Evolution Unredacted

By Dr. Anab Whitehouse

© Anab Whitehouse
Interrogative Imperative Institute
Brewer, Maine
04412

All rights are reserved. Aside from uses that are in compliance with the 'Fair Usage' clause of the Copyright Act, no portion of this publication may be reproduced in any form without the express written permission of the publisher. Furthermore, no part of this book may be stored in a retrieval system, nor transmitted in any form or by any means – whether electronic, mechanical, photo-reproduction or otherwise – without authorization from the publisher.

Published 2018

Published by One Draft Publications

"All through school and University I had been given maps of life and knowledge on which there was hardly a trace of many of the things that I most cared about ... until I ceased to suspect the sanity of my perceptions and began, instead, to suspect the soundness of the maps." – E.F. Schumacher

Table of Contents

Opening Quote – page 5

Dedication – page 9

Introduction – page 11

Section I: The Rashomon Effect

   The Evolutionary Landscape: Setting the Stage – page 19

   Critique of An Abbreviated Textbook Perspective – page 39

   A Few Lessons Related to Archaea – page 65

   Deep Solutions for the Problems of Biological Origins – page 87

   Extinction – page 135

   The Evolution of Human Beings – page 153

   Some Evolutionary Roots of Psychology – page 171

   Evolution's Black Box – page 177

Section II: Through A Glass Darkly

   I Must Be Related to John Scopes – page 195

   Complexities of a Simple Theory – page 199

   Spontaneous Mutations – page 219

   The Natural and the Supernatural – page 229

   The Metric of Ignorance – page 231

   Evolutionary Creativity – page 241

   Punctuated Equilibrium – page 249

   Intermediate Presumptions – page 263

   Questionable Facticity – page 267

   The Quality of Intelligence – page 273

   Irreducible Complexity – page 279

   Guerilla Warfare – page 307

   Disputing the Indisputable - page 319

   Quantum Uncertainty – page 323

   Replaying Life's Tape – page 333

Origins of Asymmetry – page 341

Fundamentalism – page 345

God, Science, and Evolution – page 349

Chance, Choice, and Determinism – page 365

The J. Wellington Wimpy Factor – page 377

The Making of Flowers – page 381

Finding Darwin's God – page 385

Section III: Anatomy of a Theory
Introduction – page 389

Opening Remarks – page 395

What on Earth Is Happening? – page 405

Beach Front Property – page 427

Ah, Sweet Mysteries of Life – page 459

An Ocean of Difficulty – page 493

Monkeying Around with the Containment Blues – page 527

Science of Presumption Can Be a Beautiful Thing – page 563

Transposable Conceptual Elements – page 587

Closing Arguments – page 609

Section IV: Some Evolutionary Considerations
Analysis – page 623

Bibliography – page 659

For my mentor, Dr. Baig ... who taught me, among other things, that searching for the truth is essential to being human. He also taught me how important character is to such an undertaking.

I am unlikely to ever realize the truth in the way, or to the extent, that he did. Nonetheless, the fact that after more than four decades I am still deeply engaged in trying to bear witness to the foregoing process of searching – albeit in my own way and according to my very limited capacity -- is largely due to his example.

There are no words that adequately can convey the depth of gratitude I feel for the fact that he came into my life and helped make it better than it otherwise would have been. The words that follow are mere shadows of the truths that he tried to communicate to me, and I wish I had been a better student.

Introduction

A lot of controversy, anger, and intolerance have been generated through the creationist/evolutionist debate. But, as the old Buffalo Springfield's song, 'For What It's Worth', states: "Nobody's right, if everybody's wrong", and in this debate, there is a great deal of 'wrong' that is being committed on both sides of the argument.

What follows isn't about whether one side or the other is correct with respect to what frequently amounts to a two-tiered monologue. Much more space than is occupied by the present introduction or the three sections that follow would be necessary to try to arrive at a judicious judgment concerning the tenability of any given position ... a problem that is made more difficult since there is more than one position being given expression through each side of the debate.

Instead, this essay is a comment on the apparent inability, or unwillingness, of all too many, supposedly, rational people to be interested in searching for the truth, as opposed to merely advancing any scientific, philosophical or theological perspective that they might wish to champion. My point of departure is a series of e-mails that I sent to representatives from both sides of the debate.

More specifically, twenty years ago I wrote a book entitled: *Evolution and the Origin of Life* (the contents of that book are contained in Section III of *Evolution Unredacted*). The book used a mock trial to serve as a literary device through which to explore various ideas involving modern, scientific accounts concerning the origins of life issue.

Although self-contained, the book was, at times, a fairly technical examination of a body of data drawn from a variety of sources. Among the resources that formed the backdrop against which the mock trial was to be conducted were materials dealing with: Pre-biotic chemistry, earth sciences, molecular biology, thermodynamics, cytology, and membrane functioning ... materials and information that tend to be used to lend support to an evolutionary account concerning the appearance of life on Earth.

After finishing the above book, I began to contact a number of people that I believed might be interested in the contents of the foregoing work in the hope of being able to induce some degree of discussion along certain lines that I felt – at least up to that point in time -- had been receiving insufficient attention. The people contacted were those who espoused either 'creationist' or 'evolutionist' inclinations.

While I cannot argue that the sample of responses on which this introduction is based is representative of creationist or evolutionist populations as a whole, nevertheless, there were some disturbing results -- informal and statistically questionable though these might be – that arose in conjunction with my overtures to different individuals concerning the evolution-creation conflict. Moreover, oddly enough, what was most disturbing about many of the responses involved an attitude that appeared to be held in common by many individuals from both sides of the controversy.

Perhaps, the best way to describe that to which I am alluding is to suggest that the attitude in question seemed somewhat reminiscent of the orientation of the clergy at the time of Galileo who refused to look through the telescope in order to verify whether there was any factual substance to Galileo's claims about certain aspects of the physical universe. Or, stated in another way, both creationists and evolutionists seemed to be saying: Don't bother me with facts. They only confuse the matter.

Over the years, I have tried to enter into discussion with people from both of the foregoing camps. With certain exceptions, I have found each of the camps (yes, so-called scientists as well) to be fairly arrogant, intolerant, closed-minded, and surprisingly ill-informed about a variety of issues.

I'll describe two, relatively brief, examples to try to convey what I have in mind here. One illustration comes from the 'creationist' side of things, and the other example is derived from the 'evolutionist' perspective but let me reiterate that neither side can be distilled down to a single, monolithic position so one has to reflect on what is being expressed below with a certain soupçon of intellectual caution.

In any event, when I e-mailed a variety of people who considered themselves to be part of the 'creationist' camp and informed them about my aforementioned book, I suggested their position might be enhanced if they were to refrain from trying to base their reasoning on an 'argument from design' approach because this had the effect of deflecting focus away from a variety of important, factual issues and, in the process, permitted members of the evolutionist camp to be able to alter the framework of discussion. In other words, the style in which many creationist accounts are presented tend to provide proponents of evolution with the opportunity to avoid having to deal with problematic scientific data and/or conclusions and, instead, various proponents of evolution would be able to spend all their time on critiquing a vulnerable philosophical position (i.e., arguments from design), rather than having to defend the weak underbelly of their own scientific theories concerning the claims of evolutionary theory concerning the origins of life on Earth.

Almost invariably, the responses that I received in relation to the foregoing suggestion were variations on the following theme: There is no need to remove the 'argument from design' issue from the table because they wanted to be able to show not only that evolution was incorrect but that the available evidence served to 'prove', as well, that their theological perspective had captured the truth of things. In other words, what appeared to be most important to those individuals was an underlying desire to push a theological position rather than a concern with evidential considerations ... even though, possibly, if they had restricted themselves to just critically examining the evidence, then their ultimate goal might have been better served.

At the very best, an argument from design, cannot possibly demonstrate that one theological perspective is more correct than some other such perspective. All that kind of argument can show -- even if correct -- is that somehow, order is present in the universe, and one can only speculate as to why and how that order came to be.

In fact, as some proponents of complexity theory might argue, there could be physical laws in the universe that operate in such a way that emergent structures arise out of the interaction of allegedly random systems when one goes from one level of scale to the next. So, the existence of determinant structure does not necessarily point in the direction of a theological answer ... or, so someone might wish to argue.

By insisting on a modus operandi that is, at heart, theological in nature, the creationist camp opens itself up to a whole series of issues that takes attention away from what should be the sole topic of debate: Namely, whether, or not, available evidence actually 'demonstrates' that an evolutionary account of the origins of life is tenable. When proponents of a creationist point of view permit their theological agenda to get in the way, then those individuals tend to undermine their own interests and unnecessarily complicate the discussion as well as create problems for themselves along the way.

By contrast, people from the evolutionist camp often tend to argue that modern evolution constitutes the only theory that is accepted by scientists as being reflective of the available evidence. This might be true, but it is neither here nor there as far as trying to establish whether modern evolution embraces a correct understanding concerning the origins of all species or how life began on Earth.

Across history, the majority of scientists eventually have been shown to be wrong – partly or wholly -- concerning many, if not most, of the things that they held to be true concerning the nature of reality. Although every generation of scientists tends to believe that it possesses insights into the 'truth' about nature, nonetheless, succeeding generations of scientists tend to expose the flaws in, and problems with, many facets of those earlier scientific theories and understandings.

Consequently, to say that the majority of scientists today contend that the theory of evolution is true or they consider that theory to be the only, available candidate worthy of being advanced, probably says more about the sociology of science than it does about the state of the universe. Furthermore, even if one were to agree that modern evolutionary theory is the only, available, scientific account for the origins of all species or the origins of life on Earth, this is like saying that because the police have only one suspect who they are seriously considering, then, therefore, the police's theory about how things occurred must be correct.

Among other things, there is general confusion between macroevolution and microevolution. Essentially, the former is preoccupied with population genetics, whereas the latter is about how genetic systems come into being in the first place.

Population genetics is entirely irrelevant to the origin-of-life question. Population genetics only becomes relevant when one has populations of biological or quasi-biological systems that are capable of passing on information about how to perpetuate, or generate new, viable forms and functions to subsequent generations.

Microevolution attempts to explain how biological or quasi-biological systems arise from conditions that are devoid of those kinds of phenomena. In other words, how does one make the transition from either the realm of non-living chemical systems to the world of living entities or go from organisms without certain kinds of capabilities (e.g., circulatory, pulmonary, immune, endocrine, nervous, and so on) to organisms possessing even various approximations of those capabilities.

My aforementioned book, *Evolution and the Origin of Life*, was a critique of modern scientific accounts concerning the origins of life on Earth. The book was not pushing a theological agenda, nor was it trying to prove some theological position.

Rather, in effect, the book put forth a variety of scientific data and arguments indicating that microevolutionary accounts were not even remotely close to providing an adequate, plausible, tenable theory concerning how life came into being on Earth. There were just too many, unanswered questions and too many lacunae in the theory, or set of theories, that constitutes the modern evolutionary account of the origins of life.

Oftentimes, proponents of evolution defend their position by saying: 'Well, if the theory of evolution is not true, then, what are you going to put in its place?' In truth, when we don't know something, we should admit our ignorance rather than try to force-fit facts into a theory that is, at best, fundamentally incomplete, and, consequently, tends to raise more questions than it answers.

In any event, rather than engage me in a purely factual or evidential discussion concerning the adequacy of evolutionary theory vis-à-vis the origins of life enigma, the responses I often got back from proponents of evolution were either complete silence, or some sort of critical musing that wondered how someone (i.e., me), in this day and age, who has a graduate degree, could be so scientifically unsophisticated as to not accept evolutionary accounts concerning the origins of life on Earth.

Like Galileo, I have looked at the available evidence and have come to certain conclusions that, in time, might prove to be correct or incorrect. Unfortunately, the responses of all too many advocates of evolution whom I have contacted seem to be like the clerics of Galileo's time who refuse to really look at the evidence because they are afraid, apparently, of what they might 'see'.

From an educational perspective, I am uncomfortable with either creationists or evolutionists being in charge of shaping curriculum with respect to the origins of species issue or the origin of life issue. My discomfort arises from the fact that, despite certain exceptions on both sides, I have found each side to be resistant to the idea that the fundamental commitment we have to children and to young people is to help them search for the truth rather than be force-fed preconceived, problematic doctrines that might prove to be obstacles and stumbling blocks on the way to finding truth, understanding and wisdom.

Am I saying that all of spirituality or all of science is doctrinaire and problematic? Not at all, but just because something calls itself scientific or spiritual or claims to be in the best interests of children or young people, this does not, in and of itself, automatically make it so.

Section I: The Rashomon Effect

The Evolutionary Landscape: Setting the Stage

Perhaps, nothing is uttered by most scientists and many non-scientists with a greater sense of certainty these days than that life is a function of, and arose due to, the process of evolution. However, the meme that "life is caused by evolution" might not be the slam-dunk that so many people appear to suppose is the case, and, furthermore, the foregoing kind of certainty is often rooted in ignorance about the underlying nature of what is being said with such alleged certitude.

More than 150 years ago, Charles Darwin conjectured that every modality of life that exists, or has existed, or will come to exist on (and in) the Earth has descended from one primordial life form. Approximately 113 years later Theodosius Dobzhansky, an evolutionary geneticist, wrote an essay for the March 1973 edition of the *American Biology Teacher* that bore the title: 'Nothing in Biology Makes Sense Except in the Light of Evolution."

A similar statement had surfaced nine years earlier in another piece by Dobzhansky that appeared in the 1964 edition of *The American Zoologist*. The title of that article was '*Biology – Molecular and Organismic.*'

Although Dobzhansky was a Christian in the tradition of Russian Orthodoxy, he also became a world-renowned evolutionary biologist who advocated a form of theistic evolution that he believed should be developed through the principles of science rather than received through the pages of scripture. According to him, the Bible, the Qur'an and other books of sacred teachings were very useful when it came to exploring the relationship between human beings and God, but those same works should not be, and -- according to him -- were never intended to be, treatises on science.

Echoing Darwin, Dobzhansky was committed to the idea that life arises via an evolutionary process that depends on the woof and warp of (a) unified principles of biological dynamics being intermingled with (b) different patterns of diversity. He believed that the Divine juxtaposing of biological principles of unity and diversity were what enabled evolution to make sense of the vast array of biological data that, otherwise, would remain as, apparently, disparate pieces of information.

One could point out in passing, however, that the property of being 'disparate pieces of information' is a rather relative notion. The biological information that Dobzhansky believes would be disparate if the theory of evolution were not true could still make a great deal of sense if it were considered from some other perspective, and the fact Dobzhansky has not grasped the nature of such a perspective doesn't mean that unity and diversity couldn't give expression to a mode of reality other than the evolutionary one championed by Dobzhansky.

In the aforementioned *American Biology Teacher* article, Dobzhansky stated: "I am a creationist and an evolutionist. Evolution is God's, or Nature's, method of creation. Creation is not an event that happened in 4004 BC; it is a process that began some 10 billion years ago and is still under way." As Dobzhansky pointed out in his article, it was not scripture (either Biblical or Quranic) that put forth the figure of 4004 BC as fixing the beginning of life on Earth but, rather, a 17th century figure, Bishop James Ussher, who, apparently for good measure, also specified in a 1658 publication that the great event of Creation took place between the night of October 22nd and the following day of October 23rd ... possibly feeling that specificity might be construed as an indication that his pronouncement was giving expression to the 'gospel truth'.

A contemporary of Bishop Ussher, Sir John Lightfoot -- Vice-Chancellor of Cambridge University – came to the same general conclusions as Ussher did but added that the time of the Creation event was 9:00 A.M. Moreover, Sir John apparently came to those conclusions 14 years earlier than Bishop Ussher had been able to do.

Now, as preposterous, amusing, or amazing (take your pick) as the foregoing calculations might seem, one cannot necessarily attribute the attempt to come up with precise answers to difficult questions as a function of the ignorance of 17th century scholars. After all, in an exercise of calculation that is an attempt to be even more precise than Ussher and Lightfoot had been, Nobel Laureate, Steven Weinberg, had a book published in 1977 entitled: *The First Three Minutes* in which he sought to explain what was transpiring in the universe from about 10^{-32} seconds (the end of the Planck epoch) through the next 2 minutes, 59 seconds-plus seconds following the Big Bang.

What happened prior to the 10^{-32} second mark is said to be something of a mystery because, according to many modern scientists, the laws of physics apparently were in disarray during that period of time. The idea that the laws of physics were in some sort of chaotic, broken down state in the time before the 10^{-32} mark is a long-winded euphemism for ignorance.

If we don't know what the status of the universe was prior to 10^{-32} seconds ... if we don't know what laws of physics, if any, were operable prior to that time ... if we don't know how the laws of physics suddenly became operational in the transitional period leading up to the 10^{-32} second mark, then, in some ways, the intriguing calculations of Steven Weinberg are every bit as contentious as are the calculations of Bishop Ussher and Sir John Lightfoot. All three of the foregoing individuals were trying to provide something of a temporal timeline or perspective according to what was considered to be the 'best' evidence available to each of them, but all three accounts leave much to be desired.

While the alleged nature of the unfolding of the universe within the first three minutes of the universe's existence is certainly an evolutionary theory of sorts, in the present chapter I would like to concentrate on the issue of biological evolution. However, I will return to the theory of 'The Big Bang' in a subsequent chapter.

Among other things Darwin's *Origins of Species* (The original title was longer – namely, *The Origins of Species by Means of Natural Selection or the Preservation of Favored Races in the Struggle for Life* – but the book's title was shortened for the 1872 6th edition) put forth data and arguments to support his belief that species or populations of organisms undergo a process of change, or evolution, in accordance with the principles of natural selection to which the environments in which such species exist give expression. This central notion of changes in a species brought about forces of natural selection is really not all that extraordinary although, as ensuing history has shown in dramatic fashion, Darwin's idea was interpreted as being in conflict with a variety of theological positions, and, as a result, there was considerable resistance to the foregoing theme in Darwin's initial, written foray into the issue of origins.

Slightly more controversial was Darwin's belief that <u>new</u> species could arise (the process of speciation) through the action of natural selection on a given population of organisms (i.e., a specific species). To claim that the conditions of natural selection might bring about changes in what properties of a population were most likely to be passed on to future generations is one thing (and breeders of plants and animals had been demonstrating this for centuries prior to the time of Darwin), but to argue that entirely new species could arise through such a process seemed to be pushing the envelope of credulity, and this was especially the case since quite a few theological positions that were prominent during Darwin's time presupposed that species had been fixed at the time of creation.

More controversial still was Darwin's contention that all species in existence or which had been in existence at some point in the past were derived from a common, primordial form or ancestor. For example, maybe, given the right conditions of natural selection, it might be possible for different subsets of a specific population of organisms to biologically drift apart from each other to a point where the members of those subsets could no longer interbreed with the members of the other subsets (or with the remaining members of the "mainstream" population) and, in addition, drift apart to the point where various characteristics of the larger, mainstream population might disappear altogether from one, or both, of those subsets – pushing those subsets in a different evolutionary direction and, in the process, generating new species. However, to try to maintain that all life forms evolved from a common, primordial form of life seemed – at least for many people – to push the matter of evolution beyond the pale of reasonable, plausible discussion.

Darwin's books, based on extensive years of meticulous research, were collectively quite suggestive with respect to the idea that all current life forms might possibly have arisen from a common primordial form of life. Nonetheless, not only did his books fail to definitively prove what was being suggested (this task fell to his successors), but, in addition, Darwin had no explanation for how the first primordial form of life came into being.

Although Darwin rarely wrote or spoke about the issue of primordial origins, he did, on occasion, speculate about such a possibility. For example, in a February 1st 1871 letter to his friend Joseph Hooker he wrote:

"It is often said that all the conditions for the first production of a living organism are now present, which could ever have been present. But if (& oh what a big if) we could conceive in some warm little pond with all sorts of ammonia & phosphoric salts — light, heat, electricity etc. -- present, that a protein compound was chemically formed, ready to undergo still more complex changes ... ."

Obviously, the implication of Darwin's foregoing conjecture was that a collection of the right sort of chemical elements might, somehow, come together under the right sort of environmental conditions and, somehow, form a complex compound that was, somehow, capable of undergoing still more changes until, eventually, somehow, life emerged. It would take another 60-70 years before various individuals began to try to fill in the details of the "somehows" that were left unanswered by Darwin even as the nature of those possible 'somehows' were being alluded to by him in his letter to Hooker.

There is a similar set of lacunae inherent in Darwin's contention that once a primordial form of life somehow came into being, then, all subsequent life forms would descend from that point of origin. More specifically, even if were to accept the idea that new species might arise through one, or another, collection of forces of natural selection acting on the original population of primordial organisms (assuming, of course, that such a population could, somehow, arise from a single primordial form of life), there is nothing to guarantee that the capacity to give rise to the emergence of <u>some</u> new species necessarily would lead to the rise of <u>all</u> subsequent species.

In other words, one needs to distinguish between: (1) speciation as a function of natural selection whose capacity to produce new forms of life constitutes a potential of unknown parameters, and (2) the idea of common descent from a primordial form of life. More specifically, the three-four billion history of life on Earth consists of millions, if not billions and trillions, of changes – some minor and some major -- in the forms, functions, capacities, biological components, and metabolic pathways of living organisms, and the fact that one might be able to account for some of these changes through the processes of speciation does not necessarily mean that one can plausibly account for all such changes through the kinds of speciation process that were being proposed by Darwin and that are being explored by modern evolutionary biologists.

Darwin believes (as do most, if not all, evolutionary biologists) that speciation tends to generate further speciation. Darwin also believes (as do most, if not all, evolutionary biologists) that if one were able to add up the entire set of instances of speciation that have arisen over billions of years as a result of the forces of natural selection (although, for practical empirical and methodological reasons, one might not be able to succeed in completely accomplishing such a project), then one will be able to account for all branches of the tree of life ... in other words, one will have demonstrated (or so the claim goes) that one can trace an unbroken path extending from a primordial form or species of life that, subsequently, transitioned seamlessly into other species, that, in turn and over vast swaths of time, led seamlessly to the successive generation of every single life form that ever existed in conjunction with the planet Earth.

Even if we limit our discussion to just the considerations introduced in the last several pages, it is obvious that the term "evolution" can have a variety of meanings. For instance, 'evolution' might refer to the process in which a given population of organisms (a specific species) gives expression to changes over time with respect to which set of physical and biological properties will come to enjoy the most success as a function of a given set of conditions of natural selection. Moreover, this way of rendering the notion of evolution also would include the belief that as the conditions of natural selection change, then, so too, will the character of the set of properties that are able to take advantage of the changes occurring with respect to various forces of natural selection.

A second sense of 'evolution' has to do with the process of speciation in a limited sense. In other words as environmental conditions and a given population of organisms (a specific species) engage one another, the dynamic of that engagement might lead to the generation of subsets of the population that, in time, become, among other things, reproductively isolated from one another and in the process give rise to modified or descended form(s) of the original species population that constitute the beginning of a new branch that is growing on the tree of life.

However, the extent to which such a process of speciation is capable of proceeding might be limited. In other words, while speciation does occur, there might be limits to how far it can proceed and on what 'new' possibilities might arise in conjunction with that sort of process.

A third meaning of 'evolution' concerns the limits, if any, in relation to the potential for speciation. That is, there are those who believe (and Darwin was one of these individuals) that the potential inherent in the process of speciation is, for all practical purposes, indefinitely great and, as a result, such a process has the capacity, sooner or later, to generate every form of life that has arisen since the first primordial organism arose on Earth … assuming, of course, that the forces of natural selection co-operate with, and lend support to, such changes in speciation.

Finally, a fourth notion of 'evolution' concerns the origins of life. More specifically, this sense of the word has to do with accounts of how the first primordial form of life – the first species – emerged.

Returning to the ideas of Dobzhansky, he seems to have had some strange ideas about what making sense entails with respect to the relationship among God, evolution and biology. For instance, Dobzhansky raises some rather arbitrary issues in his *American Biology Teacher* article about what God might and might not do in conjunction with the human task of trying to figure out what is going on in the universe.

More specifically, Dobzhansky seems to be of the opinion that God would not perpetrate hoaxes on, or try to deceive, or seek to fool human beings by fabricating evidence in an effort to mislead human beings concerning the origins of life or the laws governing life. While Dobzhansky might well be correct in his beliefs, his manner of reasoning doesn't eliminate the possibility that human beings can perpetrate hoaxes on themselves (e.g., the Piltdown man), as well as deceive and fool themselves, without any assistance from God, about any manner of things … including the issue of evolution.

In any event, after putting forth additional arguments, Dobzhansky comes to the conclusion that the unity and diversity of life can be explained best as a function of evolutionary processes that are shaped and molded by forces of natural selection. According to Dobzhansky, this is how God proceeded with respect to the act of creation.

I can't say that I know what God would and wouldn't do in the case of human beings, and I have my doubts about whether Dobzhansky knew such things either. I do have an intuitive feeling that I cannot expect God to operate in accordance with principles that conform to what does and doesn't make sense to me, and while I appreciate that what made sense to Dobzhansky was a function of what he believed to be the case concerning how things (such as evolution) worked in the universe, I don't necessarily have a lot of confidence in certain aspects of what made sense to him with respect to such issues.

Maybe the position outlined in the '*Nothing in Biology Makes Sense Except in the Light of Evolution*' article by Dobzhansky is correct. Indeed, previously I have stated that if so inclined – which I am not – I could accept much of what evolutionary biologists have to say about the origins of life or its descent across time, and all this acknowledgment would mean to me is that I might have to rework certain aspects of my worldview so that those features of my understanding reflected necessary "truths".

The fundamental issue is to seek and determine the nature of truth. Our belief systems need to adapt to whatever that truth turns out to be.

Nonetheless, the ideas of Dobzhansky notwithstanding, there might be other ways to account for the principles of unity and diversity to which life gives expression that need not depend on the physical principles of evolution. Moreover, just because we might not know what those ways are does not necessarily mean that the process of life on Earth is without sense ... rather, the nature of life – on a number of levels -- just might have a sense that we do not, yet, grasp ... and, perhaps, we never will.

The fact an idea helps one to make sense of things is not proof that one's sense of things is true. Truth (and proof) requires something more than meaningfulness.

For example, Dobzhansky points out in his *American Biology Article* that the idea of evolution is able to make sense out of the fact that extinction is the fate of most species that have appeared in Earth's history since environmental conditions have changed during that time and, yet, only a relatively few species have been able to successfully adapt to those changes and continue the process of descent. He goes on to assert: "but what a senseless operation it would have been, on God's part, to fabricate a multitude of species ex nihilo and then let most of them die out."

In effect, Dobzhansky is saying that if something does not make sense to him, then, it couldn't possibly make sense to God. Apparently, Dobzhansky believes that what makes sense to God should be a function of what makes sense to Dobzhansky.

The fact of the matter is -- and let us accept Dobzhansky's assumptions: That God exists, that God created life ex nihilo (whatever this means), and that God permitted most life forms to become extinct – Dobzhansky is engaged in an exercise of speculation concerning how God 'thinks' about things or how God goes about making sense of Creation. Conceivably -- and, like Dobzhansky, I am just speculating here -- God permitted so many life forms to become extinct because (a) this constituted a heuristically valuable theme on which human beings needed to reflect or meditate, and (b) perhaps the nature of creation is about constantly giving expression to new forms of manifestation while letting the old forms of manifestation become extinct after they run their course with respect to whatever role the latter played in the Divine scheme of things ... a scheme that I am not claiming to understand and a scheme that I suspect Dobzhansky did not necessarily understand either.

Evolution might be an idea that helps people like Dobzhansky to organize a vast array of biological data in order to try to make sense of that material. However, perhaps, one needs to engage in a process of critical reflection with respect to whether, or not, evolution's manner of organizing such data makes as much sense as Dobzhansky and other evolutionary biologists seem to believe.

For instance, I believe that many facets of biology make sense in the light of evolution in both of the first two senses noted previously. In other words, when one considers the changes that a given population or species undergoes across changing environmental circumstances, or when one considers the possibility of speciation as an expression of a relatively limited set of combinations and permutations that are inherent in such a population's gene pool (i.e., there are various kinds of forces and factors that place constraints on how far speciation can proceed with respect to the possible subsets of a given population), then evolution in the foregoing two senses does tend to give a unified sense to a great deal of diverse biological data.

Essentially, both of the foregoing senses of the idea of evolution are entailed by the principles of population biology. Moreover, I believe there is a great deal of evidence to support many of the principles of population biology.

However, I believe many things in biology do not make sense in the light of a sense of evolution that shines forth from the second and third meanings of evolution noted earlier. In other words, first, I have a lot of questions concerning the tenability of the idea that the potential of speciation is so indefinitely great that, given appropriate conditions of natural selection, it can account for the diversity of all life forms that have appeared over the last 3-4 billion years with respect to Earth. Secondly, I question the tenability of claims that the origins of the initial, primordial life form can be explained (as Darwin hinted might be the case in his February 1, 1871 letter to Joseph Hooker) in terms of known principles of physics and chemistry.

According to Dobzhansky's article in the *American Biology Teacher*: "Evolution as a process that has always gone on in the history of the earth can be doubted only by those who are ignorant of the evidence or are resistant to evidence, owing to emotional blocks or to plain bigotry." Dobzhansky's 'my way or the highway' sort of mentality is fairly dogmatic and resonates with the way many so-called experts propagandized the myth that there is a chemical cure – e.g., SSRIs -- for mental illness.

Such intransigence in understanding is also reflected in Dobzhansky's subsequent contention that evolution: "...is a general postulate to which all theories, all hypotheses, all systems must henceforward bow and that they must satisfy in order to be thinkable and true." Well, I suppose it is not that much of a leap to go from -- as pointed out earlier – telling God what must make sense to Divinity, to telling human beings what must make sense to them.

Evolutionary biologists often switch between referring to evolution as a fact, and/or a hypothesis, and/or a theory. However, let's reflect on this a little.

For example, currently, there is no plausible evolutionary account for the origins of life (and there will be more on this issue a little bit later during the present and subsequent sections of *Evolution Unredacted*). Consequently, one is not necessarily entitled to refer to evolution as being a fact when it comes to the origins of life issue.

Moreover, evolution is not, really, even a hypothesis when considered in conjunction with the task of trying to explain the origins of life. To formulate a meaningful hypothesis, one has to have a way of testing that hypothesis, yet, in many, if not most, respects, one can never recreate the conditions of early Earth because we do not know precisely what those conditions were, and, consequently, any hypotheses that might be postulated in this regard are entirely arbitrary and predicated on some presumed scenario concerning the conditions of early Earth.

The foregoing comments should not be construed to mean that nothing is known about whatever conditions might have been present some four-to-five billion years ago. Rather, what is being alluded to is that we don't currently possess sufficient, specific knowledge to be able to construct a reliable picture of what was taking place in any given location on early Earth.

We might know some of the general things that likely might have been happening in and on early Earth from a geological, hydrological, meteorological, and/or chemical perspective. Nonetheless, we do not know enough about how those forces were specifically interacting with one another from place to place on early Earth to be able to generate a reliable model or simulation of how protocells supposedly came into existence.

To be sure, individuals (such as Darwin in his previously cited letter to Joseph Hooker) have speculated about what the conditions on primordial Earth might have been. Furthermore, various researchers have run experiments (there will be more discussion on this later on) that were based on what those individuals believed might have been realistic conditions out of which components of the first protocell could have emerged, but there is no independent way of demonstrating that such proposed conditions are, in fact, realistic representations or models of what was the case on early Earth.

If one likes, one can formulate any number of arbitrary hypotheses rooted in speculations about the conditions of early Earth (and the prebiotic literature is replete with these sorts of arbitrary speculations). However, all one is testing are the conditions set forth in those speculations ... speculations that might have little, or nothing, to do with the realities of actual conditions 4-5 billion years ago.

We just really don't know all that much about such matters. Furthermore, so-called "educated guesses" are, first and foremost, just that – namely, guesses. In addition, 'educated guesses' leave open the question of whether, or not, one should accept all the biases, assumptions, and philosophical understandings that frame someone's notion of what it means to be "educated".

For example, as noted previously, Dobzhansky was of the opinion that individuals who did not accept the theory of evolution are "ignorant of the evidence or are resistant to evidence, owing to emotional blocks or to plain bigotry" ... and, therefore, he had a rather self-serving view of what it means to be educated. One can throw in for good measure the theory of education that was given expression by the anthropology teacher I mentioned in the introduction to this book who responded so contemptuously toward me when I had the audacity to raise a few questions in conjunction with the tenability of evolutionary theory.

All too frequently, scientists are all for skepticism, open discussion, and critical inquiry except when it comes to questioning the theories that they hold dear. It is difficult for education in any meaningful and heuristically valuable sense to take place in such an oppressive atmosphere.

So, if one cannot refer to evolution as a 'fact' or a 'hypothesis' when it comes to accounting for, among other things, the origins of life, can one refer to evolution as a theory that attempts to make sense of that issue? A theory is said to be a coherent collection of interconnected claims that are given expression through, and shaped by, an array of reasoned arguments and empirical data that have the potential capacity to account for a variety of phenomena.

Given the foregoing characterization of the notion of a theory, then, certainly, evolution is a theory. However, saying that something is a theory is not necessarily coextensive with saying that such a theory is either true or that it is necessarily even scientific.

To be sure, the theory of evolution (however one might wish to parse the term "evolution") is a relatively coherent body of interconnected claims. Furthermore, the theory of evolution does consist of a set of reasoned arguments concerning a body of empirical data. And, finally, the theory of evolution does offer an account of – although not necessarily the truth about -- why certain phenomena might have the character they do.

All theories – whether philosophical, religious, psychological, historical, or technical – consist of a relatively coherent body of interconnected claims. What makes evolutionary claims either true or scientific?

All theories – whether philosophical, religious, psychological, historical, or technical – consist of a set of reasoned arguments concerning some aspect or aspects of the empirical data of lived experience. What makes evolutionary arguments true or scientific, and what are the criteria for considering whether, or not, something has been effectively reasoned?

All theories – whether philosophical, religious, psychological, historical, or technical – purport to offer an explanation of why something is the way it is. What makes an evolutionary explanation true or scientific?

Furthermore, is it possible for something to be true but not scientific? Alternatively, is it possible for something to be scientific but not necessarily true?

During the first chapter of *Final Jeopardy: The Reality Problem Volume I*, a fair amount of space, time and words were used to point out that people who refer to themselves as scientists, or who are referred to by others in this manner, don't necessarily always know what they are talking about. Cancer treatments based on the use of Antineoplastons were – and still are – opposed by a majority of the cancer research and medical communities around the world despite the fact that Antineoplastons have been proven to be non-toxic, effective, and have successfully met the challenge of Phase III, randomized trials. In addition, SSRIs are almost universally endorsed by psychiatrists, medical doctors, and researchers despite the fact there is no proven, specific, underlying theory about what role serotonin plays in the dynamics of depression (or its treatment), and despite the fact there is considerable evidence to indicate that SSRIs are extremely toxic and, as a result, are capable of inducing various forms of 'medication madness' and discontinuation syndrome in those individuals to whom it is prescribed or administered. Furthermore, despite the existence of a significant amount of evidence supporting the idea that HIV does <u>not</u> cause AIDS -- as well as the existence of very little evidence that demonstrates that HIV does cause AIDS – the vast majority of clinicians and researchers continue to maintain that HIV causes AIDS.

The thirty-plus year campaign <u>against</u> Antineoplastons claimed to be rooted in science, but it wasn't. The thirty-plus year marketing campaign to promote SSRIs as a chemical cure for depression (and a growing assortment of other maladies) was based, supposedly on science, but this was not, and is not, the case. The thirty-plus year attempt to claim that HIV causes AIDS had its origins in an allegedly Nobel-worthy series of experiments performed in the early 1980s, and, yet, none of those experiments -- along with the hype that surrounded and permeated them -- seemed to have little to do with anything that could meaningfully be described as scientific because 'bad science' is not really science at all despite the presence of labs, experiments, technical gadgetry, and people who have credentials of one kind or another.

Science cannot exist in the absence of critical reflection. Whatever other trappings of science might be used and applied, if rigorous critical reflection is not in evidence, then, the activities taking place in the midst of such trappings is something other than science ... at best they might be referred to as being pre-scientific.

Mathematics and quantification might be necessary for science to be possible, but they are not sufficient conditions to guarantee that science will take place. Observations, hypotheses, and experiments tend to constitute necessary conditions for the existence of science, but those activities do not necessarily constitute sufficient conditions for the possibility of science to be manifested. The use of instrumentation plays a useful, if not crucial, role in the activity of science, but the presence of instrumentation is not necessarily sufficient to ensure that science will take place. Having individuals who have the credentials and/or the experience that enable those people to have facility with: Mathematics, measurement, observation, generating testable hypotheses, experimentation, and instrumentation are all necessary – but not sufficient -- conditions for the practice of science.

The Antineoplastons issue, the SSRI matter, and the 'HIV causes AIDS' affair all entailed substantial elements of mathematics, quantification, observation, hypotheses, experiments, instrumentation, and credentialed, experienced individuals. Yet, most of the people involved in those controversies were not doing science because they refused -- for whatever reasons (e.g., fear, greed, ego, power, jealousy, corruption, etc.) -- to critically engage the issues at the heart of those discussions.

Only a small number of people were actually doing science in any of those three areas of research (i.e., Antineoplastons, SSRIs, and HIV/AIDS). This is the case because only a relatively few people engaged in those areas of research were employing the necessary qualities of critical reflection to be able to ask the right kinds of questions concerning the tenability of the uses to which various modalities of mathematics, measurement, observation, hypothesis, experimentation, instrumentation, and expertise were being put.

Although skepticism plays a role in the process of critical reflection, the latter process involves much more than being willing to maintain a stance of caution concerning the veracity of various claims about the nature of the universe. One must be willing to ask questions that are intended to be something more than expressions of resistant doubt but, instead, are intended to seek out and realize the truth of an issue ... at least to whatever extent such truth can be sought out and realized.

The individual who spends his or her life committed to nothing except the practice of skepticism is not really a scientist. If there is no intention to try to ferret out whatever dimensions of truth are possible to grasp in some set of circumstances, then one is a philosophical skeptic, not a scientist.

There might be any number of questions surrounding whether, or not, one actually has grasped some sort of truth in a given situation. Nevertheless, asking questions in an attempt to be able to root one's claims in the truth in some demonstrable, substantive, fashion is a very different sort of activity than just raising questions and stating objections as ends in themselves.

The questions that are asked during the process of critical reflection should be directed toward establishing a form of understanding that is capable of engaging, and withstanding, rigorous forms of challenge concerning the quality and reliability of whatever modes of mathematics, observation, hypothesizing, experimentation, instrumentation, and expertise are employed in a given research project. With important exceptions, there is a fatal absence of the right kinds of questions, understandings, and critical reflections that is all too evident – as I feel has been demonstrated in Chapter 1 – with respect to the controversies involving Antineoplastons, SSRIs, and HIV/AIDS, and I believe there is a similar fatal absence of the right kinds of questions, understandings, and critical reflections evident with respect to certain dimensions of the evolution issue.

The theory of evolution is often said to be true because a group of scientists have come to agree on the general form of the nature of the coherency that lends sense to the set of interconnected claims and statements that give expression to a coherent hermeneutic of experience with respect to, among other things, an array of biological phenomena. However, how does agreement concerning the nature of coherency in the foregoing manner make such a theory either true or scientific since there have been many occasions during the history of scientifically oriented endeavors in which a coherent sense of things was discovered not to be true or was considered to be scientific only to be shown later to be quite unscientific as well as false?

Eliminating falsehoods is part of the process of science. Nonetheless, does advocating a theory that turns out to be false make such a theory scientific in any way other than that some individuals referred to as scientists once believed the theory to be true, or, is it the case that even though some people called scientists subscribed to the theory, it might be said that such a theory was never really scientific?

Is a hypothesis that is proven to be false, a scientific hypothesis? Aren't the waters of clarity muddied by the ambiguity that is created when someone refers to hypotheses as being scientific when, later on, they are demonstrated to be false?

Forming a hypothesis is not necessarily a scientific process. On the other hand, demonstrating that such a hypothesis is either true or false might be an expression of science ... depending on the character and quality of the demonstration.

If a group of people who are referred to as, or who refer to themselves as, scientists put forth a set of reasoned arguments concerning some set of empirical data, does their claim that the arguments are reasoned make them reasoned? Or, does their agreement that the arguments are reasoned just – possibly – a matter of unjustifiably labeling those arguments as being reasoned?

Furthermore, even if those arguments are judged as being well reasoned, does this necessarily make such arguments true or scientific? And, if those arguments are not true, then, can those arguments legitimately be described as being scientific no matter how well reasoned they might be?

If a group of people referring to themselves as scientists – or who are referred to in that manner by others – cite a theory as the explanation for why things are the way they are, does such a claim make the theory a true explanation or even necessarily make such an explanation scientific? For example, as part of the arguments put forth in his *American Biology Teacher* article, Dobzhansky provides an explanatory account concerning what, apparently, should make sense to God.

Was such an explanation scientific? And, if so, in what sense was it scientific?

Was the argument he used to substantiate his sense of things with respect to the foregoing account well reasoned? Without really giving a great deal of serious effort to critically analyzing Dobzhansky's argument, I put forth several suggestions earlier in this chapter indicating that, perhaps, his reasoning wasn't as conclusive or as well-conceived as he seemed to believe.

His way of understanding the matter made sense to him. However, wasn't the coherency to which his sense of things gave expression really anything more than a circular function of his belief system and for which he had no independent evidence to advance in support of his claim?

Many people claim that evolution is the best scientific theory to account for an array of biological data. While evolution might well be a theory, it might not really be a scientific theory except when it comes to the principles of population biology (more on this later).

However, many individuals (some of whom are scientists) want evolution to encompass more than the dynamics of population biology. Such individuals want to be able to claim that evolution is a scientific explanation for the origins and subsequent descent of all life forms.

While evolution might be a theory in the aforementioned generic sense of constituting a coherent set of interconnected statements that entail a group of reasoned arguments concerning a body of empirical data that collectively serve as a meaningful account for various biological phenomena, nevertheless, none of this makes the theory of evolution either true or scientific when it comes to both the origins of life issue or when it comes to proving that speciation is capable of accounting for all changes that can be observed (either directly in living organisms or indirectly via fossils) across the millions of species that make up the tree of life. In fact, I believe it is possible to demonstrate that the theory of evolution falls far short of having proven to be either a true theory or even a scientific theory when it comes to issues such as the origin of life.

The purpose of the present book (i.e., *Evolution Unredacted*) is not to advance a creationist perspective or an intelligent design worldview as an alternative to the theory of evolution. Rather, this book is about exploring the possibility that the theory of evolution does not actually constitute a viable account of anything in relation to either the origins of life issue or in relation to the idea that certain kinds of speciation, in conjunction with natural selection, are sufficient to explain the multiple branches that make up the tree of life.

Some people seem to think that providing an account of the origin(s) of life is an either/or issue. That is, either one must accept some version of the theory of evolution or one must accept a theory of creation or intelligent design.

However, it might be the case that neither theories of evolution nor theories of creation -- as currently conceived -- are necessarily correct or the only plausible possibilities. Perhaps Hamlet was right when he said: "There are more things in heaven and earth, Horatio, then are dreamt of in your philosophy."

By pointing out problems with the theory of evolution, this exercise in critical reflection does not automatically make me a card-carrying member of some philosophy club involving one, or another, version of creationism or intelligent design. Moreover, pointing out problems with the theory of evolution does not automatically require me to commit to any particular alternative to the theory of evolution or to a specific theory of creationism or intelligent design.

If we return to the Michelangelo approach to sculpting a statute that I alluded to earlier, sometimes it is more important to remove what doesn't belong in a given situation than it is to try to fashion a structure and, in the process, impose an arbitrary design on the materials with which one is working. Continuing to search for and, where possible, realize the truth of things is the appropriate alternative to accepting theories (such as a theory of evolution or a given theory of creation) that might be problematic in important ways.

If a given theory is problematic in the foregoing sense, then one cannot automatically assume that such a theory necessarily gives expression to a scientific theory (best or otherwise) simply because it is the only one currently available that is alluded to in those terms (i.e., as being scientific) by people who refer to themselves as scientists (or who are referred to as such by others) and, as a result, should (as Dobzhansky's previously quoted comments seem to suggest) become everyone's default position. If a given theory is problematic in important ways, then the existence of those kinds of problems is the very issue that stands in the way of the theory being considered to be scientific in any substantial, rigorous, and plausible sense.

A theory entails problems in "important ways" if one can demonstrate the existence of themes that undermine the essence or heart of a theory's sense of coherency, modes of reasoning/arguments, and/or explanations concerning the nature of the universe, or some aspect thereof. The theory of evolution is a theory that is problematic in important ways – or so it will be argued in the following pages – and, consequently, that theory is not really scientific in any substantial, rigorous, plausible, or definitive sense.

When it comes to issues like the origins of life, evolution is a theory. However, it is not necessarily a scientific theory despite the fact that it emerges in a context that has many of the trappings of a scientific-like process with respect to the use of observation, hypothesis generation, experimentation, measurement, instrumentation, and credentialed individuals.

Critique of An Abbreviated Textbook Perspective

In most, if not all, textbooks that provide an introductory overview concerning the theory of evolution -- along with many of the specifics that the authors of those books believe are in support of, or entailed by, the theory of evolution -- a person is likely to find chapters dealing with a variety of issues. The following discussion constitutes, I feel, a fairly typical synopsis from which chapter themes are often derived, developed and expanded upon according to the inclinations of the author(s).

First and foremost, the idea of evolutionary change is rooted in the dynamics of the changes taking place within a population of organisms that are collectively referred to as a species. Such a population can be described in terms of a combination of both phenotypic and genotypic properties.

A phenotype refers to a particular, observable physical trait – such as size, weight, color, anatomical features, and so on – that is given expression in an individual exemplar for the species being considered ... traits that tend to be exhibited, by most, if not all, members of a species population. Not every member of the population will necessarily manifest phenotypic properties to the same quantitative extent or in the same qualitative manner, but for the most part -- and despite the presence of some exemplars or properties in members of a population that might be phenotypically anomalous in some way – nonetheless, a set of phenotypic properties exists that tends to be characteristic of a given species and helps differentiate the members of one species from the members of other kinds of species who manifest their own unique set of phenotypic traits.

Genotype refers to the genetic capacities that help to make possible and give expression to phenotypic traits, and, as well, that have the potential of being transmitted to subsequent generations (if any) via the coding, transcription, and translation that occur in conjunction with DNA and RNA molecules. Although the genotype for a given individual tends to be fixed, the expression of different dimensions of that genotype tends to vary with changing conditions both within and without such an individual.

The gene pool (the collective set of genotypes) for the population to which an individual belongs might contain phenotypic potentials that are not necessarily included in the genotype of a given individual exemplar of that species population. Among other things, this means there might be more than one version of a given gene (known as alleles) that have the capacity to underwrite which variant of a certain phenotypic trait will be expressed in a given individual.

A particular phenotype can be the result of the gene expression that is either simple or complex. In the case of simple forms of genotypic expression, usually only one gene underlies a given phenotypic trait, while in more complex forms of gene expression, a number of genes might interact to produce a specific phenotypic trait.

Phenotypic expression also can be shaped by more than genotypic considerations. In other words, the environment in which an individual's set of phenotypic and genotypic potential is rooted can affect the way in which genotypic potentials unfold and give rise to observed phenotypic characteristics of one kind rather than another.

Generally speaking, although the environment can affect the way genotypes and phenotypes are expressed in a given organism, the environment does not usually have any impact on the nature of the properties of the genotype that are passed on. In other words, phenotypic characteristics that are acquired during the life of an organism's life cycle usually are not passed on to subsequent generations.

However, there is a growing amount of evidence indicating that the foregoing position might not be as set in stone as once thought. More specifically, there are dynamics at work involving, for example, methyl groups – referred to as epigenetic tags -- that have the capacity to affect whether, or not, certain genes will be expressed.

Histones are proteins that form the structural spools around which DNA winds itself. Depending on now tightly or loosely DNA is wound around the histone core, the expression of the genes present in such wound DNA might be easier or more difficult to express.

Each and every cell of the human body is believed to possess a unique pattern of histone and methylation activity. Consequently, methyl groups interact with DNA and can have a determinate effect on whether, or not, a given stretch of DNA will be expressed.

Changes in epigenetic tagging can be acquired during the life of an organism. For example, a poor diet might lead to methyl groups binding to DNA in ways that tend to switch off the expression of one, or more, genes.

Such epigenetic changes can be passed on to, or inherited by, offspring. Consequently, there is a sense – i.e., the epigenetic tagging of DNA by methyl groups -- in which acquired characteristics could be inherited by subsequent generations, and "epigenetics" is the field of study through which the nature and impact of such changes are explored.

The evolutionary change that occurs in a given population is a function of the transition in frequencies and proportions of genotypes and phenotypes that are brought about by the way genes are transmitted in that population and, as well, by the way in which the forces of natural selection act on those patterns of transmission over time. As the frequency and proportion of certain genotypes change, the phenotypic characteristics of that population also are likely to undergo transition.

For the most part, evolutionary change is a function of what takes place within a <u>population (or its subsets)</u> in relation to a given species. Among other things, this means that evolutionary change is not generally measured by what happens to individual members of a population but only by what happens over time to the frequencies and proportions of different kinds of phenotypes and genotypes that characterize a given population or its subsets.

Obviously, the potential for evolutionary change begins when certain kinds of changes occur in relation to individual members of a population. However, unless those changes lead, eventually, to transitions in the proportions and frequencies of such genotypic and phenotypic traits in the population as a whole (or subsets thereof), then, change of an evolutionary nature has not really occurred.

Changes in genotype frequencies and proportions are believed to come about through two primary forms of dynamic. These two modalities are known as 'genetic drift' and 'natural selection'.

Genetic drift refers to those kinds of fluctuations in the frequencies and proportions of genotypes within a small population that are brought about by what is described as a random dynamic involving various environmental forces and circumstances that result in the removal of certain genes due to either the death of individuals or the inability (for whatever reason) of those individuals to reproduce and leave offspring containing the genes in question. The disappearance of such genes is not because they, in some way, lack adaptive capacity but because the luck of the draw (i.e., random events … including mutations, a perfect storm of circumstances that are disadvantageous, "freak" accidents, etc.) did not permit them to continue.

The idea of genetic drift is intertwined with the neutral theory of molecular evolution. This latter theory contends that: (1) while a relatively small minority of mutations result in some form of advantage with respect to the prevailing conditions of natural selection and, therefore, are fixed or favored by those conditions, and (2) while other mutations result in some form of disadvantage and, as a result, are eliminated by the forces of natural selection, nonetheless, (3) the vast majority of mutations are relatively neutral in character – that is, such mutations are neither more advantageous nor less advantageous than other genetic possibilities – and are fixed or eliminated by the vagaries of genetic drift.

Natural selection is a determinate process in which given subsets of a population exhibit a superior capacity, relative to other members of the population, to, in general, survive, and, in particular, to successfully pass on those kinds of capacities to their offspring. Adaptation gives expression to the interaction between individual organisms and their environments that results in the natural selection of those organisms that have best adjusted to existing circumstances and, in the process, both survive and reproduce at rates and in ways that allow a particular set of genotypes and phenotypes to continue on in subsequent generations.

Evolutionary biologists maintain that natural selection has the capacity to alter the characteristics of an existing population through changing the frequencies and proportions of various genes that might affect the way a given phenotypic and/or genotypic property is manifested. For example as genes are combined and recombined with one another during the process of reproduction, new genotypes and phenotypes might arise, and those new phenotypes and genotypes will be acted upon by the forces of natural selection that, in turn, provide the new kids in town with the opportunity to spread through the population, and, in time, possibly alter the genotypic and phenotypic properties of the population.

Members of the same species might respond differently to different geographical conditions. Those conditions will tend to induce various dimensions of the underlying genotype to express itself in different ways over time as a result of changes in the nature of competition, together with changes in the kinds of opportunities and challenges that exist with respect to changes in geographic conditions.

Genetic differences also arise in subsets of a given species through changes in one or several genes. Many of those changes have phenotypic consequences of one kind or another, and such phenotypic consequences are acted upon by natural selection.

As a result, populations possess genetic and phenotypic variability. That variability engages changes in environmental circumstances in different ways, and under the appropriate conditions of natural selection, certain dimensions of that variability might change more quickly than other dimensions of that same variability.

Speciation refers to the process in which two or more subsets of a ancestral population arise through processes that entail sufficient genetic differentiation and/or geographic separation to bring about a genotypic and phenotypic break with, or branching from, the ancestral population. Over time, the frequency and proportion of such changes move through the newly formed subsets of the ancestral population. Moreover, those changes occur in such a way (due in large part to the existence of relative, geographic segregation) that occasional or sporadic instances of interbreeding with members of the ancestral population do not prevent the transition in genotype and phenotype frequencies/proportions from continuing to move away, to varying degrees, from the set of genotypic and phenotypic traits that characterized the ancestral population.

The processes that lead to gradations in phenotypic and genotypic differences generating speciation tend to continue across hundreds of millions of years. Eventually, out of the foregoing continuous processes, the collective series of instances of speciation will lead to the emergence of new kinds of genera, families, orders, classes, phyla, and kingdoms ... that is, taxonomic categories.

The foregoing several pages of discussion highlight the essential themes of most textbooks that deal with the theory of evolution. Those themes are: natural selection (sexual selection, kin selection, and group selection are just variations on this theme), adaptation, population dynamics, genetic drift, modalities of geographic segregation, transitions in phenotypic traits, recombinant DNA/gene shuffling, mutation, biodiversity, and speciation.

The textbooks that are being alluded to in the foregoing several pages will develop the aforementioned themes in different ways. While the vocabulary that is used to accomplish such augmentation will introduce topics involving: historical considerations, various discoveries, fossil records, modes of classification, methods of quantification, and a plethora of details based on observations, experiments, studies, and disagreements, nonetheless, all of the new vocabulary being introduced into such textbooks tends to be directed toward expanding and lending specificity to the ten, or so, central themes that give expression to the theory of evolution and that previously have been outlined (however briefly) in the present chapter.

Unfortunately, although attempts are made in those textbooks to explain various topics – for example, the diversity of life, together with the biological principles that unify such diversity, as well as the origin(s) of life -- by weaving together, in different combinations, various elements from the ten, or so, central themes of evolutionary theory, nevertheless, there are key junctures in those explanations that repeatedly disappear into an omnipresent mist of assumptions. As result, those elements are never verified or substantiated.

For example, earlier, in conjunction with providing a brief overview involving the ideas of genetic drift and the neutral theory of molecular evolution, the notion of randomness was mentioned. Genetic drift was described as being the result of some combination of chance, random events that had nothing to do with the evolutionary fitness of an organism but were just a matter of the slings and arrows of outrageous fortune that impacted on whether, or not, an organism survived or reproduced and whether, or not, a given gene survived in order to be passed on to the next generation.

What does it mean for an event to be random? There are several possibilities.

One characterization of the idea of randomness is that we do not possess (at least currently) the methods, means, or understanding to be able to trace the ultimate causes of certain terminal events – including, for example, the occurrence of what are referred to as instances of genetic drift and/or mutations. The causes of those events are indeterminate in nature, and by referring to those kinds of causes as random, we really are saying we don't know why the events occurred in the way they did.

Of course, although we don't know how a given event came to be, nonetheless, there might be a possible explanation that does account for such an event, but, at the present time, we just don't know what that kind of an explanation looks like. However, one of the <u>possible</u> explanations for this or that event has to do with another sense of the meaning of randomness.

More specifically, this alternative approach is rooted in a philosophical orientation that claims there is no ultimate purpose to the universe. As a result, events merely give expression to the dynamic interaction of a chain of forces and factors that happen to come together – for no overarching rhyme nor reason -- and give expression to this or that phenomenon.

The foregoing philosophical mode of engaging the issue of randomness comes in at least two flavors. (a) There is no determinate set of principles and forces that required a given event to have occurred but, rather, <u>independent</u> forces and principles arbitrarily engaged one another and, in the process, became entangled in such a manner that, among other things, an event or phenomenon of a certain kind took place. However, the nature of the entangled dynamic of forces and principles that did take place might have turned out otherwise if slightly different kinds of interaction had taken place at certain points along the way … slightly different kinds of interaction that might easily have occurred but, inexplicably, did not. (b) There is an <u>interdependent</u> and determinate set of principles and forces that led to the occurrence of a given event, but there is no reason or purpose underlying why such a particular set of principles and forces exists or gives expression to the universe rather than some other set of principles and forces ... this is just the way things are.

Consequently, from one perspective, randomness is just another term for ignorance. From another perspective, randomness gives expression to a philosophy concerning the ontological nature of the universe and how it supposedly operates.

Are mutations random? If so, in what sense is that the case?

Are mutations random in the sense that we do not necessarily know how they came about? Or, are mutations random in the sense that they merely constitute the outcome of a long chain of interacting dynamics that, ultimately, are arbitrary in nature and just happen.

Whether one construes the idea of randomness as a way of alluding to one's ignorance or one construes the idea of randomness as a way of referring to how one believes the universe operates, in neither case does one actually <u>know</u> what, ultimately, is transpiring in the universe ... although, obviously, one might have <u>beliefs</u> concerning those matters. On the one hand, ignorance concerning the nature of how an event came to occur is a confession that one does not know what is going on, and, on the other hand, those who advocate randomness as an inherent property of the universe are not in any position to prove that this is the case and, therefore, really have no knowledge about whether, or not, the universe is random in any sense at all and have no knowledge concerning the manner in which random events conspired to generate one set of events rather than some other set of events.

I remember reading (nearly four decades ago) a May 1975 *Scientific American* article by Gregory Chaitin entitled '*Randomness and Mathematical Proof*'. One of the central themes of the article was that while one might be able to define randomness and measure it, one could not always prove – except in certain, special cases -- that a given sequence of numbers was random.

The general ideas underlying, and associated with, Chaitin's algorithmic approach to the issue of randomness will surface again later on in the book when various issues are explored in conjunction with: black holes, thermodynamics, and mathematics. So, for present purposes, I will restrict my comments concerning the foregoing *Scientific American* article.

Chaitin maintained that one of the key differences between random and non-random sequences had to do with the issue of compressibility. More specifically, on the one hand, non-random sequences could be represented by an algorithm that provided one with a set of instructions or a formula that permitted one to generate the sequence in question but that algorithm came in a compressed form that was smaller (contained less information) than the sequence that it generated, whereas, on the other hand, a random sequence could not be compressed into an algorithm that contained less information than the sequence itself.

Conceivably, a sequence of numbers might appear to be random because one hadn't found any algorithm capable of compressing the information so that the algorithm could be expressed using less information than the sequence it generated. However, what if an algorithm were subsequently discovered that could compress the information contained in the sequence in the desired way ... that is, into a specifiable, relatively short (compared to the sequence) algorithm?

One of the reasons why a sequence might not be capable of being proven to be random is because any proof that one advanced in this respect could not eliminate the possibility that the right kind of algorithm might emerge at some later point in time. As a result, there would be a certain amount of uncertainty or incompleteness concerning such proofs.

Chaitin makes a similar-sounding point in the aforementioned article by tying his definition of randomness to Kurt Gödel's work. However, I am going to construe the idea of uncertainty/incompleteness in a slightly different direction.

For example, I can think of at least one sequence of numbers (and there are many others that are similar to it) that might prove very difficult to predict what came next in the sequence unless one knew the algorithm for producing such a sequence. That sequence of numbers belongs to π.

To some, the sequence of numbers in π might appear to be random. However, there is a determinate method for producing each succeeding number in the sequence even though the number itself is said to be infinite in character.

What happens if one uses the notion of randomness in conjunction with a theory – such as evolution -- that is said to be scientific in nature? In what way is the notion of randomness scientific?

If something actually were random (whatever this might mean), we could never prove that this was the case. If there is no possibility of proof, then in what sense is science present?

Furthermore, if one were to talk about mutations in terms of what was, or was not, compressible algorithmically, this still would leave open the possibility that someone, at some later point in time, might be able to come up with an algorithm that could account for such a mutation that was expressible as a function of some compressible, algorithmic form capable of describing the dynamics underlying a mutation that, initially, appeared to be random (e.g., as might be the case if someone discovered -- after the fact -- that a given, known chemical was a mutagen or had carcinogenic properties and played a prominent role in causing a given mutation).

Alternatively, some individuals might like to argue that the idea of randomness is one of the assumptions or postulates that one takes as given, and, then, science proceeds from there. The issue, then becomes, one of trying to account for how events of a provably determinate and functional nature arise out of phenomena that are, ultimately, said to be random in character.

For instance, one might ask: How do random events lead to determinate and functional metabolic pathways, genetic systems, or viable organisms? The modern answer -- from an allegedly "scientific" perspective -- is that the processes of natural selection and genetic drift -- along with the other set of usual suspects or central themes of evolutionary theory -- tend to shape which series of random elements will be fixed or eliminated.

However, both natural selection and genetic drift presuppose the existence of a functional system or organism upon which to operate.

Therefore, neither natural selection nor genetic drift can explain the origins of the functionality or order that they are said to fix.

The neutral theory of molecular evolution maintains that most changes at the molecular level are random events that confer no advantage or disadvantage (i.e., are neutral) with respect to fitness. Consequently, such molecular changes cannot necessarily be described as the source of new modes of functionality

Of course, the foregoing sorts of changes might affect whatever genetic and phenotypic properties are present, but they cannot do so in any way that compromises the evolutionary fitness of existing, biological functionality. Furthermore, in order to be able to provide a scientific account concerning the emergence of such new functionality, one would have to be able to show how that new kind of genotypic and/or phenotypic functionality arose through a given set of neutral changes that, on the one hand, were random and, on the other hand, did not confer any advantage or disadvantage in the process of coming together as a new kind of functional unit.

Mutations -- alleged to be random -- that are disadvantageous tend not to survive. Forces of natural selection generally (but not always or not always right away) eliminate organisms containing traits that don't function properly or capacities that do not adapt well to existing, environmental conditions.

Therefore, while disadvantageous or lethal mutations are a source of newness in a biological system or population, that modality constitutes a form of 'newness' that is destined to disappear in either the short run or the long haul. As a result, no new forms of constructive, lasting functionality arise out of those kinds of mutation.

So, the only source of constructive newness must be in the form of mutations – said to be random – that lead to a sequence of events that inexplicably acquires the capacity to function in a way that can be endorsed, reinforced, or fixed by the forces of natural selection. However, in effect, one has assumed one's conclusions by arguing that functionality arises out of random events without ever demonstrating the truth of one's claims (e.g., that the events are truly random), and this is little more than argument by assertion.

Previously, I spoke about the idea that one could not prove that a sequence of numbers was random. One could only demonstrate that one did not currently know whether, or not, there was an algorithm capable of generating that sort of sequence.

Now, it seems that one cannot prove that a new form of functionality in an organism is a product of random events. One can only acknowledge that one does not currently possess an understanding capable of explaining how functionality arose out of randomness but, instead, one must assume (due to ignorance and/or philosophical inclination) that this is the case.

The randomness of something cannot be proven. Furthermore, the idea that randomness is capable of generating order cannot be proven if something cannot be shown to be random in the first place.

After all, what one is assuming to be a random phenomenon might not be. Instead, that phenomenon might just be the result of some determinate process for which one does not, yet, know the underlying algorithm.

By making randomness a fundamental postulate for a theory alleged to be scientific, what is one actually doing? One is muddying the waters as far as being able to clearly demarcate between science and philosophy is concerned.

If one has arranged one's postulates or assumptions in such a way that one either cannot know how things have come to be the way they are, or, one must allude to unproven philosophies concerning the manner in which the universe supposedly operates, then how can one be said to be doing science? If one cannot prove the likelihood of one of the most basic assumptions underlying the theory of evolution – namely, randomness – then while one might have a theory of evolution, the theory is not a scientific one because the ultimate 'explanation' for the origins of everything in biology that has a novel, functional character rests on something other than what can be shown to be true or accurate in a scientific manner.

Sometimes, the idea of randomness plays a central role in the formation of models that might reflect the possible nature of how things work. In other words, one develops a quantitative framework for what one might expect if events were described as being of a random nature, and, then, one compares what is observed against that model. However, most, if not all, quantitative models of randomness tend to be rooted in a theory about what constitutes the criteria of being random ... criteria that tend to entail arbitrary considerations.

For instance, consider the tossing of a coin. Supposedly, there are two possible outcomes to such a tossing process.

In actuality, there are more than two possibilities. For example, a coin could be lost when it lands ... perhaps, disappearing down a hole in the ground or down a heating duct in the floor. Or, a coin might fall in a way that it ends up landing on an edge and, perhaps with the help of some object against which it leans, stays that way.

There are an indefinite variety of ways that a tossed coin might become lost or land on an edge. Nonetheless, such possibilities are ignored, and a simplifying assumption is made that limits those possibilities down to just two.

The coin can turn up heads, or it can land tails up. The likelihood that either a heads or a tails will show up on any given toss of the coin is calculated to have a probability of ½ or .5.

Tests have been carried out, and the long-term distribution of coin tosses tends to match the foregoing probability. The more tosses that take place, then the closer the statistical distribution of those coin tosses approach the indicated probability calculation.

Does such a probability calculation capture something of the dynamic of randomness? To be sure, there is an element of randomness in the sense that we don't know which side of the coin will show up on any given toss.

If we bet on the outcome, we are taking a chance that we could be wrong in with respect to the character of our guess. However, ignorance concerning an outcome doesn't necessarily make the outcome an ontologically random event.

On the other hand, the nature of the collective sequence from one coin toss to the next might be considered to form a random sequence of an epistemological character. Nevertheless, aside from the previously mentioned issue of not being able to prove that a given sequence is random in an ontological sense because of the possibility that there might be some unknown algorithm capable of producing such a sequence, there is another consideration that impinges on the judgment of randomness with respect to such a sequence.

If the sequence is truly random, why does it generate a long-term distribution pattern of ½? The law of large numbers indicates that the more trials of the coin toss that are conducted, the closer the average of those trials should come to the expected distribution value of – in the case of coin tosses -- .5, but no one has ever explained why the law of large numbers works.

If that law were explained, perhaps we would know how order comes out of randomness. Unfortunately, the law of large numbers doesn't really explain anything ... it merely describes the determinate character of the average, expected outcome of a series of events.

In other words, the law of large numbers talks about how the expected outcomes -- based on the calculation of probabilities in a given situation – tend to approach what is actually observed if enough trials are completed. That law says nothing about how, or why, the dynamics of events that seem to give expression to a so-called random sequence are able to generate a determinate result.

Why assume that the expected outcome for a coin toss is 50-50 or .5? Why couldn't it be 70-30 or 20-80?

As indicated earlier, experiments have shown that the statistical distribution for heads and tails will approach the .5 figure given enough trials. Moreover, if there were a departure from such a distribution profile, one might begin to suspect there was some force or factor that was skewing the results away from an outcome for which there seems to be no obvious reason why it should be other than it is – that is, .5.

The law of large numbers resonates with the idea introduced earlier that indicated one isn't able to prove that a sequence is random because there might be an unknown algorithm capable of generating such a sequence. The law of large numbers alludes to the existence of such an algorithm, and, in fact, indicates – at least in the case of coin tossing -- that the nature of the algorithm consists in the flipping of the same coin in roughly the same fashion for a large number of trials or times, and one will be able to produce a long-term determinate outcome with respect to the distribution of heads and tails.

The foregoing algorithm is shorter than the sequence of heads and tails that it produces – assuming, of course, that the process of tossing the coins goes on for a sufficiently long enough period of time. Thus, the coin-tossing sequence is compressible (it can be represented by an algorithm) and, therefore, is not random in nature.

So, in a sense, we know the nature of the algorithm underlying the production of a sequence of events that appears to be random and, yet, is not random because that sequence leads to a determinate result or outcome that displays an average distribution sequence that is close to that which had been expected or predicted on the basis of a probability model developed in relation to a given set of conditions. Indeed, we might argue that one can repeat the experiment as often as one likes, and although the sequence from one experiment to the next is likely to be different and will appear to be random, nonetheless, the outcome of those experiments will always end by approaching a determinate result if the sample of experiments or trials is sufficiently large.

Nevertheless, despite what we might know about the probabilities of average outcomes, we still don't know what is going on. Does a sequence of events -- that are described as being random -- produce predictable, determinate results, or is that sequence of events only apparently random but, in actuality, gives expression to a determinate set of forces that – at least for the moment -- has escaped our understanding or ability to grasp what is transpiring?

Models of probability do not necessarily describe random events. Those models are about constructing methods for calculating outcomes based on the perceived number of degrees of freedom in a given set of circumstances. When it comes to coin tosses, there are two degrees of freedom ... in the case of dice, there are six degrees of freedom (and more degrees of freedom if one uses a pair of dice) ... in the case of playing cards, there are – if one excludes jokers -- 52 degrees of freedom (or 13 degrees of freedom if one only considers the members of a given suit, or 12 degrees of freedom associated with face cards, or 36 degrees of freedom for numbered cards, or 4 degrees of freedom for aces) ... and so on.

The degrees of freedom with respect to coins, dice, cards, and the like constitute <u>framed limits</u> that are <u>determinate</u> in terms of the kind of possibilities that they allow, but the manner in which those degrees of freedom will be manifested is unknown in terms of how that dimension of being determinant will play out in any given instance, Not just anything happens, but, rather, what happens, happens in terms of the nature of the phenomenon being considered.

In addition, models of probability are predicated on the principle that there is no force or set of forces that is capable of affecting how those degrees of freedom will normally manifest themselves in a given set of circumstances. For instance, dice should not be weighted in any manner that could render some results as being more likely than other possibilities, or cards cannot be shuffled in ways that lead to a stacked deck or they cannot carry identifying marks that reveal their identity in a manner that would skew the degrees of freedom that normally govern what can occur in conjunction with a deck of 52 cards.

Probability models constitute <u>descriptions</u> of how certain phenomena manifest themselves over time. Probability models will try to accurately reflect the degrees of freedom present in such phenomena in order to be able to construct reliable methods for quantifying what will happen in conjunction with such a set of degrees of freedom in the long run.

If the law of large numbers holds in relation to those sorts of phenomena, then – given a sufficiently large number of trials -- there will be a correlation between observed outcomes and predicted outcomes. However, probability models do not constitute an <u>explanation</u> for how or why a series of seemingly random events – that is characterized by some given number of degrees of freedom -- is able to end up as a determinate result.

Now, let's shift gears a little and consider the issue of mutations. Mutations might have x-number of degrees of freedom associated with the parameters of possibility to which those mutations are capable of giving expression. Moreover, the mutations that occur in conjunction with any of those degrees of freedom might prove to be advantageous, disadvantageous, or neutral.

However, on what basis would one claim that such mutations are random in nature? If one has developed a probability model to describe and predict the possible outcomes for what might happen in relation to the degrees of freedom entailed by the process of mutation, none of those degrees of freedom necessarily constitute <u>random</u>

variables, per se, and to label them as such is an exercise in arbitrariness.

On the one hand, if we <u>don't know</u> what brings a given mutation about, then, one is not in any position to claim that the mutation is a function of random events in any ontological sense that gives expression to a provable theory about how the universe operates in accordance with allegedly random forces. On the other hand, if we <u>do know</u> what causes a given mutation, then, an individual has his or her work cut out with respect to proving that the known proximate cause of the mutation is, actually, the end result of a random conjoining of a long chain of previously unrelated events.

Neither the idea of natural selection nor the ideas of adaptation, genetic drift, geographic segregation, or speciation can, in and of themselves, account for how new functional capacities arise. All those ideas presuppose biological functionality, and the forces to which those terms allude operate in conjunction with existing biological functionality.

Speciation occurs under two broad sets of general conditions. Those conditions involve: Either some modality of geographical segregation, or the emergence of new forms of genetic variation, or a mixture of both sets of conditions.

If one, or more subsets, of an ancestral population becomes geographically segregated from that population, the physical character of the segregation might, in and of itself, induce the genotype of members of the segregated subset to manifest different phenotypic properties as a result of the: New opportunities, decreased competition, and different challenges that might be associated with the formation of an environmental niche that is brought about by the process of segregation.

Any speciation that occurs in relation to the foregoing set of circumstances does not necessarily require, or depend on, the existence of entirely new capacities. Rather, new dimensions of already existing capacities are brought into play as a function of the changed character of the dynamic between the members of the subset of the original ancestral population and the new geographical circumstances that segregates them – temporarily, partially, or permanently – from the original population.

The question, then, becomes, what are the limits, if any, on the potential for manifesting different capacities as a result of the possibilities inherent in the gene pool that constitutes the collective potential for the members of a given subset of the ancestral population? Can one suppose there are no limits and, therefore, the potential for continued speciation is indefinitely large? Or, are there determinate limits on what is possible with respect to the shuffling of genes within any given gene pool as far as the emergence of further subsets is concerned that take place in conjunction with additional instances of geographical segregation that might tap into previously unexpressed genetic dimensions of new subsets that are drawn from the subset that, in turn, had been drawn from the original, ancestral population?

The boundary conditions of speciation are shrouded in uncertainty. We often do not know what the capacity for speciation of any given gene pool is ... that is, we do not know how many previously unexpressed dimensions (capacities) of a gene pool (or its descendent gene pools) are capable of being induced to express themselves under the right circumstances of geographical segregation, and, in the process, lead to further instances of speciation.

There is nothing that is currently known which justifies assuming that the capacity for speciation with respect to any given population, or descendant subsets, is indefinitely large. At the same time, we cannot necessarily establish or determine what the precise limits are in relation to the capacity for speciation with respect to a given population or its descendant subsets.

There are several ways in which it can be said that we don't know what the capacity for speciation is with respect to any given population -- together with descendant subsets. First, we do not know what the capacity is for the process of geographical segregation to be able to induce a given gene pool to manifest the sort of genotypic and phenotypic differences over time that would generate a new species. Secondly, we do not know what the capacity of a given gene pool is with respect to generating the sort of genotypic variation (via conjugation, recombinant DNA, mutations, and/or gene shuffling) that would be capable of leading to continuous speciation given the right opportunities (such as certain kinds of geographical segregation).

Is the process of speciation capable of leading to the formation of all the species, genera, families, orders, classes, phyla, and kingdoms that make up the known tree of life? We don't know, because, as indicated earlier, we don't know what the capacity for speciation is with respect to any given population or descendant subsets.

Many textbooks on evolution provide an array of specific instances -- steeped in considerable detail -- that explore the issue of how certain kinds of speciation might have occurred. Nonetheless, one cannot use an individual case – or even a series of them – to prove that all cases of speciation, in general, must have come about in a similar fashion.

Specific, documented cases of speciation certainly are suggestive with respect to what might have gone on in relation to undocumented instances of speciation. However, the former cases do not necessarily constitute any sort of proof as far as what can, or can't be said, with respect to the process of speciation in general.

Since the tree of life first set down roots on the planet Earth, it has sprouted millions, if not billions, of branches. Every branch entails some form of speciation that carries implications for issues involving the possible origins of genera, families, orders, classes, phyla, and kingdoms.

Speciation leads to what might be termed the branching problem. Although evolutionary biologists assume that all the branches on the tree of life have been generated through the known dynamics underlying speciation (e.g., natural selection, genetic drift, geographical segregation, biodiversity), there is really very little, if any, proof concerning any of this.

The movement from branch to branch is largely a function of assumption. Speciation occurs at the branching points, but what exactly is involved in such a process is not necessarily known.

This is especially the case when it comes to the appearance of new capacities and new functions that cannot necessarily be shown to have been possible in the context of a given gene pool ... even when conjugation or gene shuffling is taking into consideration. While some new genotypic and phenotypic capacities can be accounted for by the manner in which geographic segregation induces previously untapped dimensions of a gene pool to become manifest, one cannot necessarily demonstrate that the emergence of all new phenotypic and genotypic capacities came about in that fashion.

For example, Darwin's finches give expression to the sort of speciation potential that might be contained in an ancestral population from which different subsets break off and become geographically separated from one another. Over time, and given different geographical/ecological niches, one might anticipate that different subsets of the original ancestral population of finches might eventually show up with, among other things, longer beaks, or more curved beaks, or shorter beaks, and so on.

However, one would not expect those finches to show up as giraffes, kangaroos, or T-Rexes. The potential for speciation in a given population might not be precisely known, but there are certain types of limitations that seem to circumscribe that potential.

The members of the population for a given species give expression to an array of possibilities. Nevertheless, that array of possibilities cannot give expression to just any characteristic one likes but, instead, the set of possibilities for a species (its potential for speciation) tends to fall within a range of variations on particular themes that typify such a species.

The potential for speciation of a given population is intertwined with the branching problem. If one does not know what the potential for speciation is for a given population, then, one will have difficulty accounting for how a new species arose – if it did – from such a population.

There are all manner of questions entangled with the aforementioned branching problem and the related issue concerning the indeterminate character of a given species' potential for speciation. For example, we don't know how the first protocell(s) branched off from inorganic and organic chemical reactions, and among the reasons why we don't know the foregoing, is because we don't know what the speciation potential is -- if anything -- for prebiotic interactions.

Similarly, due to the indeterminate nature of the speciation potential for the relevant population, we don't know how the capacity for DNA coding branched off from life forms with no capacity for coding DNA? In addition, we don't know how organisms with the capacity for photosynthesis branched off from organisms without such a capacity. We don't know how the capacity to generate and use adenosine triphosphate to provide energy for metabolic pathways branched off from organisms that did not possess that capacity. We don't know how active forms of membrane transport branched off from non-active forms of membrane transport. We don't know how optical handedness in the molecules of life consisting of sugars (D – Dextrorotation – optical isomers) and amino acids (L – Levorotation -- optical isomers) branched off from life forms whose sugar and amino acid molecules might have consisted of racemic mixtures as far as their optical activity is concerned with respect to the way in which such molecules polarize light. We don't know how organisms with the capacity for meiosis and mitosis branched off from organisms without such capacities. We don't know how bacteria branched off from protocells. We don't know how aerobic life forms branched off from anaerobic life forms. We don't know how multicellular organisms branched off from single cell organisms. We don't know how the Eucarya, Archaea and Bacteria Kingdoms branched off from one another? We don't know how organisms with the capacity for motility branched off from organisms without motility. We don't know how animals and plants branched off from one another. We don't know how organisms with immune systems branched off from organisms without immune systems. We don't know how flowering plants branched off from non-flowering life forms? We don't know how organisms with skeletal systems branched off from organisms without a skeletal system. We don't know how organisms with a developmental life cycle rooted in specialized cell functioning branched off from organisms without such developmental life cycle that is rooted in cell specialization. We don't know how organisms with the capacity to form hearts, kidneys, livers, lungs, pancreases, stomachs, and circulatory systems branched off from organisms without such capacities. We don't know how neurons and glial cells branched off from other kinds of cells. We don't know how organisms with the capacity for memory branched off from organisms without a capacity for memory. We don't know how organisms with endocrine systems branched off from organisms without endocrine systems. Finally, one needs to add to the foregoing considerations, the array of branching problems that arise in conjunction with issues of how consciousness, intelligence, emotion, language, and creativity arose from organisms not possessing those sorts of capabilities.

Evolutionary biologists <u>always</u> assume that the speciation potential for the relevant population in all of the foregoing cases is capable of accounting for the branching problem associated with each of the challenges noted above. However, as far as the cases cited in the previous paragraph are concerned, evolutionary biologists have not brought forth any conclusive evidence about how any of the foregoing branching problems would have been bridged by the population that is, supposedly, giving rise to the new species ... a species that has capacities not present in the previous population.

As a result, allegedly random events that, supposedly, are helping to account for speciation -- or the branching problem -- are shrouded in the mists of the unknown and, perhaps, the unknowable. Moreover, the process of speciation – along with the issue of speciation potential for any relevant population linked to the branching problems outlined in the last several pages – also are shrouded in mists of the unknown and, perhaps, the unknowable.

The branching problem encompasses both of the foregoing dimensions of the unknown – and, possibly, the unknowable. In other words, neither known forms of speciation, nor allegedly random events – considered separately or together -- can necessarily account – in any reasonable or scientific manner -- for how the transition from one branch of the tree of life to another one actually takes place, but, instead, the transitions are often bridged by assumptions that are not capable of being proven.

Conjectures abound. Unfortunately, proof concerning the truth of any of those conjectures is largely, if not entirely, absent.

The extended dynamics of population biology are capable of plausibly accounting for <u>some forms</u> of speciation, but not necessarily <u>all forms</u> of speciation (and one should keep in mind that a plausible account is not necessarily the same thing as a true account). Many of the theories that describe the dynamics of population biology can justifiably be referred to as scientific ... but only to the extent that the claims entailed by those theories can be rooted in the rigorous practices of scientific method.

The theory of evolution might well be a theory. However, it cannot necessarily be justifiably referred to as a scientific theory because the dimension of science is often missing from its conjectures concerning its proposed solutions to the branching problem that has been outlined in the last several pages.

Aside from the dynamics of population biology considered in rather narrow terms (i.e., minus conjectures, speculations, and assumptions), the theory of evolution is largely a narrative, rather than a scientific theory. That narrative is glued together with assumption upon assumption inserted at critical junctures in relation to all of the foregoing sorts of branching problems (and millions more) involving speciation and, as well, is glued together with assumption upon assumption inserted at critical junctures in relation to the idea that so many evolutionary events are supposedly of a random nature ... but a randomness that cannot be proven as such.

Theodosius Dobzhansky claims that 'nothing makes sense in biology except in the light of evolution' because the latter theory is capable of tying together a large set of what, otherwise, would be isolated, disparate pieces of biological information and showing how that theory provides a unified framework for understanding diversity. However, that sense of unity is largely a function of assumptions involving the roles that speciation and randomness are conjectured to play in the grand philosophy to which evolution gives expression.

As far as Dobzhansky is concerned, nothing makes sense in biology because he – and anyone else who thinks in the same way – was not prepared to take a sufficiently, rigorous critical look at all the ways in which evolution is not capable of making scientific sense of anything in biology unless one buys into a litany of assumptions concerning speciation and randomness ... assumptions that have not been proven and might never be able to be proven. Stated in another way, for Dobzhansky, the nature of biology is largely bereft of meaning unless one subscribes to the philosophical assumptions that subsidize the theory of evolution and render it meaningful.

Whatever science exists in conjunction with the theory of evolution is a function of what is required to establish the truth concerning the disparate observations, measurements, and experiments that Dobzhansky seems to find so devoid of meaning.

Consequently, up to a point, the framework of population biology is able to make sense of many of those variable instances of observation, measurement, and experiment, precisely because it gives expression to a methodologically rigorous way of tying together many observations, measurements, and experiments that, otherwise, would be isolated pieces of information.

The framework of population biology is not necessarily co-extensive with evolutionary theory. In fact, population biology only gives expression to one relatively small dimension of evolutionary theory.

Population biology – which, among other things, studies changes in the phenotypic and genotypic frequencies/proportions that characterize a given population over time and, as well, explores how such changes tend to hinder or help 'fitness' with respect to a given set of environmental conditions -- entails a considerable amount of science. Technical areas of study such as: Statistics, mathematics, genetics, molecular biology, and ecology are all part of the mix when it comes to exploring and developing the science of population biology.

Evolutionary theory attempts to bask in the glow of the foregoing sorts of scientific features. In the process, evolutionary theory seeks to illicitly borrow some degree of credibility from the science that takes place in conjunction with the study of population biology and, then tries to transfer that illegitimately acquired credibility to the philosophical narrative that falls beyond the horizons of population biology but lies at the very heart of evolutionary theory.

Population biology does not try to – nor does it have to -- solve the branching problem outlined earlier because population biology does not concern itself with explaining how speciation occurs. Rather, population biology takes the existence of a species as a given, and, then, seeks to explore what happens, over time, with respect to changes in the frequencies/proportions of phenotypic and genotypic properties of a species under different environmental circumstances and genetic conditions.

The theory of evolution does not make a whole lot of sense unless one can demonstrate that all the branches of the tree of life are a function of the processes of speciation that operate in collaboration with a lengthy series of allegedly random events. If there is no detailed, coherent account of speciation that demonstrates how each and every branch of the tree of life arose, then, one really doesn't have a scientific theory, but, instead, one has a philosophical narrative that is posing as a scientific theory.

A Few Lessons Related to Archaea

Some scientists never seem to learn. They are like a more sophisticated version of those times when Charlie Brown believes that he has Lucy all figured out and has come up with satisfactory answers for his anxieties about whether, or not, Lucy will pull the football back just as Charlie is trying to kick the ball.

Unfortunately, Charlie's calculations and predictions in this respect invariably turn out to be wrong. There are some relatively simple reasons for why things consistently turn out the way they do for Charlie as far as the football kicking (euphemistically speaking) issue is concerned.

Firstly, Charlie doesn't seem to have much insight into how Lucy's mind works. Secondly, Charlie is inclined to extend the benefit of a doubt (involving his own assessment of the situation) to someone that he shouldn't trust.

Similarly, many scientists don't necessarily have much insight into how the universe works and, as a result, they keep deluding themselves concerning the nature of reality through one conjecture or another. Moreover, many scientists often give the benefit of a doubt to other individuals – scientists who believe they know when they don't – and toward whom the former individuals ought to harbor a more critical perspective.

Approximately 39 years ago, a revolution began with respect to the way in which evolutionary biologists and microbiologists, among others, understood the nature of reality. As is the case with so many revolutions in science, the upheaval in understanding that began to emerge nearly four decades ago was in opposition to the biases and beliefs of an array of scientific experts and leaders who, in certain respects, conflated their ignorance with whatever knowledge they actually had.

Unfortunately, there were a lot of researchers and academics that, like Charlie Brown, placed their conceptual trust in individuals who didn't necessarily deserve that sort of consideration. At least part of the reason for such misplaced trust is that many of those researchers didn't necessarily have all that much insight into the issue under consideration, and, as a result, they permitted scientific reputations to lead them around by the nose instead of critically engaging the topic for themselves.

Up until approximately 1975, the world of living organisms had been divided into two broad Kingdoms – prokaryotes and eukaryotes. The differences between the two categories of life are fairly extensive.

Prokaryotes do not have a true nucleus (that is, there is no permeable membrane surrounding, among other things, the genetic instructions for a cell), but eukaryotes do exhibit a true nucleus. Eukaryotic cells wrap their DNA around histones (a certain kind of protein), whereas prokaryotes wind their genetic material around histone-like proteins. Eukaryotes possess multiple chromosomes, whereas prokaryotes tend to have one plasmid (often consisting in a circular strand of DNA). Mitochondria -- where biochemical processes involving energy production and respiration take place -- exist in eukaryotes but do not exist in prokaryotes. The ribosomes -- functional units that bind messenger RNA and transfer RNA in order to generate (synthesize) polypeptides and proteins -- in prokaryotes are significantly smaller than their counterparts in eukaryotes. When chlorophyll is present in prokaryotes it tends to circulate freely in the cytoplasm of the cell, but in eukaryotes, chlorophyll is contained within organelles known as chloroplasts. Organelles such as: the Golgi apparatus (which has various functions including intracellular transport), the endoplasmic reticulum (a network of membranous-like structures connected to the nucleus that plays a role in the synthesis of lipids and proteins), and lysosomes (contains enzymes that can break down various kinds of molecular structures) exist in eukaryotes, but not in prokaryotes.

There are a variety of other potential differences between prokaryotes and eukaryotes. However, the foregoing set of comparisons is adequate with respect to the current discussion.

Beginning in the early-to-mid 1970s, research by Carl Woese, a molecular biologist, strongly suggested that a third realm of life forms should be added to the previously established bi-modal classification scheme that divided up life forms into prokaryotic and eukaryotic kinds of organisms. Although the name eventually given to these newly discovered life forms was "Archaea", there was a time between 1977 and 1990 when Woese referred to them as archaebacteria.

The latter terminology might have misled some people. More specifically, the name seemed to suggest that the newly discovered life forms were some species of bacteria, but this was not the case (more on this shortly).

Around 1990, Woese began to refer to three domains of life: namely, eukaryotes, bacteria, and archaea. The designation "prokaryote" had disappeared from his manner of classifying the different kingdoms of life.

For quite a few years -- beginning in the early 1960s and before his discovery of archaea -- Woese had been trying to come up with a molecular taxonomy for life forms that would help connect known organisms to their molecular origins in relation to the formation of the first, primitive protocells. If – as evolutionary biologists maintained – inorganic and organic chemistry somehow led to the appearance of semi-functional and/or functional protocells, then all subsequent life forms should be solidly rooted in the formation of the specialized molecules that arose out of various kinds of inorganic and organic reactions that, eventually, led to the emergence of a variety of life forms.

Thus, one of the major reasons for organizing life forms into the aforementioned tripartite scheme of classification was rooted in Woese's interest in drawing the attention of scientists to some of the differences in molecular biology among various life forms. In other words, Woese wanted to develop a taxonomy for certain kinds of life forms that was based on molecular differences and that might be capable of linking life forms (both current and extinct) to their molecular past in relation to the advent of the first protocells since protocells were thought of as a set of interconnected molecular pathways that, somehow, had acquired the capacity to organize the synthesizing and degrading of various molecules in ways that helped make life possible.

Prior to the time when Woese began his project concerning the development of a system of molecular taxonomy, microbiologists and evolutionary biologists had invested considerable time in trying to discover some principle or set of principles that might point the way to arranging the bewildering array of microorganisms in an ordered, understandable manner. They had considered all kinds of physical, chemical, and metabolic properties in the search for a theme or themes that might help structure the multitude of microorganisms in an intelligible way ... but to no avail.

As a result, many microbiologists and evolutionary biologists seemed to have become disillusioned with the possibility of ever being able to make sense of the history underlying the evolution of different kinds of microorganisms and how they might have branched off from one another. The discovery of archaea might provide the sort of conceptual foothold needed to begin to make progress in the development of a molecular-based taxonomic system.

Quite a few scientists seemed to think that once the structural character of the code for DNA had been established, everything else was merely derivative detail. Woese, on the other hand, believed that more was needed in order to be able to get a better grasp of how things might have developed over evolutionary history, and, for Woese, part of the 'more' that was needed revolved around the problems involved in coming to understand how DNA coding was translated into components that could give expression to biological activity.

Ribosomes consist of an integrated set of proteins and RNA molecules that are responsible for stringing together an array of amino acids to form various kinds of polypeptides and proteins. Given the significance of the role played by ribosomes, Woese felt that these entities might cast an illuminating light on some of the possible ways in which the capacity to synthesize (to translate) polypeptides and proteins might have arisen over the course of evolutionary history.

In other words, differences in the structural character of ribosomes might imply differences with respect to evolutionary history. In this respect, Woese believed that an important key to unlocking at least part of the character of evolutionary history might come through identifying the structural character of ribosomes that were intimately involved in the process of translating DNA coding into proteins ... proteins that, in turn, could be used to lay down metabolic pathways through which an array of other kinds of biological activity might arise.

Woese concentrated on sequencing the 16s rRNA (ribosomal RNA) gene that occurs in microorganisms. These units consist of just

1,542 nucleotide bases, and, yet, they became the royal road to differentiating microorganisms from one another because each species of organism had an oligonucleotide 'fingerprint' – that is, a relatively small subsection of the nucleotide bases that constituted a unique sequence of coding for any given species of microorganism.

Woese knew from his own research that the nucleotide base pairings that underwrote 16s ribosomal RNA tended to be highly conserved in various species. Consequently, when one came across significant differences in those base pairings, one had found something that might be of considerable importance with respect to being able to develop a method for tracing or mapping the changes in various kinds of microorganisms that occurred over time.

Woese believed that the more similar the 16s rRNA oligonucleotide sequences were in relation to different species, then, the more closely (in terms of evolutionary history) their branches might be connected to one another. Alternatively, the more dissimilar the 16s rRNA oligonucleotide sequences for different organisms were, then, the more distantly related – in evolutionary terms – those organisms were considered to be with respect to one another.

While identifying unique oligonucleotide sequences might be able to help one to classify and differentiate microorganisms from one another, this capacity doesn't necessarily permit one to resolve the branching problem discussed the previous section of this chapter. In other words, establishing the fact (as Woese did) that microorganisms carried oligonucleotide signatures or markers that enabled one to classify different species of microorganisms, this mode of classification didn't necessarily account for: How any given oligonucleotide signature/marker arose in the first place, or how the transition was made from one kind of signature/marker to another.

Some of those transitions might be accounted for in a reasonable manner by means of the dynamics of speciation as understood by evolutionary biologists. However, possible transition scenarios might not always be plausible, and, consequently, one would have to go on a case-by-case basis as to whether, or not, any given proposed transition of oligonucleotide sequences made sense or, instead, began to stretch one's willingness to extend the benefit of a doubt concerning the credibility of such proposed transitions.

In any event, as Woese's catalog of oligonucleotide sequences began to grow, one of Woese's colleagues – Ralph Wolfe – began to wonder about where some of the microorganisms he (Wolfe) had been studying might fit into the molecular taxonomy that Woese was constructing, and, as well, Wolfe began to wonder where some other interesting, but little studied, microorganisms might be placed in such a taxonomy.

More specifically, Wolfe had acquired some expertise in being able to culture or grow anaerobic (environmental conditions involving no free oxygen), methane-producing microorganism (known as methanogens). Establishing such cultures was a finicky affair involving, among other things, the right combinations and amounts of nutrients.

In nature, methanogens were found in some rather unsavory environments – or so it might seem to some individuals -- such as sewage sludge and the rumens (the first stomach) of cows. Later on, it was discovered that methanogens could also flourish in the high temperatures of volcanic vents.

Wolfe also knew about the existence of other microorganisms that were found in environments of a rather inhospitable nature. For example, some microorganisms had been discovered basking in extreme conditions involving both elevated temperatures (thermophiles) and high sulfur content, while other organisms had been found in conditions characterized by high salt content (halophiles).

When the 16s rRNA oligonucleotide genetic sections of such organisms were sequenced, the foregoing organisms, along with methanogens, seemed to exhibit oligonucleotide signatures that were very different from any of the other microorganisms (mostly bacteria) that had been catalogued by Woese. In addition, these particular life forms seemed to share other characteristics that were not found in bacteria.

For example, the organisms that appeared to be un-bacteria-like displayed lipid linkages -- as well as a form of chirality with respect to the central carbon atoms in glycerol units -- that were different from what one encountered in most bacterial lipid molecules (which play important roles in the membranes of bacterial and archaea cells).

Furthermore, these apparently non-bacterial forms of life used a different kind of RNA polymerase – the enzyme that is used to convert DNA into messenger RNA – than bacteria do, and, as well, they shared a resistance to certain antibiotics (e.g., rifampicin ... which disrupts bacterial transcription – the process of instantiating DNA information in the form of RNA sequences).

The newly discovered life forms seemed to be neither fish nor fowl. That is, they didn't seem to belong to either prokaryotic or eukaryotic categories of classification.

In 1977, Carl Woese, along with George Fox (a post-doctoral student), wrote a paper that appeared in the November edition of the *Proceedings for the National Academy of Science*. Their paper discussed some of the evidence that supported their ideas about a new way of classifying life, and the two authors of the article argued that the newly discovered, non-bacterial and non-eukaryotic forms of life to which they were alluding in their paper should form a domain of their own.

The new domain of life subsequently was described by NASA, NSF (National Science Foundation), and *Newsweek* magazine as being a more ancient form of life than either prokaryotes or eukaryotes. However, such descriptions seemed to be devoid of any explanation with respect to how the transition in speciation from the new domain of life forms -- archaebacteria -- to prokaryotes was made.

After all, as indicated previously, the differences between archaebacteria and bacteria went beyond their respective oligonucleotide signatures, but encompassed, as well, major differences in, among other things, RNA polymerase enzymes, antibiotic sensitivity, and lipid formation. Consequently, there were a lot of changes for which to account before someone might plausibly claim that she or he could explain how bacteria branched off from archaebacteria ... if that is how things actually took place.

Notwithstanding the foregoing considerations, many scientists – including at least one Nobel laureate -- ridiculed the idea that a new domain of life needed to be added to the already established domains of prokaryotes and eukaryotes. However – and most unfortunately -- the criticisms directed toward Woese and Fox did not revolve around meticulously crafted scientific arguments that were rooted in observation, experiments and critical analysis.

Rather, those attacks were rooted in the stasis of entrenched ways of thinking about things. Inertial conceptual forces had been set in motion as a reaction to the Woese/Fox paper, and those forces were trying to prevent the light of a different and promising way of looking at data from gaining traction in the hallowed halls of research and academia.

People that shouldn't have been trusted on the issue -- because, at least for a time, they forgot what science actually involves -- were trusted. Furthermore, individuals who weren't willing to critically engage the evidence concerning archaebacteria on its own terms were prepared to act like lemmings and follow the nominal 'leaders' over the cliff of scientific propriety.

One giant figure in the annals of evolutionary biology – Ernst Mayr – appeared to support the work being done in conjunction with archaebacteria ... at least this seemed to be the case in the early years of that research. However, Mayr became opposed to things when Woese started to treat species of archaea as part of a formal, taxonomic system of classification that divided life forms up into three domains.

As a result, Mayr went to his grave maintaining that archaea did not form a separate domain of life forms. He felt that Woese had gone too far in relation to the molecular approach to taxonomy that was being thrown into the fray, and, yet, the criteria for what constituted going 'too far' seemed rather arbitrary and tied to unproven, pet theories about how the universe of evolution was believed to operate.

Arthur Schopenhauer once indicated that all truth goes through three stages. "First, it is ridiculed. Second, it is violently opposed. Third, it is accepted as being self-evident."

For years certainly, Woese's research was ridiculed. In addition, it was opposed ... adamantly and unpleasantly perhaps, rather than violently so. And, finally, it was accepted as being – if not self-evident – at least true.

J. Craig Venter sequenced the genome of a methanogen and compared it with both the sequenced genome of a bacterium as well as with certain oligonucleotide sections that had been derived from eukaryotic organisms. On August 25, 1996, Venter, together with some editorial personnel from the prestigious magazine *Science*, organized a press conference at the National Press Club in Washington, D.C. in order to announce the results of Venter's comparison study.

Venter had come to the conclusion that there should no longer be any doubt concerning whether, or not, Archaea constituted a different taxonomic domain of its own. There were, in fact, significant differences among the sequences for the three life forms that were being compared in his study.

Indeed, during the press conference, Venter noted that at least two-thirds of the genes sequenced in the methanogen did not resemble anything that had been observed in conjunction with either bacterial or eukaryotic life forms. Venter was confirming -- and rather emphatically expanding upon -- the research that Woese had been carrying out for more than a quarter of a century.

Unsuspectingly, Venter also was contributing to the branching problem outlined in the previous section of this chapter. More specifically, how does one account for the emergence or origin of so many genes that are unlike anything previously seen in either bacterial or eukaryotic life forms? How does one account for the transition from such a different set of genes in archaea to the ones that are observed in bacterial and eukaryotic life forms?

Woese has indicated that it might not be possible to sort out such questions and issues. The reason for this has to do with something called "horizontal" or "lateral" gene transfer.

Horizontal gene transfer does not operate in accordance with the normal mode of gene transfer – referred to as "vertical gene transfer" – in which genes are passed down to progeny via some form of reproductive process (asexual or sexual). Horizontal gene transfer involves the exchange of genes via conjugation or via the transfer of genes in conjunction with some sort of viral agent or via jumping genes (mobile segments of DNA).

Ever the maverick and original thinker, Woese had developed a perspective that ran counter to the more traditional or Darwinian idea in which all subsequent life forms (including all three domains or kingdoms) arose from a common ancestor. Instead, Woese believed that 'in the beginning' there might have been three sorts of protocells or protocell-like organisms that were immersed in a medium consisting of, among other things, many kinds of genes.

Moreover, it is even possible that the starting points for life might have been some sort of network of metabolic pathway that could have served as precursors to the emergence of protocells. In either case, genes flowed into and out of metabolic networks and/or protocells via horizontal gene transfer.

According to Woese, horizontal gene transfers, operating in conjunction with whatever protocells or networks of metabolic pathways existed in the early days of evolution, eventually led to the rise of the three domains of life forms that are known today. Nevertheless, in whatever way Woese wishes to describe his ideas, they still leave unanswered or unaddressed the issue of how functional genes of any kind arose in the first place.

Proposing that a medium existed at some point on early Earth that was replete with sequences of DNA that are referred to as "genes, is neither here nor there. Unless those 'genes' have some sort of functionality of their own and/or have a functionality in the context of a network of metabolic pathways that is capable of synthesizing one, or more, components that leads to the establishment of biological functioning of some kind, then one could exchange as many 'genes' as one likes through horizontal gene transfer, and nothing of much interest will necessarily take place.

Not just any sequence of DNA or RNA will suffice. As indicated in the foregoing paragraph, sequences must have, in some sense of the term, "functional potential", and, therefore, there needs to be an account of how functionality arises in the sea of genes that Woese is envisioning.

If one likes, one can assume that metabolic pathways made up of an interlocking set of functional genes somehow emerged. Nonetheless, one still needs to scientifically demonstrate how any of this is possible ... if not plausible.

One can assume as many things as one likes. However, at some point, the inclination to rely almost exclusively on assumptions as a means of bridging whatever seems inexplicable or problematic takes one beyond the horizons of science and into the realm of philosophy.

The issue of functionality is related to, but not necessarily coextensive with, the branching problem discussed earlier. To be sure, the branching problem requires one to explain how one moves from one kind of biologically functioning system (i.e., species) to another – somewhat different – kind of biologically functioning system (i.e., the new species that gives expression to the process of speciation).

Nevertheless, one encounters a different set of problems when one is faced with the task of trying to account for the emergence of functionality in the first operational protocell or network of metabolic pathways. In other words, accounting for how the very first species – along with the archetypal prototypic capacity for speciation – came into being entails a slightly different set of explanatory problems than does trying to account for how subsequent species arise <u>given</u> the existence of an already functional life form ... although, admittedly, there is a certain amount of overlap between the two kinds of problems.

Similarly, accounting for how a protein with an entirely new kind of functional capacity arises is a different kind of problem than trying to account for how a certain protein might have transitioned into a slightly different protein that possesses a marginally different function than did the former protein. Furthermore, trying to explain how a new metabolic pathway first became established is a different issue than trying to explain how an existing metabolic pathway might have acquired certain differences over a period of time, and, in the process, led to the formation of a new species.

For example, consider the archaea life form known as Nanoarchaeum equitans. This organism was discovered at a depth of approximately 350 feet within a volcanic vent of the Kolbeinsey Ridge, north of Iceland.

The foregoing organism is attached to the outer membrane of a variety of archaea species. Many, if not most or all, of these latter species belong to groups known as thermophiles (able to flourish in conditions of high heat ranging from 150 to 170 degrees Fahrenheit) and hyperthermophiles (flourishing in temperatures up to, at least, 235 degrees Fahrenheit) ... if not beyond.

Nanoarchaeum exists in a number of different forms. These differences seem to be a function, at least in part, of the kind of hyperthermophiles to which they are attached.

In a variety of ways, Nanoarchaeum constitute a rather strange form of organism. On the one hand, it operates without a complete set of genetic instructions.

Therefore, like a virus, it must borrow a certain amount of metabolic machinery from its host ... machinery that is needed for the synthesis of, among other things, amino acids, certain co-factors, and lipids. Furthermore, like a virus, it apparently remains dormant when not attached to a host.

Yet, Nanoarchaeum has not been classified as a virus. Instead, Nanoarchaeum is considered to belong to the domain of Archaea ... although -- since it appears to be either a symbiont or a parasite -- it is a form of archaea that had not been encountered previously.

Relatively recently, various kinds of megaviruses have been discovered ... some of which have roughly four times (2300 genes) the number of genes (563 genes) contained by Nanoarchaeum. In addition, relative to other non-viral entities, Nanoarchaeum is quite small, and, in fact, it is one of the smallest -- if not the smallest – life form ever discovered.

Based on an analysis of the amino acid sequences found in a number of its ribosomal proteins, Nanoarchaeum turns out to be quite different from many other species in the domain of Archaea. Indeed, some of its properties are so different that various microbiologists believe – but not everyone agrees -- that Nanoarchaeum might form a separate branch of life within the domain of Archaea.

On the other hand, whatever disagreements might exist in conjunction with whether, or not, Nanoarchaeum gives expression to a new branch of Archaea, there seems to be a general consensus that this species of Archaea constitutes a very ancient form of life. Some individuals believe that it might even have made up part of the root of the tree of life from which subsequent species sprang.

The foregoing possibility leads to a variety of questions. For example, if Nanoarchaeum is closely affiliated with, or is an instance of, some of the primitive precursors of later life forms, and if

Nanoarchaeum consists of a system of genes that cannot function on its own, and if Nanoarchaeum tends to remain dormant without the presence of an appropriate host, then how did Nanoarchaeum – along with the other species of archaea on which it depends -- come into existence.

By being one of the smallest -- if not the smallest -- life forms known to humankind, Nanoarchaeum might only possess an incomplete set of 563 genes, but, nevertheless, they are functional genes. The origin of that sort of functionality needs to be explained, and this amounts not to one question, but 563 of them ... in fact, additional questions (at least 490, 885 of them) will arise and lead beyond the foregoing number (563) as one tries to account for how the hundreds, if not thousands, of nucleotide base pairs that make up each of those 563 genes came to have sequences that, when translated and transcribed, formed functional units.

The foregoing issues, problems, and questions are not inconsequential. This is because the aforementioned figure – 563 genes – is close to what some evolutionary biologists consider to be around the minimal number of genes needed for life to be self-sustaining and, therefore, might, or might not, be intimately caught up with the origin of life issue.

Aside from the considerations noted during the last several paragraphs, there are other sorts of questions and problems that tend to emerge. For example, if Nanoarchaeum life forms came into existence before, say, the aforementioned hyperthermophiles, and, as such, constitute some sort of predecessor to the latter species, then, how did the hyperthermophiles arise? On the other hand, if the hyperthermophiles were first up on the tree of life, then one must try to account for a form of life that has many more genes than Nanoarchaeum does (and, consequently, is capable of generating many more questions) and, as well, there will be an additional litany of questions about how Nanoarchaeum evolved subsequent to the hyperthermophiles.

Earlier, in passing, entities referred to as megaviruses were mentioned. The size of some of these megaviruses – in terms of the number of genes and the number of base pairs they encompass, as well as in terms of the fact that they are big enough to be seen with a light (rather than with an electron) microscope – dwarfs the size of Nanoarchaeum and are even are larger than some forms of bacteria.

The genome of the larger of two megaviruses found in the ocean near Chile and in a fresh-water lake in Australia contained as many as 2.6 million base pairs. The genome of Nanoarchaeum contained less than one-fifth (490, 885) the number of base pairs carried by the larger of the two aforementioned viruses.

When the base pairs of the megaviruses (referred to as Pandoraviruses) were sequenced, something very intriguing was discovered. More specifically, only between 7 and 15 percent of the base pairs matched up with anything in the databases that catalogued sequenced base pairings.

Since, for the most part, the foregoing base pairing sequences didn't match up with known base pairing sequences this means there are millions of questions surrounding how such differences arose. Those questions have to do with trying to figure out how millions of base pairings came together to form 2300 functional genes ... and why certain genes necessary to make the megavirus into an autonomous, self-perpetuating life form were missing.

One of the smallest -- yet fully autonomous -- species of bacteria that exists is quite common in the oceans of the world. Its name is Pelagibacter ubique.

It consists of 1,389 genes and 1,308,759 base pairs. This makes it more than twice as large as Nanoarchaeum, but only half the size of some Pandoraviruses.

Having a substantial number of functional genes doesn't, in and of itself, necessarily confer life. Pandoraviruses have more genes and more base pairs than Pelagibacter, but the former is not considered a form of life – at least not of an autonomous kind – while Pelagibacter is classified as a bacterial life form.

Nanoarchaeum has less than half the genetic material (both in terms of the number of genes and the number of base pairs) that is contained in the bacterium, Pelagibacter ubique. In addition, Nanoarchaeum does not carry the full complement of genes needed to code for proteins, lipids, co-factors, and so on, and, yet, unlike

Pandoraviruses, Nanoarchaeum is still considered a life form belonging to the domain of archaea.

Where and how one draws the line that separates the living from the non-living does not seem to be a straightforward function of either the number of genes, the number of base pairs, and whether, or not, a given entity is fully autonomous. Nonetheless, no matter how one defines the line of demarcation between life and non-life, one still has to account for how genes with some degree of functionality arose out of thousands, if not millions, of base pairs ... because although functional genes obviously exist in species from the domains of archaea and bacteria – both of which are described as being made up of living species -- nevertheless, functional genes also exist in megaviruses such as Pandoraviruses that are considered to be non-living entities.

How does functional order arise out of an ocean of 'genes' that are not necessarily functional to begin with ... unless, of course, one arbitrarily – and, therefore, without proof -- supposes that at least some of those 'genes' have a functional capacity? Even given an ocean of at least some genes with a degree of functional capacity, how does the horizontal transfer of genes bring about ordered systems of biological functionality?

Various microbiologists seem to feel that the discovery and study of Archaea – and, perhaps, megaviruses -- gets us all closer to arriving at an understanding concerning the origin(s) of life. However, evolutionary biologists don't seem to have any means of separating the wheat from the chaff when it comes to trying to answer any of the foregoing sorts of questions in a rigorous and a reliable fashion.

Saying -- as Woese does -- that protocells or networks of metabolic pathways arose in an ocean of "genes" -- where horizontal gene transfer was common -- doesn't address any of a variety of basic questions in a very specific manner. Instead, 'explanations' – if one can call them that -- are so saturated with assumptions of one kind or another that trying to claim that any such network of assumptions gets us closer to understanding the origin of life is a lot like trying to claim that landing on Pluto gets us closer to reaching the Andromeda galaxy ... which in a sense might be true, but not in any way that makes much of a difference as far as reaching Andromeda is concerned.

A reader should not infer from the foregoing comments that I believe horizontal gene transfer doesn't occur in 'real' life or that such transfers don't play important roles in the lives of microorganisms. For example, one might note that since 1988 the Hawaii Ocean Time-series research project (HOT) – together with related research projects elsewhere in the world – acquired microorganisms and viruses from ocean samples collected at depths ranging 40 to 13,000 feet.

The researchers have sequenced the genomes of those samples. Among the millions and millions of base pairs that have been catalogued in relation to those samples, a multitude of new genes (extending into the thousands) have been discovered.

More importantly, at least for present purposes, those researchers uncovered a great deal of evidence supporting, if not proving, the idea that an extensive amount of horizontal gene transfer occurs among the microorganisms and viruses that made up the samples collected. However, demonstrating that horizontal gene transfer currently occurs in viruses and microorganisms – and, as well, has occurred in the past – does not really explain how the capacity for horizontal gene transfer arose originally, nor does it account for how the genes that are being horizontally transferred acquired their initial functionality millions – perhaps billions -- of years ago ... nor does the current existence of horizontal gene transfer explain how the functional genes that have been horizontally transferred became incorporated into the genetic programming of the entity to which the genes have been transferred ... nor does it account for how the capacity to incorporate genes from other organisms came into being so that those transferred genes could become appropriately modified to become adaptive, functional units within the cellular ecology of the latter organisms.

The foregoing issues point in the direction of something that, potentially, has considerable importance. More specifically, those considerations suggest – as do many other considerations in this chapter -- that the theory of evolution is not a scientific theory missing a few, inessential details. Rather, it is a theory that is missing almost all of the foundational components needed to explain and demonstrate the specific character of the dynamics that, supposedly, are at the heart of evolutionary change.

The theory of evolution pretends to be a scientific theory. However, when it matters most, that worldview, again and again, resorts to the use of unproven – and, perhaps, unprovable -- assumptions, speculations, or conjectures in an attempt to provide the underpinnings for purported explanations concerning a vast array of questions and problems that the theory should be able to address in a plausible manner if were truly scientific in character ... but does not do so.

One of the features that typify philosophical activity has to do with the inability of such a process to be able to demonstrate – in an independent and rigorous manner -- the truth of many, if not most, of its claims concerning the nature of reality. Yet, when the theory of evolution manifests the same sort of inability to prove its essential claims, nevertheless – and, perhaps (given the nature of philosophical bias), not so mysteriously -- that theory retains its alleged scientific status.

Consider the following. At the heart of the dynamics of many extremophiles (organisms capable of surviving and flourishing in extreme physical conditions), are proteins with specialized properties of functionality. These proteins have the capacity to assist organisms to adapt to extremes of, among other things, acidity, alkalinity, salinity, radiation, heat, cold, and pressure.

For instance, thermophiles and hyperthermophiles possess certain kinds of proteins that exhibit enhanced hydrophobic (water resistant/avoidant) properties along with an elevated capacity for a variety of electrostatic interactions that are needed to help lend stability (through, among other things, packing and folding activities) to the metabolic pathways that must operate in the midst of conditions involving high temperatures (up to, at least, 235 degrees Fahrenheit). Now, since proteins with a certain amount of hydrophobic properties and capacity for electrostatic interactions exist in most cells, the problem becomes one of explaining how thermophiles and hyperthermophiles acquired the ability to push – in a functional manner -- the cellular envelope with respect to such properties and capacities.

The general issue being alluded to in the foregoing paragraph also applies to halophiles ... that is, organisms which have the capacity to survive and flourish in conditions involving high saline content. In other words, while the specialized proteins that help make halophiles possible tend to exhibit a high negative surface charge -- as a result of: The presence of an increased number of acidic amino acids, together with the insertion of certain kinds of peptide linkages at appropriate junctures -- nonetheless, other cells that cannot tolerate conditions of high salinity also possess proteins with some degree of acidic amino acid content, and, so, one wonders how the right set of specialized proteins (or the underlying base pairing) arose in halophiles that enabled them to deal with conditions of extreme osmotic stress brought about by the presence of a high saline content in the environment in which the halophiles exist.

Similarly, psychrophilic organisms – ones that survive and flourish in temperatures near, or below, freezing – possess certain proteins that exhibit reduced hydrophobic properties, as well as display a reduced electrostatic charge on the surface of those proteins, and, in the process, helps such organisms to adapt to cold temperatures. How did proteins with these kinds of characteristics arise?

The specialized proteins that are key to thermophilic adaptation are different (in terms of hydrophobic properties, acidic amino acid content, electrostatic interaction on their surfaces, as well as properties of folding and packing) from the specialized proteins that are key to halophiles and psychrophiles. Indeed, they are all different from one another in various ways.

Similar sorts of differences extend to other kinds of extremophiles that inhabit conditions of high pressure, radiation, acidity, and alkalinity. In certain cases, however, there is some degree of overlap with respect to various amino acid sequences and base pairings since some organisms that can exist in, say, conditions of high saline content also exhibit the capacity to survive in conditions of high alkalinity, or organisms capable of existing in conditions of low temperatures also often tend to display an ability to deal with conditions of high salinity.

Areas of overlap notwithstanding, there still are different kinds of specialized proteins that play various kinds of roles in all of the foregoing cases ... although in certain instances involving multiple conditions of extreme environments (e.g. low temperatures and high saline content, or high temperatures as well as a high acidity), some of these proteins might play a secondary role rather than a primary one. Consequently, at some point the existence of proteins with the foregoing sorts of specialized functions and capabilities have to be accounted for as far as questions involving the origins of those proteins are concerned.

There are many conjectures that might be offered with respect to such matters. Maybe, for instance, the transition to proteins with a specialized capacity for functioning was gradual (Many, if not all, of the following comments also could be directed toward the idea of non-gradual transitions as well.).

If the transitions were relatively gradual, there are many pathways that might have been made such transitions possible. Nonetheless, if a person advances a theory of transition concerning the origin of a given form of extremophilic protein, then, one would like to know not only the precise route that was taken to make the transition from some kind of non-extremophilic protein to an extremophilic one, but, as well, one would like to know the nature of the dynamics at each step along that route.

In addition, one would like to know how the ancestral proteins emerged that, allegedly, predated the appearance of the specialized, extremophilic proteins. One can point, if one likes, to any number of possible transitions from the base pairings underlying one kind of protein to the base pairings underlying some kind of subsequent, specialized, extremophilic protein, but, at some point one is going to have to explain how the first protein arose that was part of the original branch that, eventually, led to the evolutionary branches on which specialized, extremophilic proteins are found.

Moreover, one cannot suppose that the <u>functionality</u> which might arise from the fact that some given sequence of thousands of nucleotide base pairings came together in just the right way to be selected by natural selection is capable of accounting for why one such sequence, rather than some other sequence, occurred. Natural selection acts upon such sequences <u>after the fact</u> and <u>not before the fact</u>.

In other words, natural selection might be able to help explain why a given set of genes -- with certain capabilities – survives or flourishes in a given set of environmental circumstances. However, natural selection cannot necessarily explain how those genes or a sequence of base pairings came to exist in the first place.

Therefore, when one engages issues such as trying to account for the origins of specialized, extremophilic proteins (or their supporting metabolic pathways) an individual is always engaged in matters that tend to transcend the horizons of the idea of natural selection. One is engaged in a rather mysterious realm that the theory of evolution attempts to explain away through, among other conceptual devices, use of the notion of random mutations (see the following 'Deep Solutions ...' section of this chapter for a discussion concerning some of these other conceptual devices being alluded to in the foregoing).

If mutations constitute the ideational bridge that is intended to 'explain' the movement and dynamics along some proposed pathway of gradualness, then, what caused which mutations to happen at what points and in which sequence? If someone objects that such questions cannot be answered, then, to whatever extent they cannot be answered, then, to that extent one does not have a scientific theory.

Moreover, if someone claims that the mutations were random – and, therefore, inexplicable in character -- then, one should be ready to acknowledge that this sort of claim doesn't really explain much of anything. Instead -- as pointed out earlier in this chapter during the brief discussion that revolved about the idea of randomness -- such an account tends, at best, to presuppose its own truth ... something that, scientifically speaking, is not really an appropriate thing to do.

How does one prove or demonstrate what the nature of the sequence of events was that made a protein with specialized, extremophilic capabilities possible? How did the requisite kinds of nucleotide sequences come together to underwrite such a capability, and how did the requisite base pairing coding for the associated metabolic pathways come about in a manner that would be able to arrange for supplying the right kinds of specialized proteins at the right times and in the right amounts and in the right places?

One could imagine many scenarios for how such a complex, integrated set of events might have come about. Imagination is capable of many things.

Nonetheless, imagination does not always give expression to, or lead one toward, the truth of things. In fact, we often can image what is false much more readily that we can imagine what is true.

Many so-called scientific journals, books, and academics are often filled with conjectural imaginings of all kinds as a way of alluding to the possible significance of a given set of observations or experiments. Yet, most of those imaginings disappear as quickly as they arose because they lack the necessary, substantive properties that can tie them to reality in anything more than what ultimately proves to be a tangential -- if not asymptotic -- manner.

The process of conjecturing and speculating about the nature of reality can be a useful exercise because it helps to stimulate further research and critical reflection. However, the content of those conjectures and speculations does not become scientific until one can rigorously demonstrate that such content gives expression to the truth or can play a substantial role in helping to lead one to the truth.

Deep Solutions for the Problem of Biological Origins

In addition to the notion of random mutations (and, some of the problems surrounding this notion have been touched upon previously), there are a variety of other terms that appear in some of the literature that seeks to outline and explore the theory of evolution. Like the idea of "random mutations", these other terms are used in ways that create the impression that something is understood when this is not necessarily the case.

For instance, some people refer to the chaotic properties of various kinds of biological or chemical systems in which small changes in initial conditions can lead to unpredictable results. One could accept such a statement without necessarily being any closer to understanding – in specific, provable ways -- how, say, non-living systems turn into living systems, or how non-extremophilic proteins transition into extremophilic proteins, or how metabolic pathways come into existence.

The earlier reference to chaotic properties can be replaced by an array of other terms such as: 'spontaneous', 'self-organizing', 'far from equilibrium conditions', 'self-criticality', and 'emergence'. In each case, a term is used that is intended to serve as a means of explaining how some given structure, property, activity, network, or capacity arose in a given set of circumstances that didn't contain such structures, properties, activities, networks or capacities prior to a certain point in time ... or prior to some given threshold being reached.

Thus, far from equilibrium conditions generate dissipative structures. The properties of such structures could not have been predicted on the basis of the existence of far from equilibrium conditions on their own.

Or, interacting components of the right kind spontaneously lead to self-organizing systems. The possibility of systems with those sorts of capacities could not have been predicted knowing just the nature of the properties of the individual components involved prior to the point of being brought into an interactive dynamic with one another.

Or, complex systems give rise to emergent properties. The nature of these latter properties could not have been predicted before the conditions necessary for such complex systems occurred and attained a certain level of self-criticality.

No one who has any degree of familiarity with what has been taking place in science over the last 50 years, or so, would deny that there are an array of circumstances in which far from equilibrium conditions are capable of generating unanticipated forms of dissipative structures ... or, in which certain kinds of systems do organize themselves -- seemingly spontaneously – in unexpected ways as a function of the forces and elements present in those systems ... or, in which various kinds of properties inexplicably emerge out of systems exhibiting complex sorts of behavior. Instances of all of the foregoing scenarios have been demonstrated on numerous occasions.

Nonetheless, demonstrating the reality of the foregoing sorts of phenomena does not prove or force one to conclude that any given process for which one does not have a ready explanation concerning how such events are possible must be the result of some chaotic, complex, spontaneous, self-organizing, or emergent dynamic that automatically generates what cannot be otherwise explained. For example, unless one can show scientifically that a particular set of inorganic and organic interactions is capable of spontaneously organizing itself in a way that leads to life as an emergent property when certain thresholds of self-criticality have been reached in the context of complex systems behavior -- where initial conditions are of considerable importance -- then, all of the foregoing terminology constitutes little more than a bunch of buzz words that purport to explain things but, in reality, do nothing of the sort.

To claim that life is due to a sequence of phenomena that are built on layer after layer of spontaneously emerging properties arising out of systems that have become – and are continuing to become -- increasingly complex as a result of the accumulation of, and ensuing interaction of, the foregoing sorts of emergent properties might be a meaningful way of engaging a great deal of data, but it is entirely too vague to be of any scientific value.

Furthermore, claiming that since we seem to have no other explanation for how life arose, then, life "must have" arisen through such an inexplicable -- but determinately emergent -- set of processes might be an interesting conjecture. Nevertheless, unless one can explicate the inexplicable in demonstrable, provable ways, then, there is no science present ... just conjecture.

Many scientists believe that the structural character of the universe gives expression to, and is the result of, a set of natural laws capable of being discovered through a reiterative process that rigorously pursues observation, experiment, and critical analysis in an attempt to produce a coherent, consistent, accurately reflective portrait of some facet of reality. Consequently, scientists tend to believe that the universe and all it encompasses – including life – must be the result of some set of natural forces and principles that have the capacity to generate, among other things, the phenomena we experience.

In other words, life is considered to be an inevitable product of the interaction of chemical and physical laws. Given the right set of chemical ingredients, forms of energy, kinds of forces, environmental conditions, and sufficient time, then, according to the foregoing way of thinking about things, the emergence of life will occur.

However, an on-going problem for scientists is that they have had a heck of a time trying to figure out what the right set of chemical ingredients, sources of energy, forces, environmental conditions, and so on are. Indeed, to date, scientists have not been successful – not even remotely so -- in their attempts to show that life is, in fact, an inevitable outcome that is rooted in the interaction of a determinate set of naturally occurring physical/material events.

Many scientists believe there are three general steps that lead to the dance of life. First, one throws into the evolutionary pot an array of carbon-containing molecules, together with an assortment of other kinds of inorganic molecules that can spice things up. These molecules might have arisen on Earth, and/or they might have come to Earth via asteroids and comets, or they might even somehow have found their way to Earth from somewhere in the cosmic void.

Secondly, scientists presume that the structural character of one's pot is sufficiently complex that it permits a variety of processes to take place that are capable of: Bringing together, concentrating, and assembling the molecules initially present in such an evolutionary pot. This complexity is believed to extend to the structural properties of the interior of the pot that needs (if the theory is to have a chance of being correct) to consist of the right sort of surfaces, textures, and minerals, to be able to help catalyze and compartmentalize an array of molecular reactions that, supposedly, lead to the emergence of complex molecules such as: Proteins, nucleic acids, lipids, and carbohydrates.

Thirdly, the dynamic of interacting molecules within the right kind of evolutionary pot eventually establishes networks of metabolic pathways that are able to sustain themselves, compete for resources, and self-replicate. Such systems are believed to give expression to a set of characteristics that have varying degrees of capacity for survival and, as a result, forces such as natural selection and genetic drift begin to push and pull populations of organisms in different directions as a function of the interaction between the capabilities of those sorts of organisms and the degree to which environmental conditions lend support to, or are antagonistic to, those capacities.

Aside from the many problems that are entailed by the foregoing tripartite narrative (and there will be more discussion concerning such problems shortly), people often get bogged down with trying to determine when, exactly, life emerged during the aforementioned three-step process. Some people identify the beginning of life with the appearance of the first systems that were capable, in some sense, of self-replication via RNA and/or DNA.

Other individuals believe the beginning of life is synonymous with a capacity to establish metabolic pathways. Still other individuals refer to the capacity to form semipermeable membranes as marking the emergence of life. And, finally, there are those who believe that appropriate combinations involving all of the foregoing capabilities are necessary for life to exist.

I believe the foregoing kinds of considerations are rather premature. Before one even addresses the issue of 'what is life?' one must account for how the order necessary for underwriting the emergence of capacities -- such as: Self-replication, metabolism, and semipermeable membrane formation -- arose out of a assortment of interacting carbon-containing molecules and a variety of other inorganic molecules. The primary issue is not a matter of figuring out how to differentiate life and non-life, but, rather, the primary issue is a matter of trying to determine how functional order arises out of circumstances comprised of elements and forces that do not, on their own, exhibit such functionality or order.

To begin with, there are 80-90 years of collected data indicating that various kinds of molecules occurring in living organisms are capable of being fashioned in the laboratory under the right kind of experimental conditions. However, there is little, or no, evidence indicating that the chemical interactions occurring on early Earth went about their business in the manner in which laboratory experiments suggested might have been the case.

Are the laboratory experiments being alluded to in the foregoing paragraph rather suggestive? Of course, they are ... but being suggestive is not proof of anything.

By their very nature, experiments require the organizational capacities of one, or more, experimenters in order for those experiments to be able to take place. Experimenters bring materials together in a specific manner (e.g., amounts, sequence, length of time, and conditions) and ensure that those materials are subjected to a certain set of events within an environment that is highly regulated.

When things are done in the foregoing manner, various consequences follow. But, what happens when materials and forces are left to their own devices, sans experimenter ... will the same kinds of consequences that are observed in the laboratory also occur?

Maybe! However, the issue is whether, or not, those sorts of consequences will inevitably occur independently of the ordered conditions of a laboratory experiment.

Experiments can provide a proof of concept – that is, experiments can demonstrate that certain kinds of consequences are possible and follow from certain kinds of conditions. Nevertheless, there is no guarantee that the natural world will necessarily give expression to the conditions and circumstances that are necessary for 'interesting' kinds of consequences to emerge.

For example, let's consider a classic experiment conducted back in the early 1950s, under the guidance of Harold Urey – a Nobel Prize winning chemists -- by Stanley Miller, a second-year graduate student. Among other things, the tabletop simulation of early Earth conditions had a 5-inch, 300-milliliter glass flask that was two-thirds full of water that supposedly represented the ocean.

Depending on circumstances, ocean water comes in a variety of forms. The saline content of that water can vary, as can its temperature, pH value, and mineral content ... all of which can impact the rates and character of whatever sorts of reactions might take place in that kind of a medium.

Moreover, on early Earth, surface waters would have been bathed relatively continuously in a certain, unknown amount of ultraviolet light. As a result, whatever reactions might have taken place in the liquid medium also might have been quickly degraded as a result of the presence of that ultraviolet light ... and the impact of ultraviolet light on organic molecules is only part of the broader problem of photolysis in which the presence of light has the capacity to degrade the reactants and products of various reactions.

In addition, ocean water would have been subject to tidal forces, currents, and storms of varying intensities. How tides, ocean currents, and storms might have affected chemical reactions is a further set of considerations that need to be factored into one's analysis of the possible significance – or lack thereof – of the Miller/Urey experiment.

The bottom line is that what might take place in a flask filled with water that is hooked up to other experimental equipment is not necessarily indicative of what might take place on early Earth. No matter what geological period one is considering, ocean water is a much more variable and complex medium than is 'ordinary' water.

The aforementioned 300-milliliter flask of water was hooked up to a 10 inch, 5-liter flask filled with a number of gases that are fairly reactive – namely, hydrogen (H_2), methane (CH_4), and ammonia (NH_3). In addition, the latter flask contained two metal electrodes that were intended to serve as the experimental counterpart to lightning strikes.

Opinions concerning the composition of the atmosphere on early Earth have gone through a number of fairly significant changes since the Miller/Urey experiment. For example, seven, or so, years, after the Miller/Urey experiment had been completed, the evidence from additional geological and geochemical experiments/analysis tended to indicate that the atmosphere of early Earth consisted, to a large extent, of carbon dioxide and nitrogen.

Unlike hydrogen, methane, and ammonia – which are reactive gases – carbon dioxide and nitrogen are less reactive than the gases used in the Miller/Urey experiment. What is more important, however, is that the updated version of the atmosphere of early Earth was quite different from the composition of the atmosphere envisioned by Miller/Urey.

Consequently, what might happen in a flask containing a set of reactive gases does not necessarily have much relevance with respect to what might have happened in the actual early Earth atmosphere. The latter atmospheric environment might have contained different gases that were less reactive than the Miller/Urey experimental set-up.

Furthermore, other than involving electricity, I'm not quite clear about how the electricity delivered through two metal electrodes is much like what happens when lightning strikes. The sparks in the Miller/Urey experiment involved 2-4 watts of energy, whereas lightning strikes deliver the equivalent of approximately 8000 watts.

Moreover, the experimental sparks were fairly regular and in the same area. Lightning strikes, on the other hand, are sporadically intermittent and tend not to regularly visit the same, confined area again and again.

Unless, of course, one wishes to include the interfacing of the Catatumbo River with Lake Maracaibo in Venezuela where lightning strikes occur up to 280 times per hour, ten hours a day, and between 140 and 160 nights of the year. And, if one did wish to factor such possibilities into the matter, one would introduce an array of problems for the reactants and products of chemical reactions that would arise in any context involving that kind of a constant barrage of powerful, electrical discharges.

One of the problems being alluded to in the foregoing paragraph is that whatever chemical reactions might have been helped along with one lightening discharge might very well have been disrupted or destroyed with subsequent lightning strikes. No one has performed experiments simulating the Catatumbo River/Lake Maracaibo conditions, so, it is hard to determine what might, or might not, have taken place under those sorts of conditions.

The two flasks in the Miller/Urey experiment were linked up with one another via glass tubing. At one point along the tubing, a condenser section was set up, and the flask containing the water was heated on a continuous basis by a relatively low intensity source of energy that was intended to simulate the condition of evaporation that was believed to be present on early Earth.

Once the Miller/Urey experiment started, the experiment was permitted to run over several days. After a few days, the formerly clear flask water began to become yellowish in color, and, as well, the area of the electrodes was exhibiting some blackish residue.

Miller subjected the residue and the water to chemical analysis. His primary tool in this aspect of the experiment was paper chromatography -- a process that helps to differentiate chemical molecules from one another – and Miller discovered the presence of glycine ($C_2H_5NO_2$), the least complex member of the amino acids that make up the proteins of life.

Miller re-ran the experiment. This time he let it proceed for a week, and, as well, he turned up the heat in the flask containing the water so that the latter slowly boiled.

At the end of seven-day experimental period, Miller again used the process of paper chromatography to separate out whatever molecules might be present in the water. He discovered the presence of a wide array of organic molecules, including quite a few amino acids.

The foregoing is all very interesting. However, the Miller/Urey experiment also raises a lot of questions above and beyond the problems already noted in earlier comments.

For instance, the experimental apparatus was sealed and involved a continuous circulation of chemical components that were regularly exposed to: Relatively low-intensity, electric sparks; conditions of condensation; and being passed through boiling water. Why should one suppose that conditions on early Earth also consisted (in part or whole) in a similar sort of environment that involved: Materials being sealed off from the rest of the world; continuous circulation of the same components; regular doses of low-intensity electric sparks;

regular and consistent conditions of condensation, as well as boiling water ... all for a period of no more than a week?

To the foregoing question, one can add several other issues. For example, whatever chemical residues accumulated – under questionable conditions -- over a period of a week, likely would be subjected also to the continuous, degrading actions of a wide variety of hydrological, ultraviolet, photolytic, and other environmental forces (e.g., acidity, alkalinity, etc.).

What would survive from an interacting set of synthesizing and degrading forces is anybody's guess. One cannot necessarily assume that, over time, the forces of synthesis would necessarily overpower the simultaneously occurring forces that served to undermine and degrade whatever the forces of synthesis might have brought forth.

On May 15, 1953, Miller's two-page article – *'A Production of Amino Acids Under Possible Primitive Earth Conditions'* – was published in *Science*. While the foregoing title was technically correct, what might possibly have been the case on early Earth is not necessarily how things actually were back then, and, therefore, the title of Miller's article is also potentially misleading if what he considered to be 'possible' didn't accurately reflect actual, early Earth conditions.

Since the 1953 paper was released, a great many other experiments have been conducted that were able to demonstrate how, under certain conditions, different molecules that played important roles in the biology of life could be produced experimentally. For example, in 1960, by heating a concentrated solution of hydrogen cyanide (HCN), John Oró was able to synthesize considerable amounts of adenine – one of the five nucleobases from which nucleotides are formed as well as being a central component in adenosine triphosphate (a major source of energy for many biological reactions).

The Oró experiment is very suggestive in relation to the origin of life issue. That is, the experiment is suggestive provided there were concentrated solutions of hydrogen cyanide on early earth, and those solutions were heated in just the right way, for just the right amount of time, and were not subsequently subjected to any of the forces (e.g., water, light, acidity, alkalinity, and temperature) of molecular degradation that have been present on Earth from a very early time.

Later on in the 1960s, Leslie Orgel demonstrated that if one froze water rich in organic molecules, then, as water crystals grew in the freezing water, this had the effect of concentrating the organic molecules that remained in solution within the portions of water that had not, yet, frozen. Therefore, Orgel prepared dilute solutions of hydrogen cyanide and slowly lowered the temperature of the solution to -20 degrees Centigrade.

Orgel's experiment led to the production of small amounts of highly, concentrated HCN. Moreover, over a period of weeks and months, the HCN molecules were observed to establish linkages involving up to four HCN molecules.

Oró's experiment (outlined earlier) required concentrated solutions of HCN to be <u>heated</u> in order for adenine to be synthesized. How did the concentrated brine of HCN produced by freezing dilute solutions of HCN in Orgel's experiment come to be sufficiently heated for the right amount of time (and the process of thawing, for example, might not entail sufficient heat) to yield small amounts of adenine?

Well, one way of responding to the foregoing question is to hypothesize that heating might not have been required. Ten, or so, years later -- in the mid-1970s -- Stanley Miller and several colleagues repeated the Orgel experiment.

When the initial portion of this replicated experiment was completed, the researchers stored the flasks in a freezer, waited more than twenty years, and, then, proceeded to analyze the contents of those flasks. They found a fair amount of adenine had been produced while in a frozen condition.

Low temperatures tend to slow down the rate of reactions with respect to the synthesis of various molecules. However, given a long enough period of time, cold, freezing conditions will not necessarily inhibit the formation of more complex molecules from taking place.

However, what If Orgel's experimental solution contained other kinds of organic molecules as well as HCN, how would this have affected his results? If those solutions were: Acidic, alkaline, exhibited a high saline content, and/or contained various assortments of minerals suspended in solution, how would any of these added factors affect Orgel's experiment?

How likely is it that one might find water-based solutions on early Earth that contained dilute amounts of only HCN? Such solutions might be possible, but how likely were they?

Any answer that one gives to the foregoing questions will be fairly arbitrary. This is because we don't actually know what the conditions for any particular part of early Earth actually were even if we might know what some of the general conditions were that prevailed at that time.

Over the years, a wide variety of prebiotic experiments sought to fill in the gaps with respect to how a variety of molecules of importance to the origin of life might have arisen under "possible" conditions with respect to primitive Earth. However, as was the case with the Orgel, Oró and Miller experiments, just how likely any of those possible conditions might have been is not known with any high degree of certitude.

Furthermore, even if one were to accept-- for the purposes of argument -- the idea that all of the "possible" conditions described in a whole set of different experiments might have reflected actual conditions in different areas of early Earth, there was still another, major hurdle to get over. How can one be sure that all of the different sets of conditions (e.g., temperatures, pH conditions, energy sources, chemical materials, atmospheric conditions, degrees of concentration, and so on) that were needed -- according to an array of experiments -- to produce different kinds of molecules important to life would necessarily have taken place in close proximity to one another?

After all, one might grant – and this would require a person to overlook quite a few problems and questions -- that the conditions established in various experimental set-ups could have reflected actual conditions in different parts of early Earth. Nevertheless, how did all of the molecules synthesized under an array of variable, experimental conditions come together in one small area – say, the size of a cell -- in order to be able to form functional, metabolic pathways?

No scientist (or group of scientists) has been able to do a single, self-contained experiment that simultaneously: (1) Simulated all of the conditions said to be necessary for producing the array of molecules essential to life as we know it, and, then, (2) observed the products of those differential conditions proceed to self-organize into functional, integrated, metabolic pathways that were capable of underwriting the existence and complexity of the simplest of living organisms. Therefore, even if one grants the possibility that individual components important to the existence of life were, somehow, synthesized on early Earth – and this is, by no means, a foregone, scientifically proven conclusion – nevertheless, there is no explanation for how all of those components came to be functionally organized in one place, no bigger than a cell.

Furthermore, one cannot take the mass of data concerning the prebiotic conditions of early Earth that have accumulated over decades of so-called research and try to claim – with a straight face – that it all gives expression to the best scientific theory we have concerning the origin of life issue. The fact that a bunch of scientists – some of whom won Nobel Prizes – spend time in a laboratory, formulate hypotheses, conduct experiments, draw conclusions, and come up with this or that equation, does not mean that what they have done constitutes science or is scientifically viable.

Many scientists have developed theories concerning the origins of life. Many religious people have developed theories concerning the origins of life.

Many scientists criticize the latter individuals because the religiously inclined have no viable, provable account of how the dynamics of life came into being even though such people use terms like 'creation science' and 'intelligent design' in order to give the appearance of having put forth a scientific theory of some kind. What is appropriate for the goose is also appropriate for the gander.

Therefore, since scientists have not been able to put forth any viable, provable account of how the dynamics of life came into being through purely physical/chemical means, then, despite the fact that scientists use terms like the 'science of evolution' and the 'scientific method', scientists are really no further ahead in the origin of life explanation lottery than religious people are. There is no 'best available scientific account of the origin of life' because there is no science in this area that is capable of demonstrating itself to be reliable.

Just because people refer to themselves as scientists and spew forth a lot of hypotheses, speculations, conjectures, opinions, and experimental results, this doesn't automatically render what they say and do in this respect to be of any scientific value. Nor does their position as scientists automatically award them scientific bragging rights with respect to people who are religiously inclined and have immersed themselves in activities called creation science and intelligent design.

If one doesn't know the truth of things, then, irrespective of what phrases might be used involving the word "science" or "scientific", one is ignorant. If one doesn't know the truth of something, then it becomes an exercise in foolishness to try to claim that one unproven, allegedly scientific theory/account is better, more scientific than some other unproven, allegedly scientific theory/account.

All attempts to scientifically account for the origin of life – whether through creation science or "mainstream" science – are equally inept and riddled with an array of problems. There is no "best scientific account" concerning the origin of life ... there is just ignorance all the way around.

Educators – the sort of people Dobzhansky was addressing in his previously discussed, *American Biology Teacher* article ('*Nothing in Biology Makes Sense Except in the Light of Evolution*') -- who want to teach a theory concerning the origin of life or that – purportedly – deals with the nature of speciation as being scientific -- when such theories entail little more than ignorance at virtually every crucial juncture -- seem to be under a misunderstanding when it comes to the process of education. Ignorance is ignorance, and it shouldn't be packaged as being anything other than ignorance.

There is no such thing as scientific ignorance. There is just ignorance.

What we don't know when it comes to a scientific account of the origin of life is close to 100%. The "best scientific account" that we have concerning the origin of life is that we have no idea how, or if, the origin of life can be demonstrably explained in terms of known scientific laws and principles.

Proponents of evolutionary biology have gone to court on many occasions defending the idea that students are, in effect, being educationally abused when those students are forced to take courses in the public-school system that are imbued with the biases and misconceptions of the proponents of creation science and intelligent design. However, the fact of the matter is that students are no less educationally abused when they are forced to take courses in the public-school system that are imbued with the biases and misconceptions of the proponents of evolutionary biology.

Biases and misconceptions are just that. Propagating bias and misconceptions as being anything other than what they are is not an exercise in learning how to do science, or learning how to become scientific.

If educators want to teach science in science classes, then assist students to develop the sort of critical understanding that allows them to be able to differentiate the wheat from the chaff with respect to the search for truth concerning the origin of life issue. Educators need to assist students to learn, on the one hand, about the lacunae, problems, unknowns, missteps, and unanswered questions that saturate the whole field of origin of life research, as well as many facets of the field of evolutionary biology, and, on the other hand, educators need to assist students to learn that at the present time there is no scientific theory concerning the origin of life that is even remotely viable.

To try to do anything else in a classroom (whether elementary, high school, college, or university) would constitute an exercise in educational abuse. Unfortunately, the judges who issue decisions concerning cases involving the proponents of evolutionary biology versus the proponents of creation science or intelligent design biology don't seem to even understand the nature of the issues about which they are making legal judgments ... judgments that will affect the lives of millions of students.

Over the last 40 years, or so, a new approach -- with respect to trying to provide a scientific account concerning the origin of life issue -- has gained a certain amount of traction in at least some scientific circles. This new approach is referred to as the 'hydrothermal hypothesis.'

By way of background, one of the sticking points for a lot of origin of life theories up until the 1970s circulated around the issue of water.

It was an inconvenient truth that the presence of water tended to resist and/or undermine certain, essential, chemical reactions (that are important to life) from proceeding vary far, if at all ... water is Janus-like in its capacity to both help facilitate as well as help undermine a variety of chemical reactions.

Another possibility, however, began to bubble to the surface of consciousness beginning in the late 1970s. Jack Corliss, an oceanographer, took the submersible, research vessel Alvin to the bottom of the ocean, and in the process, he discovered incredible networks or ecosystems of life that were flourishing in conditions of tremendous pressure, no sun light, and the very high temperatures that exist in various undersea volcanic vents.

When water is subjected to high pressures (say, a kilobar – a thousand atmospheres -- or more), together with sufficiently high temperatures (say, 175 degrees Celsius, or more), then, the dielectric constant of water goes down. In many ways water becomes like an organic solvent under these conditions.

As a result, water tends to behave very differently under the foregoing sorts of conditions than it does at much less extreme temperatures and pressures. Perhaps, therefore, the physical and chemical differences that manifest themselves in water under conditions of high pressure and high temperature might be able to permit certain kinds of chemical reactions to proceed that might not be able to take place when placed in water at the sort of temperatures and pressures that exist in many places on the surface of the Earth or in the top several hundred feet of the ocean.

For example, consider pyruvate (CH_3COCOO^-) -- a source of energy that, among other things, helps to subsidize many reactions taking place within one, or another, metabolic pathway. When glucose, a six-carbon atom, is split, first into two pyruvate molecules due to the presence of the right kind of catalytic enzyme, and, then, subsequently, the pyruvate molecules are split into still smaller molecules – again, due to the presence of the right kind of enzyme -- energy is released along the way, and this energy is used to help advance various reactions that will not occur spontaneously ... that is, on their own.

In addition, if one combines pyruvate, carbon dioxide, and the right kind of enzymatic catalyst, one can generate a molecule known as oxaloacetate [HO$_2$CC(O)CH$_2$CO$_2$H]. This latter molecule has further uses within certain metabolic pathways of living organisms.

However, when pyruvate is left to its own devices in water – that is, without the presence of the right sort of catalytic enzymes – it tends to break down into smaller molecules. These latter molecules contain only one or two carbon atoms, and a result cannot be combined with carbon dioxide to produce the four-carbon molecule, oxaloacetate ... at least not at normal room temperatures and pressures, and not without the presence of an appropriate kind of catalyst that can speed up the reaction rates of such chemical components, and not in the presence of water.

What would happen if one were to combine water, pyruvate, as well as carbon dioxide and subject those ingredients to various conditions of high temperature (say, 150 to 300 degrees Celsius) and high pressure (say, 500 to several thousand atmospheres)? Would one obtain molecules of oxaloacetate or anything else of interest to the origin of life issue?

Harold Morowitz and Robert Hazen -- along with the assistance of Hat Yoder and George Cody -- undertook the foregoing experiment ... or, at least, a facsimile thereof. They placed water and pyruvate (both liquids) into a gold tube -- the size of a long grain of rice – and introduced carbon dioxide gas into the tube via the way of a chemical known as oxalic acid dihydrate (H$_2$C$_2$O$_4$·2H$_2$O) that decomposes into carbon and water at temperatures above 100 degrees Celsius.

The open end of the gold tube into which the various chemical reactants had been introduced was sealed up using a complex process involving a carbon-arc welder, a graphite rod and liquid nitrogen -- at a temperature of -196 degrees Celsius that was used to keep the other end of the gold tube sufficiently cool so that this would help prevent the pyruvate – which tends to be fairly volatile under certain conditions -- from boiling away when the carbon-arc welding process went about its sealing business at the other end of the tube.

The small, rice-sized, gold tubes (there were three of them) were, then, placed within a complex arrangement consisting of: a platinum holder, nickel metal cylinder (served as the electric furnace), ceramic filler rods, ceramic end caps, thermocouple wires, all packed within a white aluminum powder. The foregoing arrangement was, then, attached to a steel plug that is capable of retaining pressure while, simultaneously, providing a means of insulating an assortment of wires through which an electric current flows that controls the amount of temperature/heat being applied to the gold tubes.

Finally, all of the above is sealed within a metal container. This container is capable of withstanding a pressurized gas (argon) being pumped into the contraption and once pumped in will subject the gold tubes to the same sort of pressure.

The pressure selected for the experiment was two kilobars or 2000 atmospheres. The temperature was set at 250 degrees Celsius and was controlled by a computer.

The whole, experimental set-up was permitted to run for several hours. Supposedly, the conditions established through the experiment were intended to simulate the conditions that might be found several miles down in the ocean along one, or another, volcanic vent.

Following the aforementioned two-hour experimental period, a combination of gas chromatography and a mass spectrometer was used to analyze the contents of the gold tubes (after they were opened). Those contents didn't reveal the presence of oxaloacetate – as the researchers thought might be the case -- but the contents did contain thousands of other molecules of various descriptions.

Among the molecules that were synthesized were alcohols and sugars. In addition, they discovered complex molecules that contained dozens of carbon atoms ... some of which formed the sort of branching structures and rings that are similar to various kinds of branching and ring structures that are found in living organisms.

The researchers drew certain conclusions from their experiment. More specifically, among other things, they felt they had demonstrated – in a proof of concept sort of manner – that hydrothermal conditions of great pressure and high temperature were capable of generating a vast array of molecules that might have played various roles along the way toward the prebiotic origins of life.

There are a number of problems that permeate the foregoing experiment. For example, where does one discover hydrothermal conditions – except, perhaps, in a laboratory – such that a set of circumstances lasts for only several hours and takes place under very carefully controlled, sealed conditions of temperature and pressure in which a chemically inert gold tube presses in on the primary chemical reactants without permitting any outside agents (of a possible reactive nature) into the reaction chamber?

What would have happened if the experiment had gone on for a thousand or a million or a billion years rather than for two hours? We don't know, and, yet, the former set of possibilities is far more likely than a two-hour experiment, so, really, what does the experiment outlined above actually teach us?

After all, if you pressure-cook certain foods for several hours at an elevated temperature, you might get a tasty meal. If you cook the same dish for several thousand, million, or billion years, the meal might not be so tasty … or suggestive with respect to the origin of life issue.

Even assuming there were real-world hydrothermal conditions that provided a niche within which certain chemical ingredients could be completely sealed for, say, a two-hour period, what happens to those contents once the container is breached and its contents are released into the ocean waters that are circulating through a given volcanic vent? We don't know because – for a variety of reasons (not the least of which is an inability to maintain control over a wealth of variables in such a situation) -- no one has performed that kind of an experiment.

What happens if the innermost, sealed container in actual, non-laboratory based conditions does not consist of a soft, relatively inert material such as gold? Will we get the same results?

Or, approaching issues from a slightly different direction, let's take the experiment at face value. One sets up an experiment, and one gets some interesting and unexpected results.

What does any of this 'unexpectedness' do for the origin of life issue? How will all of the unexpected molecules and molecular fragments fit into an attempt to explain how life might have arisen from such a concoction?

Could life have arisen from some arbitrary set of ingredients subjected to some arbitrary set of conditions? Maybe, but, to the best of my knowledge, Morowitz, Hazen, Yoder, and Cody, didn't discover life in their gold tubes, and, therefore, one is left to ponder and critically reflect on the possible significance, if any, of what they discovered.

The researchers were expecting one thing – some molecules of oxaloacetate – and got a whole lot of something else. They seemed to believe that what they actually got was some sort of emergent set of properties, and, perhaps, conducting other kinds of similar experiments -- at different temperatures or pressures, and with different ingredients -- might lead to an array of additional sets of emergent properties ... that is, entities that were not anticipated prior to running such experiments but that showed up, nonetheless, and entailed some interesting possibilities.

When one performs an experiment and that experiment does not yield the results one expected, one hasn't stumbled upon emergent properties. Rather, one has come across evidence pointing to one's ignorance concerning the nature of the forces and principles that are likely to be operative with respect to the dynamics of a given set of conditions that have been set in motion by one's experiment.

Claiming that those sorts of allegedly emergent phenomena might, somehow, lay the basis for constructing a provable account of, or explanation for, the origin of life seems rather strained ... to say the least. The only emergent dimension of such an experiment is that one comes to learn some things that one didn't know before.

If one piles emergent properties upon emergent properties upon emergent properties (and so on indefinitely) one doesn't necessarily end up with life. One might end up, however, with some new facts ... facts that might, or might not, have something of relevance to disclose with respect to the origin of life issue.

John Holland is one of the individuals who helped bring the field of emergent modeling into prominence. He used computer algorithms – that is, a programmed system or network of operational rules -- to simulate various phenomena.

He believes emergent properties can be shown to be a function of the kinds of selection rules one uses to model the phenomenon out of which such properties emerge. Moreover, he believes that the degree of complexity inherent in some, given emergent phenomenon might be closely related to the number of lines of programming code that are needed to faithfully simulate or reflect the properties of such a phenomenon.

In his 1998 book: *Emergence: From Chaos to Order*, Holland indicates that the idea of emergence is so complex that formulating a concise definition for the phenomenon is unlikely to take place. Furthermore, he admits that he has no concise definition to offer in conjunction with the notion of emergent behavior.

Nevertheless, Holland's perspective in relation to the problem of trying to concisely define the phenomenon of emergence could be quite prescient. After all, there really might be a realm of emergent phenomena that do not necessarily share a set of overlapping, operational selection rules, and, therefore, such phenomena tend to resist being reduced down in a way that could be encompassed by any kind of concise definition that would be capable of capturing the variability and complexity of those sorts of phenomena.

On the other hand, Holland's opinion about whether, or not, the complexity of emergent phenomena renders them resistant to concise definition might be steeped in a certain amount of confusion about what emergent phenomena actually are ... or are not. In other words, can one – or should one -- automatically assume there are an array of special dynamics that give expression to something called "emergent phenomena," or is the idea of "emergent properties" just a catch-all term that tends to camouflage the presence of considerable ignorance with respect to how various things work in the universe.

For example, as previously noted, Harold Morowitz and Robert Hazen held a tentative hypothesis that if one subjected water, pyruvate, and CO_2 to sufficiently high pressures and temperatures one might be able to produce molecules of oxaloacetate despite the absence of an enzyme to help facilitate the process. However, their experiment produced something quite different than the molecules of oxaloacetate that had been anticipated as possible outcomes if one ran the experiment at issue -- namely, the experiment yielded an array of thousands of unanticipated molecules.

What were the dynamics underlying the differences between what was expected and what actually occurred? Were there some sort of special, emergent dynamics that were taking place or did the researchers just not understand how things work under certain conditions?

What was actually happening in the sealed rice-sized gold tubes containing water, pyruvate, and carbon dioxide that were being heated to 250 degrees Centigrade, as well as being subjected to 2000 atmospheres of pressure? The fact of the matter is we don't know.

Was the heat primarily responsible for the synthesis of unexpected molecules? Was the pressure primarily responsible for the production of the unanticipated? Was the cooling down period that preceded opening the gold tubes primarily responsible for yielding outcomes that had not been predicted? Was the combination of heat, pressure, time, and cooling down primarily responsible for what took place, and why weren't the researchers able to predict such an outcome? We don't necessarily know how pressure, temperature, and certain ingredients interact with one another across various ranges of values.

Do molecules that are inert under "normal" conditions remain so under more extreme conditions? Are their various thresholds of pressure and temperature that if surpassed will give expression to certain kinds of phenomena that, currently, we do not understand? What underlies such thresholds and why do they occur at some junctures and not others? What are the limits, if any, that might exist in relation to the interaction of pressure, temperature, and various substances across a range of values? What impact does the amount of time that transpires during the experiment have on how pressure, temperature, and molecules combine together to generate products?

I'm not sure that I see any emergent phenomena that are taking place in the Morowitz-Hazen experiment. I do see an awful lot of unanswered questions and considerable ignorance concerning the physics and chemistry of what is transpiring inside the gold tubes during the experiment … questions and ignorance that all tend to revolve around not knowing why what was observed to happen in the experimental outcome was able to take place.

To refer to the unexpected and unanticipated as giving expression to the dynamics of emergent phenomena doesn't really explain anything at all. In fact, such a way of talking might constitute little more than a certain kind of magical thinking in which causal attribution is assigned to a hypothetical entity – namely, emergence – whose actual, specific dynamics cannot be verified ... and that might not even be an actual phenomenon (just a way of descriptively referring – if rather vaguely -- to phenomena whose internal dynamics lie beyond the horizons of our understanding).

Furthermore, the capacity to develop a computer algorithm – as Holland and others have done -- to simulate a phenomenon doesn't necessarily – in and of itself – prove that the phenomenon being modeled or simulated is a function of the sort of operational or selection rules that are contained in the algorithm. Such simulations/models only demonstrate that, to varying degrees of accuracy, a computer can mimic certain <u>behavioral</u> properties by means of a given algorithm that has been set in motion by a working computer that has the capacities needed to run the algorithm successfully.

A psychopath can mimic the emotional behavior of 'normal' people. However, the psychopath did not necessarily generate his or her own behavior in the same way that normal people generate their emotional behavior.

A painter can simulate, with considerable accuracy and attention to detail, some of the visual properties of a scene of nature or the external characteristics of a person. However, the manner in which a painter arrives at her or his terminal juncture (i.e., the existence of a finished picture) is not the way in which nature arrives at its terminal juncture (i.e., the existence of the natural phenomena or person being painted).

A computer simulation might be able to model some of the behavioral properties of certain real-world phenomena. This does not necessarily mean the processes (computer algorithms and real-world phenomena) underlying the respective surface behaviors are the same.

The possibility that one might not be able to predict what a computer algorithm will generate if given enough time is not an expression of emergence. It is a statement of ignorance concerning how the dynamics set in motion by that computer algorithm will unfold over time.

For human beings, trying to follow the dynamics of the foregoing processes is too complicated. There are too many variables for a person to be able to keep track of simultaneously so that an individual can understand what is happening from second to second in real time.

For a computer, the issue of understanding an algorithm is irrelevant. All the computer does is to run the program for a specified time (both the parameter of running and stopping are specified by something other than the computer), and things end up wherever they end up. However, if one were to ask the computer to predict the outcome of the algorithm prior to the program being run, the computer would be in no better position (without running the program) than a human being is as far as giving a reliable prediction is concerned because ignorance has central prominence in both cases.

So-called emergent properties are as 'mysterious' to a computer as they are to human beings. In both cases, neither the computer nor the human being is capable of predicting how, respectively, a given algorithm or comparable real-world dynamic will unfold over time.

There is no emergent phenomenon going on in either case. There is just ignorance about the character of the outcome and how such an outcome arises from the dynamics that are inherent in a computer program or a real-world context.

When someone says that life is a quintessentially emergent phenomenon, what is that person trying to say? Generally speaking, individuals say this sort of thing when they don't understand how life is possible but wish to be able to continue to believe that various physical laws (both known and unknown) are capable of coming together and giving rise to life in ways that his or her current understanding does not grasp.

In other words, such people tend to believe that somewhere, somehow, the right combination of forces and elements came together within the right sort of circumstances and conditions to be able to give rise to some kind of living protocell. Thus, life is an emergent property of the interacting combination of the right set of unknown: Forces, elements, circumstances and conditions.

Every time someone – either with a computer on in a lab – is able to run an experiment leading to unexpected or unanticipated results that appear to be somewhat suggestive in relation to origin of life issues, then, those kinds of results are considered by some individuals to constitute evidential support for the possibility that life also arose through a similar process of unexpected and unanticipated outcomes.

For some individuals, emergent properties and emergent behavior seem to become the answer to every unknown issue involving the origin of life. However, such 'answers' never explain or account for anything in specific detail, and, consequently, just how certain kinds of underlying dynamics are capable of generating life is always left unaddressed, and the resulting gap in understanding is papered over by using the term: 'emergent behavior'.

Emergence is not a scientific term. It is a philosophical one, and it entails many of the same kinds of ambiguities and arbitrary assumptions that characterize any number of philosophical positions.

Furthermore, dressing up the idea of emergence in scientific clothing doesn't make the concept any more rigorous. For example, one can talk all one likes about how: Far from equilibrium conditions are capable of creating conditions involving the flow of energy which dissipate that flow in unexpected and unanticipated ways and, in the process, gives rise to certain kinds of ordered structures of energy flow that are quite different from what takes place near equilibrium conditions ... and, if one likes, one can quantify the whole description with lots of spiffy equations and mathematical expressions.

Nonetheless, however scientifically valuable such accounts might be in conjunction with describing or modeling an array of phenomena, there is no, or little, transfer value when it comes to explaining the origin of life. In other words, there is no scientist (or group of scientists) who has (have) come up with a far from equilibrium scenario that reliably and demonstrably accounts for precisely how the dissipative structures that constitute different life forms (or the dissipative structures inherent in various kinds of metabolic pathways) arises in various kinds of far from equilibrium conditions.

Before moving on, the reader [(?) ..., readers (?) -- see Introduction] should understand that there are two broad kinds of hydrothermal vents. They are referred to as black smokers and white smokers, and each kind of vent system gives expression to different sets of chemical and physical conditions.

Black smokers arise in connection with volcanic activity in the depths of the ocean. The 'black smoke' is not actually smoke but consists of a acidic mixture of metal sulfides and seawater heated to temperatures of around 400 degrees Centigrade under tremendous pressure from the ocean depths in which black smokers exist (one of the deepest, if not the deepest, black smoker discovered to date resides in the Cayman Trough, a little more than 3 miles below the ocean surface).

The material surrounding the channel-way that rises up through the black smoker chimney system is made from various kinds of sulfur minerals. This material has precipitated out from the heated, metal sulfide solution that is churning up through the black smoker chimney.

Black smokers increase in height as a result of the continuing precipitation of the aforementioned sulfur minerals. The growth rates of the chimneys built up from the precipitants vary with conditions, but those structures can reach heights of several hundred feet before beginning to fall apart after 20,000 years, or so.

White smokers, unlike black smokers, are not a function of volcanic activity. Instead, the interaction of seawater with mantle-derived rocks releases energy in the form of heat along with a variety of gases, including: Hydrogen, hydrogen sulfide, carbon dioxide (which, given the right conditions, can lead to the formation of methane), and nitrogen (which, given the right conditions, can lead to the generation ammonia).

The foregoing heated solution is not nearly as hot as the seawater mixture that churns up through black smokers, but the solution in white smokers does contain a great many electrons in the form of reduced reactants. White smokers are alkaline in character.

Moreover, while white smokers sometimes form chimney-like structures similar to black smokers, white smokers more often form complex, interconnecting structures made of materials precipitating out of the 'white smoke' rising from and through such structures. The white color of the "smoke" that emanates from white smokers comes from the calcium, silicon, and barium compounds contained in the hydrothermal mixture associated with such structures.

The arrangement of bubble chambers and compartmentalized units in white smokers are roughly the size of living cells. They form extended, interconnected, microscopic, networks of porous materials.

The existence of each kind of smoker has inspired various researchers to conjecture that life might have arisen in one or the other form of hydrothermal vent system. The minerals, gases, pH conditions, surface structures, and compositional materials that are associated with the respective smokers are considered by various researchers to be ideal "breeding" grounds for an array of chemical reactions that might lead to life.

The amazing, but different, ecosystems that are found living in harmony with each of the respective smokers have suggested to some individuals that, perhaps, life arose as a function of the physical and chemical conditions present in one, or the other, kind of smoker. The task then becomes a matter of showing how life could have arisen in the underlying physical and chemical conditions associated with one of those two kinds of smokers … or, perhaps, both.

While acknowledging the differences between the two aforementioned sorts of smokers, much of what is said in this section of the present chapter, is directed toward the <u>general</u> kinds of problems that are likely to be encountered by both modes of smokers. For example, irrespective of whatever the particular physical and chemical conditions of a given smoker might be, if one hopes to develop a plausible theory concerning the origin of life, one must account for how various modes of order arise in those different kinds of conditions that are capable of establishing an array of interacting, metabolic pathways that will perform the functions that are able to initiate and sustain living organisms.

In short, the physics and chemistry of each smoker are different. Nevertheless, the ultimate problem facing both of the smokers is the same – namely, how does one induce a set of basic physical and chemical reactions to form, first, more complex biomolecules and, then in turn, to assemble such biomolecules into functional, self-sustaining metabolic pathways before they disassemble under the onslaught of a variety of forces involving: temperature, water, pressure, pH values, unfavorable thermodynamic conditions, changing geological conditions, and competing chemical reactions.

Stanley Miller – whose classic 1950s experiment was discussed earlier – believed that life must have formed somewhere within the area between a few centimeters and several hundred meters, or so, relative to the surface of oceans/lakes. This area is known as the Photic Zone, and it represents the depth to which light – considered by many evolutionary biologists to play a primary role in the origin of life – will penetrate in a given body of water.

Miller and others were critical of the hydrothermal vent hypothesis that maintained life might have formed near the bottom of the ocean along volcanic vents. Among other reasons, Miller and many other similar-minded researchers felt that the heat from such vents would have destroyed more molecular precursors to life (e.g., amino acids and ribose sugars) than they would have created.

Various experiments have been conducted in an attempt to shore up some of the perceived weaknesses (such as the Miller criticism noted in the preceding paragraph) that are associated with the hydrothermal vent hypothesis. For instance, in the period spanning 1999 and 2000, Jay Brandes designed a number of experiments to try and address a few of the theoretical problems with which the hydrothermal vent hypothesis was faced.

One of the experiments Brandes performed involved the amino acid leucine [$HO_2CCH(NH_2)CH_2CH(CH_3)$]. This is an important biomolecule (that is, a molecule known to play various roles in living organisms).

When leucine is subjected to temperatures of, say, several hundred degrees Centigrade, under conditions of elevated pressure, leucine tends to decompose fairly quickly (in a few minutes). However, when leucine is exposed to the foregoing sorts of conditions in the presence of an iron-sulfur mineral known as pyrrhotite, the amino acid is able to survive for a number of days.

Pyrrhotite is significant because it is a fairly common component in oceanic volcanic vents. Moreover, while the means through which pyrrhotite is able to help prevent the breakdown of leucine seems to be somewhat elusive, the implication of the Brandes experiment is as follows: One cannot automatically assume that the biomolecules that might arise in hydrothermal vents will necessarily and automatically decompose when exposed to conditions of high temperature and pressure since there are various kinds of minerals existing in those vents – such as pyrrhotite -- that might be able to help stabilize those biomolecules and extend their molecular lives.

Findings from other experiments also have suggested that if amino acids can establish strong bonds with various kinds of minerals, they might have a better chance of remaining intact for a longer period of time than in the absence of such mineral bonds. The implication of this research is that, perhaps, there were conditions in hydrothermal vents that facilitated boding between various kinds of minerals and different amino acids, and, in the process, helped preserve the molecular identity of those amino acids in extreme conditions.

Whatever truths are entailed by the foregoing sorts of experiments and research, they sound somewhat strained when it comes to trying to account for the origin of life. Simply because one can demonstrate that the life of a molecule might be extended for a short period of time under certain circumstances, this does not necessarily have any relevance to what might have actually happened on early Earth or what needed to happen on early Earth if the origin of life is to be explained purely in terms of the laws of physics and chemistry.

The fact there is evidence to show that something could have happened does not mean that this is what actually did happen. A lot of prebiotic experiments and research seems to resonate with the words that the Marlon Brandon character, Terry Malloy, voiced in the movie: *On the Waterfront* – namely, "I coulda' been a contender."

A lot of things could have been but are not. It remains to be seen whether various prebiotic pretenders turn out to be bums or real contenders. However, for the most part a lot of those researchers just seem to be caught up in their own fantasies of what "coulda' been" or should have been or might have been if things worked the way their ideas claimed was possible.

Does extending the life of a molecule from a few minutes to a few days really appreciably change anything as far as accounting for the origin of life is concerned? Do we know whether, or not, all the biomolecules of life regularly established bonds with various minerals and that these sorts of arrangements lasted sufficiently long to enable the biomolecules to enter into reactions with other biomolecules while still subjected to the extreme conditions of hydrothermal vents? Do we know whether, or not, the existence of the biomolecule-mineral bonds would have interfered with the ability of the attached biomolecules to interact with other biomolecules?

Unless all of the foregoing questions – plus many others -- can be answered in definitive terms, one is not necessarily dealing with something of scientific value as far as the origin of life issue is concerned. The aforementioned experiments and research do not constitute evidence in favor of a scientific account for the origin of life because we really don't know what, if any, relevance those findings have with respect to the actual conditions in any given hydrothermal vent on early Earth.

For example, hydrothermal vents tend to exist in very unstable, geological conditions. How long do vents last in such unstable conditions?

During the first year, or so, of this century, the submersible vehicle Atlantis was involved in the discovery of an alkaline vent system located on an underwater mountain known as the Atlantis massive, roughly 9 miles from the Mid-Atlantic ridge. Some of the vents found there were nearly 200 feet tall.

One of the white-smoker, alkaline vents found on the Atlantis massive was dubbed the 'Lost City'. It has been estimated to have been venting for 40,000 years – twice the length of time usually associated with the life-span of black-smoker, acidic hydrothermal vents that had been discovered many years before.

20,000 to 40,000 years is not a very long period of time for nature to work with and through which to catalyze the basic molecules of life and, then, assemble them into some network of metabolic pathways. Even if one were to arbitrarily add on several hundred thousand years to the life span of white smoker vent systems (and there is little, or no, evidential basis for extending the life-span of various kinds of smokers in this way), one still needs to assume a great deal to suppose that, somehow, the first prototypes of life emerged in such vent systems.

One might be willing to concede that some minerals could extend the lifetime of certain biomolecules under fairly extreme conditions of temperature and pressure. Nevertheless, we really don't know if any given hydrothermal vent would be around long enough for the foregoing sort of possibility involving mineral-biomolecule bonds to be able to make any difference as far as the origin of life is concerned.

Research also has been done which demonstrates that a variety of relatively common minerals – such as the sulfides of copper, iron, nickel, zinc or cobalt (as well as the oxides of some of the foregoing minerals) – have the capacity to promote (catalyze) the addition of carbon atoms to other molecules under certain conditions ... one of which involves elevated temperatures. Thus, mineral-rich hydrothermal vent systems might be excellent sources for the building of more complex biomolecules as carbon atoms are added to, among other possibilities, hydrogen molecules.

If one considers the fact that there are, and have been, tens of thousands of deep ocean ridges in the oceans of the world that are peppered with various kinds of hydrothermal vents, and, then, if one throws in millions of years of time through which the mineral-rich hydrothermal vents will be permitted to do their work of preservation and catalysis, then, someone might – and there are those who have – come to the conclusion that an abundance of biomolecules of varying degrees of complexity must have been formed in and around hydrothermal vents. Seemingly, one is off and running – perhaps taking a lead – in the explanatory races with respect to the origin of life issue as one imagines all manner of metabolic pathways that might emerge as different combinations of minerals worked their catalytic magic in the hydrothermal vents ... or, so, the theory goes.

Even if one were to grant each and every possibility outlined in the last few pages – and, for a variety of reasons, I am not inclined to do this because, among other things, far too many unproven assumptions are necessary to make the hydrothermal scenario work -- none of the granted possibilities, either alone, or in combination with one another, accounts for how functional order arises out of the morass of biomolecules that might have been generated through hydrothermal vents.

The assumption is made that given so much biomolecular material with which to work, then, surely, functional, metabolic pathways must have arisen again and again. The problem is that there is absolutely no proof such an assumption is rooted in reality.

To be sure, some of the prebiotic experimental and research data are highly suggestive. There are interesting speculations. There are promising conjectures. There are intriguing possibilities ... but there is absolutely no scientific proof that the hydrothermal vent hypothesis is true.

One can imagine whatever one likes, but that is all one ends up with: imagination. It is an exercise in magical thinking in which someone supposes that because he or she believes something must be true, then, this is the way reality must be.

As far as the hydrothermal vent hypothesis is concerned, in order to have any chance of demonstrating a truly scientific explanation for the origin of life, one must show that hydrothermal vents will produce functional proteins, lipids, nucleic acids, and carbohydrates. This has not been done.

Moreover, in order to have any hope of demonstrating a truly scientific explanation for the origin of life, one must show that hydrothermal vents will produce workable metabolic pathways capable of performing not just a few, minor biological functions but everything that is necessary for such pathways to be able to continuously sustain themselves over a period of time. This has not been done.

Furthermore, in order for there to be some possibility of demonstrating a truly scientific explanation for the origin of life, one must show that the metabolic pathways that do arise (assuming they do) will form integrated networks that are able to replicate and pass on such capabilities in a manner that permits additional, independent, integrated, biologically functioning networks of metabolic pathways to become established. This has not been done.

Maybe the day will come when one, or more, individuals will be able to successfully meet all of the foregoing three challenges. However, today is not that day.

The hydrothermal vent hypothesis is not a scientific hypothesis. Rather, it is a conjecture in search of scientific proof.

One could assemble a library of scientific facts concerning the hydrothermal vent conjecture ... and, indeed, there are many large library-like collections containing that kind of technical material. However, unless one can show how a given library of facts can prove the truth of the hydrothermal vent conjecture, one doesn't have a scientific theory.

Instead, one has a conjecture to which various scientific facts have been attached. This situation is somewhat akin to the way certain prebiotic research has indicated that a biomolecule sometimes can become bonded to a mineral that might help to prolong the life of the former biomolecule through an unknown mechanism and, therefore, with no real understanding of whether, or not, those facts are actually capable of sustaining the lifetime of the conjecture for any length of time.

Claiming that a library of scientific facts is <u>consistent</u> with the hydrothermal vent hypothesis does not make that conjecture either true or scientific. In order to be able to assess the relevance of such claims, one needs to critically examine the nature of the 'consistency' that is being claimed in order to understand in what way, if any, an allegedly scientific fact is capable of establishing a viable, demonstrable, concrete bridge between the hydrothermal vent conjecture and a sustainable account of the origin of life.

As intimated a page, or so, back, there are no scientific facts that are said to be consistent with the hydrothermal vent conjecture/hypothesis that are capable of establishing a viable, demonstrable, concrete bridge between that conjecture/hypothesis and a sustainable account of the origin of life. Consequently, whatever scientific facts are claimed to be consistent with the hydrothermal vent hypothesis or conjecture are of an entirely inessential kind because they cannot prove what needs to be proved as far as the origin of life issue is concerned.

If one needs to travel from Boston to Seattle, and one finds oneself in Atlanta, then, being in Atlanta might be considered by some to be entirely consistent with the character of the stated journey. Nonetheless, one's presence in Atlanta also tends to raise a lot of questions about, whether, or not, one will ever make it to Seattle or whether one even knows where one is going or what one is doing.

The same sorts of questions tend to arise in conjunction with the idea of consistency when considering the hydrothermal vent hypothesis. If one needs to reach the destination of having a viable account for the origin of life, and one is wandering around a library of scientific facts trying to figure out if, or how, any of those facts will enable one to travel from the hydrothermal vent hypothesis to the truth, one's status is sort of like the situation described in the previous paragraph.

In other words, one started out on one's journey in Boston (the hydrothermal vent hypothesis). Now, however, one finds oneself in Atlanta (the library of scientific facts) on the way, possibly, to Seattle (the truth), and, unfortunately, claims of consistency don't possess a whole lot of value under such circumstances because they don't necessarily get one any closer to one's destination ... however interesting and intriguing the possibilities in Atlanta might appear to be.

A key metabolic pathway in living organisms is the citric acid cycle. This is also known as the TCA (Tricarboxylic Acid) cycle, as well as the Krebs cycle.

This pathway consists of a handful of relatively small compounds made up of just three molecules: carbon, hydrogen, and oxygen. The molecules are: (1) oxaloacetate, (2) citrate, (3) isocitrate, (4) α-ketoglutarate, (5) succinyl-CoA, (6) succinate, (7) fumarate, and (8) malate. In addition, two-carbon atoms from pyruvate are inserted into the beginning of the cycle in conjunction with acetyl CoA.

During the course of the TCA cycle a variety of molecules are produced that are important building blocks for the synthesis of other biomolecules ... including amino acids, sugars, and lipids, as well as molecules that serve as a source of energy in the form of ATP (in bacterial cells and plant mitochondria) and GTP (in animal mitochondria). However, each step of the cycle requires the presence of a different enzyme that catalyzes a specific reaction by helping to rearrange the bonds and relationships among the carbon, hydrogen and oxygen molecules involved in the cycle with the help of coenzymes or cofactors (NAD – nicotinamide adenine dinucleotide – and FAD – flavin adenine dinucleotide) that accept or donate electrons at certain points in the cycle.

In the mid-1960s, some microorganisms were discovered that ran the foregoing metabolic pathway in reverse. Not surprisingly, this newly discovered process was referred to as a reverse citric acid cycle.

What was surprising, however, is that a certain point in the cycle, citric acid split up into a molecule of oxaloacetate and acetate, and, in the process, opened up the possibility for an additional metabolic cycle to be established provided that a few modifications were made in relation to the acetate molecule. Consequently, the reverse citric acid cycle seemed to constitute a metabolic pathway that had a potential capacity for self-replication.

Some people entertained the idea that the reverse citric acid cycle might have been closely related to the first metabolic pathways contained in primitive protocells. Among other things, this possibility had the virtue of giving expression to a self-replicating process that – at least in principle – doubled its potential with each completion of the cycle.

Of course, there was still the problem of having to account for, among other things, the origins of the 9 enzymes and several coenzymes that made the cycle possible. However, if one returns to the topic of the sulfide and oxide minerals that were explored earlier in this section, then perhaps, those minerals -- along with other components that either circulated in the hydrothermal vents and/or were part of the structure of those vents – might have been able to help facilitate (i.e., catalyze) different steps in the cycle and, therefore, helped sustain the cycle until the right sort of enzymatic proteins came into existence that would be able to introduce added efficiency and speed to various reactions taking place within the reverse citric acid cycle.

The idea of sulfide minerals serving as interim catalytic-like agents in protocells is rendered somewhat more plausible by the fact that at the core of a variety of modern enzymes are groupings of sulfur, iron, or nickel molecules. Perhaps, modern enzymes somehow arose from the simple beginnings of sulfur, iron or nickel sulfide minerals when aspects of the latter became incorporated into amino acid complexes that resulted in a protein with enzymatic properties.

Unfortunately, all of the foregoing possibilities are really little more than speculation. For instance, no one has shown that an array of sulfide minerals (and/or oxide minerals) has the potential to be arranged in just the right sort of sequential way and with just the right amount of sufficient catalytic activity to make a reverse citric acid cycle work at all ... let alone within a plausible time frame for such a system to be able to survive and replicate amidst conditions involving extreme temperatures and pressure as well as existing in an environment that is not necessarily all that stable from a geological point of view.

Moreover, no one has shown how sulfide minerals with catalytic properties were able to transition to proteins with catalytic properties. In other words, how does a person go, on the one hand, from: (1) A metabolic-like system regulated by a sequence of conveniently placed sulfide minerals within a compartmentalize niche of some given hydrothermal vent, and, on the other hand, to: (2) A nucleic acid based system of coding that gives rise to a metabolic pathway that is regulated by enzymes that contain cores involving iron, nickel, or sulfur molecules?

George Cody, Robert Hazen, and Hat Yoder of the Carnegie Institute group conducted a number of experiments in the latter part of the 1990s that were intended to study the behavior of citric acid in conditions of high temperature and pressure. These conditions were intended to simulate what might have happened in the vicinity of various hydrothermal vents on early Earth.

Cody analyzed the results of those experiments. He determined there were two kinds of reaction cycles that tended to take place under the conditions specified by the experimental design, and he labeled them alpha and beta pathways.

The alpha pathway begins with citric acid breaking down into acetate and oxaloacetate molecules and, therefore, mimics what also can be observed to occur in the reverse citric acid cycle. However, under the experimental conditions, the foregoing oxaloacetate further degrades into pyruvate plus carbon dioxide, and, then, the pyruvate molecules decompose into acetate.

Irrespective of whatever combination of reactants and minerals were used in the experiments, the researchers could not get pyruvate to generate oxaloacetate. Consequently, the alpha pathway could not even get the reverse citric acid cycle started, let alone find a way to move on to the other steps of the reverse citric acid pathway.

The beta pathway discovered by Cody seemed more promising because it was rooted in a carbon dioxide produced series of successive reactions that yielded molecules with five, four, and three carbon atoms respectively. However, with one exception (aconitate) these carbon-containing molecules were different kinds of five-, four-, and three-carbon molecules than the ones that characterized the reverse citric acid cycle.

Cody also uncovered some evidence indicating that the beta-pathway that sometimes occurred during the experiments he was analyzing had the ability to be reversible in the presence of nickel sulfide. This suggested that the beta-pathway might be able to form a closed metabolic loop.

However, there was an absence of certain other kinds of evidence in the foregoing experiments. More specifically, there was no data that showed that the beta-pathway gave expression to a metabolic potential that might be able to generate the sort of biomolecules that are synthesized in either the citric acid cycle or the reverse acid cycle ... biomolecules that play important roles as building blocks with respect to the synthesis of additional biomolecules – such as amino acids, sugars, and lipids -- which are intimately involved with the process of life as we know it.

Furthermore, Cody's analysis did not appear to demonstrate that the beta-pathway associated with the experiments his group ran was capable of producing ATP or GTP ... a key source of energy in biological systems. In fact, given that the molecules that showed up in the beta-pathway were mostly different from the molecules found in the reverse citric acid cycle, the very molecules that made up the beta-pathway might constitute an obstacle with respect to the formation of either ATP or GTP.

For example, the molecule, succinyl-CoA, plays an essential role within the citric acid cycle by helping to bring about the synthesis of ATP and GTP. Succinyl-CoA accomplishes this through holding on to the energy generated during the oxidation of the α-ketoglutarate molecule by means of a thioester bond. When the latter bond is hydrolyzed, the path has been cleared for the synthesis of ATP and GTP.

It is uncertain whether, or not, any of the molecules in the beta-pathway analyzed by Cody might be capable, under suitable circumstances, of preserving energy in the same way that succinyl-CoA does. If none of the molecules in the beta-pathway is capable of achieving this step, then a very important element is missing from the beta-pathway even if -- as Cody feels might be the case -- that pathway was capable of forming a closed metabolic loop.

A few more problems can be added to the foregoing considerations. One set of such problems is similar to an aforementioned issue.

For example, let's assume one begins with something like the beta-pathway that possesses nickel sulfide to serve as a catalyst of sorts. Given such a starting point, one must be able to account for how a nucleic acid base pairing system -- that encodes for proteins with enzymatic capacities that might have some nickel atoms at their core -- arises from the aforementioned, non-nucleic acid base pairing starting point.

There are still more problems or questions that need to be explored in conjunction with the beta-pathway possibility. For example, what about the other metabolic pathways that will be necessary in order to be able to take functional advantage (i.e., to help a given protocell to survive) of the molecules that are produced via the beta-pathway?

How did these other metabolic pathways come into existence? How did they become integrated with the biomolecules that are supposedly being synthesized through the beta-pathway?

A metabolic pathway that operates in the way that the reverse citric acid cycle does is not sufficient unto itself as an explanation for the origin of life. It needs to be augmented by, and integrated with, a variety of other metabolic pathways, that, in turn, are involved with still other metabolic pathways ... and this principle of additional, integrated, complementary metabolic pathways applies to the beta-pathway scenario as well.

Moreover, all of these complementary metabolic pathways need to be set in a context that consists of an appropriate sequence of the right kind of sulfide and oxide minerals (or other, alternative, catalytic agents) that are capable of making such metabolic pathways functional. Even if one were to grant that the beta-pathway formed closed metabolic loops, this is not enough ... in other words, that loop must be demonstrated to be both functional (i.e., capable of producing useful biomolecules), as well as connected to other metabolic pathways that can make use of what is being synthesized through the beta-pathway.

None of the foregoing issues were part of the Cody analysis of the beta-pathway. Therefore, if the beta-pathway is to be considered a viable candidate with respect to explaining the origin of life, then a lot more works needs to be done.

Moreover, as it stands -- and quite apart from the problems arising in conjunction with the need to account for additional metabolic pathways to complement the beta-pathway – the beta-pathway idea is missing some important ingredients. Among other things, and as indicated earlier, one doesn't even know whether the beta-pathway has any biologically relevant functionality with respect to the molecules that are produced through it when citric acid is subjected to conditions of extreme temperature and pressure.

It is not enough to show that a given pathway might form a closed loop. One also has to be able to demonstrate that the pathway has a potential functional value for helping to account for the origin of life.

Apart from – but also related to – the beta-pathway, there are a number of other issues that need to be addressed. For example, assuming that the reverse citric acid cycle (or something very much like it) might have been one of the first metabolic pathways to become established in the times of early Earth, how was the transition made from the reverse citric acid cycle to the 'normal' citric acid cycle?

Any number of possibilities might be advanced to address such a problem. However, there is no definitive evidence to show that any particular one of those possibilities is the right one (i.e., that it accurately reflects what happened).

An even more important issue has to do with the emergence of the DNA code? No one knows how this came about, and there aren't even any reasonable conjectural candidates for consideration.

The hydrothermal vent hypothesis maintains that metabolic pathways might have arisen in conjunction with various sulfide and oxide minerals that populated those vents. Aside from the previously noted problem of having to come up with a credible scenario for how an array of such minerals came to be arranged in just the right way and with just the right sort of catalytic activity to be able to give rise to functional, metabolic pathways, one must also be able to provide a credible account for how such prototypes transitioned into a coded set of nucleic acid base pairs that was able to incorporate the metabolic information contained in the sulfide mineral based metabolic pathways despite the fact that the system of coded nucleic base pairs seem to have nothing to do with arrays of sulfide and oxide minerals that have a structure and composition that is quite dissimilar from nucleic acid base pairs.

If one cannot explain how the transition from sulfide mineral-based metabolic pathways to nucleic acid-based metabolic pathways was accomplished, then one is left with several problems. The first problem is the huge hole that exists in any origin of life account that might be associated with such an inability to bridge the aforementioned transition issue.

The second problem is as follows: One will have to consider the possibility that the DNA/RNA coding system arose entirely independently of whatever primitive protocells might have formed that are based on the sorts of sulfide/oxide mineral metabolic pathways that often are envisioned to have arisen in conjunction with the hydrothermal vent hypothesis. Among other things, such a possibility suggests that all of the prebiotic research connected with trying to show how biological systems might have been given their start through various kinds of inorganic and organic chemistry that made use of networks of sulfide and oxide mineral pathways of catalysis in the context of hydrothermal vents is relatively worthless because none of that research really explains how the nucleic acid base pairings came to contain the sort of information that made the former sort of functionally integrated metabolic pathways possible.

In short, the degree of difficulty with respect to the problem of accounting for the origin of life via physics and chemistry has just doubled. One not only has to explain how functional sulfide mineral-based metabolic pathways arose, but, one also has to explain how quite different nucleic acid base pair systems arose that were able to independently solve the same set of problems involving metabolism that had been at least partially solved by the sulfide mineral approach to forming functional metabolic pathways that were facilitated by the presence of catalytic agents.

As a side note, and before moving on to other issues, all of the difficulties that saturate sulfide/oxide mineral–based accounts concerning the rise of metabolic pathways also befuddle various theories (e.g., Graham Cairns-Smith) which claim that certain kinds of mineral-laden clays could have served as a catalytic medium that might have brought about metabolic pathways through which important biomolecules might have been synthesized. If one analyzed the previous discussion concerning the role that sulfide minerals might have played in the hydrothermal vent hypothesis approach to the origin of life issue, and one substituted the word "clay" whenever terms such as "sulfide minerals" or "oxide minerals" appeared, all of the problems that have been pointed out with respect to the latter terms (e.g., sulfide minerals) would carry over to the clay-based theories.

Whatever strengths and properties might be associated with the idea of clay serving as a template, of sorts, for the origin of life, those strengths and properties are not enough to overcome the problems permeating that idea ... problems that already have been raised in conjunction with the sulfide/oxide mineral-based theories. Not the least of such problems is the relative dearth of evidence that is capable of demonstrating precisely how clay-based theories were able to generate the sort of functional metabolic pathways that are needed to provide a viable means of accounting for the origin of life.

If life is not to be a one-and-done proposition, there must be a way of storing information that contains instructions for generating an integrated set of structural and dynamic properties that not only are needed for survival but which also can be transferred in a manner that permits similar, but independent units of instructional storage to arise. Some individuals – such as Sidney Fox – have tried to account for the foregoing sort of informational storage process through the notion of proteinoids (protein-like entities).

Proteinoids consist of a sequence of amino acids that are synthesized through various chemical and physical conditions ... conditions that, given enough time, will lead to the formation of chains of peptides (i.e., proteinoids) that are theorized to have the sort of properties that -- when arranged in appropriate sequences -- can synthesize nucleotides. In turn, these nucleotides serve as the storage units for an array of operational instructions that are not only necessary for a protocell to be functional in a biological sense but, as well, are necessary for the transmission of that information in a way that can assist other protocells to arise, function, and enable the same sort of information to be passed along down a line of molecular, if not cellular, descent.

Potentially, amino acids are capable of linking up with one another in an enormous array of possibilities that extend far, far, far beyond even a realm of arrangements that entails hundreds of trillions of combinations. The vast number of such combinatorial possibilities seems to pose a rather significant problem for anyone who might want to provide a rational, credible account for how a set of, say, 20 amino acid combinations from amongst an array of such enormous possibilities came together in just the right way to be able to synthesize the right kind of nucleotides that, in turn, would come to form a sequence of base pairs that were capable of leading to the synthesis of the same set of amino combinations from which similar sequences of nucleotides could be synthesized again and again.

There is exactly zero proof indicating that any of the evolutionary scenarios concerning the origin of life is capable of explaining in a convincing manner how the foregoing mass of combinatorial possibilities was able to give rise to the sort of functional network of metabolic pathways that are necessary to account for even the simplest forms of life. Of course, one could assume that everything somehow happened in a just-so way, but assuming one's conclusion in such a fashion tends to be a sign of the presence of magical thinking rather than the presence of a rigorous, critically reflective, methodical process of science.

Hope, however, springs eternal ... even in the minds of people who consider themselves – or are considered by others – to be scientists. So, let's consider a few more ideas.

In 1982 Thomas Cech and Sidney Altman uncovered the existence of ribozymes. Ribozymes consist of RNA molecules that not only are able to store information but, as well, are able to catalyze certain kinds of biochemical reactions.

The discovery of ribozymes offered a possible way to eliminate a dilemma with which origin of life theories had been confronted prior to 1982. More specifically, although everyone conceded that both proteins and DNA/RNA were necessary for life, no one could figure out a plausible account for which of the two ingredients come first.

Did the origin of life process start with DNA/RNA and subsequently lead to the synthesis of proteins? Or, alternatively, did the origin of life process begin with proteins (as Fox and a few others maintained) and this, in turn, led to the synthesis of DNA/RNA?

Ribozymes appeared to resolve the foregoing problem rather nicely. A biomolecule had been found that, seemingly, might be able to take on the roles of both proteins and DNA.

While, in principle, ribozymes appeared to have explanatory potential with respect to addressing the aforementioned chicken-and-egg priority issue in relation to the origin of life, there also were some outstanding questions swirling about that notion. For example, even though ribozymes were capable of catalyzing some reactions, did ribozymes have the capacity to catalytically facilitate all manner of reactions?

If the answer to the foregoing questions is no, then, there are determinate limits to the catalytic properties of ribozymes. Depending on the nature of such limits, ribozymes might not constitute as big a treasure trove with respect to be able to provide a credible evolutionary account for the origin of life as some theorists might have hoped.

Ribosomes (not to be confused with ribozymes) consist of a complex integration of proteins and strands of RNA. In combination, the foregoing two components of ribosomes assist cells to bring about the assembly of proteins.

Initially, scientists believed that the proteins in ribosomes were primarily responsible for the sort of catalytic activity that facilitated the linking up of amino acids with one another during the formation of various proteins. Eventually, however, research determined that ribosomal RNA, not proteins, played the lead role in the assembly of proteins.

There was a further tantalizing piece involving the origin of life puzzle that complemented the discoveries involving the role of RNA in ribozymes and ribosomes. RNA nucleotides (or closely related molecular structures) are found in some coenzymes that play important roles in, among other reactions, the citric acid and reverse citric acid cycles that were discussed earlier in this chapter.

Finally, research also has uncovered the existence of what are sometimes referred to as 'riboswitches'. These are segments of RNA found in messenger RNA that are capable of regulating some aspects of gene expression by turning certain genes on and off through the way in which their conformational shapes change when binding to various kinds of molecules.

Many people today are familiar with the term "junk DNA". This phrase was introduced in 1972 by Susumu Ohno, a Japanese geneticist, as a way of referring to the fact that only about 2% of the human genome consists of genes that actually code for proteins, whereas the other 98% of the genome consists of, apparently, useless, nucleotide residues.

The traditional picture of protein synthesis was that a given gene codes for, and is transcribed into, messenger RNA. Messenger RNA helps bring about the assembly of the protein that was specified by the gene that led to the appearance of messenger RNA.

In 2001, a more complex picture began to emerge. Among other things, researchers discovered that something called "microRNA" (sometimes consisting of as few as 22, or so, nucleotide sequences) was capable of binding to various segments of messenger RNA, and, as a result, the microRNA was able to modulate the activity (or expression) of the messenger RNA molecules.

In addition, researchers discovered that what had been considered to be useless or junk DNA was coding for the tiny sequences of RNA known as microRNA. These small, coded segments of RNA were performing a vast array of regulatory functions within cells.

For example, consider the protein myosin that plays a major role in orchestrating the activity of heart muscles. Researchers uncovered the fact that microRNA sequences were tucked away in one of the introns associated with the production of myosin.

An intron is a nucleotide sequence that is removed by RNA splicing during the translation of messenger RNA into – in this case -- a myosin protein. The intron that is removed during RNA splicing was considered to be useless or junk RNA that for 'reasons' lost in the distant past had, nonetheless, been retained and still was able to code for the transcription of such sequences ... and, as a result, the segment of the DNA sequences coding for the useless messenger RNA was considered to be a junk form of base pairings as well.

Yet, lo and behold, the intron was not useless. It contained information that helped regulate the activity of myosin in a variety of circumstances.

Among other things researchers found that such microRNA sequences helped heart muscles to respond in various ways to the presence of, among other things, thyroid hormones. In addition, researchers discovered that as the nature of the microRNA changed, then so too did the manner in which heart activity was regulated also change.

An obvious question that emerges in conjunction with the foregoing findings is the following one. How did all of the regulatory information become embedded within the DNA genetic sequences so that it could be removed from the messenger RNA sequences in the form of introns and, then, subsequently be released to perform regulatory functions in conjunction with whatever protein had been assembled and according to whatever set of conditions happen to prevail at the time the protein was assembled and began to go about its functional business?

Both the hydrothermal vent hypothesis and the prebiotic soup conjecture concerning the origin of life have no plausible, credible, evidence-based way of responding to, or accounting for, the emergence of regulatory order in living systems. Consequently, since order plays such an essential, defining role in making life what it is, neither the hydrothermal vent hypothesis nor the prebiotic soup conjecture really give expression to a viable scientific, evolutionary theory for the origin of life.

Furthermore, neither of those perspectives constitutes the best available scientific theories of the origin of life. This is because neither of those perspectives gives expression to a scientific theory in any meaningful sense of the word.

They each consist of little more than speculations, assumptions, and pieces of isolated, disconnected, and highly questionable data. Such data might have been derived through scientific means, but this is not sufficient to qualify the ideas that make use of such data as being scientific in nature.

As has been shown throughout this chapter, those pieces of data cannot withstand any sort of rigorous critical analysis. After the dust of such a process of considered, critical reflection clears, neither the hydrothermal vent hypothesis nor the prebiotic soup conjecture has been able to plausibly and credibly demonstrate how the pieces of data that have been gathered together over more than 60 years of extensive research would be capable of permitting one to bridge the huge gap that separates, on the one hand, the hydrothermal vent hypothesis or the prebiotic soup conjecture from, on the other hand, a coherent, detailed, consistent, evidence-based account concerning the origin of life.

The inability of science (on so many levels) to generate a successful theory concerning the origin of life issue does not leave one with the best available scientific account for the origin of life. The failure of science in this respect leaves one with no scientific theory at all.

The presence of scientists does not necessarily render a theory scientific. The presence of experimental research conducted by scientists does not necessarily transform the pieces of data that come from such research into a scientific theory. The writing and publishing of an array of articles, books, and essays that are steeped in scientific jargon, terminology, and technical calculations does not necessarily mean that the subject matter of those books, articles, and so on constitutes a scientific theory.

Something is scientific when one can demonstrate -- through the use of reason and empirical data -- that the claims being made in the name of science are capable of being defended in a way that demonstrates that the reasoned, evidence-based system of understanding underlying those claims is able to accurately reflect, to varying degrees of specificity and predictability, those facets of reality to which the claims allude. Evolutionary theories concerning the origin of life have not been able to satisfy – even in minor ways -- the foregoing challenge, and, therefore, those theories are not scientific ... they are just theories, hypotheses, and conjectures, and they do not deserve being assigned the label of "scientific".

Notwithstanding the foregoing considerations, ribozymes, ribosomes, riboswitches, and microRNA represent independently derived forms of support for the idea that, perhaps, RNA should be at the heart of any origin of life scenario that sought to explain how certain capacities – that is, storing information, transmitting it, and handling whatever catalytic activities might be necessary to facilitate such storage and transmission activities -- were possible.

As appealing as all of the foregoing facts concerning RNA sound (and a set of theories known as the 'RNA world hypothesis' were constructed and updated through such facts), nevertheless, there are a variety of problems inhabiting and threatening the RNA world hypothesis. To begin with, having a potential means to store and transmit information involving operational instructions for setting up metabolic pathways that can be catalyzed in appropriate ways is one thing.

However, the origin of such operational instructions is quite another matter. The existence of ribozymes, ribosomes, riboswitches, microRNA, or any other RNA-based capacity does not explain how the set of operational instructions that coordinate and regulate the activity of ribozymes, ribosomes, riboswitches, microRNA, or other RNA-based capacities came into existence. Ribozymes, ribosomes, riboswitches, and microRNA only have biological value when they are capable of operating at a time and in a way and in a place and for a duration that is capable of producing what is needed for cells to be functionally viable.

Therefore, without an appropriate script and directorial oversight, ribozymes, ribosomes, riboswitches, and microRNA are somewhat limited in their capacities to explain origin of life issues. They are sort of like a group of actors who, individually, might possess certain acting talents but, nonetheless, if those actors are left to their own devices to give expression to an array of arbitrary actions, then those actors will not necessarily be able to produce a qualitatively coherent, sensible, and functional film (i.e., explanatory account).

Furthermore, despite decades of trying to find a plausible way of generating RNA molecules from simple precursors under various kinds of simulated early Earth scenarios, no scientist or group of scientists has been able to successfully synthesize RNA. In addition, no one has come up with a plausible mechanism (either in the context of some variation on the hydrothermal vent hypothesis or in relation to some version of the primordial soup scenario) for inducing RNA molecules to link up with one another under early-Earth-like conditions.

To varying degrees, scientists understand how ribozymes, ribosomes, riboswitches, and microRNA work in functional cells. However, scientists have little, or no, understanding about how those RNA-related components came to be organized in a way that, along with other factors, gave rise to functional cells.

[In passing, I find it interesting that Stanislaus Burzynski -- the person who discovered Antineoplastons and their possible role in the etiology of cancer had stumbled upon the discovery of a group of small peptides that seemed to have a regulatory function with respect to preventing various kinds of cancer from being able to gain a foothold in individuals. One wonders if there might be a variety of microRNA segments that code for such peptide sequences.

Perhaps cancer is caused by a variety of factors (carcinogens) that inhibit the expression of those sequences of microRNA that are responsible for the presence of Antineoplastons in healthy individuals.

By using antineoplastons as a treatment for cancer patients, Dr. Burzynski might have been introducing countermeasures for an acute or a chronic problem involving the expression of those segments of microRNA that, normally speaking, regulate the generation and activity of Antineoplastons in healthy individuals.]

| Evolution Unredacted |

Extinction

The vast majority of organisms that have appeared on, or in, the Earth at one time or another have become extinct. Various species are going extinct on a regular basis, and this tendency gives expression to a background rate against that existing life forms play out various possibilities with respect to forces of natural selection that determines whether, or not, such life forms will become part of that background rate.

The average lifespan for species, in general, is approximately 5 million years. However, the lifespan for any given species might be as little as a hundred thousand years, or as long as 15 million years.

The 'normal' background rate of extinction seems to run around 10 to 20%. In other words, out of every 100 species, 10 to 20 of the members of the larger set of species will become extinct over a million-year period ... which works out to be roughly .00001-.00002% of existing species per year.

Occasionally, a type of extinction occurs that deviates substantially from the aforementioned background rate. These are events involving mass casualties resulting in the disappearance of numerous species within – geologically speaking -- a relatively short period of time.

Extinction events might be caused by an array of conditions. Among such possibilities are: Massive volcanic eruptions, relatively rapid changes in climate, large meteor impacts, the release of considerable quantities of methane from hydrates (methane that becomes entangled within a crystalline form of water and, in the process, forms a structure that is similar to ice), and so on –

An array of evidence collected over many years indicates that as many as 17 relatively minor, kinds of mass extinctions have taken place since life first appeared on Earth. For example, there were many large mammal species that became extinct by the end of the last ice age, 10,000 years ago.

On the other hand, there have been, at least, five 'events' involving mass extinctions that are considerably larger than the minor forms of extinction being alluded to in the last paragraph. These major instances of mass extinction usually encompass at least 40% -- if not more -- of the life forms existing at a given time.

The mass extinction events for which evidence exists are as follows. The first 'event' happened in the late Ordovician period some 440 million years ago; (2) a second 'event' occurred in the late Devonian period roughly 370 million years ago; (3) a third 'event' took place near the end of the Permian period approximately 250 million years ago; (4) a fourth 'event' occurred near the end of the Triassic period around 200 million years ago; and, finally, (5) the so-called KT 'event' occurred near the end of the Cretaceous period some 65 million years ago (the 'K' refers to the Greek word for chalk – kreta – which is commonly found in rock strata that tend to mark the boundaries of the Cretaceous period, and 'T' refers to the Tertiary period that followed the Cretaceous period). A mass extinction 'event' might consist of just one dynamic (such as large scale volcanic eruptions), or that kind of event might involve several kinds of dynamics that interact with one another.

There might well have been additional events entailing mass extinction. However, clear-cut evidence for such possibilities is missing due to problems involving a lack of fossils, along with various difficulties associated with being able to establish precise dates for such events.

Nonetheless, there is evidence that at least one other mass extinction event apparently took place in Precambrian times, some 560 to 550 million years ago. Initially discovered in the Ediacara Hills of Australia -- but subsequently found among the fossils in many other locations throughout the world -- there were a variety of animals (including the first fossils that could be seen with the naked eye) that existed in Precambrian times. However, during the latter portion of the Precambrian period, those animals disappeared from the fossil record.

The mass extinction event that occurred during the latter stages of the Ordovician period seemed to involve mostly marine fauna of one kind or another. The available evidence concerning that event suggests it might have been due to relatively rapid changes in climate as tropical-like temperatures were replaced by much cooler conditions.

The second mass extinction event for which there is a fairly substantial amount of evidence took place late in the Devonian period. Instead of consisting of one event, this mass extinction event seemed to be the result of a series of extinction events that occurred approximately 360 million years, or so, ago.

Armored fish, aquatically skilled cephalopods, and other organisms became extinct as a result of whatever events were going on during those days. Opinions on the nature of the cause of the extinctions vary, but one of the possibilities mentioned is that Earth might have been hit by a fairly sizeable asteroid or comet late in the Devonian period.

The largest of the foregoing five events took place near the end of the Permian period, 250 million years ago. It might have wiped out 90%, or more, of all life forms that were in existence at the time.

Many species involving both flora and fauna were involved in the foregoing mass extinction event. Furthermore, the species that were affected inhabited both land (e.g., insects, reptiles, plants, and amphibians) as well as water environments (e.g., most of the marine groups that had been dominant during the Paleozoic era disappeared or were severely decimated during this event).

The fourth mass extinction event occurred fairly late during the Triassic period. This event seemed to affect mostly aquatic life forms, but many water-based families of species became extinct at approximately the same time.

As arrays of species were disappearing during the late Triassic mass extinction event, there were other species that arose in, and around, the time of the Triassic mass-extinction event. For instance, quite a few modern groups – such as mammals, turtles, and crocodiles – began to appear at this time.

The KT 'event', which took place approximately 65 million years ago, eliminated an incredible array of species … including the life form (dinosaurs) that had dominated the Earth for roughly 135 million years. The reign of dinosaurs had begun during the Triassic period and extended until the end of the Cretaceous period when the Earth was hit by a sizable asteroid in Chicxulub, Mexico on the Yucatan peninsula.

However, there were many species of life other than dinosaurs that disappeared as a result of the KT event. A great deal of research indicates that as many as three-quarters of all plant and animal species extant at the time of the event soon disappeared.

The impact of the aforementioned asteroid created conditions that were somewhat similar to those that various scientists have claimed might arise in conjunction with a 'nuclear winter' scenario. Among other things, under such circumstances, the atmosphere would have become filled with so much debris due to the impact of an asteroid, as well as ensuing fires, and so on, that the light of the sun likely would have become blocked out for a substantial period of time (perhaps, years), and this could have eliminated a variety of species in several ways.

For instance, not only would some species die out due to the relatively rapid drop in temperature that would occur in the aftermath of the asteroid strike, but, in addition, plankton and plants would have become unable to perform photosynthesis since light from the sun likely would have had difficulty penetrating a debris-filled atmosphere. The extinction of plants and plankton would lead, in turn, to the demise of whatever species relied on such plants and plankton as a food source, and, this, subsequently, would lead to the disappearance of whatever forms of life fed on the consumers of plants and planktons ... etc., etc., etc.

Let's briefly review the time frame for the five mass extinctions for which evidence exists. The first in a series of five mass extinction events took place 440 years ago. Following another 70 million year period, a second mass extinction occurred. 120 million years later, life forms underwent a third mass extinction that eliminated up to, at least, 90% of all species. 50 million years further down the temporal road, a fourth mass extinction occurred. Finally, a fifth mass extinction took place approximately 135 million years later in which three-fourths of all forms of plant and animal life disappeared.

In other words, within a period of 185 million years, life on Earth was substantially extinguished in vast numbers including one instance of mass extinction that eliminated at least 40% of life forms existing on Earth at that time, as well as second mass extinction event that was calculated to have wiped out between 90-95% of all life forms, along with a third mass extinction event that extinguished three-fourths of all life forms existing at the time. Moreover, during that 185 million year period, there were 6 additional substantial, but much more limited, extinction episodes.

Normally speaking, when many people speak about evolution, what they have to say is couched in terms of what might transpire over many hundreds of millions of years. Nonetheless, a variety of extinction events – both major and minor -- tend to induce one to critically reflect on what might have been going on within an extraordinary period of 185 million years ... a period of time in which life was reduced down to 10% of its former number of species, and, then, over the next 120 million years, life forms were again reduced by three-fourths of their numbers ... plus whatever mass extinctions occurred as a result of the event that occurred near the end of the Triassic period ... along with a number of other minor – but still substantial – instances of mass extinction.

How did life recover sufficiently from being nearly extinct some 250 million years ago to becoming sufficiently robust that it was able to withstand another major extinction event 50 million years later? How did new life forms rise like Phoenix from the ashes during that 50 million year period?

Alternatively, how did that which happened with respect to life, between 65 million years ago and the present time, occur? In other words, how does one go from a point in which three-fourths of life has been wiped out to the present state of biodiversity?

Following each mass extinction event, a variety of new flora and fauna appeared on the scene – such as when mammals, crocodiles, and turtles emerged in, and around, the time of the mass extinction of the late Triassic period, or in relation to the rapid radiation of animals with shells that occurred following the mass-extinction event associated with the Precambrian period.

The foregoing data tend to indicate that, theoretically speaking, one no longer has the usual luxury of having hundreds of millions of years to work with in order to try to account for how various evolutionary events might have taken place. Instead, one is dealing with time frames of 50 and 65 million years respectively.

Of course, 50-60 million years is still a very long time. Nonetheless, the time within which the recovery of life must take place and, in the process, give expression generate many new forms of life is considerably truncated from time frames consisting of hundreds of millions of years.

Furthermore, one cannot necessarily pinpoint the place, time, or circumstances when a given species first shows up. The fossils found in geological strata, together with various methods of dating, might be able to provide a general framework for the appearance of a given species, but the precise time, place, circumstances, and means through which a species became a thing unto itself appears to be relatively hidden (and this is part of the branching problem discussed in a previous section of this chapter).

In addition, the problems surrounding the emergence of new life forms following mass-extinction events can be intensified somewhat relative to the aforementioned time frames of 50-65 million years. Consider the following.

The Cambrian explosion -- or radiation – that predates all five of the mass extinction events -- began approximately 540 million years ago and lasted for roughly 20 million years. During that relatively brief period of time, the general body-plan for many of the major phyla of modern metazoans, or members of the Animal Kingdom, came into existence, and, as well, there was considerable diversification of other kinds of organisms such as phytoplankton.

In addition, the fossil record indicates that certain kinds of complex organisms arose during the Cambrian explosion that appeared to be unlike any phyla existing today. Obviously, these sorts of organisms subsequently became part of the background extinction rate.

Many evolutionary biologists -- since, and including, Darwin -- tend to agree that during the Cambrian explosion the phyla for all modern animals seemed to simultaneously emerge in the fossil record within a relatively short period of geological time. The problem that arose from such an acknowledgement had to do with the need to explain how so much diversity emerged – relatively speaking -- so quickly.

Since the time of Darwin, some researchers (including Darwin himself) suspected that the incomplete nature of the fossil record might contain a great of information concerning the nature of the explosion ... that is, if such fossils had been discovered rather than missing. In other words, the perspective advocated by a variety of individuals inclined toward the theory of evolution suggested that if the fossil record were to become more complete, not only would the evidence needed to explain the explosion have been readily available, but, as well, such data would have been able to demonstrate that the evolutionary branching process entailed by the Cambrian explosion was relatively uniform and gradual.

The foregoing contention might well be true. On the other hand, one could say something very similar in conjunction with almost any issue for which there is a relative death of evidence capable of supporting whatever one believes might be the truth concerning such an issue.

To be sure, the absence of evidence does not necessarily constitute evidence of absence. Nevertheless, an absence of evidence does not constitute any form of positive evidence either.

Stephen Jay Gould and Niles Eldredge developed an alternative approach to trying to account for evolutionary phenomena like the Cambrian explosion. The two paleontologists believed the fossil record contained very little evidence supporting Darwin's belief that evolution occurred through a process of gradualism -- that is, there seemed to be very little, overall, phenotypic change exhibited across the geological history occupied by the fossils for a given species.

This period of relatively limited, net, evolutionary change is known as a condition of stasis or a stage of equilibrium. However, from time to time, that condition of stasis or equilibrium is punctuated by periods of change in the fossil record.

These periods of change entail a process of speciation. New species arise within the context of small populations that have been separated geographically, ecologically, or in some other way from the ancestral population

This process of speciation is related to Ernst Mayr's founder effect notion. Small populations are moved from one adaptive peak (defining one species) to another, different kind of adaptive peak (defining a new species), by – somehow – moving through a valley in which such adaptive transitions tend to run up against forces of natural selection of one kind or another.

At some point, the newly minted species comes back into the picture via the presence of fossils. Consequently, if one just looks at the surface evidence (fossils), the emergence of the new species might appear to be strange and involve a seemingly inexplicable transition in the fossil record, when, in reality (or so the theory goes) the process of speciation has been perfectly 'normal' but took place off-stage (i.e., without fossil evidence).

Not all evolutionary biologists believe that the sort of speciation entailed by the theory of punctuated equilibrium is necessary in order for adaptive, evolutionary change to be able to occur. Such individuals believe that phenomena like genetic drift and/or an array of mutational events are capable of bringing about adaptive changes independently of the process of speciation outlined by Eldredge and Gould.

Moreover, the evolutionary biologists to whom I am alluding in the foregoing paragraph tend to claim that genetic constraints of one kind, or another, often prevent transitions in morphological character (i.e., evolution) from taking place. From the perspective of those theorists, genetic drift and/or mutational events are necessary before substantial phenotypic transitions (of an evolutionary nature) will be able to occur due to the manner through which genetic drift and/or mutations, of one kind or another, overcome previous genetic constraints.

Nevertheless, whether one believes that crucial evidence has disappeared in the mists of incomplete fossil records, or one advances a theory of punctuated equilibrium that revolves around the possibility of a certain kind of speciation process, or one maintains that adaptive change can arise through the phenomenon of genetic drift and/or a sequence of mutational events, there is an absence of any proof which shows that one plausibly can account for the apparently sudden emergence of life forms like the ones that seemed to occur during the Cambrian explosion.

All one has is a certain amount of data mixed in with an array of conjectures, assumptions, and hypotheses concerning those events. The dimension of proof is entirely absent.

Because technical terms and phrases -- along with a few equations – are often sprinkled among the conjectures, speculations, and assumptions, the aforementioned positions appear – at least to some individuals -- to be scientific in nature. Nonetheless, no real science is present.

As far as the theory of evolution is concerned, none of the essential dynamics have been proven. Nothing of a critical nature has been substantiated. Nothing of a fundamental nature has been confirmed. Nothing has been demonstrated as being likely to be true.

At best, whatever might have been proven, substantiated, confirmed or demonstrated tends to be entirely a function of surface phenomena. None of the deep, dynamic principles that are capable of bringing about the cause of so-called evolutionary change or bringing about the cause of the order that is manifested through the surface phenomena associated with what is alleged to be evolutionary change have been brought into the light of understanding.

Like the fossil evidence that might potentially exist but has not, yet, been discovered, so too, data consisting of direct, observational evidence involving the actual dynamics of punctuated equilibrium, genetic drift, and mutational events all occur off-stage, so to speak, and are unavailable to us except in indirect ways that rely more on the process of interpretation, assumption, and conjecture than they do on the presence of real concrete evidence.

Similar kinds of problems tend to permeate the periods of recovery that follow each of the periods of mass extinction. Within a fairly short period of time following such events – geologically speaking – there often appear to be Cambrian-like explosions of life forms that seem to come into view in relatively inexplicable ways.

How does one explain such phenomena? Ideas such as: an incomplete fossil record, or punctuated equilibrium, or genetic drift, or a set of mutational events – whether considered individually or collectively – are not scientific explanations for what occurred during the various explosion of life forms that followed mass extinction events, but, instead, those ideas are allusions to the possibility of explanations that, unfortunately, lack the presence of anything more concrete than various kinds of experimental data and research that are suggestive or interesting without being conclusive or compelling … that is, there is an absence of any semblance of proof in conjunction with the aforementioned ideas.

Those ideas – taken individually or collectively – might be correct. However, there is no proof this is the case.

How does an idea become scientific when there is no proof that such an idea actually accounts for what it purportedly explains? In the absence of proof certain ideas might provide one with a hermeneutical understanding that possesses a kind of meaningfulness that helps make sense of some facet of reality (as was discussed previously in conjunction with Dobzhansky), but what is meaningful and what is true are not necessarily coextensive.

The time-frame issue becomes even more acute when one comes to the matter of human evolution. Instead of talking in terms of 50-65 million years as in the case of some of the mass extinction events, or speaking in terms of the 20 million years associated with the Cambrian explosion, the time frame for human evolution supposedly occurs over a period of 2-3 million years.

Some evolutionary biologists wish to extend the foregoing 2-3 million year period by an additional 3-4 million years -- and more will be said about this in the next section of the current chapter. However, whether one is speaking in terms of a time frame lasting 2-3 million years, or one is talking in terms of period of time lasting 6-7 million years, one still is dealing with a theory that claims that an incredible array of complex phenomena took place with a relatively short period of time – indeed, apparently, such events took place within a frame of time that is significantly shorter than any other time frame in evolutionary history as far as the emergence of significant new capabilities is concerned.

More specifically, during a period of time covering anywhere from two to seven million years (which is still 13 to 18 million years less than the time frame for the Cambrian explosion and 32 to 47 million years less than the time frame for the recovery of life following the late Permian and KT extinction events respectively), evolutionary biologists claim that very complex capacities involving: Language, creativity/inventiveness, reason/logic, insight, problem-solving, various kinds of genius (e.g., artistic, musical, mathematical, or mechanical), memory, imagination, reflexive consciousness, spirituality, hermeneutical activity, morality, and the like came into existence. There are libraries filled with conjectures concerning all of the foregoing phenomena, but what is missing from those documents is: Compelling evidence that the individuals producing such documents know – in specific, demonstrable terms -- how any of the foregoing capabilities came about; or, compelling evidence that the authors of those documents know how the origin of life came about; or, compelling evidence indicating that such individuals know how new forms of life arose following mass extinction events.

Of course, someone might counter with the possibility that the rudiments of intelligence, reason, logic, language, creativity, memory, morality, imagination, insight, reflexive consciousness, and so on might have begun to take root in various earlier species, and, if this is the case, then trying to shrink the time frame in the manner that is being suggested in the previous paragraph is quite misleading. This, certainly, is a possibility.

However, the underlying problems don't really disappear in conjunction with such a counter proposal. Instead, the character of the problems that must be explained merely shifts a little.

Firstly -- and <u>assuming</u> that the foregoing possibility is correct -- one must be able to account for how any of the rudimentary capacities for logic, language, and so on were initially able to arise prior to the appearance of hominid-like creatures. To offer a date, or related evidence, for <u>when</u> those kinds of capabilities might have first begun to emerge is not enough.

One must be able to provide a detailed, concrete account of what the nature of the dynamics was that led to the emergence of even the most rudimentary, primitive forms of those abilities. No such evolutionary account exists.

Secondly, one must be able to account – in specific, concrete terms -- for the dynamic history of transitions or transformations that led from the rudimentary editions of the aforementioned capabilities to their modern, human counterparts. No such evolutionary account exists.

Finally, whether one considers the time-frame for the emergence of advanced cognitive/mental abilities to be between two and three million years, or one broadens that time frame somewhat and contends that the emergence of those sorts of cognitive/mental capabilities started between six and seven million years ago, one is not really specifying precisely when, or how, any of it took place. In other words, the mental capabilities at issue didn't necessarily take two to three million years or six to seven million years to evolve ... those time frames merely mark the period of time within which – we know not where, what or how – something happened.

Contending that advanced cognitive/mental capacities emerged during a two-to-three million year period or maintaining that they emerged during a six-to-seven million year period is not necessarily the same as, or equivalent to, the claim that it took two to three million years or six to seven million years for those capabilities to evolve. Since we don't know how – or precisely when – such capabilities emerged, we really have no idea how long it took for any of those abilities to appear on the scene.

To be sure, the idea that some sort of significant set of gradual steps was necessary to produce a complex phenomenon seems to be somewhat easier to wrap understanding around with respect to how something might have happened than is the notion that events might have taken place in some non-gradual manner. However, as pointed out earlier, irrespective of whether things took place relatively gradually or relatively quickly, establishing a time frame of whatever length of time doesn't actually tell us when or how something happened – only that whatever happened, happened somewhere within that period.

For instance, the late Permian mass extinction event was followed by a 10 million year fossil gap in the oceans of the Earth. This gap extended into the early Triassic period.

The rich, multi-mile-long reefs consisting of large walls made from, among things, the remnants of coral life – systems of reefs that were prominent during the Permian period -- had all disappeared. Moreover, up until the present time, paleontologists have been unable to discover any evidence within the geological strata covering ten million years relative to that period of history -- involving both the extinction event of the later Permian period and the early part of the ensuing Triassic period -- which suggests the presence of any kind of reef structures during that time frame.

Apparently – at least according to the fossil record -- all that survived in the world's oceans following the late Permian mass extinction event were five species of shelly organisms consisting of: Four bivalve creatures (Unionites, Claraia, Promyalina, and Eumorphotis), along with one kind of brachiopod (Lingula). On land, one reptilian form (Lystrosaurus) appeared to dominate life (constituting as much as 95% of all life forms).

Nothing else – as far as visible life forms are concerned -- appears to be present for the next 10 million years. Then, slowly, evolutionary events seem to begin to pick up speed as life moves into the middle part of the Triassic period.

The foregoing data appears to indicate that evolution went on a holiday for 10 million years. What happened?

There is always the possibility that paleontologists simply haven't yet been able to discover the evidence that is out there somewhere waiting to be found and, if discovered, would provide the proof that evolutionary changes constituted an on-going, robust set of phenomena in the oceans of the world following the late-Permian mass extinction. This is a possibility but it is not a scientific one.

Such a possibility becomes scientific when one has the necessary evidence to back up that kind of a claim. Until the time when the necessary sort of evidence is forthcoming, the aforementioned possibility is merely a non-scientific conjecture or speculation.

In passing, the reader might wish to note that paleontologists have found all manner of fossils (non-marine deposits) elsewhere in the geological strata corresponding to the early Triassic period. However, these plentiful findings are limited to just a couple of species: Clararia and Lystrosaurus.

If the 10 million year gap is genuine – that is, it constitutes reliable evidence indicating that evolution was at a standstill – one is presented with several puzzles. What prevented evolution from taking place, and what got it going again?

Given – i.e., assuming -- that the post-extinction environment had been so toxic and obstacle-riddled that life – let alone evolutionary change – was not possible, then, how did five species of shelly creatures, plus a reptile, manage to survive and, given that they survived, why didn't they evolve? Moreover, if evolution was not possible in the post-apocalyptic period following the late-Permian mass extinction, then what was it that had to change in order for life to once again begin to evolve? And, finally, once conditions conducive to the dynamics of population biology began to appear, what was it that actually happened so that things – evolutionarily speaking – could once again begin to move in more diversified directions?

If 90-95% of life forms on Earth had become extinct due to the late-Permian event, whatever remained is likely to have been scattered in the form of relatively small populations. Small populations tend to limit the variation that is available to the gene pool of such species, and this raises several problems.

To begin with, how does a population of limited variability find a way to survive for 10 million years despite – presumably -- changing conditions? This is not to say that such a question cannot be answered, but, currently, we lack sufficient evidence -- concerning both the precise ecological conditions of the early Triassic period, as well as the capabilities of the few species that were living in those conditions -- that would be needed to address that kind of a question with any degree of compelling credibility.

Secondly, given the likelihood of such limited genetic variability, how did the capabilities arise that permitted the relatively few species of life existing in the early Triassic to begin to evolve in relation to variable conditions of natural selection? Moreover, why did it take 10 million years for such variable capabilities to emerge?

What kind of a system of genetic drift and/or series of mutations would enter into stasis for 10 million years, and, then, suddenly (relatively speaking) begin to become active again? Did allegedly random events of either variety (i.e., genetic drift and/or mutations) suddenly stop occurring for that period of time, and, if so, why did this happen?

Seemingly, evolutionary theory is a lot like Archimedes's notion when he is alleged to have claimed words to the effect of: "Give me a place to stand, and I will move the Earth." It is all about leveraging the assumption that there is a place where one can stand and through which one can accomplish what one claims is possible.

Similarly, evolutionary theory is largely a function of looking for a place to stand (the evidence) from which one can uncover (move) the weighty conjecture concerning the nature of so-called evolutionary change. However, finding the requisite standing place in relation to evolutionary theory is, in many respects, as elusive as realizing Archimedes thought experiment might prove to be.

Many people – including quite a few scientists -- will argue that theology and religion are not scientific in nature because those systems of thought can't prove their assertions or because theology and religion have no reliable, intersubjective means through which to uncover the kind of empirical data that is needed to be able to advance a compelling and demonstrable case for any of their claims concerning the nature of reality. Fair enough.

However, the epistemological status of evolutionary theory appears to be very much like that of religion and theology. The former system of thought can't prove any of its assertions concerning the underlying cause of the sort of changes that occur in life forms over time which are said to be evolutionary in nature, and, furthermore, evolutionary theory doesn't seem to be rooted in any reliable, intersubjective means through which to uncover the kind of empirical data that is needed to be able to advance a compelling and demonstrable case for any of its claims concerning the nature of reality.

Everyone agrees that things – including life – change. Nevertheless, no one has any proof capable of being agreed upon by the vast majority of individuals that the reason(s) why things change is (are) because the nature of reality is 'X'.

Yet, the hermeneutical musings of scientists – which are devoid of proof when it comes to evolution -- are said to be scientific in nature, while the hermeneutical musings of theologians and people of religion are said to be non-scientific in nature. This seems to be a distinction without a difference.

Some people who are inclined toward an evolutionary perspective concerning the nature of reality (including life) might wish to argue that one must become a scientist in order to truly understand the extent to which the theory of evolution is capable of proving itself.

Similar arguments have been -- and continue to be -- advanced by theologians and proponents of this or that religion.

In other words, the sort of argument that sometimes emerges from the evolutionary and religious perspectives is that a person can only understand the nature of the truths that are given expression through a particular belief system by becoming the right sort of technical expert within the context of that framework. When one acquires such expertise, one will be able to see the truth of things.

This is an exercise in framing. One's understanding is being shaped, organized, and manipulated to accept a certain point of view as being true quite apart from whether, or not, there is any way to independently show that what is being said concerning the nature of reality is true in the way such perspectives claim is the case.

Scientists tend to insist that theologians and people of religion play the game of evidence and proof according to strict rules. There are no presumptive freebies permitted in such a game.

Every claim must be backed up with proof. And, this is as it should be.

Amazingly, however, people who advocate an evolutionary point of view apparently do not believe they are required to play the aforementioned game by the same set of strict rules of proof and demonstration. People who are inclined toward an evolutionary perspective tend to refer to their claims as being scientific without ever having to prove the scientific character of those claims.

Such individuals consider what they believe to be scientific even though what they believe constitutes a system of thought that is largely incapable of demonstrating the truth of any of its essential claims concerning the nature of reality or how reality supposedly works. The proponents of evolution continuously grant themselves all manner of presumptive freebies in relation to underlying assumptions – such as randomness – but insist that this is an entirely different matter than when theologians and people of religion try to assume their way through this or that claim.

In evolutionary theory, every branch -- as well as the trunk and the underlying root system -- of the tree of life is held together via assumptions. One cannot conceptually move from a prebiotic root system to the trunk of the evolutionary tree of life without making a huge number of assumptions, nor can a person theoretically move from that trunk to a given branch, nor can an individual hermeneutically move from one evolutionary branch to another branch without assuming that changes – considered to be largely random in nature (either via a series of just-so mutations or the vagaries of genetic drift) – come about in such a way that the dynamics of those changes cannot actually be observed but must be assumed to have occurred in the way they were claimed to have occurred.

The Evolution of Human Beings

As recently as the late 1990s – less than 20 years ago – the mainstream evolutionary account concerning the emergence of human beings ran somewhat along the following lines. At some point prior to 4.4 million years ago, the initial member of the life forms that are referred to as hominins branched off from various kinds of primates, and, then, approximately a little over two million years later, the genus, Homo, arrived on the scene.

Hominins refers to a group of species and genera that are considered to be more closely related to human beings than they are to chimpanzees and bonobos (sometimes referred to as pygmy chimpanzees). The basis of this relationship of closeness involves, among other features, varying degrees of: Exhibiting an upright posture; being bipedal, as well as having a larger brain relative to chimpanzees and bonobos.

In addition, the hip/pelvis region of hominins was much shorter and more bowl-shaped than that of apes … a feature that helped stabilize bipedal movement as well as assisted hominins to stand in an upright position. There were also various characteristics involving leg length and the type of bones in the feet that tended to differentiate (to a degree) various members the hominins from chimpanzees and bonobos.

The hominins encompass a variety of genera with which Homo sapiens, along with a number of other human-like species (very broadly construed), have been grouped for purposes of comparison, and those genera include: Homo, Australopithecus, Paranthropus, Ardipithecus, and Kenyanthropus. Currently, the Homo genus classification consists of at least eight species: Homo habilis; Homo rudolfensis; African Homo erectus (also known as Homo ergaster); Homo erectus (from Asia); Homo neanderthalensis; Homo floresiensis; Homo heidelbergensis, and Homo sapiens.

Between, on the one hand, the advent of the first hominin more than 4.4 million years ago and, on the other hand, the rise of the genus Homo several million years later, there were additional hominin species that appeared on the scene -- including at least six species of Australopithecus, two species of Paranthropus, and several editions of

Ardipithicus. What any of these life forms have to do with one another is uncertain and the subject of a great deal of debate.

The, now famous, "Lucy" (discovered in Ethiopia in 1974) is a member of the species Australopithecus afarensis. Her species stumbled onto the evolutionary scene roughly 3.2 million years ago and survived for about 900,000 years before becoming extinct.

A hundred thousand, or so, years later – somewhere around the three million year mark -- the Paranthropus group of hominins begins to show up. One or another species from this group managed to survive for a little less than several million years before disappearing.

Several hundred thousand years later, roughly around the two million year mark, the earliest versions of the Homo genus begin to arise. Of the eight editions of the Homo genus that we know about, only one species – Homo sapiens – still survives.

According to the late 1990's, mainstream version of events, hominins did not begin to leave the African continent until about 1 million years ago. These hominins migrated into various areas of the world and began to give rise to a variety of species in the Homo genus.

For example, according to the 1990s version of human evolution, Homo neanderthalensis became established in Eurasia and appeared to flourish for several hundred thousand years. Eventually, that species became extinct when – from the perspective of the predominant view of the late 1990s – that species was completely supplanted some 28,000 to 30,000 years ago (possibly through combat, competition, or both) by the smarter, tool-making, symbol-manipulating Homo sapiens.

A variety of evidence uncovered during the last 20 years has changed the foregoing picture substantially. For example, fossil research from Dmanisi, in the Republic of Georgia, suggests that hominins might have left Africa (around 1.78 million years ago) nearly three-quarters of a million years earlier than the roughly one million years ago that was believed to have been the case in the late 1990s, and, in addition, that migration might have been accomplished completely independently of the Homo genus that, previously, had been thought to have begun the African exodus.

Moreover, additional research conducted on the Indonesian island of Flores could push the aforementioned exit of hominins back even further than the evidence discovered in Dmanisi, Georgia. Moreover, the small brain and body of the species Homo floresiensis found on Flores suggest that this organism might have descended from an earlier species of Australopithecus or something similar to the Australopithecus.

In addition, the Flores data indicates there were versions of the genus Homo that had survived at least another 13,000-15,000 years beyond the period 28,000 to 30,000 years ago when Homo sapiens supposedly replaced Homo neanderthalensis. In other words, dating-data indicate that there are hominin fossils from the island of Flores that place Homo floresiensis in that locality as late as 17,000 years ago ... 13,000 to 15,000 years after Homo neanderthalensis allegedly became extinct.

Another pocket of data, based on fossils found in the Djurab Desert, indicates that hominins might have first arisen in the vicinity of Chad rather than in East Africa. In addition, the Djurab evidence suggests that the first hominins might have appeared some two million years earlier than previously thought ... pushing back the origins of hominins to approximately seven million years ago.

Furthermore, relatively newly discovered evidence in Malapa, South Africa by Lee Berger – a paleoanthropologist at the University of Witwatersrand in Johannesburg, South Africa -- is changing perceptions about where the Homo genus actually might have begun. Such research raises the possibility that the Homo genus could have first emerged in the south of Africa rather than in eastern Africa as earlier believed.

As well, views about the species Homo neanderthalensis and its relationship with Homo sapiens also have undergone a substantial transformation over the last 20 years of research. For example, evidence has been discovered indicating that Neanderthals seemed to have had some ability to make tools, and the members of this species also appeared to exhibit a capacity for some degree of symbol-based traditions that were reflected in the systems of jewelry, feathers, and paint that adorned their bodies.

Moreover, whereas in the late 1990s experts believed Homo neanderthalensis and Homo sapiens did not interbreed, nevertheless, more recent analysis of DNA samples indicates otherwise. Collectively speaking, anywhere from 3 to 20% of Homo neanderthalensis genes might have been passed on to various populations within the genus of Homo sapiens ... including some genes that might have helped confer a certain amount of enhanced immunity.

Current evidence also indicates – at least to some individuals – that the history of hominins does not necessarily tell a story in which one species or genus replaces another in some sort of linear fashion. Rather, the evidence suggests that a number of different hominin groups might have overlapped somewhat and, in the process, interacted with one another to an unknown degree, and if this is the case, then, sorting out which – if any – particular group begat another becomes a much more difficult task.

For example, the earlier picture of human evolution maintained that Australopithecus – which began to show up in fossil remains found in southern Africa during the 1920s – was supplanted, eventually, by the taller, larger-brained species Homo erectus that showed up in Asia (Java and China) and that eventually – supposedly -- evolved into Homo neanderthalensis, followed by Homo sapiens. Thus, at a certain juncture in mainstream evolutionary thinking, Australopithecus, Homo erectus, and Homo neanderthalensis were all considered to be part of the direct lineage leading to Homo sapiens.

The discovery of fossils by Louis and Mary Leakey in Olduvai Gorge in Tanzania, East Africa initiated a process of re-thinking the evolution of hominins. Part of this re-conceptualizing of hominin history was rooted in an ability to date the geological strata in which fossils were found through independent means (e.g., magnetic and volcanic data) that permitted researchers to establish roughly accurate starting and ending points concerning the rise and fall of various hominin species.

Moreover, an array of newly discovered evidence indicated that hominins did not necessarily form a sequence of organisms – with one kind of hominin life form succeeding from a previous species of hominin -- but, instead, different kinds of hominin sometimes overlapped with one another. For instance, data indicated that two different genera -- Homo habilis and Paranthropus boisei – contemporaneously inhabited the same region of East Africa for thousands of years.

Whether, or not, the two foregoing genera were directly ancestral to Homo sapiens is uncertain. Whether, or not, the two aforementioned genera engaged in some degree of interbreeding similar to what occurred with Homo neanderthalensis and Homo sapiens is also unclear.

Since the findings of Mary and Louis Leakey began to move thinking about the evolutionary history of hominins in different directions, a variety of evidence has arisen indicating that as many as – possibly -- six species of the Homo genus were extant at various times during the last one hundred thousand years. To what extent any of those species interacted or interbred with one another is unknown, and, as a result, we are faced with the possibility that there might have been a multiplicity of lineages underlying the Homo genus.

Consequently, the question of who was related to whom -- if at all -- and in what way -- if at all -- makes reconstructing the history of hominins much more difficult. Evolutionary connections – or possible connections – no longer seem to be as straightforward and linear as once appeared to be the case.

Some of the foregoing issues might be addressed more completely as new hominin fossils are uncovered. However, relatively speaking, hominin fossils have been difficult to unearth (and, this is also the case in conjunction with chimpanzees and various other African apes ... all of whom have a relatively impoverished fossil record).

Nevertheless, many paleontologists find comfort in the fact that only about 3% of the land area that is encompassed by Africa has been scoured for, among other things, hominin fossils. Many researchers believe a much larger sample of land mass will have to be explored before anyone can claim that the fossils that have been found can be said to constitute a fairly representative sample of evidence as far as the evolutionary history of hominins is concerned.

When one throws in the findings from the islands of Flores and Java in Indonesia, along with the fossils discovered in Zhoukoudian, China, together with the newly discovered treasure trove of fossils associated with Malapa, South Africa, one realizes that there could be a great many more pieces of the puzzle involving human evolution in particular, as well as hominin evolution in general, that are out there somewhere, waiting to be found. However, whether, or not, such pieces of the puzzle will be found or actually are out there waiting to be found is, at this time, unknown.

In the meantime, there are a number of questions that should be raised. For example, the species Homo floresiensis that was found on the island of Flores in Indonesia and survived until approximately 17,000 years ago was small-brained and, possibly, linked to some Lucy-like exemplar from one, or another, of the various Australopithecus genera found in East Africa, and, therefore, one might ask: What did either Homo floresiensis or some progenitor form of Australopithecus have to do with Homo sapiens?

How did Homo floresiensis get to the island of Flores? Where did they come from? Who were their direct ancestors?

Even if one were to uncover fossil evidence that provided a much more robust evidential lineage that linked some form of Australopithecus to the rise of Homo floresiensis, what implications – if any -- would this have for the origins of Homo sapiens? For example, how was the transition made from the small brain of Homo floresiensis to the much larger brain of Homo sapiens, and why should one be forced to suppose that Homo floresiensis and Homo sapiens have any common connection whatsoever?

We do not know where Homo floresiensis came from. Although the possibility exists that this species might have had some evolutionary connection (still unproven) with a small-brained ape-man or ape woman, Australopithecus, found in East Africa, the origin of Homo floresiensis is an on-going mystery.

The reasons are fairly clear about why Homo floresiensis is classified as a hominin. It possesses a variety of anatomical characteristics (such as being bipedal and having a capacity to stand upright, as well as a few other features) that seem to place it in closer evolutionary proximity to different hominins (including Homo sapiens) than to either chimpanzees or bonobos.

However, why consider Homo floresiensis to be part of the human family? Even if this species is linked to a form of Australopithecus in East Africa that suggests hominins might have exited Africa more than two million years ago, what, if anything, does this have to do with the evolution of Homo sapiens?

One could raise similar questions in conjunction with the seven million year old fossils found in the Djurab Desert in Chad. Those fossils might be hominin in nature, but what is there about that discovery that demonstrates they are direct, or even indirect, relations of Homo sapiens?

The discoveries in the Djurab Desert might be able to push back the history of <u>hominins</u> several million years. However, why automatically assume that those fossils also push back <u>human</u> evolution several million years as well?

Unless one can demonstrate determinate evolutionary links between the Djurab fossils and Homo sapiens, one really has no basis for claiming that the former fossils require researchers to extend the evolutionary history of Homo sapiens by several million years. The general category of hominins is one thing, and the particular category of Homo sapiens might be quite another thing.

Something is considered a hominin because of how that life form relates more closely – across an array of anatomical characteristics – to human beings than such organisms relate to chimpanzees and bonobos. To say that something is hominin does not necessarily render it human in some sense despite the presence of whatever anatomical similarities it might share with human beings.

For instance, Homo neanderthalensis and Homo sapiens interbred. Nonetheless, ancestral origins of both Homo neanderthalensis and Homo sapiens are something of a mystery.

Both of those species might have arisen from Homo erectus (aka Homo ergaster) that first appeared – as far as current fossil evidence indicates – between 1.9 and 1.6 million years ago. On the other hand, the newly uncovered fossils from Malapa, South Africa might, or might not, indicate there was some alternative ancestral path to either Homo neanderthalensis, or Homo sapiens, or both.

However, what, if anything, do the foregoing possibilities have to do with hominins in general? Did Homo erectus descend from some form of Australopithecus or Paranthropus or Ardipithecus? Did the Malapa, South African life forms descend from Australopithecus or Paranthropus or Ardipithecus, and, if not, from what did they descend?

Currently, we don't know the answer to any of the foregoing questions. Consequently, it seems premature to conflate the history of hominins with the possible history of Homo erectus, and/or Homo neanderthalensis, and/or Homo sapiens.

Approximately 2.9 to 2.4 million years ago – roughly around the time when the Lucy line of Australopithecus became extinct – two new life forms (both quite different from Lucy's Australopithecus family) showed up in the fossil record. One of those life forms belonged to the Homo genus, while the other life form belonged to the genus Paranthropus.

The genus Paranthropus is a member of the hominin group. However, it does not appear to be part of the ancestral tree of Homo sapiens ... again underlining the fact that not all members of the hominin group are necessarily human in some essential or fundamental sense of the term -- despite the presence of characteristics that incline researchers to consider them to be closer, in a certain sense, to human beings than to chimpanzees and bonobos.

The other member of the hominin group that appeared on the scene about the same time as the Paranthropus genus is considered to be the very first exemplar of the Homo genus. The general body form of that life form possessed certain features that, to a limited degree, are somewhat reminiscent of Homo sapiens.

In addition, this alleged ancestor of Homo sapiens also had a much larger brain than anything that preceded it and was capable of making various kinds of very simple tools. Nonetheless, the brain-size of this species was much smaller than that of Homo sapiens or even the species Homo erectus that appeared roughly a million years later after the aforementioned founding member of the Homo genus appeared on the scene.

Between 1.9 and 1.6 million years ago, Homo erectus (aka Homo ergaster) arose. The body features of this species were virtually indistinguishable from Homo sapiens, and, as indicated in the last paragraph, the brain-size of Homo erectus was larger than the life form that arose approximately a million years earlier and that is thought to have gotten the Homo genus its start.

Did the first exemplar of the Homo genus descend from some edition of the Australopithecus genus that became extinct around that time? We don't know.

Did Homo erectus descend from the foregoing, groundbreaking form of Homo genus? We don't know?

Did either Homo neanderthalensis or Homo sapiens descend from Homo erectus? We don't know.

Home erectus is believed to have migrated out of Africa and, eventually, populated various parts of Asia and Europe. Did Homo neanderthalensis interbreed with Homo erectus – much as Homo sapiens interbred with Homo neanderthalensis – but, nonetheless, was a separate species that had arisen in some way independent of Homo erectus -- as also might have been the case in relation to Homo neanderthalensis and Homo sapiens?

How did the larger brain-size (relative to, say, Australopithecus) of the first member of the Homo genus arise? How did the still larger brain-size of Homo erectus emerge during the million years that separated those first two members of the Homo genus?

We don't know the answer to either of the foregoing questions. Consequently, the genus Australopithecus might not have anything to do with the origins of the Homo genus, even though Australopithecus is considered to be a member of the hominin group.

In addition, there is little, or no, evidence indicating that the genus Paranthropus has anything to do with the origins of the Homo genus, and, yet, Paranthropus is considered to be a member of the hominin group. Moreover, the genus Ardipithecus seems even less likely (due to its even more ancient pedigree) to have anything to do with the origins of the Homo genus (although it might have some evolutionary connection to Australopithecus), and, yet, Ardipithecus is considered to be a part of the hominin group … although, to be sure, this issue is not without its share of controversy.

For example, in 2009 a 4.4 million year old, fairly intact skeleton was discovered in the Afar region of Ethiopia by a group of researchers led by Tim White. The remains were designated as the species ramidus in the genus Ardipithecus and were given the nickname, Ardi.

Ardi was a mixed mosaic of physical characteristics. More specifically, Ardipithecus ramidus exhibited anatomical features that were conducive to both traveling through the trees (e.g., long, curved fingers; a divergent big toe; relatively flat feet), as well as features that would have aided bipedal movement (e.g., the backward flexibility of minor toes, along with a certain degree of stiffness in the foot).

In short, Ardi suggested that the presence of anatomical features that might have facilitated climbing and arboreal locomotion didn't necessarily preclude the possibility of the simultaneous presence of other anatomical features that might have been conducive to some degree of upright posture and bipedal movement. A life form could be considered to be hominin even though there were some ape-like anatomical features that were present.

Again however, while the foregoing considerations indicate there might be compelling reasons for extending the definition of what constitutes a hominin, nonetheless, such an extended way of characterizing hominins might have little, or nothing, to do with determining the origin(s) of human beings. This is especially the case if evidence cannot be found – and none has been discovered to date due to a relative lack of fossil evidence -- which demonstrates that Ardipithecus ramidus is some sort of direct (if distant) antecessor to human beings.

Does Ardipithecus ramidus have anything to do with the rise of Australopithecus anamensis, which, in turn, might have possible progenitor links with Australopithecus afarensis (Lucy)? We don't know, but even if it did, the jump from Australopithecus afarensis to the Homo genus is a fairly big one (and the difference in brain size forms only one part of the explanatory chasm existing between the two genera).

As pointed out earlier in this section, some researchers believe that Homo floresiensis might have had some sort of connection with Australopithecus afarensis (Lucy). Nonetheless, even if one were to assume such an unspecified connection, there is little, or no, evidence to indicate that Homo floresiensis had any direct connection with the rise of Homo sapiens.

Three years after Ardi was discovered, another set of fossil-remains was unearthed ... also in the Afar region of Ethiopia. The remains were found in 2012 at a site called Burtele – approximately 48 kilometers from where Ardi was discovered -- and consisted of eight small bones that belonged to a foot and, perhaps not surprisingly, was referred to as the Burtele species.

The Burtele foot has been dated as being contemporaneous with Lucy, but that foot is also quite different and more anatomically archaic than anything found in Australopithecus afarensis. While the big toe of the Burtele foot appears to indicate that the species to which the foot belongs is a hominin of some kind, nonetheless, there are other features of the Burtele foot that are more reminiscent of Ardi than Lucy ... that is, there are features associated with the Burtele foot that seem to be consistent with some degree of arboreal locomotion as well as with a degree of bipedal motion.

Did the species to which the Burtele foot belongs arise from Ardipithecus ramidus? We don't know, but even if there is a connection, of some kind, between Ardipithecus ramidus and the Burtele foot species, we don't know what, if anything, any of this has to do with the origin (s) of Homo sapiens.

The foregoing issues are rendered even more complex when one takes the idea of homoplasy into consideration. Homoplasy refers to situations in which different species acquire similar characteristics independently of one another.

More specifically, two species, separated by several million years, might each be associated with a certain amount of evidence indicating that they both possessed some degree of capacity with respect to being bipedal. Nonetheless, one cannot automatically conclude that the two species are evolutionarily linked together because both species might have acquired the capacity to be bipedal independently of each other.

Notwithstanding the foregoing considerations, the last common ancestor between hominins and chimpanzees is estimated to have existed between six and ten million years ago. Moreover, that last "common ancestor" is likely to be situated in the context of a population rather than as a function of a single individual.

A population would permit an array of the "right" combination of hominid genes to align themselves in a subset of that last common ancestor population. This subset of the common ancestor population could branch off subsequently from the rest of the larger population.

Even if one assumes that all of the foregoing is true, none of those "givens" establishes what the nature of that last common ancestor population might have looked like, or, even more importantly, how the genes necessary for hominin-like characteristics arose in that population and came to be aligned in some sub-set of that common ancestor population. Furthermore, even if one were able to establish what the nature of the last common ancestor population might have looked like – at least in general terms – this only gets us as far as the rather amorphous collective referred to as hominins.

The origins of: Homo sapiens, Homo erectus (Asian not African), Homo floresiensis, Homo neanderthalensis, and Homo Heidelbergensis continue to be shrouded in mystery. The relationships of various members of the Homo genus with one another also are largely shrouded in mystery.

Among some researchers, there is speculation that Homo heidelbergensis might be the predecessor of either Homo neanderthalensis, or Homo sapiens, or both. In turn, Homo heidelbergensis – again, rather speculatively – has been linked with some, unknown antecessor of the Homo genus, and the foregoing unknown antecessor of Homo heidelbergensis is conjectured to have arisen from some unknown member of Homo ergaster (African) or Homo erectus (Asian).

The foregoing family tree might turn out be correct. At the moment, however, the possible family tree is constructed from little more than assumptions, speculations, and conjectures.

Researchers maintain that the members of the hominin group are all more closely related to human beings than those members are related to chimpanzees and bonobos. However, beyond this, we really don't know, or understand, very much.

Evolution Unredacted

All humans are members of the hominin group. Nonetheless, not all members of the hominin group are necessarily human in any essential way.

Contrary to what some researchers are suggesting, the origin of Homo sapiens might not date back from six to ten million years. Furthermore, those life forms that are hominin and which do date back from six to ten million years might have little, or nothing, to do with the origins of Homo sapiens.

To claim, as some evolutionary biologists do, that human evolution covers a period of, at least, from six to ten million years seems to be -- potentially at least – somewhat misleading. Such a claim assumes that being human – rather than being hominin-like – began from six to ten million years ago, and, yet, there is no proof that this is the case.

There is no concrete, detailed explanation for how different members of the hominin group acquired the similarities that make them more similar to human beings than it makes them similar to chimpanzees and bonobos. Furthermore, there is no concrete, detailed explanation for how or when different members of the hominin group branched off from one another … if they actually did branch off from one another.

Homo floresiensis has a brain size that is much smaller than most other members of the Homo genus. However, that species has enough of the right sort of other physical characteristics that permit it to be classified as a member of the Homo genus.

The brain size of Homo floresiensis suggests that it might have evolved from some form of Australopithecus. Nonetheless, two different genera are being linked here, and, therefore, one has to provide an account of how Homo floresiensis acquired all of the properties that make it a member of the Homo genus rather than some kind of Australopithecus.

Moreover, to say that floresiensis belongs to the Homo genus doesn't necessarily make that species human in some sense. In fact, one is confronted with the question of what, exactly, does it mean to be human.

In terms of gross, physical properties, there are a number of features that differentiate human beings from other members of the hominin group. Among those properties one finds the following characteristics: short toes; arched feet; strong knee joints; enlarged femur head; short, broad pelvis, long, flexible waist; barrel-shaped rib cage; low shoulders; twisted humerus; strong wrist; long, opposable thumb; forwardly placed opening for spinal cord; chin; small canine teeth; and a large brain.

According to the theory of human evolution, the foregoing characteristics did not arise all at once like Athena allegedly arose fully formed from the head of Zeus. Those features were supposedly acquired at different points in evolutionary history.

For instance, seven million years ago, small canine teeth and the forwardly placed opening for the spinal cord arrived on the scene. Roughly 3 million years later (at the 4.1 million year mark), strong knee joints were acquired. Around 3.7 million years ago, short toes and arched feet came into being. Approximately 3.2 million years ago, the long thumb and short, broad pelvis showed up. Two million years ago, the twisted humerus and low shoulders appeared on the scene. A hundred thousand years later – 1.9 million years ago – long legs, enlarged femur heads, and a long, flexible waist arrived. 1.6 million years ago, the barrel-shaped rib cage arose. Several hundred thousand years later, strong wrists were acquired. Approximately one million years ago, a large brain emerged, and 800,000 years later, the modern chin evolved into place.

How any of the foregoing features came into being is unknown. How all of the aforementioned features collectively found their way into Homo sapiens is unknown.

The story of human evolution – along with the story of evolution in general -- might be somewhat like the phi phenomenon in psychology. In this perceptual illusion a sequence of flashing lights is perceived as forming one continuous motion. In point of fact, the illusion consists of a series of separate events that are interpreted to give expression to continuity.

In the phi phenomenon a given flashing light is not what causes a subsequent bulb to flash. Moreover, the sequence of flashing lights is not set in motion by, or caused by, the first light that goes off during the sequence.

Each instance of a light flashing is a separate event that occurs within the context of a timed sequence in which only one light at a time is flashing. Nonetheless, the sequence and timing of a series of such separate events appears to create the illusion of continuous motion.

This is the same sort of phenomenon that is at the heart of motion pictures. When a sequence of static images of a certain kind is flipped at, or run with, sufficient speed, an observer experiences a sense of continuous motion or action, when, in reality no such action is present in any single image ... there is only a series of static images.

Similarly, in evolution, one encounters an illusion that is created by a sequence of images (the moment to moment dynamics of a population) that supposedly appear to give rise to subsequent species. However, with the exception of potentially limited cases such as Darwin's finches (touched upon in an earlier section of this chapter), there is nothing that evolutionary theory can point to in the way of hard evidence (as opposed to conjectures) that is able to demonstrate how the trunk and branches of the tree of life were <u>all</u> produced through the process of speciation ... and not just that <u>some</u> of those branches involved cases similar to Darwin's finches that emerge in the context of the principles of population biology.

Assumptions are made – e.g., random mutations, and/or genetic drift, and/or punctuated equilibrium, and/or speciation – that, allegedly, connect a given species with subsequent ones. Nevertheless, the point when the transition is made from one species to another via natural processes is rather ill defined and seems more like an illusion created by a series of discrete events than it has been demonstrated to constitute a process of continuity.

Evolutionary arguments – including the ones involving human evolution -- seem to be somewhat like a film maker trying to claim that static images or individual frames are actually connected to one another in some sort of <u>causal</u> manner, and, therefore, those images or frames are not really static and independent from one another but are linked in some mysterious fashion as a dynamic function of the images themselves. Of course, the static images of the filming process are connected together because that process permits one image after another to be collected, stored, and run but the connection among those frames, relative to one another (rather than as a function of the filming process), is sequential not causal.

Moreover, in a film, there are people – such as the editor, director, and producer – who help shape the sequence in which individual frames are spliced together in order to give the impression that a certain set of actions has taken place. However, in evolution, whatever takes place through processes of genetic drift, mutation, punctuated equilibrium, and speciation are considered to be entirely independent of what occurred before, or what happened after, those sorts of 'evolutionary' events.

Natural selection plays the role of editor, director and producer. Nevertheless, natural selection is not interested in generating one kind of action sequence rather than some other kind of action sequence.

More specifically, natural selection doesn't worry about whether, or not, a given set of physical, chemical, geological, hydrological, and atmospheric factors will interact in such a way at a given point in time and space that those factors are capable of enabling a given life form or metabolic pathway or biomolecule to be able to survive rather than becoming extinct. In short, the forces of natural selection are described as operating quite independently of whether they are conducive to the creation, continuation, or extinction of some given precursor to life or some given form of life.

Consequently, events that are allegedly of an evolutionary nature might actually be static images that are entirely independent from one another. Yet, advocates of evolution seem to want to insist there is a connection among those static images that is caused by genetic drift, mutations, punctuated equilibrium, and/or speciation and, therefore, evolution constitutes a process of continuity rather than discontinuity with respect to the transition of one species to another … but this might be an illusion of perceptual understanding rather than the actual nature of reality.

One can grasp how a filming and editing process can result in a sequence of film that gives expression to a continuous story that makes sense. Nonetheless, one has more difficulty grasping how the process of natural selection can result in a sequence of living events that gives expression to a continuous story that makes functional sense.

In the filming/editing process the presence of human intention on the part of the editor/director/producer, together with the intelligence of the observer, is responsible for the sense of order that is contained in the sequence of images. In natural selection there is no intention or intelligence that is present, and, consequently, one has a bit more difficulty trying to figure out how the interaction between, on the one hand, the forces of natural selection and, on the other hand, the random, independent events of genetic drift and/or mutation are responsible for the sense of order that emerges again and again across species – from the beginning of life until the present time.

To be sure, one is able to understand how the forces of natural selection that are present at a given point in time and space might act on a chemical/physical system and, in the process, permit that system to continue on because there is a set of compatibilities between the properties of that system and the characteristics of the forces of natural selection that are engaging those properties. Nevertheless, one has a harder time understanding how random, independent events are capable of continuously providing just the right sort of features to feed into the dynamics of the forces of natural selection so that life is able to arise and, then, radiate out in a diverse array of functional forms for some 3-4 billion years.

At heart, evolutionary theory appears to be something of an illusion. Not only do the dynamics of speciation seem, as previously outlined, to be engulfed in a phi-like phenomenon, but, as well, the manner in which allegedly random events generate the functional order that makes any given species – including humans -- capable of adapting to prevailing conditions of natural selection also seem illusory in nature as well.

Some Evolutionary Roots of Psychology

Not too long ago I watched a TED talk (TED is an acronym for 'Technology, Entertainment, and Design') by Alison Gopnik. Dr. Gopnik has done some very interesting research in conjunction with learning and development -- research that makes her an important part of the trend in psychological sciences over the last 10-15 years that has altered the way in which many people (both professional and lay people) think about some of what goes on in the mind of young children (say, 1-4 years of age).

In a variety of ways, infants and young children are much more sophisticated explorers of their universe than many people give them credit for. Indeed, in some ways, young children might be better and more open explorers than adults are.

Unfortunately, all too many adults are socialized out of realizing some of their inherent potential for learning and development via the very process of schooling that many people assume is how human beings maximize their capacities for learning about the world. There are many ways in which schooling interferes with and undermines the process of learning as children are induced – through techniques of undue influence -- to accept an educational institution's view of some given issue ... such as the theory of evolution.

With the foregoing considerations in mind, several points seem worth mentioning in conjunction with Dr. Gopnik's TED presentation. To begin with, she seeks to place her work in an evolutionary context that, in and of itself, is unremarkable since many researchers in psychology do the same sort of thing these days.

Nevertheless, at certain points during her talk, Dr. Gopnik refers to neurochemistry, neurotransmitters, and so on, as if the mere use of that sort of terminology fully explains what is going on in the brain or how the brain and mind are connected, when, in point of fact -- as is also the case in relation to evolutionary theory -- no one has shown in a rigorously empirical manner how either neurochemistry or neurotransmitters came into existence or how they are able to generate: consciousness, thinking, reasoning, logic, memory, creativity, understanding, and so on.

To be sure, various aspects of neurochemistry are correlated with mental functioning. However, correlation is not causation, and until the precise causal steps are nailed down, then, reducing mind to brain constitutes a bit of myth making, not science.

A second point concerning Dr. Gopnik's presentation involves the work of Thomas Bayes. Bayes was an eighteenth-century mathematician who invented a form of statistical thinking that is capable of leading to improved descriptions of a system based on a computational technique that incorporates new data into one's calculations ... calculations that are able to improve, to a certain extent, upon some initial probability model with which one began in relation to the system or situation being explored by an individual.

Dr. Gopnik suggests that young children are capable of running Bayesian-like computations in order to work out which hypothesis concerning an aspect of reality is more likely to be true based on their experiential interaction with such an aspect of reality ... something that even adults might have difficulty working out -- at least this would be the case if adults were required to use and apply the mathematical properties of Bayes' theorem to arrive at an answer.

However, one might respectfully suggest that although on the surface there might be certain parallels between Bayesian probability methods and the manner in which children try to work their way through various possible solutions to a problem, it does not follow that children are engaged in some sort of Bayesian computation ... any more than an outfielder in baseball necessarily uses calculus to track down a fly ball. Yes, there is a process of reasoning and logical analysis taking place in the mind of the child, but this does not necessarily mean that Bayesian mathematical methods are being employed ... although one is entitled to say there is, at the very least, an analogical relationship between what children are capable of doing and Bayesian statistical techniques.

In fact, one might speculate that Bayes original idea was a specific, concrete, creative application of the sort of mental capacity to which Dr. Gopnik refers in her presentation as being present in children. In other words, it is our inherent capacity to learn from experience and, in the process, update our understanding of such experience that might have served as the inspiration for Thomas Bayes theorem and that reflected some of what had been taking place in his mind when he went from an initial understanding of something, and, then transitioned to an improved version of that idea through the incorporation of new experiences by means of a mathematical model.

In short, Bayes might have worked out a formal, mathematical model that captured, in a limited way, certain facets of the aforementioned more general capacity to be able to learn from experience by incorporating what we learn into our previous understanding. As such, Bayes theorem is an analog for a cognitive process that takes place in human beings – including children – but those cognitive processes transcend Bayes theorem even though that theorem does reflect certain aspects of what occurs during the process of seeking to understand some given phenomenon or dimension of experience.

Even if one were inclined to accept Dr. Gopnik's idea that children operate in accordance with Bayesian probability functions, one is still faced with a considerable conundrum. How did human beings acquire the capacity to think in that manner? What were the specific, evolutionary steps that made that sort of capacity possible?

The idea of evolution appears to be used by Dr. Gopnik as sort of a convenient, but very vague, background, rhetorical prop through which to frame her audience's understanding with respect to how things might have come to be the way they are. Supposedly -- or, so the evolutionary story goes -- we got to our present level of cognitive ability through evolution, and, yet, no one -- including Dr. Gopnik -- ever provides a detailed account of how those kinds of capacities actually came into being.

Everything is run through the presumptive lenses of evolutionary interpolation and extrapolation. In the process of framing things in the foregoing manner, understanding becomes steeped in mythological-like elements.

I do not say the foregoing as someone who seeks to advance either a creationist position or some sort 'intelligent design' notion. Instead, I say what I do as a hardnosed empirical skeptic who, like Cuba Gooding in the movie: 'Jerry Maguire' is saying: "Show me the money."

If one cannot produce the blow-by-blow empirical account of how things came to be the way they are (and the pages of this book suggest that the proponents of evolution cannot accomplish this in any credible fashion), then one is not talking about science. Rather, one is dabbling in philosophy while seeking to leverage the halo-like effect of the term: "science'.

Evolutionary theory might guide much of modern thinking in a variety of areas – especially in relation to psychology. Unfortunately, a great deal of that thinking is rooted in the sort of speculative philosophy and assumptions that cannot be proven and, consequently, is not rooted in real science ... even as evolutionary theory seeks to clothe itself in scientific jargon in order to give the impression of being scientific without having to meet the standards of actual substantive rigor.

Many people, of course, might respond to the foregoing by saying words to the effect: 'Well, of course, everyone admits there are many lacunae in evolutionary research, but it is the best available scientific theory to account for a wide array of phenomena ... indeed, if one rejects evolutionary theory, then with what do you propose to replace it?"

The foregoing is like a prosecutor saying: "Well there is very little actual, concrete evidence indicating that the person we have in custody is responsible for the crime with which he is being charged - although there is considerable circumstantial evidence and, as well, there are many expert witnesses who are willing to testify, according to their biases, that the right individual is in custody) -- but, gee, since there is no other viable suspect, why don't we just go along with the idea that the guy we have in custody is guilty ... after all, do you have anyone who would serve as a better suspect?"

A person doesn't have to offer up an alternative theory that explains things better than evolution does. One only has to understand that the available evidence does not support or justify holding on to the suspect of evolution simply because that suspect is the only entity our state of ignorance and limited imagination can conjure up to account for the questions at issue.

Unlike my evolutionist friends, I am not afraid to say that I do not know what the truth of the matter is. Notwithstanding such an acknowledgement, nevertheless, when one looks carefully at various accounts concerning the origins of life, or even the origins of novel, biological capabilities, existing evolutionary accounts leave one deeply dissatisfied. There is almost no intellectual rigor (despite the presence of scientific sounding jargon) present in those arguments, and I see no reason why I, or anyone, should adopt a theory that is so steeped in a cloud of unknowing and, yet, simultaneously, assume that evolutionary theory constitutes good science ... because this is just not the case.

Talking with many individuals who are advocates of the theory of evolution is like interacting with a bunch of K-street, Washington lobbyists who yammer away trying to induce people to support their grandiose, but rather empirically shaky and self-serving ideas. Being convinced of the truth of something is not necessarily the same thing as being correct concerning that to which one is so passionately attached.

Unfortunately, if one should express some sort of resistance to the marketing campaign of the evolutionists (as I am doing now), then look out, for the proverbial stuff is likely to hit the fan. Labels and epithets often soon follow -- such as: 'That person is anti-science'; or, 'that individual is a 'luddite''; or, 'such people are standing in the way of intellectual progress', or 'that person is hopelessly irrational' -- when all one is doing is pointing out (concretely and not theoretically) that there is a wealth of empirical and conceptual problems that beset the theory of evolution across an array of issues -- starting with the 'origin of life' matter, and extending into such topics as: the origins of consciousness, reason, logic, memory, creativity, morality, cognitive development, and so on.

Evolution's Black Box

In engineering and science, a black box is at the epicenter of an unknown set of processes. One can talk about what arises out of those processes as an output, and one can talk about what some of the inputs might have been that are fed into that box and, to varying degrees, might have helped shape what transpires within that box, but the actual character of the dynamics of the black box that makes such outputs possible is a mystery.

At the heart of evolutionary theory are a series of black boxes. For example, consider the DNA code.

In 1953 James Watson and Francis Crick worked out the general, double helical structure of the DNA molecule. Within the context of that helix, they knew guanine and cytosine paired off with one another to form some rungs in the helix, and, as well, they knew that adenosine and thymine linked up together to form other rungs in the double helix.

In their April 25 letter to *Nature* magazine -- which introduced their discovery to the world -- they also intimated that the aforementioned pairing arrangements might serve as the basis for a copying system in which either of the strands making up the helix could serve as a template for the generation of the other, complementary strand. What the two researchers didn't know at that time was just what any given sequence of bases actually meant, and, in fact, it would take another ten years before an answer, of sorts, could be offered in relation to the meaning of the DNA code.

The term "of sorts" is used in the foregoing paragraph because the answer that took ten years to work out concerned discovering what the sequences meant. That answer, however, had no clue how such a system of coding came into being.

There are two questions swirling about the DNA code. One question concerns the nature of the code, while the other question has to do with how that code came into existence, and scientists largely have been preoccupied with – and only have answered -- the first question.

Let's take a look at the first question noted above. What is the code?

There are four different bases in DNA. There are 20 amino acids.

How are the two kinds of molecules related? What is the "meaning" of any given base with respect to the generation of amino acids?

A one-to-one correspondence between a base and an amino acid doesn't work. There are too many amino acids, and, therefore, one base pair could call for any one of five, or more, amino acids, and, as a result considerable confusion would enter into the process of translating a given base into the particular amino acid that was needed to help form this or that protein.

If the DNA code consisted of a pair of bases, only 16 amino acids could be formed. In other words, if the code was a base doublet, then any one of four bases could appear in the first position, and, as well, any one of four bases could appear in the second position, for a total of 16 possibilities.

There would appear to be four possibilities too few to form the necessary 20 amino acids, and, as a result, a certain amount of confusion would be present. Such confused understaffing might tend to undermine the precision-oriented nature of living organisms.

Eventually, several individuals (Sydney Brenner and Francis Crick) demonstrated that the DNA code – whatever it might turn out to be – had to consist of at least three bases. However, if any of the four bases could occupy any one of the three coding positions, then, this would lead to 64 possibilities (4 x 4 x 4), and this seemed to give more than three times as many possibilities as were needed to code for just 20 amino acids.

Quite a few suggestions were put forth during the next ten years in an attempt to identify the precise nature of the relationship between nucleic bases and amino acids. Eventually, researchers discovered that there appeared to be several levels of coding taking place.

On what might be considered the most outward level of coding, there was a degree of redundancy or degeneracy built into the DNA code. This meant that while some of the three-letter nucleic base words coded for just one amino acid, nonetheless, in another case six different three-letter nucleic base combinations coded for the same amino acid (e.g., leucine), and in still another instance, three, three-letter nucleic base combination coded for the same amino acid (e.g., isoleucine), or in a number of cases, sets of four three-letter nucleic base combinations coded for the same amino acid (e. g., valine, serine, proline, threonine, alanine, arginine, and glycine)

Moreover, there were three three-letter nucleic base combinations that didn't code for an amino acid. Instead, they represented stop signals.

The foregoing, several paragraphs outline the general structure of the code. In other words, they are part of the description that deals with what the code is.

What has not, yet, been explained is how the code came to be the way it is. For instance, how did a three-letter base combination come to mean "stop", and how did any given three-letter nucleic base combination come to 'mean' one, rather than another, amino acid?

Some researchers were not content with knowing how the code worked. They wanted to understand what processes led to the code assuming the form it did.

During their journey of discovery, they learned that different factors seemed to be associated with each of the three nucleic bases that made up any given three-letter nucleic combination. For instance, consider the first nucleic base letter of any triplet or codon ... some individuals felt it might code for much more than originally had been believed.

More specifically, in a cell, amino acids are synthesized in several different ways, and each of these ways begins with simple, molecular precursors. Research indicated that there seemed to be a relationship between the first nucleic acid base-letter that made up a given triplet (codon) and the identity of the precursor that began the process of synthesizing the amino acid being coded for by that DNA triplet sequence.

For instance, pyruvate – which, among other things, helps get the Krebs cycle started – is a precursor for the synthesis of certain amino acids. Researchers found that all of the amino acids that have pyruvate as a precursor are coded for by a three-letter nucleic acid codon that begins with the nucleic acid base 'T' or thymine.

The foregoing relationship is very intriguing and interesting. Nonetheless, such a relationship doesn't really solve the underlying puzzle: How did the DNA code come to be the way it is?

In other words, how did the nucleic acid base thymine come to mean that the three-letter codon for which it was the first nucleic acid base-letter would code for one of the amino acids that uses the precursor pyruvate in its synthesis? Why didn't codons starting with thymine code for a precursor that initiated synthesis for a different kind of amino acid?

How did thymine come to "mean" or stand for pyruvate? What was the nature of the dynamic linking a nucleic acid base (consisting of a five-carbon sugar, a phosphate group, plus a nitrogenous base) and pyruvate (CH_3COCOO^-)?

At what point in the evolutionary process was it determined that any given amino acids would be represented by this or that nucleic acid triplet and, in addition, determined that the composition of that triplet would begin with a nucleic acid base that specified the identity of the precursor that would initiate the synthesis of the amino acid being coded for by such a triplet? How did this sort of determination come about?

If the origins of life <u>are</u> rooted in black smokers, white smokers, and/or some Stanley Miller kind of scenario, and, as a result an interconnected set of metabolic pathways arose – in a manner that is not currently understood -- that were capable of initiating and sustaining life, then, how did the information contained in such a arrangement get transferred to a sequence of three letter codons? Moreover, how did that information get incorporated into the sequence of codons in a way that not only stipulated which triplet would stand for which amino acid but did so in a way that specified that the first letter of the codon would identify the precursor that was necessary for the synthesis of the amino acid being coded for in the triplet.

The foregoing, mysterious conspiring of events sounds even more preposterous than the idea that Francis Crick (the Noble Prize winning scientist who, along with James Watson, had established the basic helical character of DNA) came up with in an attempt to explain why the DNA code was the way it was. More specifically, Crick put forth an idea that he referred to as: "directed panspermia".

According to Crick, life on earth was the result of a seeding process in which a bacterial life form of extraterrestrial origin had been introduced into the planet Earth at some point in the distant past. Moreover, according to Crick, the seeding process was intentional and conducted by some sort of alien intelligence.

Crick's idea is more of an evasion than it is an explanation. Even if his idea were correct, it still doesn't explain how extraterrestrial bacteria or intelligent, alien life forms came into being.

Crick's conjecture notwithstanding, one is still left with two problems: (1) Accounting for how the DNA coding process acquired its system of linking nucleic acid base triplets with amino acids; (2) accounting for how the DNA coding process acquired its system for linking the first nucleic acid base letter in a given triplet to the identity of the precursor that would help initiate the synthesis of the amino acid being coded for by that triplet.

If the origins of life <u>are not</u> rooted in the prebiotic chemistry of black smokers, white smokers, and/or Stanley Miller-like scenarios, then how does one account for the millions of DNA sequences that would have to arise in order to be able to code for – mean, stand for, represent – different amino acid combinations in the form of peptides or proteins that played central roles in helping to make this or that metabolic pathway possible? In addition, how did the DNA coding system acquire the ability to have the first nucleic acid base letter in any triplet code for -- mean, stand for, or represent -- the precursors that are needed to synthesize the amino acid being encoded?

No matter how one would like to proceed with respect to trying to account for the origin of life, one is faced with a deep mystery, puzzle, or problem. On the one hand, an individual can start with various scenarios involving prebiotic chemistry that -- in a way that is not currently understood – came together in a manner that eventually was able to transfer information about metabolic pathways to, or incorporated that information into, a DNA coding system. Or, on the other hand, a person can begin with some sort of scenario in which there is an accumulation of millions of nucleic acid base sequences over millions of years that somehow – in a way that is not currently understood -- came to give expression to a DNA coding system in which certain triplets came to represent specific amino acids, and, as well, those triplets came to give expression to a DNA coding system in which the first nucleic acid base letter in a triplet identified the precursor that was to be used to help initiate the synthesis of the amino acid being encoded by that triplet.

Neither of the foregoing two possibilities is any better equipped to explain or account for the origin of life than is Crick's notion of directed panspermia. None of the foregoing three possibilities constitutes a scientific explanation.

Just as certain research has pointed out the intriguing relationship (whose origins are a complete mystery) between the first nucleic acid base letter of a given triplet and the identity of the precursor that helps to initiate the synthesis of the amino acid being coded for by that triplet, similar research also has indicated that there appears to be a connection between the nature of the second nucleic acid base letter in a given triplet and the degree to which the amino acid being coded for through that triplet is soluble in water. More specifically, researchers have discovered that five out of six of the amino acids that are most insoluble (hydrophobic) in character have 'T', or thymine, as the second nucleic acid base letter for a given triplet, while all of most water soluble (hydrophilic) amino acids are coded for by DNA triplets that have 'A', adenine, as the second nucleic acid base letter in the triplet that is coding for such amino acids.

Once again, the puzzle of origins rears its ugly, inexplicable head. How did the nucleic acid base letter 'A' – adenine – come to code for, mean, or represent an amino acid that has the property of being highly hydrophilic when 'A' is the second nucleic acid base letter in a triplet coding for that amino acid? How did the nucleic acid base letter 'T' – thymine – come to code for, mean, or represent an amino acid that has the property of being highly hydrophobic when the second nucleic acid base letter in a triplet coding for that amino acid is 'T'?

In short, how did a nucleic acid base letter come to determine whether the amino acid being encoded would be hydrophilic or hydrophobic? How did the positioning of a nucleic acid base letter within a triplet come to mean, stand for, or represent the solubility of the amino acid being encoded?

One also might ask why DNA triplets didn't code for lipids or carbohydrates, instead of coding for amino acids? The short answer, of course, is that this is just the way things are.

Nonetheless, the foregoing short answer doesn't really account for how and why DNA triplets got connected with amino acids rather than lipids or carbohydrates. This is especially the case given that the basic molecules that make up nucleic acids, amino acids, lipids, and carbohydrates are quite different from one another, and, consequently, there is no obvious reason why nucleic acids should be linked to amino acids rather than lipids or carbohydrates … the arrangements that are in place with respect to the link between nucleic acid base triplets and amino acids seem to be rather arbitrary in character.

The third nucleic acid base letter in a DNA triplet is, to a great degree, fairly degenerate in character. Another way of referring to this state of affairs is to say that the third nucleic acid base letter often tends to be devoid of useful information … that is it is information free.

For example, consider the amino acid, glycine. A triplet coding for glycine is GGG (guanine times three).

However, the final nucleic acid base letter 'G' in the triple guanine codon could be 'A' – adenine – or 'C' – cytosine – or 'T' – thymine, and each of those triplets would still code for glycine. The identity of the nucleic acid base letter holding down the third position in the DNA triplet doesn't seem to matter.

Yet, the third nucleic acid base letter in a DNA triplet does matter in certain instances. For example, the amino acids tryptophan and methionine are encoded, respectively, by TGG and ATG, but if the final 'G' in either of these triplets is changed to 'T' – thymine – or 'A' – adenine – or 'C' – cytosine – one will not be coding for the same amino acid but, rather, a different amino acid is being encoded or a 'stop' signal is being indicated.

One finds similar triplet specificity when it comes to stop codes. TAA, TAG, and TGA are all stop codons, and, yet, if the character of the nucleic acid base letter occupying the third position in the triplet is altered to some other nucleic acid base letter, one will get an amino acid and not a stop signal.

There are eleven triplets among the 64 possible DNA combinations of the four nucleic acid bases that are a little less concerned with the identity of the third nucleic acid base letter than, say tryptophan or methionine are, but, nonetheless, those eleven triplets will not permit just any nucleic acid base letter into the third position of the triplet. For example, phenylalanine will permit either 'T' – thymine – or 'C' – cytosine – in the third position of the DNA triplet and either triplet will code for phenylalanine, but if the third position of the triplet is occupied by 'A' – adenine -- or 'G' – guanine – one will get the amino acid, leucine, not phenylalanine.

How did TTT and TTC come to stand for phenylalanine but not leucine? How did TAA, TAG, and TGA come to 'mean' stop rather than some amino acid? How did TGG and ATG come to represent tryptophan and methionine respectively?

How did the third position in a nucleic acid base triplet come to be significant in some instances but not others? How did the third position in a nucleic acid base triplet become semi-important in some cases (for example, phenylalanine, tryptophan, histidine, glutamine, and asparagines – to name a few) but not in other cases (e.g., leucine and glycine)?

Some researchers have proposed that the primordial code was a function of doublets (two nucleic acid base letters), and, at an unknown point during the process of evolution, there was some sort of codon capture dynamic that turned the doublet code into a triplet code. When such a switch-over occurred is not known, nor is it known how such a transition in coding took place, nor is it known how the initial doublet code came into being ... if any of this is the way things actually began.

Natural selection might account for why a given coding system was endorsed due the survival value that was entailed by such a system once it came into existence. Nonetheless, natural selection does not account for how either a doublet code or a triplet code came into existence or how the former (doublet) coding system transitioned later into a triplet coding system – if this is what took place rather than just being a conjecture.

DNA coding is one of many black boxes occupying the heart of evolutionary theory. No one knows how that coding system came into existence. No one knows how nucleic acid base letter triplets came to 'mean' amino acids rather than, say, lipids, or how nucleic acid base letter triplets came to mean one amino acid rather than another. No one knows how certain nucleic acid triplets came to mean 'stop' rather than stand for an amino acid of one kind or another. No one knows how some DNA triplets became very particular about the nucleic acid base letter occupying the third position in a codon while other triplets were less fussy (or not fussy at all) with the nucleic acid base letter that occupied the third position of a codon.

There are all manner of inputs that have been conjectured as – possibly – having helped shape the evolutionary process through which the system of DNA coding might have come to assume its central place among biological systems organisms on Earth. There are all manner of outputs that have been described as having arisen through the coding system of DNA.

Nonetheless, the origin(s) of the DNA coding system are steeped in mystery. Those origins entail a dynamic that is a total black box as far as evolutionary theory is concerned, and, as a result, at the present time there is no way to account for those dynamics in a scientific way.

In 1883 Andreas F.W. Schimper conjectured that, originally, chloroplasts might have been part of a symbiotic relationship between nonphotosynthetic cells and certain kinds of photosynthetic bacteria. Chloroplasts are organelles that are found in the cytoplasm of both plants as well as algae, and those organelles contain molecules of chlorophyll pigments, along with various enzymatic proteins, that make photosynthesis possible as well as make possible the production of ATP – adenosine triphosphate -- one of the primary mediums of energy currencies in cellular life.

Schimper believed that over a period of time the symbiotic relationship between the nonphotosynthetic organism and certain bacteria transitioned into a permanent arrangement. As a result, two organisms began to function as one life form when various metabolic pathways of the two organisms were integrated while other metabolic pathways possessed by one or the other of the two organisms fell by the wayside.

Approximately forty years later, Schimper's idea was broadened to include mitochondria ... the double-membrane organelle found in the cytoplasm of eukaryotes (life forms that – unlike prokaryotes -- possess a true nucleus together with a number of cytoplasmic organelles such as the Golgi complex, lysosomes, and the endoplasmic reticulum). In other words, some biologists conjectured that mitochondria (which, among other things, are responsible for the production of energy-containing molecules in cells) originated in a symbiotic relationship between bacteria (purple bacteria to be specific) and some form of protoeukaryote (a primitive form of eukaryote) and, then, over time, the two life forms merged into one organism.

The underlying idea came to be known as endosymbiosis. This referred to a process in which some kind of protoeukaryotic life form would ingest bacteria (cyanobacteria – or some ancestor -- in the case of chloroplasts and purple bacteria – or some ancestor -- in the case of mitochondria) that established a symbiotic relationship (i.e., one from which both organisms derived benefit) and, eventually, that symbiotic relationship becomes transformed, somehow, into just one organism as the different kinds of bacteria became dedicated organelles -- i.e., chloroplasts or mitochondria – within a larger protoeukaryotic life form.

The idea of endosymbiosis was largely rejected and ignored for more than 70 years. However, during the 1960s, research revealed that chloroplasts and mitochondria are semiautonomous organelles that are capable of dividing on their own, synthesizing their own proteins, and they also contain DNA, mRNA (messenger RNA), tRNA (transfer RNA), as well as ribosomes (small particles consisting of rRNA – ribosomal RNA – and proteins that engage in protein synthesis).

With so many semiautonomous capabilities present in chloroplasts and mitochondria, the possibility of endosymbiosis no longer seemed to require such a large leap of imagination. Mitochondria and chloroplasts appeared to share many of the characteristics of various bacteria, and, so, if certain kinds of bacteria were ingested by a larger protoeukaryotic form of life, and this ingestion was followed by the establishment of some kind of symbiotic relationship, and, then, finally, the two life forms became integrated over a period of time, there seemed to be a plausible set of steps through which endosymbiosis might have taken place ... a theory that was more fully developed by Lynn Margulis.

The theory of endosymbiosis presumes that prior to the incorporation of certain kinds of bacteria into protoeukaryotic organisms, life forms were anaerobic – that is, such organisms relied on a form of respiration in which nutrients are converted into useful forms of energy and materials by moving electrons around through metabolic pathways that were centered on molecules other than oxygen. Organisms were required to operate in the foregoing manner because there was very little oxygen in the primitive atmosphere of early Earth.

At some point (between one and two billion years ago), certain forms of bacteria (known as cyanobacteria) acquired the ability to use pigments – such as chlorophyll – to capture energy from certain wavelengths of light and, then, transform that light energy into a form of chemical energy that was capable of subsidizing a variety of metabolic pathways through which an array of biomolecules were synthesized and subsequently used to sustain life processes. Prior to the advent of cyanobacteria, bacteria would have had to use some other molecule – such as molecular hydrogen or hydrogen sulfide -- as an electron donor (rather than chlorophyll) in the process of photosynthetic respiration.

The process of photosynthesis involving chlorophyll is a fairly complex process that aside from sunlight also uses water that is a very stable molecule and, therefore, resists giving up any of its electrons. In photosynthesis water is first broken open by orienting the water molecule in just the right way so that its electrons can be engaged one by one, and, in the process, oxygen is released.

Next, photosystem II – comprised of P680, the pair of chloroplast molecules that constitute one of several reaction centers – removes electrons from the aforementioned oxygen-generating process when that reaction center is activated by light of the right wave length (680 nanometers). The electrons that have been captured through photosystem II are, then, shunted down an electron transport system and along the way those electrons are used in the synthesis of ATP (a carrier of energy in the form of phosphoanhydride bonds) before being transferred to photosystem I – comprised of P700, the pair of chloroplast molecules that constitute the second of two reaction centers that are activated by light of the right wavelength (700 nanometers).

This latter photosystem re-energizes the electrons involved in the process of electron transport before passing them on to NADPH (nicotinamide adenine dinucleotide phosphate), a coenzyme that accepts several electrons (and a proton) from photosystem I. NADPH subsequently becomes involved in a metabolic pathway that activates carbon dioxide and converts the latter molecule into sugar. The foregoing process is known as oxygenic photosynthesis.

Aside from all of the other useful results that arise out of this sort of photosynthesis (and that were outlined in the previous paragraph), one of the most eventful dimensions associated with it revolves around the oxygen that is released as a waste product. As oxygen was released into the atmosphere of the Earth 1-2 billion years ago, this gradually led to the disappearance -- for the most part -- of the materials (like hydrogen sulfide or dissolved iron) that certain anaerobic organisms used to survive. Consequently, these organisms found ecological niches that tended to be devoid of oxygen.

Nevertheless, somewhere along the evolutionary way, organisms capable of aerobic respiration arose, and, as a result, were able to oxidize glucose to carbon dioxide and water by using oxygen as an electron acceptor. In the process, a considerable portion of the energy released during those reactions was conserved in the form of ATP.

Once protoeukaryotic organisms arose that were capable of aerobic respiration, the stage was set for the appearance of eukaryotes that, in part – according to the theory of endosymbiosis -- involved the ingestion of purple bacteria or cyanobacteria that would become, respectively, mitochondria and chloroplasts. However, before bacteria could be ingested by a protoeukaryote, the latter organisms had to develop a capacity for endocytosis that constitutes a way of consuming extracellular materials – such as bacteria – through the infolding of a plasma membrane that is, then, pinched off as a membrane-bound vesicle containing whatever was taken in through this means.

There are at least 150 different kinds of eukaryotes that have within them diatoms (e.g., phytoplankton), photosynthetic organisms, and other small organisms living as endosymbionts within the larger organisms. The cell walls of the ingested life forms might have been stripped away and, as well, some of the cell structure of the ingested organisms might be whittled down in various ways, but what remains still continues to function ... at least for a time.

For example, various marine slugs have chloroplasts contained within some of the cells that line the digestive tract of such slugs. These chloroplasts come from the algae being eaten by those slugs, and they continue to operate their photosynthetic equipment for quite some time following ingestion.

However, these ingested chloroplasts do not divide or grow. And, within a few months, they stop functioning.

The process through which these chloroplasts are permitted to survive for a period of time rather than becoming completely disassembled during digestion is not known. Furthermore – and, perhaps, related to the foregoing point -- why this arrangement occurs in some marine slugs and mollusks but not in other kinds of mollusks is also unknown.

Given that such chloroplasts only survive for a time and cannot grow or divide suggests that certain integrating events have to occur in order to make the condition of endosymbiosis permanent. Consequently, the fact such symbiotic relationships between certain kinds of marine slugs and green algae can be established is likely only one of a number of steps that are necessary in order for a complete process of endosymbiosis to become a reality. What those steps are is not known.

There have been a variety of comparisons between, on the one hand, the rRNA sequences in chloroplasts and mitochondria, and, on the other hand, the rRNA sequences in various kinds of bacteria. The bacterial rRNA base sequences that match up most closely with chloroplasts are cyanobacteria, while the bacterial rRNA base sequences that seem most closely related to mitochondria are purple bacteria, and, therefore, those rRNA comparisons would seem to lend credence to the idea that, at some point, cyanobacteria – or a close relative -- were involved in the origin of chloroplasts, while purple bacteria served as the ancestral origin for mitochondria.

Nevertheless, while the existence of such rRNA comparisons is suggestive with respect to the idea of endosymbiosis serving as the basis for a possible evolutionary account concerning the origin of chloroplasts and mitochondria, those comparisons don't necessarily constitute proof for the idea of endosymbiosis. While rRNA comparisons do show a degree of similarity between two sequences, those similarities don't really reveal how the two things being compared came to share that similarity.

Evolutionary biologists, of course, believe that the rRNA sequence similarities in the things being compared indicates there was an evolutionary process that led from cyanobacteria to chloroplasts, just as there was an evolutionary process that led from purple bacteria to mitochondria. They just can't tell you what was involved in the nature of that evolutionary process.

Francis Crick might claim – if he were with us today – that an alien intelligence could have genetically engineered the bacteria, as well as the chloroplasts and mitochondria of eukaryotes, using similar methods, on the one hand, with respect to chloroplasts and cyanobacteria, and, on the other hand, in relation to mitochondria and purple bacteria because in each instance similar functional requirements were in effect.

Why should an alien intelligence have to re-invent the wheel? Similar designs are used in the respective cases because they serve similar functions.

Of course, back in the 1980s Francis Crick would have had no idea how any of the foregoing might have been accomplished by an alien intelligence. However, the existence of such ignorance would have placed him on even terms with his evolutionary colleagues.

Is the evolutionary account any better than Crick's notion of directed panspermia? Is it necessarily any simpler?

In order for the theory of endosymbiosis to be plausible, there are quite a few questions that have to be addressed in a satisfactory manner. For example, how did anaerobic organisms come into being? How did photosynthetic organisms involving molecular hydrogen or hydrogen sulfide as electron donors come into being? How did chlorophyll-based photosynthetic organisms arise? How did the transition from anaerobic respiration to aerobic respiration come about? How did eukaryotic life forms arise – with their true nucleus, and an array of organelles (e.g., Golgi complex, lysosomes, endoplasmic reticulum, endosomes -- both early and late) ... organelles that are not found in bacteria? How did the capacity for endocytosis come into existence? How did endosymbiosis become established ... that is, how did an ingested bacterium lose some of its functionality, while retaining other capabilities, and how did that former-bacterium become integrated into the cellular functioning of the larger organism?

There are no concrete, step-by-step, demonstrable answers to any of the foregoing questions. Every one of those questions is rooted in an evolutionary black box of unknown dynamics.

How did the DNA coding for each of the foregoing steps come into being? One might suppose that various genes were passed around through the process of conjugation (exchange of genetic material between two organisms) that led up to one, or another, of the foregoing steps, but where did the genes (and whatever capabilities they entail) come from?

At some point, the genetic buck has to stop. One has to be able to explain how any given biological capability became instantiated in genetic information.

Genetic <u>information</u> can't be passed around through the process of conjugation, insertion, and splicing until it has been raised to the status of information (having genetic meaning) from its previous condition of being genetic noise (being genetically meaningless). In addition, a scientific account must be given for the origins of the capabilities that are instantiated in the DNA sequences that are responsible for: anaerobic respiration, aerobic respiration, photosynthesis (involving chlorophyll, hydrogen sulfide, or whatever), cyanobacteria, protoeukaryotic life forms, endocytosis, endosymbiosis, and so on.

There are no scientific accounts that show the step-by-step process through which genetic noise becomes genetic information, or the step-by step process through which: anaerobic respiration, aerobic respiration, photosynthesis, cyanobacteria, protoeukaryotic life forms (including all the characteristics that distinguish them from bacteria and archaea), endocytosis, or endosymbiosis become encoded in DNA base sequences. Everything that permeates the foregoing issues is ensconced in conjecture, speculation, assumptions, and a great deal of ignorance.

For example, no one knows how the five basic dimensions of the process of photosynthesis came into being. No one knows how water was selected to be a source of electrons. No one knows how either photosystem I or photosystem II came into existence or why those systems came to revolve around wavelengths of 700 and 680 nanometers respectively. No one knows how all the steps involved in photosynthesis came to be organized and integrated into a functional metabolic pathway that could be fed into other functional metabolic pathways. No one knows how ATP came to play such a central role as a source of electrons, or how the production of ATP came to be incorporated into the process of photosynthesis.

Where is all the science in the evolutionary theory of, in this case, photosynthesis? The science resides in some (but not all) of the inputs and in some (but not all) of the outputs.

Nonetheless, the dynamics of the evolutionary process itself remains locked in a black box surrounded by ignorance. There is no current scientific understanding that is capable of provably accounting for what has taken place, or is taking place, as a function of the dynamics contained within the black box referred to as evolution.

Section II: Through A Glass Darkly

I Must Be Related To John Scopes

When I was a freshman in high school, I gave several presentations on evolutionary theory for a science class in which I was enrolled. My older brother recently had withdrawn from college and had returned home bringing with him, among other things, a biology textbook that I considered to be of interest.

Based on material I found in that text, I drew several illustrations depicting certain facets of evolutionary history. I added some commentary, and, then, approached my science teacher to see about presenting those efforts to some of my fellow science students.

He agreed. Over the next several classes, I delivered my findings, illustrations, and comments.

Given that the foregoing events took place around 1958, the topic of evolution did not appear in many, if not, most high school textbooks used during that era, including the science book that was being used in our school. Perhaps the teacher was quite happy that the topic of evolution was going to be introduced by a student rather than him, and, perhaps, this is the reason why he gave the green light for my presentation.

I remember the other students in the class responded to the talk in a very underwhelming manner. For whatever reason, they seemed uninterested in what was being said, and, certainly, one of the reasons why they might have expressed little enthusiasm for what I did is because what I did – or tried to do -- was not all that interesting.

At the time I never considered the possibility that the apparent lack of interest displayed by the other students toward my presentation might be because I was crossing some lines that the other students – or, perhaps, some of them -- considered sacred. My failure to consider such a possibility might have been because, at the time, I, myself, was a church-going individual who never had heard any of the murmurings concerning the evolution-creationist controversies that had arisen in many parts of the country and that were exemplified, to some extent, by the 1925 prosecution of a Tennessee biology teacher, John Scopes, for lecturing about the tenets of evolution rather than about the theology of creation.

I really wasn't taking sides in the matter, because, for the most part, I was quite ignorant about those issues. I simply found the material in my brother's college textbook to be of interest and wanted to share it with other students.

In retrospect, I wonder why my teacher let me give the presentation. I have no idea what he did, or didn't, know about evolution, or the creationist-evolutionary controversy, or the Scopes trial.

He might have been as ignorant as I was with respect to the whole set of conceptual, religious, and cultural currents that swirled about the various disputes between those who were proponents of evolution and those who were advocates for a creationist position of some kind. Or, he might have his own ideas concerning those matters and was prepared to take a risk, of one kind or another, by letting me proceed with such a topic despite possibly knowing that the material was sensitive and, potentially, explosive.

To the best of my memory, the issue of evolution didn't arise again in that particular science course. Or, if it did, the presence of that sort of material did not lead to any kind of cause célèbre among the students or in the rest of the community that might have colored it with the sort of emotional overtones that would have caused memories of the affair to linger in my mind.

Fast-forward more than 50 years later to Section I of the present book. In Section I, I stated that if an individual were so inclined, then such a person could accept evolutionary theory pretty much in its entirety and stipulate – without self-contradiction – that evolution was the means through which, over billions of years, God went about introducing changes to life forms on Earth (and, perhaps, elsewhere as well). During the foregoing discussion, I further stipulated that I was not inclined to proceed in such a fashion because I believe there are a variety of substantial problems that permeated evolutionary theory and, as a result, I did not feel evolutionary theory constituted the slam-dunk that many scientists, and other like-minded individuals, seem to suppose is the case and, then, I proceeded to put forth a variety of ideas, arguments, , evidence, and resource material in support of the foregoing claim.

| Evolution Unredacted |

I would like to introduce some further considerations in support of the foregoing perspective. This additional material explores some of the work of Kenneth Miller.

Professor Kenneth R. Miller is a relatively rare individual among those who are proponents of evolutionary theory. He has found a way that he believes is capable of reconciling his religious faith with the sciences of evolution.

Nevertheless, Professor Miller has not shied away from engaging various people of faith who, for one reason or another, have tried to take issue with the edifice of evolutionary theory. He has accomplished this through: (1) Public debates with a number of individuals who are well-known advocates of positions built around notions of 'creationist science' and 'intelligent design'; (2) several high-profile court cases that have helped shape the fate of the evolutionary-creationists controversy within the context of public education in America, as well as (3) a series of high school, biology textbooks that he co-authored with Joe Levine that have been used to teach millions of teenagers about, among other things, the commonly accepted view of many scientists concerning the nature of the relationship between biology and evolutionary theory.

I have read several works by Dr. Miller, including *Finding Darwin's God: A Scientist's Search For Common Ground Between God and Evolution*, as well as: *Only A Theory: Evolution and the Battle for America's Soul*. The ensuing discussion gives expression to some of my critical reflections concerning material that can be found in the foregoing two works, but most of the commentary in the present section tends to focus on the first (*Finding Darwin's God*) of Dr. Miller's two previously mentioned books.

I feel his point of view presents readers (and I might be optimistic here in using the plural form of "reader") with an important opportunity to learn about a variety of important issues. For, not only does a critical engagement of positions held by Professor Miller have potential ramifications for the issue of evolution, but, as well, many of his arguments offer a chance to explore various possibilities concerning the nature of one's relationship with Being.

Complexities of a Simple Theory

Professor Miller indicates that his first direct encounter with Darwin occurred during the summer interregnum that bridged the period between his last year of high school and his first year of college. Along with a number of other books that he engaged during that period, he read *On the Origins of Species* and, for the most part, found it to be a rather boring work and relatively pedestrian in its mode of argument.

According to Dr. Miller, Darwin's position in *Origins* can be reduced down to a small set of simple premises. Those premises give expression to a few common-sense claims.

First, (1) considerable variation has been observed to exist among domesticated animals and plants. (2) Wild life forms – both among animals and plants – display a range of variation similar to that of domesticated species, and, in fact, the variability found in wild life forms is so great that, sometimes, scientists have had, and continue to have, difficulty determining where one species ends and another species begins. (3) Life is synonymous with a struggle for survival, and that struggle is most intense within the context of a single species because all the members of a given species are competing for precisely the same set of resources that are considered to be crucial to their individual survival. (4) Variation is preserved and enhanced through the process of natural selection when successful individuals (i.e., those that are able to survive the struggle for existence and, therefore, have the opportunity to generate offspring) automatically pass on to their progeny an array of variations in conjunction with already established characteristics, as well as in relation to new possibilities.

The foregoing, commonsense premises are introduced and developed during the first four chapters of Darwin's book: *On the Origin of Species*. The remainder of that work branches out to investigate an array of topics – ranging from: Paleontology and classification, to: Embryology and behavior, as well as a variety of other issues, in order to provide readers with an outline of how the principles that are introduced during the first four chapters of his book might play out in conjunction with a number of other topics.

One has no difficulty acknowledging that substantial variation exists in the worlds of domesticated species as well as amongst forms of wild life. However, one might pause (at least I did) amidst the sense of uncertainty that tends to sweep over one as one is required to acknowledge – as evolution demands -- that all the variation in life forms that have ever arisen were, without exception, a function of the inherited recombination or reordering of genetic possibilities that were already present in various predecessors.

The foregoing issue revolves about the dynamics of variation. What are the origins of variation and how do new variations arise and are there any limits to how far the potential of variation in a species can proceed?

Darwin's book was entitled: *On the Origin of Species*. However, in passing, one might wish – and I do so wish -- to raise questions about whether the process of variation that is at the heart of the aforementioned book is also capable of explaining how the evolutionary distances that mark the transition from one kind of species – e.g., some primitive form of protocell – to the emergence of: Domains, kingdoms, phyla, classes, orders, families, and genera were actually traversed.

Given time and under appropriate conditions of natural selection, are the variations and potential for variation that were in the first life-forms capable of underwriting the transitions in variation that would have been necessary for various domains, kingdoms, phyla, classes, orders, families, and genera to be able to come into existence following the emergence of life? Moreover, can one suppose that the variations among molecules prior to the appearance of the first protocell would have been sufficient to underwrite the origins of life under the right kind of circumstances?

Are variation and natural selection sufficient to account for all the forms that life assumes? Perhaps one also needs to posit the existence of some sort of ordering principle that is capable of generating new kinds of variation?

For example, many scientists – including Professor Miller – often resort to invoking the idea of random or chance events to account for, among other things, the origin of new modalities of variation. Frequently, however, but not necessarily always, this amounts to being nothing more than a case of assuming one's conclusions and, in the process, hermeneutically framing the situation.

How does one know that an event is truly random? What does one even mean when one refers to an event as being random?

Some people claim that an event is random when one cannot produce an algorithm that is capable of generating such an occurrence. To some extent, this makes randomness a function of one's ignorance concerning the nature of how reality operates since just because one is not aware of the existence of an algorithm that is capable of producing a given phenomenon, this doesn't preclude the possibility that such an algorithm actually does exist and simply has not, yet, been discovered.

In certain respects, citing randomness as an explanation for something is similar to citing God as an explanation for something. Both approaches often – but not necessarily always -- are steeped in considerable ignorance, and the existence of those sorts of accounts might demonstrate little more than the tendency of human beings to confabulate ways of describing experience that seek to make the unknown sound more familiar and better understood than actually is the case.

Daniel Dennett, a philosopher, considers evolution to be a dangerous idea. He believes its danger – at least in part – has to do with the way that a very simple idea – namely, evolution by natural selection -- seems to be capable of weaving an amalgam of issues involving: Cause, effect, purpose, meaning, life, time, and space into a unified set of laws concerning the nature of reality, and, in the process, by-passes the need to make any reference to the existence of God.

To be sure, the idea of evolution by natural selection does provide one with an opportunity to develop a law-like framework for exploring, discussing, and attempting to explain an array of fundamental issues involving cause, effect, purpose, meaning, and so on. However, that idea is only dangerous if, on the one hand, it is false and someone accepts it as true, or, on the other hand, that idea is true and someone resists acknowledging the reality of that truth because it threatens some sort of cherished, but delusional, understanding concerning the nature of reality.

Apparently, Dennett believes that the idea of evolution by natural selection is dangerous because it induces – or forces -- people to consider the possibility that God does not exist. However, as noted earlier, one could, if one liked, acknowledge the truth of evolution and, without fear of self-contradiction, simultaneously endorse God's existence merely by supposing that evolution gives expression to one, or more, of God's natural laws for affecting change in the physical world.

If an individual were to acknowledge the truth of evolution, then that person might have to modify certain of his, her, or their beliefs concerning the nature of God's relationship with reality. However, accepting the truth of evolution by natural selection does not necessarily require an individual to jettison ideas concerning God's existence.

For example, Thomas Jefferson maintained that the nature of the Creator was such that various forms of life would never be permitted to become extinct. Consequently, he believed that if one were to encounter fossils that were dissimilar to any currently existing forms of life, this only meant that living exemplars of that fossil were present elsewhere on Earth and had not, yet, been discovered.

Later, when machines powered by the steam engines first designed by James Watt began to remove earth in order to lay down rail lines or excavate coal to stoke the fires of industry, a wealth of fossil remains were forthcoming, many of which did not seem to have any living counterparts despite the fact that there was a rapidly diminishing number of regions on the Earth's surface that had not been explored. Clearly, evidence was mounting that the possibility of extinct life forms was not as inviolable a principle as people like Jefferson had once assumed.

If Jefferson had lived to witness the foregoing discoveries, he very likely would have been forced to admit that the idea of extinction was something that, in fact, had been permitted by the Creator. However, changing his ideas about what the Creator would, and would not, permit, carried no necessary implications for his belief about the Creator's existence.

Similarly, a person could engage any number of factual findings of science and, as a result, be required to modify his, her, or their ideas concerning the nature of God's relationship with the Universe. Nonetheless, none of the foregoing sorts of changes would require a person to necessarily conclude that God did not exist but, rather, merely would require a person to acknowledge that God's nature or

God's nature with the Universe was different than what the individual previously had supposed to be the case.

Science changes to accommodate new empirical realities. There is no reason why religion cannot do the same.

Toward the close of the 18th century, William Smith – an English surveyor and canal builder – began to realize that various kinds of geological formations had regularities to them that showed up in, among other things, the canal excavations that he was overseeing. Subsequently, Smith also realized how various kinds of fossils were uniquely associated with certain kinds of rock formations, and, consequently, he was able to use the presence of those fossils to identify the nature of the rock formations in which the fossils were located.

For Smith, fossils were a way to map geological history. Others saw the potential in Smith's findings as a basis for mapping changes in life forms across transitions in geological structure.

An enhanced version of Smith's early work was published in 1815. It gave empirical impetus to what would become known as the 'principle of faunal succession.'

The foregoing principle indicated that fossils were preserved in a particular order within geological formations and could be used to establish a record for history of life on Earth. For example, recent geological formations contained fossils that were similar to, if not largely the same as, many existing life forms, while geological formations of a more ancient vintage contained fossils that were increasingly different from the life forms with which we are familiar today.

The fossil record demonstrated that only had life changed over time, but, as well, many of those changes had been substantial. Furthermore, a recurrent theme amidst those changes was that many life forms had become extinct.

Prior to Darwin, a number of individuals (e.g., William Smith in England, Georges Cuvier and Etienne Geoffroy Saint-Hilaire in France) had established that the theme of change was indelibly written into the geological and biological record of life on Earth. Darwin was among the first individuals (and one might wish to consider Alfred

Wallace and, possibly, Jean-Baptiste Lamarck, as well, at this point) to offer a plausible scientific account concerning what might have made such biological change possible, and while plausibility does not guarantee truth, it helps to push inquiry in more rigorous directions that, over time, might be improved upon.

If a person were to examine the fossil record and critically reflect upon it, the individual will encounter evidence indicating that during the Proterozoic era, more than 3 billion years ago, prokaryotic life forms existed. Prokaryotes are single-celled organisms that do not possess: (1) A membrane-bound nucleus; (2) mitochondria (an organelle responsible for processes such as respiration and energy production); as well as (3) a number of other kinds of organelles that are found in eukaryotic forms of life (e.g., Golgi apparatus -- involved in such processes as intracellular transport of materials; lysosomes – organelles containing various enzymes capable of acting on, and degrading, different kinds of polymers that exist in the cell; or, chloroplasts – the structures within which photosynthesis takes place).

Prokaryotes are generally divided into two domains: Bacteria and archaea. Archaea or archaebacteria are similar in size and structure to bacteria, but they possess a form of molecular organization that differs significantly from that exhibited by bacteria.

Cyanobacteria are prokaryotic organisms that generate energy by means of photosynthesis, and, in the process, release oxygen. There are other forms of bacteria that have the capacity to generate their energy through one form, or another, of chemosynthesis ... that is, processes that often are carried out in the absence of sunlight and that generate energy through the oxidation of inorganic molecules.

The latter organisms are divided into two categories. There are chemoautotrophs and chemoheterotrophs.

Chemoautotrophs generate energy by oxidizing inorganic compounds (e.g., elemental sulfur, molecular hydrogen, hydrogen sulfide, ammonia, or iron). The vast majority of chemoautotrophs are extremophiles (organisms that live in conditions of high: Acidity, alkalinity, salt concentration, temperature, and so on), or chemoautotrophs also include forms of archaea and bacteria that live in other kinds of extreme environments.

Chemoheterotrophic organisms oxidize various organic compounds (e.g., lipids, carbohydrates, and proteins). In the process, they synthesize ATP (adenosine triphosphate), and this serves as a source of energy for those organisms.

Prokaryotic organisms also can be divided into anaerobic and aerobic life forms. Anaerobic organisms tend not to be able to tolerate the presence of oxygen, while aerobic organisms require oxygen to be present in order to be able to thrive.

No one knows whether the first life forms that arose during the Proterozoic era were some manner of prokaryotic organism or, instead, constituted a non-prokaryotic protocell progenitor that failed to leave any fossilized remains. If the former possibility is the case, then, one encounters the problem of having to account for the origins of prokaryotes from prebiotic beginnings, and if the latter possibility is the case, then, one is confronted with the problem of needing to account for the precise nature of the dynamics that made the series of steps possible that give expression to the transition from protocell to prokaryotic organisms.

The origins of both anaerobic and aerobic life forms are also immersed in mystery. Did they arise independently of one another – and, if so, how -- or was there a sequence of transitional steps leading from anaerobic to aerobic organisms, and, if so, what were those steps?

Similarly, no one knows how chemoautotrophs – organisms that generate energy by oxidizing inorganic materials -- or cyanobacteria -- organisms capable of photosynthesis – came into being. Did they have origins that were independent of one or another – and, if so, what are the sequence of events in each case that led to their emergence.

Did the two foregoing kinds of organisms share an evolutionary history? If so, what are the transitional steps that led from one form of life to the other?

Did archaea and bacteria arise independently of one another, or is one of those life forms descended from the other? No matter how one chooses to answer the foregoing questions, one is faced with the task of providing an account for either the origins of those life forms or an account that explains how the transitional journey from one life form to the other took place.

To date, no one has supplied specific, concrete, verifiable responses in relation to any of the foregoing questions and issues. The origins of: Protocells, bacteria, archaea, cyanobacteria, extremophiles, chemoautotrophs, chemoheterotrophs, as well as anaerobic and aerobic organisms, along with their possible evolutionary interconnections with one another, are all (to borrow from Winston Churchill) riddles wrapped in mysteries inside enigmas.

Many scientists and philosophers scoff at religiously inclined individuals who are unable to provide, upon demand, step-by-step proofs concerning, for example, God's existence, the nature of miracles, and how God runs the universe. Yet, none of those same scientists and philosophers are able to overcome a similar sort of inability with respect to providing, upon demand, step-by-step accounts concerning the origins of life or the precise nature of the transitions that bridged the differences between, say, protocells and prokaryotes, or archaea and bacteria, or chemoautotrophs and cyanobacteria, or anaerobic and aerobic life forms ... and the list of unknowns that permeate evolutionary theory can be extended almost indefinitely.

Why should scientists and philosophers be exempt from the same requirements of rigor, scrutiny, and step-by-step explanations that they wish to apply to everyone else? Surely, what is good for the goose is also good for the gander.

Yet, many evolutionarily inclined scientists and philosophers seem to feel that general notions – such as the idea of "random mutations" -- constitutes a fully adequate way of filling up the many lacunae – a few of which have been noted above -- that populate evolutionary theory. One can easily extend this sort of argument.

For instance, as noted previously, prokaryotes lack certain features that are found in eukaryotic organisms. Among other characteristics, prokaryotes do not have a bound nucleus, nor do they possess organelles such as the Golgi apparatus, lysosomes, chloroplasts, and mitochondria.

Prokaryotes do possess various protein-based micro-compartments exhibiting functions that, in some ways, are similar to some of what takes place in the organelles of eukaryotic organisms. Consequently, many people feel that the foregoing sort of compartmentalized processing areas in prokaryotic organisms are the forerunners for the organelles that arise in eukaryotic life forms, but, so far, no one has been able to establish the nature of the specific sequence of transitional steps that would take one from prokaryotic to eukaryotic life forms or from prokaryotic micro-compartments to fully functional eukaryotic organelles.

Various scientists refer to the notion of endosymbiotic processes or symbiogenesis as a way to provide an explanation for the sort of dynamics that might have made the transition from prokaryotic micro-compartments to eukaryotic organelles possible. The theory of endosymbiosis was first proposed by the Russian botanist, Konstantin Mereschkowski, between 1905 and 1910, and, further refined by Lynn Margulis in 1967.

According to the broad features of the foregoing theory, the organelles that are manifested in eukaryotic organisms are due to a number of prokaryotes that, initially, were independent from one another, but became absorbed into, or engulfed by, some larger prokaryote and, over time, the contingent arrangement became integrated or harmonized in certain ways and, therefore, led to the formation of a stable, symbiotic dynamic in those organisms. The problem with the foregoing is that although Margulis does provide evidence indicating that symbiosis of various kinds does occur in nature, and although she puts forth an outline for how such a condition of symbiosis might have arisen about 1.5 billion years ago, nonetheless, her theory is more suggestive than anything else.

In other words, things could have happened in the way she describes in her research. However, there is no concrete proof that what she theorized might have happened long ago actually did occur in the way she indicated, and, as well, she is a little vague on the precise nature of the transitional steps that permitted the aforementioned symbiotic relationships to become integrated and harmonized with one another to such a degree that the components of the system are no longer recognizable as separate entities that have a symbiotic relationship with one another but, instead, are considered to be the unified components of a cell.

Let's consider a number of other instances in which evolutionary theory is a little vague on the details. We can begin with certain life forms that arose at some point after nucleated life forms emerged.

More specifically, in 1946, fossil remains were discovered amidst the Ediacaran Hills of Australia indicating that approximately 600 – 700 million years ago life acquired the capacity to exist in multiple-cellular formats. These fossils are among the earliest representatives of metazoan – or multi-cellular – life forms.

There are a variety of mysteries surrounding the biota or life forms that were indigenous to the Ediacaran Hills. For instance, there is considerable uncertainty about whether, or not, those life forms were progenitors of later life forms or constituted an evolutionary dead end.

If the Ediacaran organisms were predecessors of later life forms, then, one would not only like to know how those organisms arose from earlier eukaryotic life forms, but, as well, one would like to know the nature of the specific steps that enabled those organisms to make the transition from single cell to multi-cellular forms of life. On the other hand, if the Ediacaran life forms were an evolutionary dead end, then, one must find an alternative way to account for the origin of subsequent editions of metazoan organisms that is independent of Ediacaran life forms.

The Cambrian period follows the Ediacaran era. The former period occurred approximately 541 million years ago.

The fossils associated with the Cambrian period contain exemplars from most of the animal phyla (a taxonomic form of classification that falls below Kingdom but above Class) that exist today. The term "explosion" is sometimes used in conjunction with the Cambrian period because of the diversity of the life forms that are found among the fossils dating from that time.

Even if one were to assume that the relative absence of fossils for progenitors of Cambrian life forms merely indicated that the right sort of conditions had not been present to create a fossil record concerning those progenitors and, therefore, the absence of evidence in this regard should not be interpreted to constitute evidence of absence, this still leaves one with a variety of questions. For example, as long as one does not have the requisite sort of fossils with which to work, then, one only can speculate about what specific sequence of events might have marked the process of transition through which different phyla arose from earlier predecessors.

Notwithstanding the foregoing considerations, most biologists -- based on fossil evidence and a variety of other considerations -- tend to agree on the basic features that characterize the succession of life forms that proceeds from the Cambrian period onward. The following material provides a general overview concerning the foregoing framework of agreed-upon succession

More specifically, starting with the Cambrian period, some 550 million years ago, corals and shellfish – among certain other life forms -- arose. Approximately 480 years ago, the first species of fish arose, but their bony offspring were not present for another 100 million years. About 380 million years ago, amphibian life forms emerged. Around 340 million years ago, reptiles came into existence, and perhaps 80 million years later, dinosaurs began to roam the Earth. In the vicinity of 210 million years ago, mammals show up, and approximately 55 million years later, birds begin to proliferate.

Evolutionary scientists note that as each of the aforementioned new forms of life first appears in the fossil record, those organisms often tend to exhibit a variety of features that resonate more with their predecessors than with later instances of that kind of life form. For example, the first amphibians tend to resemble fish more than they resemble subsequent instances of amphibians, and the first reptiles seem more amphibian-like than reptile-like, just as the first mammals appear to have more in common with reptiles than they do with later forms of mammals.

Many people believe the foregoing information demonstrates that the succession of life forms displayed in the fossil record is the result of a gradual process of change over time. More specifically, they believe that one or more species of fish gradually transitioned into an amphibian form of life, and, in turn, one, or more of those amphibians gradually transitioned into a reptilian form of life, and, in turn, one, or more, of the reptilian organisms being alluded to gradually transitioned into a mammalian form of life.

From an evolutionary point of view, one understands from whence the fishy aspects of amphibians might have come, and one has a sense of why amphibian characteristics might be present in early forms of reptiles or why reptilian features might show up in primitive forms of mammals. What is less readily evident is the precise causal nature underlying the emergence of the amphibian characteristic that distinguish amphibians from fish, or the origins of the reptilian features that differentiate reptiles from amphibians, or the derivation of the mammalian properties that make a life form a mammal rather than a reptile.

Darwin and his conceptual descendants have encountered a certain amount of difficulty when attempting to plausibly explicate the specific nature of the dynamics that supposedly generate various novel: Features, properties, structures, functions, metabolic pathways, and behavioral capabilities that differentiate, say: Fish from amphibians, or amphibians from reptiles, or reptiles from mammals. Although the general notion of "random mutations" often is advanced to account for the emergence of novel characteristics in various forms of life, nevertheless, the term tends to be little more than a catch-all phrase that is completely lacking in specificity, and, therefore, doesn't really so much constitute an explanation as much as it gives expression to an attempt to camouflage the unknown with the appearance of understanding.

Claiming that the existence of similarities among a group of fossils suggests the presence of ancestor-descendant relationships is one thing. However, claiming that the existence of differences among a group of fossils demonstrates that "descent with modification" (Darwin's term) is taking place doesn't really clarify what is transpiring because the actual nature of the process of modification remains steeped in mystery.

For instance, Darwin spoke about the fossil remains of Glyptodon, a large, armored animal that once roamed the terrain of South America and, based on certain anatomical similarities (such as the presence of an armored body), seems to be related to the modern armadillo that also is indigenous to the Western hemisphere. According to Darwin, the foregoing sort of similarity, along with the existence of numerous other instances of similarities between ancient fossils and modern life forms that are found in relative physical and geographical proximity to one another, tend to give expression to the idea of ancestor-descendent relationships as well as the principle of "descent with modification'.

According to Darwin, Glyptodons are the ancient predecessor of armadillos. However, even though Glyptodons and armadillos are allegedly related to one another, nevertheless, they are not the same life forms, and, therefore, despite the presence of a number of similarities, they also differ from one another.

Even in the case of similarities, there are differences. Depending on the degree of difference entailed by those similarities (and I realize that some people might experience a brain cramp trying to parse the notion of similarities with differences), one might wish to raise a few questions concerning the source of those differences. Are those differences merely expressions of variation within a population, or is something else involved?

What is the source of variation in any given case? Can one automatically suppose that all variation is a function of various kinds of combinatorial phenomena involving random events within the genetic potential of a given population (assuming, of course, that one could definitively determine that such processes were purely random in character)?

On the other hand, the presence of truly significant differences – i.e., beyond the differences that are associated with the foregoing sorts of similarities -- tends to make one wonder, despite the presence of the aforementioned sorts of similarities, not only whether, or not, an ancestral-descendent relationship actually exists, but, as well, such differences make one wonder how those differences came into being. The greater the degree of those differences, then the more one tends to wonder about the source of those differences and whether they give expression to the "normal" dynamics associated with the potential variants that are inherent in any given population ... in this case, either Glyptodons or armadillos.

Did variations in the genetic potential of the ancestral Glyptodon population underwrite the differences between Glyptodons and armadillos? Did variations in the genetic potential of the armadillo population underwrite those differences? Were those differences a combinatorial function involving variations of both of the foregoing possibilities? Or, were those differences a function of some other kind of phenomenon?

Are there limits to how variable any given population can become? Can one necessarily assume that every difference that shows up in a subsequent species – say armadillos -- is tied to an earlier species – say Glyptodons -- by virtue of ancestral-descendent relationships and, therefore, demonstrates the process of "descent with modification"?

What is the nature of the specific set of sequential events that mark the transition from Glyptodons to armadillos? Even if one knew the entire sequence of modifications that were responsible for generating the differences between Glyptodons and armadillos – and, in fact, no human being knows what that sequence is -- one wouldn't necessarily understand what was responsible for the occurrence of any of those modifications.

In fact, someone could invent a device that permitted one to observe and record each and every modification as it occurred. Nonetheless, the foregoing sort of invention still wouldn't necessarily enable a person to know what was responsible for the occurrence of any given instance of modification.

Was the modification due to a transcription error of some kind during the process of replication? If so, what caused that sort of error to occur?

What set of events led to the occurrence of such an error? What brought about that set of evens at a certain point in time?

Were the aforementioned changes due to the modifying impact that various kinds of, say, radiation had on an organism's biological system over a period of time? What caused that radiation to engage the organism in one manner and at a certain time and place rather than another?

Let's go back to Schrodinger's Cat that prowled a few of the pages in Volume II in the *Final Jeopardy* series and ask: What caused a quantum system – for instance, an unstable set of atoms within a closed chamber – to decay at one instant in time rather than another. Quantum mechanics provides one with a methodology that is capable of generating descriptions that exhibit considerable precision concerning the possible behaviors of those particles under an array of circumstances (including when an atom might decay) but quantum mechanics – at least as currently understood -- can't explain what, if anything, causes one particle, rather than another, to decay when it does ... and the collapse of the wave function doesn't so much explain what is going on as much as that event itself stands in need of an explanation.

During the Foreword to this book, I mentioned the work of Morton White with respect to the complex nature of causality. One doesn't have to spend a great deal of time thinking about those kinds of issues before one realizes that one can easily become lost in the multiplicity of causal details that tend to surround any given event or sequence of events.

During our lives, we develop various hermeneutical scenarios in an attempt to capture – correctly or otherwise -- what we consider to be the most important causal features of any given situation. For Darwin, one such scenario entailed the idea of descent with modification, but this doesn't necessarily take us very far because – as previously noted -- we still aren't really sure what forces make those modifications possible.

For instance, are the underlying causal forces associated with any given instance of modification random or are they non-random in nature? Do random events actually occur, and how would one know that those events are really random? Or, if the events are non-random in nature, what kind of non-random phenomenon is present?

Are the dynamics surrounding 'descent with modification' a function of laws that are self-contained and independent of Divinity? Or, do the phenomena entailed by 'descent with modification' constitute laws that give expression to the presence of Divine intentions?

Professor Kenneth Miller disagrees with those individuals who claim that Darwin's proposed mechanism for evolution was vague. Dr. Miller feels that the any given species originates by means of the existence of variations that subsequently were subjected to different forces involving, for example, changes in: Climate, geographical migration, natural disasters.

Several examples are given by Professor Miller to help clarify the foregoing point. For instance, he introduces his readers to genus Rhizosolenia (single-celled diatoms that are capable of photosynthesis, and, as well, are characterized by, among other things, silicate cell walls of fairly intricate design) and, then, proceeds to explore how two instances of that genus – namely, Rhizosolenia bergonii and Rhizosolenia praebergonii – came into being.

According to Professor Miller, one can map the fossil remains for specimens of Rhizosolenia across a period of approximately 1.6 million years and observe, first, an increase in the variation of the ancestral species followed by a branching phenomenon that gives rise to two distinct lines of descent. The emergence of the new species of Rhizosolenia took approximately several hundred thousand years to become complete.

If a person – such as Professor Miller – feels inclined to refer to the foregoing instance of speciation as an instance of descent by modification, one can take note of his use of terminology, and, still, ask whether, or not, anything has been shown to have occurred in the context of Rhizosolenia that is qualitatively different from population dynamics and, therefore, deserving, in some sense, of being called "evolutionary" in nature. Variation, change, descent, and natural selection are all present in the process of speciation that is being documented with respect to Rhizosolenia, but one has difficulty observing the presence of any sort of specific mechanism in his overview that would be capable of explaining in concrete, step-by-step dynamics how truly evolutionary forms of life arose.

How, for example, did archaea come to be different from prokaryotic life forms (or vice versa)? Or, how did eukaryotes arise from prokaryotic life forms (and the notion of symbiogenesis as presently understood is far too general with respect to the etiological portrait it paints)?

How did anaerobic organisms descend (if they did) from aerobic organisms? Or, how did cyanobacteria arise (if they did) from chemotropic life forms, or how did multi-celled organisms arise from single-celled organisms?

Professor Miller alludes to a broadening of variation or diversity that occurred during the evolutionary history of Rhizosolenia.

However, he doesn't provide an account detailing the causal precursors for that process of broadening that are capable of being verified.

Of course, one could posit a variety of natural, material possibilities that might have brought about the foregoing sort of broadening in variation. Nonetheless, one can't prove that any of those possibilities – operating in isolation or in concert with one another – were responsible for the broadening of variation to which Miller refers.

Secondly, although climate change, natural disaster, predation, and so on could all – individually or collectively – provide an explanation for why speciation occurred, we still are none the wiser with respect to understanding the concrete character of the branching process that actually took place. A possible mechanism is not necessarily the same thing as an actual mechanism.

Therefore, Professor Miller's account of speciation doesn't really add any degree of specificity to Darwin's perspective. One still doesn't know what the actual nature of the steps were that gave rise to either the broadening of diversity in the Rhizosolenia population or the kind of speciation branching process that have been observed in conjunction with Rhizosolenia fossils.

One has been given a narrative. However, one has to navigate through an assortment of assumptions and speculations in order to be able to traverse that account.

Over the last 150-plus years, the vast majority of scientists have increasingly resisted, if not rejected, the attempts of religiously inclined individuals to claim that the natural history of organisms such as Rhizosolenia reflect the handiwork of God. Those scientists indicate that an evolutionary account involving, among other things, concepts such as the notion of 'descent with modification' is far superior to what can be provided through a theological framework, and, yet, there are significant lacunae that exist in conjunction with both approaches (evolutionary and creationist) to the available evidence concerning the concrete character of the details underlying observed changes in life forms.

Professor Miller moves on from his exploration of Rhizosolenia and begins a discussion concerning the origins of human beings. He indicates that just as one can map changes in the cell wall of Rhizosolenia in order to provide data points through which one can trace the branching process of speciation in that genus of organism, so too, one can map changes in certain regions of the brain in various species that are considered to be ancestral to homo sapiens and, thereby, acquire an understanding of the evolutionary history for humankind.

For instance, cranial capacity is one of the properties used to distinguish between Homo sapiens and possible ancestral predecessors. A convention in classification has been established that uses a cranial capacity of 600 cc (cubic centimeters) to form a line of demarcation.

More specifically, hominid-like organisms that exhibit a cranial capacity higher than 600 cc are generally classified as belonging to Homo sapiens. Those hominid-like organisms that possess a cranial capacity that is less than 600 cc tend to be placed within the genus of Australopithecus.

On the basis of fossil evidence, we know that approximately 3 million years ago, a hominid-like life form existed on Earth that exhibited a cranial capacity of about 400 cc existed. This organism survived for roughly 1 ½ to 2 million years.

Somewhere around 2 million years ago, changes occurred in conjunction with the diversity and distribution of the hominid-like fossils that have been unearthed. One of the changes being alluded to that took place across an interval of several million years among hominid-like fossil remains involves the property of cranial capacity, and this encompassed a set of changes that increased smoothly from 650 cc to 1,500 cc.

Professor Miller maintains that the foregoing changes in diversity and distribution led to a split in speciation. The two hominid-like species co-existed for a while, but, after a time, one of the two lines came to an end, while the other line – which led to modern humankind -- continued on.

According to Professor Miller, the foregoing pattern of change is consistent with what evolution predicts will occur. However, precisely what Professor Miller has in mind when he makes the foregoing sort of claim is not very clear.

Is he saying that if an impartial, intelligent, scientifically literate (whatever that might mean), alien observer came to Earth some 2 million years ago, then, that individual would necessarily predict that the hominid-like species with a cranial capacity of 400 cc that the alien encountered on Earth would begin to diversify in variation and distribution and, in time, would give rise to creatures exhibiting a smooth set of increases in cranial capacity that would run between 650 cc and 1,500 cc, and that such changes would lead to a branching in speciation in which one of the lines would become extinct while the other line would persist? If Professor Miller were arguing along the foregoing lines, then, what would be the basis for making those kinds of predictions?

Populations increase their diversity on some occasions while at other times the diversity within a given population might decrease. There are instances in which diversity is followed by speciation, and there are times when this does not occur.

How could the foregoing alien observer know that a hominid-like population on Earth consisting of individuals that have cranial capacities of 400 cc will diversify or not? How can the aforementioned alien observer be certain that the direction of diversification in variation will result in an increase of cranial capacity rather than a decrease?

How would such an observer know that the hominid-like organism with a cranial capacity of 400 cc is going to persist rather than become extinct? Moreover, how does the aforementioned alien observer know that increases in variation will lead to a process of speciation that will be followed by the extinction of the hominid-like organism with the lesser cranial capacity and the survival of the hominid-like organism with the greater cranial capacity?

In addition, nothing has been said, so far, to demonstrate that increases in cranial capacity will necessarily give rise to qualitative changes in functioning and behavioral capabilities? So, even if one were to grant that increases in cranial capacity were likely (and there really is no reason to make this sort of concession), one still lacks grounds for arguing that the proposed increases will necessarily lead to a branching process in which the hominid-like organism with the greater cranial capacity will be more likely to survive since there are many species – both currently and in the past – that possess very little cranial capacity and, yet, have managed to survive.

Spontaneous Mutations

Professor Miller indicates that as a cell biologist he tends to be rather mystified by those who are skeptical about the process of evolution because the latter individuals seem to miss the obvious in conjunction with understanding how the issue of novelty – that is, variation -- arises naturally within a population. More specifically, among other things, cellular biology has established that, for the most part, genes code for proteins, and since proteins play such a central role in shaping the structural, metabolic, and functional properties of a cell, then, changes affecting proteins will have an impact on the extent to which cells thrive or become extinguished in any given environment.

Change the proteins within a cell, and the structural, metabolic, and functional properties of that cell will tend to change as well. Change the proteins within a cell, and one will affect the character of the sorts of variations that will be exhibited by the cells within any given population.

Consequently, spontaneous changes, or mutations, to the genetic capacity of a cell will have the potential to affect the fitness of that cell in conjunction with various environmental contexts. Furthermore, according to Professor Miller, because spontaneous mutations are able to delete, alter, duplicate, rewrite, and invert any, and all parts, of the genetic code, then, those modalities of change are capable – at least in theory -- of generating any genetic sequence that has ever been observed.

Given Professor Miller's foregoing perspective, one might inquire about what is meant when he and other scientists refer to the notion of "spontaneous" changes or mutations? Generally speaking, this means that few, if any, of the individuals who use those terms are sufficiently knowledgeable to be able to delineate the specific identity of the forces of causality that allegedly led to a given change in genetic structure taking place.

For example, Professor Miller mentions, in passing, an item that appeared in the periodical *Science*. The magazine report describes how Gregory Petsko, a biochemist, was having difficulty inducing a particular bacterium to produce more copies of a certain protein.

Professor Petsko indicates that at some point an event occurred in his lab that affected just one of the cells in the bacterial population being studied. More specifically, a certain gene migrated to a position that was thousands of base pairs away from where Professor Petsko's team had placed the gene, and, as a result, the new arrangement brought about precisely the change in the bacterium that was needed to enable the organism to increase its production of the protein in which Professor Petsko was interested.

Professor Petsko refers to the foregoing event as a completely random or chance occurrence. However, the truth of the matter is that he has no idea what caused the event to take place, and, he is using terms such as "random" and "chance" to confabulate his way through the discussion.

Furthermore, just because such an event – whatever its actual nature – is capable of generating a serendipitous outcome on one occasion doesn't mean that any, and all, outcomes are capable of being brought about by the same sort of phenomenon. The fact that certain kinds of rare events have been observed to occur under various circumstances is not an adequate basis for trying to claim that the nature of the universe is such that it serves as a randomized cornucopia from which any and all possibilities flow ... in relation to evolution or in relation to anything else.

To say that a given mutation assumes the form of a deletion, inversion, duplication, and so on does not provide an account of what caused those changes to occur. The use of those terms is a way of describing the outcome of an underlying causal process without specifying the nature of the dynamic to which that causal process gives expression.

In other words, use of the term "spontaneous" is another way of saying that one is not in an epistemological position to demonstrate how a change occurred, but, instead, one is merely noting that such a change did occur. Moreover, even if one were to maintain that a given mutation were due to, say, the presence of some ionizing force or an error in transcription, one would not necessarily have definitively identified the underlying cause of the mutation since one still would have to address the issue of what caused the ionization or transcription error to occur, and so on ... perhaps indefinitely.

Secondly, an individual might be able to take any genetic sequence she, he, or they liked and construct a complex algorithm to account for how that sequence might have arisen from some earlier state. Nonetheless, the fact one can conceive of such an algorithm doesn't prove that some, subsequent, genetic sequence necessarily arose through the kind of process that is encapsulated within the algorithm that was constructed, any more than a person can argue that because one is able to specify an itinerary for how someone might have made a trip, then, this necessarily shows how some individual did make that trip.

Thirdly, and, perhaps, most importantly, even if one were to agree that an underlying dynamic existed which utilized processes of deletion, duplication, inversion, and so on to be able to rewrite the instructions for generating any protein one wished to consider, nevertheless, the ability to produce an array of new proteins does nothing to account for the source of organization that is necessary to be able to arrange proteins into functional pathways. Change is not just about the identity of the proteins that are integral to cell functioning, but, as well, change is also a matter of, among other things, when, where, how much, and in what order those proteins are going to be produced.

There is a multiplicity of anabolic and catabolic feedback loops in which all proteins are immersed. Being able to account for the existence of the proteins associated with those mechanisms does not automatically mean that, therefore, one also understands how those associated feedback systems came into existence: Simultaneously with, before-the-fact of, or after-the-fact in relation to the emergence of certain proteins.

Next, Professor Miller offers a brief overview concerning the issue of antibiotics and antibiotic-resistant drugs. He describes how antibiotics have the capacity to disrupt the dynamics through which bacteria grow by inhibiting the production of an enzyme that is critical to the cross-linking process that takes place during a bacteria's construction of new portions of its growing cell walls and, thereby, renders bacteria vulnerable to the inward pressure of the water that surrounds bacteria within their host organisms.

Professor Miller also refers to the capacity of an increasing number of bacteria to be able to counter the dynamics of antibiotics, and in the process generate antibiotic-resistant strains of bacteria. What Professor Miller doesn't indicate is how those sorts of resistant strains of bacteria have acquired the ability to counter the presence of antibiotics.

Is the foregoing capacity to resist antibiotics the result of a newly emergent "spontaneous" mutation, or set of mutations, that, in some fashion, has equipped a bacterium to interfere with, or evade, the way in which antibiotics disrupt a bacterium's capacity to enhance the latter's cell walls during growth? Or, is the foregoing capacity for resistance an indication that the potential for resistance actually existed all along within a given population of bacteria and merely needed the right sort of environmental circumstances to allow that variant to flourish?

What is at the heart of a bacterium's capacity to resist the presence of antibiotics? How – in specific, concrete steps -- did such a capacity or capacities arise?

Is the capacity to resist the presence of antibiotics a complicated process that would require quite a few changes in genetic coding in order for a newly established property of resistance to be able to emerge? Or, is that capacity the function of a relatively simple process that could be generated with just a few "spontaneous" modifications in the underlying genetic code?

If the foregoing process of acquiring resistance is a complicated affair, then, evolutionary theory is faced with the problem of needing to provide a complex, step-by-step account concerning the emergence of such a capacity in order for scientists to be able to reasonably claim that they actually understand what is transpiring. If, on the other hand, the property of resistance is easily acquired, then, one doesn't need evolutionary theory to account for such a capacity but, instead, one could use principles that are inherent in population biology to provide an explanation.

Irrespective of whether the capacity for resistance is complicated or simple, one still must grapple with the issue of whether, or not, those sorts of changes are a function of random or non-random forces. To claim that the acquisition of a capacity for resistance to antibiotics –

whether simple or complex in nature – is due to random events is to filter the discussion through the colors of an assumption that cannot be proven to be true, and to claim that the acquisition of the capacity for resistance – whether simple or complex in nature – is due to non-random events requires one to list all of the causal factors that are responsible for the emergence of that sort of an ability, and such a chain of causality tends to fade into a cloud of unknowing that often permeates those kinds of discussion.

Professor Miller goes on to talk about a breakthrough in health care that occurred in 1996 when scientists were able to engineer an artificial, molecular structure that was capable of blocking the action of an HIV-protease enzyme that was considered to play a central role in the etiology of AIDS. Individuals who were given the foregoing protease inhibitor displayed a marked improvement in AIDS-related symptoms, and the discovery seemed to usher in a new era of medical treatment.

Despite the early promise of the foregoing engineered compound, nonetheless, after a few years of continued use, AIDS-related symptoms began, unfortunately, to re-emerge in patients who were being treated with the protease inhibitor. Viral variants had begun to appear that were immune to the action of the artificial compound.

According to Professor Miller, viral strains were arising that were no longer susceptible to the capacity of the engineered compound to interfere with the functions of a critical, viral protease enzyme is because the virus believed to cause AIDS. Dr. Miller claims this is because the virus in question exhibits a high rate of mutation (and one might recall from *Final Jeopardy, Volume I*, that there are a variety of troublesome questions that continue to swirl about the actual nature and identity of the HIV virus).

More specifically, the reverse transcriptase enzyme supposedly employed by the HIV pathogen to help regulate the process of copying genetic instructions is relatively "sloppy" (Professor Miller's term) in the manner in which it goes about its activities. As a result, forms of the critical protease enzyme arose that were no longer susceptible to the action of the protease inhibitor that had been engineered by scientists.

Apparently, Professor Miller didn't consider the possibility that the reverse transcriptase enzyme is not being "sloppy" when it operates in the way it does. Rather, that sort of inconsistency might have been selected precisely because of its capacity to generate variation within a given viral population and, thereby, produce variants – within certain degrees of freedom – that are able to escape from various environmental threats posed by, for example, an engineered protease inhibitor.

From a viral point of view, the upside of a genetic system that is loose – within certain degrees of freedom – about how it operates is that it has the capacity to produce a wider range of variants, and, therefore, increase the likelihood that some of those genetic experiments will prove to be successful in a given environment. The downside of such a genetic system is its potential for running a lot of combinatorial experiments that turn out to be unsuccessful.

No one knows how the genetic coding came into being that makes possible the reverse transcriptase enzyme. As a result, one has difficulty determining whether the aforementioned sloppiness is by design or the result of a series of serendipitous, but random, events.

Professor Miller seems to consider random, chance mutations to be a virtually endless source of variations that are capable of creatively resolving all manner of environmental challenges irrespective of whether those problems are external or internal in nature. Yet, the very nature of random dynamics is steeped in uncertainty, and consequently, one has no way of knowing whether, or not, purely chance events might be capable of underwriting all of the modifications and changes that have taken place on Earth over billions of years.

Even if a person were to assume that chance events were capable of giving expression to a certain kind of creative solution with respect to a particular set of circumstances, an individual cannot plausibly argue that, therefore, not only are all events necessarily governed by the phenomena of chance, but, as well, chance events are capable of eventually generating whatever creative responses might be needed to yield a particular outcome. The issue is rendered more complex when we realize that none of us can be certain that any given event should be considered to be random because, for the most part, we don't know the full complement of causal forces that make that sort of an event possible.

Seeking to buttress his arguments concerning the potential role of chance in evolution, Professor Miller outlines the 1998 experimental work that was performed by Ronald Breaker and Adam Roth at Yale University. The two scientists began with a set of DNA sequences that were said to be random in nature.

While the foregoing sequences might not have been ordered in any particular fashion, they were far from random sequences. They were chosen by the two scientists because the molecules were made of DNA rather than lipids, carbohydrates, or some other molecular structure, and, in addition, they were not only specifically paired with the amino acid histidine, but, as well, they were put through a laboratory process that created DNA sequences of variable length.

Furthermore, the DNA-histidine structures were subjected to 11 cycles of mutation and selection. One has a certain amount of difficulty understanding why a person might consider a process to be random that consists of 11 cycles of mutation and selection that are being conducted within the context of a controlled, laboratory environment.

The end result of the foregoing 11 rounds of mutation and selection was a DNA molecule that was capable of cleaving RNA molecules. In other words, Breaker and Roth had created a protease-like DNA enzyme.

Such enzymes are not found in nature. So, the two scientists were able to accomplish in a couple of days what nature could not do on its own for more than three billion years.

What implications does the foregoing scenario carry for the notion of evolution? Virtually none I believe!

For the most part, neither DNA nor RNA does all that well in the wild. When such molecules are placed outside the protected environment of a living cell or laboratory and subjected to the presence of hydrolysis, photolysis, acidity, alkalinity, thermal degradation, and an array of other forces, then, those two molecular structures tend to degrade fairly readily.

Even if one were to suppose that random forces governed the natural world, there is virtually nothing -- if not entirely nothing -- that is of a random nature in the cited experiment. The very nature of an experiment is to control the conditions under which it proceeds, and that is precisely what Professors Breaker and Roth did in the experiment being described by Professor Miller.

To be sure, there are degrees of freedom that characterize the parameters within which the Breaker-Roth experiment unfolds. For instance, the DNA sequences with which the experiment begins are permitted to be of variable length.

However, the foregoing property does not make those sequences random in nature. Instead, that property merely establishes – in a non-random fashion -- one of the conditions that will govern the experiment.

According to Professor Miller, the term "evolution" has at least two senses. On the one hand, it refers to a historical framework through which the present can be linked to the past by means of ancestral-descendant relationships that exemplify a process involving "descent with modification," while on the other hand, evolution also constitutes a theoretical framework that seeks to show how a set of naturally occurring components (such as variation, natural selection, genetics, competition, environmental contingencies, random events, speciation) interact with one another to form a process of dynamics that, over time, is capable of bringing about changes in the structure, metabolic pathways, behavior, and other properties of various organisms.

Evolutionary theory (the second sense of the term evolution) is the straw that stirs the drink of evolutionary history (the first sense of that term). Evolutionary theory supposedly provides the details and fine-tuning that describes how the heart of evolutionary history – namely, "descent with modification" – works.

However, when one examines the nature of the details and fine tuning that allegedly are entailed by the mechanism of evolution (the second sense of the term noted above), one discovers – as has been pointed out over the last thirty pages (as well as in *Volume I* of the *Final Jeopardy* series and in *Evolution and the Origin of Life*) – there actually is a complete absence of the sorts of details and fine tuning that one has been led to expect will be present. One doesn't learn how life began, nor does one learn how (and the following examples are just a very small sampling from a much larger population): Chemotrophs, bacteria, archaea, cyanobacteria, anaerobic or aerobic organisms, metazoans, Ediacaran biota, or Cambrian period life forms came into existence.

Instead, one is provided with a generalized formula for change that might, or might not, be true in any given instance. Supposedly, random mutations lead to changes in variation among different genes whose viability is shaped – individually and collectively -- by an array of possible forces that constitute the process of natural selection, and, over time, the foregoing process will generate all life forms that have, and will, emerge on the planet Earth and elsewhere in the Universe.

The foregoing scenario is offered as an explanation, but in reality, it is little more – and, perhaps, not more -- than a narrative. In other words, evolution – in both its historical and theoretical sense -- gives expression to a hermeneutical framework that organizes an array of data and confers an interpretation on that information without simultaneously being able to prove the truth of that understanding or perspective.

Just as those who are religiously inclined have difficulty lending much specificity to their creationist accounts and, as a result, are forced, by a lack of knowledge, to be content with various generalized pronouncements concerning the actual dynamics of creation, so too, those who are inclined toward evolution have difficulty lending much specificity to their own accounts concerning the origins of various life forms and, as a result, are required to be satisfied with generalized pronouncements concerning the actual dynamics of evolution.

One understands how those who are inclined toward an evolutionary account of life believe that one can account for the origins of all organisms by constructing a dynamic that consists of the right mixture of: Random events, environmental contingencies, genetic variation, speciation, competition, co-operation, combinatorics, and natural selection. The problem is that in a litany of critical cases (some of which have been discussed previously) evolutionists are unable to demonstrate precisely that mixture of the foregoing components is the correct one or how – or even if – such a mixture came to be established at any given time during evolutionary history.

There is no set of step-by-step explanations possessed by creationists that show the precise manner in which all life forms came into being. Similarly, there is no set of step-by-step accounts that evolutionists can offer which are capable of revealing precise details about how all life forms came into being.

One can acknowledge that phenomena such as: Variation, heredity, competition, co-operation, speciation, mutation, environmental contingencies and natural selection all are real. One just can't demonstrate – at least given what is currently known -- that the foregoing array of forces actually was causally responsible for the origins of the multiplicity of life forms that have arisen on Earth across more than three billion years.

The idea of randomness is an ontological assumption concerning the nature of the universe. Most people – including those who are inclined toward evolution -- have no definitive way of proving that the universe is inherently random.

The idea that genetic variation plus mutation in some primitive chemotropic prokaryotic life form is sufficient – given enough time and the right environmental circumstances – to eventually yield the sorts of potential for variation in subsequent life forms that can be acted on by mutational forces to generate, over time, the entire repertoire of variants among all populations of past and current species constitutes a chain of reasoning concerning the origin of life forms that is not conceptually sustainable ... at least for the present. Indeed, understanding concerning the source, nature, and parameters of variation in the numerous populations that are encompassed by the history of life on Earth is beset with considerable ignorance, and, therefore, the subject of quite a lot of arbitrary speculation.

The Natural and the Supernatural

The foregoing comments are not intended to imply that natural phenomena are incapable of accounting for the origins of life forms. Rather, those remarks allude to the possibility that we don't fully understand the dynamics of natural phenomena, and, therefore, our ways of engaging an array of issues might be vulnerable to a variety of conceptual missteps that arise due to our lack of understanding concerning the nature of reality.

For example, many people – both among creationists and evolutionists -- would like to establish (if only for the sake of argument) a line of demarcation between the natural and the supernatural. Such people are often inclined to claim that the physical/material world operates in accordance with natural phenomena, whereas God operates in accordance with supernatural principles, and, therefore, in order for God to be able to affect change in the universe, supernatural principles – which are supposedly totally different from what takes place in the natural world -- must somehow interact with, or intervene in, natural processes, but no one has been able to figure out how dissimilar phenomena might interact with one another.

However, if God is responsible for the universe having the character and properties it has, then, there is not necessarily any distinction to be drawn between the natural and the supernatural. This is because the natural gives expression to the intentions of Divinity, and, consequently, does not constitute a reality that is other than a manifestation – whatever the physical properties might be – that is made possible through a supernatural presence.

As such, electromagnetism, the weak force, the strong nuclear force, and gravitation -- along with whatever other forces are in the universe but yet to be discovered by humankind – could all be natural expressions of a Divine Presence that made those phenomena possible much like a computer programmer makes the varied phenomena of a virtual world possible by arranging sequences of 0's and 1's to give expression to different properties and qualities. The aforementioned forces are simultaneously both natural and supernatural because, on the one hand, if not for the presence of God, then, those forces would not exist, and in this sense they carry the imprint of a supernatural presence (that is, a presence existing prior to the emergence of what might be referred to as "natural"), and, on the other hand, in as much as those phenomena help give expression to the realities of the everyday world, they serve to give expression to the principles that govern the world of sensory experience from which our sense of the physical, natural universe is derived.

The foregoing comments are not intended to convey a pantheistic notion of God. Creating the conditions, principles, and laws that govern natural phenomena is not necessarily the same things as BEING those natural phenomena, any more than an architect who creates the plans for a building and, then, translates that design into concrete structures in accordance with certain codes of construction can be said to be the edifice that is being built even though there is an intimate relationship between the two.

In addition, one shouldn't suppose that the last seven paragraphs are part of a subtle effort to sneak in a presumption concerning God's possible existence. Instead, the comments of the last several pages refer to a logical possibility that seems to have been overlooked by many people on both sides of the creationist/evolutionist divide, and, consequently, quite apart from the ontological issue concerning whether, or not, God exists, those comments have relevancy to a variety of epistemological and hermeneutical considerations that need to be reflected upon as one tries to struggle to establish the truth concerning the nature of one's relationship with Being.

When critically reflecting on the foregoing sorts of issues, one doesn't have to adopt any particular theory concerning the origin of various kinds of life forms. In other words, one doesn't have to espouse either a creationist or evolutionist conceptual position.

One merely is taking cognizance of the fact that both creationists and evolutionists – especially those who have difficulty grasping anything beyond their own perspective -- tend to commit a variety of mistakes with respect to the way they think about various issues, and one of those mistakes (as pointed out above) is to assume that one must draw a distinction between the natural and the supernatural when, possibly, no such distinction needs to made.

The Metric of Ignorance

Another kind of mistake that sometimes surfaces during the point-counterpoint dialectic that characterizes many arguments between evolutionists and creationists concerns the issue of intelligent design. More specifically, representatives from both of the foregoing, warring parties often use ignorance – disguised as something other than what it actually is -- as a metric for trying to make sense of why reality is the way it is.

For example, Professor Miller (who believes in God but is a proponent of evolution) wants to know why an Intelligent Designer would go about fashioning organisms for the Galapagos Islands (Approximately 650 miles off the west coast of Ecuador) that are found nowhere else in the world but are similar to – yet, according to creationists, ancestrally independent from -- life forms found in South America, or why would an Intelligent Designer fashion organisms for the Cape Verde Islands (about 355 miles to the west of Senegal) that are found nowhere else in the world but are similar to – yet, according to creationists, ancestrally independent of -- life forms found in Africa? Dr. Miller believes that a simpler – and much better explanation – for the foregoing facts is to suppose that a small number of founding species from the respective main lands (South America in the case of the Galapagos Islands and Africa in the case of the Cape Verde Islands) were able to reach the two island groups and, then, new species began to emerge on those islands as a result of the impact that isolation and other environmental factors had on inducing changes in variation within different populations of life forms that subsequently were shaped by the process of natural selection on those islands.

According to Dr. Miller, the only defense that creationists have for maintaining that all of the foregoing life forms are the result of independent instances of intelligent design rather than evolution is to argue that such an arrangement is merely the way that God chose to do things. In other words, creationists apparently are not privy to the Divine mode of reflection that results in things being arranged in one fashion rather than another – and this might well be the case -- and, as a result, Dr. Miller believes that creationists are not able to provide a satisfactory response to the query he advances with respect to wondering why an Intelligent Designer would create similar life forms for, on the one hand, the Galapagos Islands and South America and, on the other hand, the Cape Verde Islands and Africa but, nonetheless, do so in a fashion that makes what happens at each location genetically independent of what transpires at any other relatively close region.

Professor Miller goes on to wonder why an Intelligent Designer's creative powers of imagination would be so feeble as to render the Designer incapable of generating an array of life forms on the Galapagos and Cape Verde Islands that are substantially different from, rather than similar to, the organisms that are found on, respectively, South America and Africa. Professor Miller seems to suppose that an Intelligent Designer would necessarily always desire to exercise Its imaginative capacities and could not possibly, for any reason (at least any reason that Dr. Miller can conceive of), create life forms that were similar to one another in geographically proximate locations but which did not share a common ancestry of any kind.

Professor Miller seems to think that if he can't imagine why an Intelligent Designer might have proceeded in one fashion rather than another with respect to the manner in which life forms are arranged and organized on the Earth, then, this must mean that evolution offers the only viable alternative. This would be like Gary Kasparov claiming that Deep Blue (the IBM chess-playing computer) couldn't possibly be executing a winning strategy if the former world champion was incapable of imagining why the computer was arranging the pieces on the chess board in one way rather than another.

Just because Professor Miller doesn't understand why, or just because creationists seem unable to offer an explanation for why, an Intelligent Designer would have done things in one way rather than another doesn't mean there couldn't be any number of reasons for why an Intelligent Designer – if such a Being existed – might have decided to arrange life forms in one manner rather than another with respect to the Galapagos Islands, South America, the Cape Verde Islands, and Africa. Both Dr. Miller and creationists seem to be largely ignorant about – that is, they appear to have little, or no, knowledge concerning -- the nature of the Divine Mind, and, yet, that ignorance seems to be playing a central role in their respective thinking processes concerning why the world is the way it is because they are thinking on the basis of arbitrary speculations rather than actual knowledge.

Professor Miller's belief that there might have been a few species that found their way to the two island groups and, over time, began to generate a variety of species that were similar to, but distinct from, various species that existed on the respective main lands (South America and Africa) is a plausible perspective. All his position lacks is a specific, step-by-step account of how the original mainland species came into existence, journeyed to the islands, and proceeded to produce the sorts of variation in the island founding populations that, in time, led to speciation of one kind, rather than another, taking place.

Creationists argue that things are the way they are because that is the way God chose for them to take place. Evolutionists argue that things are the way they are because that is the way evolution unfolded.

Creationists do not seem to have access to the nature of the Mind of the Intelligent Designer and, if this is the case, then they cannot provide specific reasons for why things have been arranged in the way they are observed to be. Evolutionists do not have access to the nature of the step-by-step events and dynamics that take one from the origins of some mainland species to the emergence of similar, but different species on various islands, and, therefore, evolutionists cannot provide a specific account that provides the details about how things came to be the way they are on, for example, the Galapagos and Cape Verde Islands.

According to Professor Miller, proponents of the form of creationism known as intelligent design believe that the species that arise on the Galapagos and Cape Verde Islands must constitute independent creations. The reason why this must be the case is because – according to the proponents of intelligent design -- one cannot logically maintain that an Intelligent Designer might have fashioned some species through acts of intelligent design but, then, decided to switch over to a process of evolution in conjunction with other species ... apparently, the Intelligent Designer is the sort of Being that does things in an all or none fashion.

Once again, ignorance has become the metric for measuring, analyzing, and evaluating reality. Contrary to what some proponents of intelligent design might believe, there is no logical principle that prevents an Intelligent Designer from fashioning some species through the act of creation while generating other species through the dynamics of variation, isolation, and forces of natural selection.

Similarly, with respect to proponents of evolution, there is no necessary natural principle that automatically precludes the possibility that certain founding species might be a function of the creative process of some Intelligent Designer while various, subsequent, derivative species might arise through the worldly dynamics of variation, isolation, and forces of natural selection. Professor Miller seems to think that it is silly for proponents of intelligent design to suppose that God thinks in accordance with an either/or modality of logic that requires everything to be a function either of creation or evolution but could not be a mixture of the two, and, yet, Professor Miller seems to be just as silly when he appears to argue that evolution precludes the possibility that some species might arise through the clever planning of an Intelligent Designer that made provisions for the possibility of a variety of degrees of freedom through which created organisms might generate subsequent life forms by means of the dynamics present in population biology.

One should not interpret the foregoing remarks to indicate that I am seeking to advance, or agree with, some ideology or theology concerning the form of creationism known as intelligent design. Once again, the purpose that motivates the present discussion has to do with providing opportunities for being able to become aware of various kinds of problematic logic that are employed by some of the proponents for both evolution and creationism.

One cannot effectively seek the truth concerning the nature of one's relationship with Being as long as one is entangled in flawed thinking and misleading assumptions. Consequently, however one decides to proceed with respect to issues of evolution and creationism, one's efforts should be as free as possible from the sorts of influences that are going to become obstacles on the path to acquiring a viable understanding of reality, and, therefore, the previous discussion has tried to provide some food for thought concerning those sorts of efforts.

Professor Miller indicates that fossils give expression to a concrete, real-world record that bears witness to the fact that new species have emerged constantly over time. He goes on to ask -- and, then, answer – a question concerning the possible nature of the process that would enable new species to be able to come into being through the medium of intelligent design.

Dr. Miller states that he can think of only two possible ways of answering such a question. He believes that one way in which new species might come into existence would be through some sort of magical puff of smoke, while a second possibility concerning the origins of new species is if a "... new organism is born as the apparent offspring of another species but is so distinctive genetically that it becomes the founding member of a completely new species."

As far as the first possibility noted above is concerned (i.e., the puff of smoke edition), rather than admit that he has no idea how God might go about the process of creating life forms – if, indeed, that is what God does – Professor Miller obfuscates matters by reducing the process of creation to a sarcastically charged 'magical puff of smoke'. In short, his ignorance assumes concrete form in the guise of a 'magical puff of smoke' that is intended to dismiss the possibility of creation with a catchy turn of phrase and without ever having to offer any actual evidence to substantiate the truth of what he claims other than his ignorance concerning the matter.

Professor Miller adds further obscurity by means of the second possibility noted previously when he introduces the idea of organisms that come from a given species but do so in such a distinctive genetic form that they are able to become founding members of a new species. What is actually meant by the notion that something seems to arise from another species but does so in such a way that it entails sufficient genetic differences to be able to serve as the founding member of an entirely new species?

How did those distinctive genetic features arise? What caused them?

Professor Miller doesn't offer any hints as to how one should go about answering the foregoing two questions. Therefore, in its own way, the notion of being "distinctive genetically" is as devoid of meaningful content as is the phrase "puff of smoke," and, consequently, as has often been said in the realm of computing: Garbage in, garbage out.

A little later on in his discussion, Dr. Miller indicates that the idea of intelligent design impels one to claim that the past is immersed in a process of magic in which life forms emerge "out of nothing". If God exists and God has the capacity to create, then, obviously, life does not emerge out of nothing but, instead, emerges out of the creative powers of a Divine Presence ... and all that such a Presence entails.

Where once there was no-thing or no material/physical substance, nevertheless, following the act of creation, some-thing comes to exist. More specifically, God thinks of a form with various kinds of qualities, says to it "Be" and it takes on, or is translated into, some sort of ontological character in the visible universe through the properties of that Divine command or directive ... perhaps somewhat like a computer programmer who imagines a form of one kind or another and, then, arranges a series of 0's and 1's to bring that form to virtual life

If the foregoing is what transpired in the universe, then even though a person might not be able to understand how any of the foregoing processes are possible, nonetheless, this failure of understanding does not automatically preclude such a scenario from being possible or even real. As Arthur C. Clarke one said: "Magic's just science that we don't understand yet," and, consequently, the nature of creation *might* just be a form of Divine science that is indistinguishable from magic to those (which includes most human beings) who are uninitiated in the science that underlies, and makes possible, the phenomenon (in this case, creation) that currently is not understood.

Earlier, a possibility was put forth indicating that the natural and the supernatural do not necessarily give expression to separate domains. If one were to consider creation as one of the sciences known to, or invented by, God, then, this would be an example of how the natural and the supernatural co-exist, so to speak, since the Presence of Divine intentions is being made manifest in the material world through a dynamic in which God weaves together the limiting and enabling properties of physical principles (such as the four basic forces) in a manner that gives expression to phenomenological structures displaying certain kinds of properties.

Professor Miller states that new species don't appear to be continuing to issue forth from a magical puff of smoke as, supposedly, was the case in the past. He goes on to ask what he seems to believe is a rhetorical question by inquiring why the emergence of new life forms seems to have come to a halt.

The implication of his question appears to be that new life forms no longer emerge out of a magical puff of smoke because new forms of life were never a function of some magical puff of creative smoke. However, given that Professor Miller seems to have no insight into whether creation occurs or, if it does, how it occurs, one should not be surprised that he seems to be at a loss with respect to being able to imagine why creation of new life forms might have stopped and, therefore, he only can wonder why this is the case since such wondering is all that his ignorance concerning the subject permits him to do.

One should not construe the foregoing comments to be a backhanded attempt to illicitly lend ontological reality to the idea of God. Rather, the focus of those comments is intended to draw attention to the fact that Professor Miller makes pronouncements concerning the nature of God or the Mind of God that are not backed up by any evidence except his inability to speculate about such matters in ways that are shaped by his lack of knowledge – or ignorance – concerning those issues.

Later on Professor Miller once again boldly leaps into the abyss of ignorance when he proceeds to speculate about how an Intelligent Designer would, and would not, behave. More specifically, he refers to all the forms of organic remnants to be found in biological organisms that Stephen Jay Gould referred to as "the senseless signs of history" that hearken back to the existence of previous ancestral-descendent relationships.

For example, Dr. Miller describes how human embryos form a yolk sac early on during the process of development. Although the foregoing sac is empty in human beings, similar sacs found in reptiles and birds contain a yolk that is filled with nutrients that help subsidize the growth of reptilian and avian embryos during the early stages of development.

Human beings are mammals that operate in accordance with a different system of nutrient delivery in conjunction with a developing embryo. For mammals, the placenta of the pregnant mother serves as the source from which the embryo draws nourishment.

Therefore, in mammals there is no need for a sac that contains a nutrient-rich yolk. According to Professor Miller, the sac that emerges during the early stages of mammalian development is nothing more than an artifact that has been partially carried over to mammals from a by-gone ancestral era when egg-laying reptiles first came into existence prior to the advent of mammals.

Professor Miller might be right in his assessment of the foregoing situation. But, then again, the perspective being outlined in the previous paragraph might just be an expression of his ignorance concerning those matters.

Suppose someone were to ask Dr. Miller to provide a step-by-step account that explained how egg-laying reptiles first came into existence possessing a sac that contained a nutrient-rich yolk capable of nourishing a developing reptilian embryo, and, then, over time, had that system of delivering nutrients to a developing embryo replaced with a placental arrangement through which the embryo was nourished. Could he answer such a question?

No, neither he nor any other proponent of evolution is able to provide the sort of a detailed account that is being alluded to in the previous paragraph. Instead, what those individuals might offer is: Firstly, a listing of all of the fossils that evolutionists believe have populated the ancestral lineage between egg-laying reptiles and mammals, and, secondly, some sort of vague, general commentary concerning the notion of random mutations and how those kinds of alterations in different aspects of the genome brought about precisely the sort of "descent with modification" that could be acted upon by natural selection and yield a succession of life forms that would become fossilized in time and lead toward the emergence of the founding member of the class of Mammalia that, in turn gave rise to – in one, or another, ordered sequence – Prototheria (e.g., platypus), Metatheria (marsupials such as kangaroos, koalas, possums, and wombats), and Eutheria (placental mammals such as cats, bats, whales, and human beings).

In addition, Professor Miller and his evolutionary colleagues might throw in some information about the process of speciation. This would be done in order to provide something akin to a proof of concept demonstration in the sense that if one can show that speciation occurs in certain cases, then, as many evolutionists have argued, one is justified in jumping to the conclusion that all life forms must have arisen in that same manner, and, one is, thereby, provided with a smooth chain of ancestral lineage from egg-laying reptile to mammals.

Now, maybe, the foregoing scenario is just the way things happened during evolutionary history. However, one is not necessarily being petty, juvenile, or unnecessarily argumentative to wonder about the degree of credibility that should be assigned to an explanation that depends, in such a fundamental way, on a set of random processes consisting of millions of alterations being able to come together in just the right order and sequence to be able to successfully underwrite the numerous transitions that led from egg-laying reptiles to the founding member of Mammalia.

Natural selection might be able to lend a helping hand to shore up such an explanation by identifying – after-the fact -- those mutations that work and, in the process, are preserved so that they are available to be utilized by subsequent members of the ancestral lineage, Nonetheless, natural selection is not capable of generating the millions of before-the fact, "just so" mutational changes that give rise to the organisms on which natural selection operates, and, as a result, one is implored to accept -- on little more than an empirically challenged form of faith – the idea that, again and again, completely random sequences of mutations came together to form functional metabolic pathways, organelles, proteins, and structures that were subsequently endorsed by the forces of natural selection.

Quite frankly, the foregoing perspective sounds an awful lot like the confabulated stories that often are espoused by individuals who fall under the influence of medication psychosis after being brought out from a medically induced coma. To the people who are uttering those kinds of confabulated ideas, everything might sound reasonable and plausible, but to anyone who is an impartial, objective bystander, the story sounds rather delusional in character.

I find it rather strange that although Professor Miller believes in God he seems unwilling to acknowledge the possibility that God might do things for reasons that Dr. Miller either does not grasp or that cannot readily be understood in the absence of the requisite information and/or insights. The yolk sac that is found in mammals during the early stage of developments might be an evolutionary remnant of an egg-laying reptilian ancestor of human beings, but it might also be a sign of some other dimension of Divine creativity or metaphysical reality that has not, yet, been disclosed to Professor Miller or the rest of us.

I do not know what the truth of the matter is concerning the significance of the yolk sac, just as I do not know what the truth is with respect to all of the other "senseless signs of history" to which Stephen Jay Gould alluded that are present in various species of organisms. Professor Miller has offered his take on the matter, but irrespective of whether, or not, his interpretation is correct, I am not inclined to permit his inability to imagine – or his lack of knowledge concerning -- what other significance the yolk sac might entail, to become the tipping point for whether, or not, I accept evolutionary theory especially given that so much of evolutionary theory is steeped in unanswered and, in many cases, unanswerable questions.

Evolutionary Creativity

Professor Miller seeks to counter those creationists who try to argue that the impact of mutations (which proponents of evolutionist see as the engine of variation) is always deleterious and, therefore, as Dr. Miller notes in passing, such individuals are inclined to point toward the issue of mutations in a dismissive fashion because they believe it constitutes a problematic dimension for evolutionary theory. As a result, Professor Miller alludes to some of the many successes associated with the idea of random mutations that can be witnessed in virtually any hospital in the world.

More specifically, Professor Miller refers to the increasing number of pathogens that are exhibiting resistance to an expanding list of antibiotics. He feels that antibiotic-resistant bacteria demonstrate that, indeed, random mutations are capable of conferring evolutionary benefits ... at least for the bacteria that acquire that sort of resistance.

He extends the foregoing argument by providing an overview for a set of experiments that led to results that Professor Miller believes lend considerable force to the importance of random mutations in the process of evolution. The experiments in question were conducted by a group of scientists at Harvard Medical School that operated under the supervision of E.C.C. Lin.

The researchers were interested in trying to discover what makes proteins vulnerable to damage from the presence of oxygen. To explore this issue they set up conditions in their laboratory that they felt might favor the emergence of a strain of proteins that would be able to thrive to some extent – or, at least, hold its own – in an oxygen-rich environment.

Dr. Lin's experiment selected a form of bacteria that had the capacity to produce a certain kind of enzyme during the process of fermentation that takes place in the absence of oxygen. The longer version of the protein's name is: L-1,2-Propanediol:NAD^+ 1-oxidoreductase, and its shortened name is ProNADO.

Since ProNADO goes about its work in the absence of oxygen, one is not entirely surprised to discover that the protein is susceptible to damage when oxygen is present. Despite that vulnerability, the researchers placed the selected bacteria in an oxygen rich environment and, then, attempted to induce the bacteria to grow by adding a relatively small organic molecule, propanediol, to the culture as a source of food.

In order for the foregoing bacteria to be able to exploit propanediol as a source of nourishment, the ProNADO protein had to be present to help convert the small organic compound into molecular forms that could be metabolized. However, since the researchers didn't have a clue about how proteins might acquire the capacity to resist the presence of oxygen, the Lin-led scientists were not in a position to be able to construct an oxygen-resistant protein.

Consequently, they decided to explore the possibility that unguided evolution might be able to provide a solution to the experimental problem. As a result, they just let the experiment continue to run.

After a certain amount of time passed (the time required for nearly several hundred generations of bacterial reproduction to occur), the researchers discovered that mutant forms of the organism had emerged in which the gene coding for ProNADO had been turned on permanently. This occurred during two separate experiments.

In one of the two foregoing experiments, the ProNADO enzyme enhanced its ability to resist the presence of oxygen by roughly 40% during the period (approximately 40 minutes) in which, normally speaking, ProNADO becomes completely oxidized. During a second experiment, mutant bacteria also emerged that exhibited an enhanced capacity (approximately 60%) with respect to its ability to resist the presence of oxygen over the period of time (40 minutes) in which ProNADO is normally totally oxidized.

Following their experiment, the scientists analyzed the genetic structure of the mutant genes. They discovered that the changes that made the foregoing enhanced resistance capacities possible were quite simple.

In one instance, the seventh amino acid in the ProNADO protein chain was changed from the amino acid, isoleucine, to another amino acid, leucine. While during the other successful experiment the amino acid was changed from leucine to valine in the eighth amino acid of the chain of amino acids that constituted the mutant ProNADO protein.

Furthermore, at a certain point in the experiment, the researchers genetically modified the strain of bacteria being employed by inserting both of the foregoing changes into the genome of the target organism. When this took place, the capacity of the mutant form of ProNADO to resist the presence of oxygen for 40 minutes increased by another 20%, reaching 80% effectiveness.

Professor Miller claims that the foregoing set of experiments demonstrates the potential of random mutational events to be able to provide functional ways of overcoming previously existing evolutionary obstacles. Apparently, undirected evolution had generated a solution that had escaped the grasp of a number of scientists.

To begin with, the fact that an organism has the capacity to produce variants that exhibit new properties under certain circumstances does not necessarily mean that all changes that have been observed to occur in organisms across more than three billion years have come about in a similar way. There is a considerable amount of inductive reasoning that is present in evolutionary theory.

Consequently, in order for Professor Miller's foregoing point to be considered viable, then virtually every aspect of evolutionary change must be a function of similar sorts of random events. Yet, if even one instance of change is not the result of those sorts of random processes, then, Dr. Miller's argument becomes vulnerable to the problematic ramifications that ensue from the presence of what amounts to black swan-like events.

Secondly, Professor Miller has not eliminated the possibility that the genetic dynamics that gave rise to mutant forms of the ProNADO enzyme during the Linn experiments might have been a function of some sort of non-random set of events. For example, he appears to be assuming that the processes of change taking place in the genomes of bacteria are devoid of intelligence and, therefore, he seems to believe that bacteria, and, perhaps, other forms of life as well, are incapable – at least within certain degrees of freedom – of being able to run experiments that bring about alterations in DNA and, thereby, make possible different arrangements of amino acids.

When Gary Kasparov played against Deep Blue, he asked for permission to examine the algorithms that regulated the computer's chess playing strategies because he was having difficulty figuring out the reasoning process that directed the computer's chess moves. The sought for permission was denied, and, as a result, the computer's logic and "thought" processes remained alien to Kasparov.

Conceivably, there might be some mode of intelligence present in biological organisms that, among other things, enables those life forms – at least within certain degrees of freedom – to explore alternative ways of engaging, or interacting with, existence by means of altered genetic programming. If this is the case, then, one can't automatically assume that the intelligence being employed will necessarily make sense to human beings ... as Deep Blue's mode of "intelligence" remained alien to Gary Kasparov, so too, the modality of "intelligence" that could be present in life and, among other things, might be responsible for enabling genetic programming to be modified in various ways – but not necessarily indefinitely – could very well continue to seem alien to human beings.

Furthermore, if such a form of intelligence were present, it wouldn't necessarily "wish" to disclose its presence. As was the case with respect to the operators of Deep Blue, that kind of information might be considered proprietary, or that kind of information could be treated as crucial to biological integrity and, therefore camouflaged in some manner to protect its continued viability.

For example, an alien-like form of intelligence could hide in plain sight by cloaking itself within an array of data points that concealed the presence of that intelligence. As such, a variety of genetic alternations could take place, but only some of them -- not enough to register as being statistically significant -- might give expression to intelligent behavior, and, therefore, statistically speaking, one would not be able to distinguish intelligent behavior from random-like events.

If the foregoing sort of genetic intelligence were present in different forms of life, several possibilities might account for its presence. On the one hand, that intelligence might have emerged at some point during evolutionary history – in a way and form that, currently, is unknown -- and begun to introduce directed changes into the genome in order to generate variations that were capable of probing different aspects of the environment in the search for effective ways to engage various ontological contingencies and, as a result, become likely candidates for survival through the forces of natural selection.

Alternatively, God could have been responsible for the changes that Professor Miller considers to be random. Although this might mot be the case, Dr. Miller really has no non-arbitrary way to rule out that kind of a possibility.

Moreover, a principle akin to what baseball commentators often refer to as "defensive indifference" might be applicable here. In other words, just as a baseball team that is winning a game by a substantial margin sometimes will permit base runners from the opposing side to steal second or third base without trying to throw the runner out because the advance of those runners is not likely to change the outcome of the game, so too, God might permit organisms to experiment within certain limits because those sorts of changes would not appreciably affect the outcome of the game of life.

Professor Miller observes, in passing (bottom of page 102 in *Finding Darwin's God*) that 99 % of God's creations have become extinct. He uses that statistic to suggest that if God is the one who is serving as an allegedly intelligent designer, then, God would not appear to be all that adept at creating the sort of "perfect" organisms that could prove themselves to be capable of evading extinction as individuals – such as Professor Miller – might anticipate in conjunction with what he seems to suppose an all-powerful, omniscient God could and would do.

Once again, Dr. Miller is letting his ignorance or biases get in the way of being able to grasp certain possibilities ... possibilities that are not necessarily true but possibilities that are logically present in the situation being explored. More specifically, he appears to be presuming that if God were the One Who is creating life forms, then, surely, God would create perfect organisms – the kind that never need to change and that are perfect in their operations and, as a result, will never become extinct.

However, what if God's purpose for creating life was not to create perfect life forms but, rather, God's purpose was to create life forms that were perfectly suited to serve some other purpose of creation. For example, suppose, for unknown reasons, God wanted to demonstrate to all organisms that came into existence that life was ephemeral, risky, and contingent on conditions that were, in many respects, beyond their control, then, creating organisms that -- in the short term or the long term – would become extinct might be one way of helping the foregoing purpose to become clear.

In any event, irrespective of whether the processes governing change within the genome of an organism are intelligent or random, one must be able to show that there are no limits to what changes are possible at any point in time if one is going to try to argue that the changes that have taken place among organisms during more than three billion years of transitions were a function of the capacity – whether directed or not – of life forms to produce the sorts of changes that would enable those organisms to generate all domains, kingdoms, phyla, classes, orders, families, genera, and species that have been observed in fossilized or non-fossilized forms. Less abstractly, Professor Miller has not, yet, demonstrated that the sorts of changes that took place in the previously discussed Lin experiment are typical – in conjunction with either the property of simplicity or in terms of kind – with respect to all of the changes that have occurred in life forms across more than three billion years.

In other words, one can acknowledge the findings of the Lin experiment without simultaneously being forced to commit oneself to the idea that the foregoing research proves that all of life – from first protocells to human beings – acquired novel proteins, metabolic pathways, organelles, structural features, and so on through the same sort of process – that is, via random mutations (i.e., descent with small modifications). The Lin experiment only demonstrates that under certain circumstances, slight changes in the genome of various organisms can lead to the emergence of new kinds of functionality, but one should note that just because those sorts of small changes can occur, this says absolutely nothing (i.e., neither ruling the possibility out nor ruling the possibility in) about whether, or not, there might be some underlying intelligence that is responsible for those changes, and, therefore leaves open the possibility that the changes which have occurred with respect to life on Earth are not necessarily random in nature.

Contrary to what Professor Miller claims, the Lin experiment does not show that evolution works. What that experiment demonstrates is that genetic changes do, sometimes, occur within organisms and, sometimes, those changes are beneficial in nature.

The Lin experiment is consistent with the idea that random mutations have the capacity to lead to constructive changes in genetic coding and, as such, could become, under certain circumstances, a driving force for the emergence of novel forms of variation that are capable of shaping the properties of a given population of organisms and, therefore, affect (i.e., be evolutionary in nature) the way the forces of natural selection engage that population. However, the Lin experiment is also consistent with the possibility that God or some other form of intelligence has the capacity to serve as a non-random driving force for the emergence of an array of novel forms of variation (not necessarily indefinite in nature) that affect the way the forces of natural selection engage the population of organisms exhibiting those changes.

One might note in passing that Escherichia coli, a form of bacteria, consists of 4,639,221 letters written in the script of DNA. The Lin experiment involves only a few of those letters, and one can't help but wonder about how the strain of bacteria that was used in the Lin experiment might have come into existence in the first place ... that is, how did the millions of other DNA letters that comprise that bacteria become organized to give rise to a functional life form consisting of nearly 4,300 genes.

The Lin experiment is intriguing due to the way in which it demonstrates the sort of wiggle room that is potentially present in life. Nevertheless, those kinds of degrees of freedom are truly miniscule – almost to the point of vanishing -- when compared to the complexities embodied by the task of trying to account for how even a relatively simple organism like E. coli – or any bacteria with thousands of genes – first came into existence.

The potential for change that exists in a fully functional life form like the one in the Lin experiment is possible precisely because of the degrees of freedom that appear to be built into that life form (e.g., the capacity of the ProNADO enzyme to be turned on or off) through the presence of a complex, integrated set of feedback mechanism to which thousands of genes contribute. Being able to account for a change of functioning with respect to the foregoing sort of mature genetic system is – as the Lin experiment shows – a challenge, but it pales in comparison to the challenge of having to explain in a step-by-step fashion how that kind of bacteria first became possible or how the precursors for that form of bacteria became possible, and, quite frankly, the Lin experiment really doesn't have much to offer with respect to illuminating the latter sort of issue except as an extremely vague, iffy, and speculative suggestion.

Punctuated Equilibrium

Let's pursue another line of inquiry concerning evolutionary theory that might help to complement the Lin research. This involves the idea of punctuated equilibrium.

Punctuated equilibrium is a theory first introduced by Stephen Jay Gould and Niles Eldredge in a 1972 paper that raised questions concerning the viability of a neo-Darwinian model of change rooted in the notion that evolution was a function of the gradual accumulation of relatively small modifications over long periods of time. More specifically, Gould and Eldredge pointed to a set of facts that is commonly acknowledged among evolutionary biologists – namely, the fossil record seemed to be characterized by long periods of stasis interspersed with periods encompassing the relatively sudden disappearance of life forms that were, in turn, followed by the instantiation of new species in the fossil record that were both substantially different from the organisms that had disappeared earlier but were, as well, clearly related to those same, extinct life forms.

In short, according to Gould and Eldredge, the available fossil evidence seemed to run contrary to the idea that life had evolved by means of a continuous accumulation of small, gradual modifications over time. Consequently, they proposed the idea of punctuated equilibrium that, instead, called for long periods of relative, evolutionary quiescence (relative equilibrium) followed by fairly rapid periods of speciation (punctuated activity) ... trends they believed were reflected in the fossil record.

The higher rates of evolutionary change implied by the theory of punctuated equilibrium seemed to be more consistent with the fossil record than were neo-Darwinian models which maintained that evolutionary change was very gradual in nature. Therefore, the problem that the idea of punctuated equilibrium poses for traditional evolutionary theory requires scientists to be able to reconcile the traditional notion of Darwinian gradualism (relative stasis or equilibrium involving slight changes) with what seemed to be the sudden emergence of new species (punctuated activity) through, possibly, non-Darwinian (non-gradual) means.

Despite various concerns, evolutionary biologists began to warm up to the theory of punctuated equilibrium throughout the next several decades. However, despite its ascendancy, the issue remained controversial because no one seemed to understand the precise nature of the dynamics that might be at the heart of the process of punctuated equilibrium.

Enter, stage right, David Reznick. During a discussion exploring the possible meanings and significance of terms such as "microevolution" and "macroevolution", Professor Miller describes the 1981 guppy (Poecilia reticulata) research of David Reznick -- a biologist – that according to Dr. Miller carries some important implications for evolutionary theory and the idea of punctuated equilibrium.

In the wild – that is, beyond the protective walls of a home, aquarium system -- guppies have to deal with the existence of predatory enemies just as many other life forms are required to do. In Trinidad, one of the enemies that guppies encounter is the cichlid, a perch-like tropical fish that seems to have a bloodlust-like hunger for the flesh of guppies.

Reznick noted that cichlids were wreaking havoc on a population of guppies that swam about in a pool beneath a certain waterfall. He and his fellow researchers decided to move some of the guppies in that pool to another upstream location that was situated at some distance from the pool below the waterfall.

Once the transfer was complete and the guppies had settled into their new home, Reznick began observing different facets of the lifecycle of the newly protected population. Those observations continued for 11 years.

Among the findings noted by Reznick as he began comparing the guppies that remained in the original, predator-laden pool with the guppies that had been moved to an upstream location free of predators, were differences involving relative sizes and the amount of time required to reach sexual maturity. The upstream guppies took longer to reach sexual maturity than did the downstream guppies, and, in addition, at sexual maturity the upstream guppies were larger than the downstream guppies.

The foregoing considerations possess potential evolutionary significance. More specifically, since the number of eggs that a female guppy can produce increases with size, anything that permits female guppies to experience a longer and greater growth period prior to reaching sexual maturity tends to increase the likelihood that female guppies will be able to produce more eggs and, therefore, this property enhances the chances of those fish to be able to leave behind an increased number of offspring.

The foregoing changes were considered to be heritable changes. In order to establish a baseline for comparison purposes with respect to quantitative changes involving the aforementioned features, the researchers also raised guppies under laboratory conditions.

Reznick and his colleagues were interested in studying the rate of genetic change among guppies that were being raised under different conditions. Evolutionary biologists use the metric of "darwins" to measure the rate of change in a given physical feature over time.

A 'darwin' refers to heritable properties that change by a factor of 2.718 over a million-year period. Reznick applied that metric to various guppy properties.

The Reznick researchers recorded changes to members of different guppy populations that were located in several streams in Trinidad as well as a population of guppies that were being maintained under laboratory-controlled conditions. Those measurements ranged between 3,700 and 45,000 darwins.

One doesn't begin to understand the potential significance of the foregoing results for evolutionary theory until one compares them with the rates of change that appear to characterize various fossil remains. For example, consider ceratopsian dinosaurs.

Ceratopsian dinosaurs were a quadrupedal herbivore that existed during the Cretaceous period and are considered by evolutionary biologists to constitute a group of organisms that are characterized by a fairly rapid rate of evolutionary change. When one measures the differences in size among various fossilized representatives from that group, one comes up with a figure of 0.06 darwins.

Depending on the guppy properties being measured, Reznick and his colleagues were observing rates of changes in guppies that were running between 10,000 and 10,000,000 times higher than the rate of change that could be observed in conjunction with the fossil record left by ceratopsian dinosaurs ... organisms that were considered to be characterized by, relatively speaking, a fairly high rate of evolutionary change. Reznick and his fellow researchers concluded that under the right set of circumstances the rate of change exhibited by guppies in the wild was more than capable of accounting for what seemed to be taking place in the fossil record.

Professor Miller broadens the thrust of the foregoing argument by referring to a variety of other studies involving lizards and a number of birds (e.g., sparrows and finches). According to Dr. Miller those studies also entailed a relatively rapid rate of genetic change that was at least a thousand times faster than the rates of change required by evolutionary mechanisms such as punctuated equilibrium.

The aforementioned research concerning guppy life in Trinidad, along with the additional rate-studies mentioned by Professor Miller that were mentioned above, appear to offer a way out of the previously noted dilemma that punctuated equilibrium supposedly poses for neo-Darwinian models concerning evolutionary change. In other words, the foregoing research – involving, among others, the work of Lin as well as that of Reznick -- appears to indicate that given the right set of circumstances, microevolutionary events consisting of random mutations are capable of underwriting both gradual and rapid rates of evolutionary change.

Punctuated equilibrium can be understood to involve either, on the one hand, a process of <u>instantaneous origination</u> followed by periods of relative stasis or, on the other hand, a process consisting of the <u>relatively rapid accumulation of slight changes</u> followed by periods of relative stasis. The Reznick guppy research, together with similar work involving finches, sparrows, and lizards, tend to provide a way to account for abrupt evolutionary change in terms of the rapid accumulation of slight changes under favorable conditions.

As such, evolutionary change would continue to be Darwinian in nature. That is, although evolutionary change still would consist of a process involving a continuous series of slight modifications, nonetheless, sometimes, under the right sort of circumstances (as the Reznick research demonstrated), such a series of slight modifications might be able to accumulate fairly quickly within a given population and, thereby, generate a rate of evolutionary change that was capable of being reconciled with the periods of rapid speciation that were entailed by the notion of punctuated equilibrium.

Professor Miller feels the foregoing considerations have the capacity to rebut those who try to argue that evolution lacks a proven mechanism capable of subsidizing change or that such a mechanism would be too slow to be able to reconcile the data from fossils which indicates that both long period of stasis and rapid periods of speciation are juxtaposed next to one another. Furthermore, according to Dr. Miller, the foregoing mechanism of evolutionary change is not theoretical but is real and, consequently, can be observed to be taking place throughout the sort of research that had been conducted by, among others, David Reznick and his colleagues.

The darwin is a measure that seeks to map the changes that take place in conjunction with certain physical properties over a given period of time. To utilize such a measure, one requires a baseline against which to compare changes for various populations.

By its very nature, the darwin is a metric that does not appear to lend itself well to being able to serve as a measure for, or establishing an index of some kind for, novel properties ... that is, properties that emerge but do not constitute variations on already existing themes. For instance, suppose one were considering various differences between certain kinds of anaerobic and aerobic organisms, and noted that at a certain point in biological history most primitive forms of anaerobic life forms didn't possess the capacity to generate the enzyme ProNADO (first mentioned in conjunction with Lin's research approximately 11 pages ago), but later on, there were some organisms – although not necessarily fully anaerobic – that did possess the capacity to produce ProNADO and, therefore, might be able to function effectively in the presence of oxygen under certain circumstances (e.g., when the ProNADO gene is switched on, as well as when the seventh and eight amino acids in the sequence of DNA molecules coding for ProNADO called for certain amino acids – e.g., isoleucine or valine – to be transcribed rather than others – e.g., leucine).

One question that might arise in conjunction with the foregoing scenario is fairly obvious. What were the specific steps that permitted organisms to traverse the genetic distance from a life form that did not have the capacity to generate ProNADO (and I do not mean that this capacity was present but just not switched on) to an organism that did possess the ability to produce ProNADO in a form that permitted the organism to function in the presence of oxygen?

There is no readily identifiable base line against which to measure the foregoing sort of transition. At one point in time, the coding for ProNADO did not exist, and at another point in time, the coding for ProNADO did exist, but, unfortunately, one has no set of starting reference points that permit one to determine the rate of change for the emergence of the ProNADO molecule over time ... at least not in terms of darwins.

Of course, if one were able to demonstrate that the coding for the ProNADO enzyme was traceable to changes in a certain sequence of DNA molecules that originally was non-functional but, over time and after numerous mutations, came to constitute the coding for the ProNADO protein, then, one could set the original non-functional sequence of DNA molecules as a base line and calculate the darwins that measured the rate of change among those DNA molecules that were marking the transition from organisms without the capacity to produce ProNADO to organisms that did possess that ability. However, to date, no one has come up with a viable way to be able to set a base line in the foregoing manner and, thereby, provide a means of calculating the number of darwins, or the rate of change, that are associated with the aforementioned sorts of transitions in DNA molecules over time.

Consequently, if one wanted to establish a measure for the rate of change with respect to circumstances in which one lacked a means of establishing a concrete base line of comparison for calculating the rate of change of a given property over time, then, using the darwin-metric in such a situation seems to be problematic. Therefore, although the Reznick research might provide a way for calculating the rate of change with respect to certain properties (e.g., size, age of sexual maturity, and so on) it does not necessarily offer a means of mapping the rate of change in conjunction with the emergence of new properties (e.g., an enzyme that did not previously exist).

Darwins are a metric that measure rates of change for properties that can be clearly defined. These changes are quantitative in nature and -- due to the absence of fixed reference points that can be used as a baseline -- do not seem to be capable of capturing the rate of change for newly emergent qualitative changes in cellular organization, metabolic pathways, organelles, structural features, behavioral capabilities, enzyme function, and the like.

The guppy research conducted by David Reznick and his colleagues indicates that the raw material – i.e., variation -- for potential speciation is clearly present in some populations. His work demonstrates that the rate of change for an array of properties is sufficiently fast that such a rate is capable of accounting for some forms of abrupt speciation that are entailed by the fossil record.

Therefore, Reznick's research provides a provisional means for reconciling the idea of punctuated equilibrium with the contention of Darwin – as well as many other evolutionary biologists – that the nature of evolution or descent with modification takes place through the accumulation of small changes. This is accomplished by demonstrating that there are circumstances (such as the guppies that were moved to a location in Trinidad that was relatively predator free) when the accumulation of small changes can take place very quickly, and, consequently, a <u>potential</u> for rapid speciation is present.

A high rate of change, as measured by darwins, indicates that certain features of a given population are undergoing a series of transitions on a fairly regular basis and, as a result, this constitutes a possible driving force for at least some forms of rapid speciation (more on this shortly) that are suggested by the fossil record (i.e., the dimension of punctuated equilibrium that involves rapid speciation). Nevertheless, despite a high, overall rate of change, the individual changes that actually occur are small, and those small changes do slowly accumulate over time and broaden the variation being expressed by a given population and, thereby, serve to anchor the population's tendency to remain – in evolutionary terms -- fairly inactive over long periods of time (the quiescent dimension of punctuated equilibrium) until a point in reached when some mutant variant in the population begins to assert its capacity to better exploit existing conditions and, therefore, becomes ascendant while other variants become extinct.

However, Reznick's research only applies to certain kinds of speciation. His work is applicable to circumstances in which there arise variant forms of existing properties (such as the length and shape of a bird's beaks, as was the case with Darwin's finches) that have the capacity to exploit a given ecological context in ways that are different from, or more effective than, what previously had taken place and, consequently, over time, might have led to the appearance of a form of life that, among other things, is no longer capable of interbreeding with similar life forms and, therefore, is said to have undergone the process of speciation.

Nonetheless, instances of speciation that involve the emergence of qualitatively novel features (such as the previously discussed ProNADO enzyme) that are not merely a matter of mutant variations involving already existing properties within a given population do not seem to be subsumable under, or entailed by, Reznick's aforementioned research. Even if the rate of change for a given property consists of a very high number of darwins, this does not necessarily establish a basis for being able to account for the emergence of truly novel features in a given population ... features such as: When a capacity for life arose, even though, earlier, only abiotic activity took place; or, when a capacity for chemosynthesis first became established after no such capacity existed previously; or, when the capacity for photosynthesis emerged even though no such prior capacity existed; or, when the capacity for multicellular life became possible, even though no such ability existed earlier in time; or, when the capacity for a eukaryotic mode of life arose despite the fact that only prokaryotic forms of life had existed heretofore.

In addition, one should keep the following point in mind. Just because Reznick's research offers a possible means of reconciling (a) some forms of rapid speciation that seem to be present in the fossil record with (b) the Darwinian principle that evolution involves the accumulation of small changes over time, nonetheless, establishing that something is possible is not necessarily the same thing as proving that what is possible is, in fact, what took place.

The rate of change that is measured by the darwin metric shows not only what has occurred in a given set of circumstances, but, as well, suggests what might have happened in other circumstances as well. However, this is nothing more than a suggestion until evidence is forthcoming which is capable of verifying that what took place in a different set of circumstances seems to be governed by a process similar to what transpired in conjunction with the guppy research of Reznick and his colleagues.

The same sort of point can be made in relation to the notion of mutation rate, a more generalized and amorphous way to give expression to the rate of change idea. In other words, even if one can establish that under certain circumstances, a given population appears to be characterized by a high rate of mutation, this does not, thereby, demonstrate that a particular series of mutations actually occurred but, instead, serves only to establish the possibility that certain kinds of mutation might have occurred.

The fewer the number of mutations that are needed to be able to account for some evolutionary transition, then, within certain limits, the more plausible such an account will be. The greater the number of mutations that are needed to be able to account for some evolutionary transition, then, the less plausible that sort of account tends to become.

At a certain point during his critical exploration concerning various aspects of the Eldredge/Gould theory of punctuated equilibrium, Professor Miller asks the question: "Do new species appear so suddenly that they require a mechanism above and beyond the ordinary processes known to take place in genetics and molecular biology?" Following an analysis of some research that had been conducted by Stephen Jay Gould in conjunction with several other scientists, Dr. Miller answers that question in the negative.

More specifically, Dr. Miller states that *Cerion* -- a land snail found on the Bahamian island of Great Inagua -- had been the focus of a fair amount of research by Gould. Professor Gould, along with David Woodruff, compiled an array of *Cerion* shells, and on the basis of the fossil evidence that had been left behind, the two researchers concluded that fairly recently – geologically speaking -- the land snail had undergone speciation.

Some fossil deposits discovered by Gould and Woodruff showed a total absence of *Cerion rubicundum* (which exists today) and, instead, those deposits were made up entirely of remnants left from the largest form of *Cerion* – namely, *Cerion excelsior* – that has now become extinct. The fossils collected by Gould and Woodruff appeared to establish a set of transitions in form (e.g., involving the size and shape of shells) that extended smoothly from *C. excelsior* to *C. rubicundum*.

The appearance of smooth changes in shell size and shape that seemed to mark the transition from the soon-to-be extinct edition of *Cerion* (i.e., *excelsior*) to its extant, modern counterpart (i.e., *rubicundum*) was substantiated through the findings of additional research that subsequently was carried out quite a few years later by Gould and a geochemist, Glenn Goodfriend. Using: (1) A methodological technique of analysis that measures the changes that take place over a period of time in the properties of amino acids found in the shells of *Cerion* organisms, as well as (2) radiocarbon dating technology, Gould and Goodfriend were able to establish a rigorously empirical chronological order for Gould's original collection of fossils ... an order which demonstrated that the transition from the older, extinct species (i.e., *C. excelsior*) to its modern, existing descendent (i.e., *C. rubicundum*) constituted a series of smooth transitions in the size and shape of fossilized shells.

Professor Miller concludes that the foregoing research demonstrates that modern species do not arise in a magical puff of smoke. Rather, he maintains that the Gould/Goodfriend research illustrates how species (in compliance with Darwinian principles) arise as the result of gradual, smooth changes in a population until, at some point (due to various genetic and environmental contingencies), the population bifurcates to give rise to several forms of species ... one kind of organism that continues on into the future, as well as another life form that often – but not always -- ends up becoming extinct.

From a geological perspective (in terms of fossil deposits), the organism that survives might seem to have emerged abruptly because it is found in deposits that do not contain the extinct form of the organism (or vice versa). Nevertheless, when one applies various kinds of analytic methodology to the available data (as Gould and Goodfriend did in their research), the appearance of abruptness is placed in geological perspective, and, as a result, that which seemed, on the surface, to be abrupt is found, instead, to operate in accordance with Darwinian principles that call for the accumulation of small, gradual changes over time (i.e., descent with modification).

However, as was pointed out earlier in conjunction with the Reznick guppy research, the Gould/Goodfriend findings do not necessarily justify Professor Miller's belief that new species always – and necessarily -- operate in accordance with the natural processes that are described by genetics and molecular biology, and, therefore, do not emerge abruptly. Although speciation might occur fairly smoothly in conjunction with changes that are reflected in relatively easily identifiable properties such as the size and shape of a snail shell, nonetheless, Gould's and Goodfirend's research doesn't necessarily carry any implications for cases that are characterized by certain kinds of novel properties (e.g., aerobic capabilities rather than anaerobic capabilities, or chemotropic properties instead of abiotic characteristics, or eukaryotic qualities rather than prokaryotic feature, or photosynthetic capabilities instead of chemotropic abilities, or multicellular features rather than single-cell properties).

Even in terms of the emergence of new species that might be explicable by means of Darwinian principles, there are still a number of issues that need to be addressed. For example, were the changes in DNA coding that led to differences in the size and shape of *Cerion* shells simple or complex?

The simpler those changes in DNA coding are, the more plausible a Darwinian-based explanation of speciation becomes. The more complex those changes in DNA coding are, then, depending on the degree of complexity that is involved, the less plausible a Darwinian-based explanation becomes ... not because those changes weren't small and gradual but, rather, because the whole sequence of changes that are necessary to bring about transitions in certain properties (such as the size and shape of a snail shell) becomes increasingly improbable as the number of the changes begins to increase in order for phenotypic properties to become manifest.

The foregoing considerations lead to a second issue. Can one presume that changes in the size and shape of, say, a snail shell were the only modifications that led to a bifurcation of species?

Conceivably, changes in shell size and shape might, in and of themselves, interfere with the ability of organisms -- that, ostensibly, seem to be members of the same species – to be able to procreate. If so, then, one has a readily available means for explaining why organisms from the same species reach a point in which they no longer can procreate because of the way in which the size and shape of their shells interfere with that process.

On the other hand, if the size and shape of, for example, the shells of some species do not constitute a barrier to procreation, one must begin to look elsewhere to discover why two organisms that once, supposedly, were part of the same species are no longer able to either procreate or create fertile offspring. This possibility would seem to suggest there could be a variety of other changes to the DNA coding that accompanied changes in the size and shape of shells that might be more responsible for the breakdown in the capacity to interbreed that leads to the bifurcation of a species -- a process that is an integral part of the phenomenon of speciation -- and, if so, then, this sort of possibility would seem to indicate that the number of changes necessary for speciation to occur goes beyond the modifications that underwrite changes in the shape and size of shells.

If the latter sorts of additional changes in DNA coding are relatively few in number (i.e., beyond that which is needed to account for changes involving, say, the size and shape of shells), then, the plausibility of the Darwinian model remains relatively unaffected. However, if the number – and, perhaps, complexity -- of changes in DNA coding that are associated with the breakdown of interbreeding begin to climb, then, so too, will the number of questions concerning the precise nature of the speciation process and whether, or not, one can adequately and plausibly account for those changes in purely Darwinian terms ... that is, as a function of the random accumulation of small modifications over time.

Another way of engaging the foregoing issue is to raise a question concerning the nature of punctuated equilibrium. How punctuated must a change in equilibrium be before it no longer readily lends itself to a Darwinian model of gradual descent by means of small modifications?

On the one hand, a person has little difficulty in grasping how the sorts of changes documented by Lin (involving the enzyme ProNADO), Reznick and his colleagues (guppies), or Stephen Jay Gould and Glenn Goodfriend (*Cerion* land snails) could be understood to be instances of punctuated activity that are fully in line with Darwinian principles of descent with modification. On the other hand, an individual has considerably more difficulty understanding how the sorts of changes that are involved in the transitions: From abiotic systems to living protocells, or from chemotropic organisms to photosynthetic life forms, or from anaerobic to aerobic organisms, or from bacteria to archaea (or vice versa), or from prokaryotic to eukaryotic life forms, or from single-celled organisms to metazoans, or any number of other transitions that differentiate domains, kingdoms, phyla, classes, orders, families, and genera from one another (and, therefore, do not exhibit just a few sequences of DNA that are unlike one another) constitute instances of punctuated activity that can be demonstrated to be completely reconcilable with Darwinian principles of descent due to the accumulation of small modifications over time.

Intermediate Presumptions

At a certain point in his defense of various facets of evolutionary theory, Professor Miller explores several possibilities concerning events that might have led to the emergence of vertebrate life forms. For instance, many evolutionary biologists believe that *rhipidistians*, a freshwater fish that swam about in the waters of the Upper Devonian period approximately 350 million years ago, were an early ancestor of four-footed land vertebrates or tetrapods.

Dr. Miller attempts to fill in some of the transitional history that supposedly leads from rhipidistians to tetrapods by citing several scientific discoveries. One finding took place in 1991, while the other discovery occurred in 1998.

Several British researchers made the earlier of the two foregoing discoveries. Those individuals came across a well-preserved fossilized specimen of *Acanthostega gunnari*, a fish-like tetrapod, containing some intriguing details.

For instance, the fossil revealed that the fish-like tetrapod possessed an internal system of gills. This was significant since no other known amphibian possessed such a gill system.

Acanthostega gunnari was capable of breathing underwater by means of its gills. The organism also was able to breathe on land via its lungs.

The fossil exhibited features that combined properties of both fish and land vertebrates. As a result, Professor Miller considers the fish-like tetrapod to constitute an intermediate form that helps mark the evolutionary transition between fish and land vertebrates.

There is a substantial lacuna inherent in the foregoing perspective. More specifically, if one assumes that the gills in *Acanthostega gunnari* were inherited from, say, *rhipidistians* or their ancestors, one also needs to be provided with a step-by-step account that indicates how the lung system arose in land vertebrates.

Notwithstanding the fact that if one pushes the issue of origins sufficiently far back in evolutionary history, one is going to have to provide a step-by-step account of how gills first came into being, for the moment one might be willing to entertain the idea that the fish-like aspects of *Acanthostega gunnari* – i.e., the gills – could have a plausible explanation since one can claim – as evolutionary biologists do – that the capacity for gills was inherited from the line of ancestors extending back to *rhipidistians*. However, what is land-like in *Acanthostega gunnari* – i.e., the lungs – seems to lack a comparable backstory.

Acanthostega gunnari might very well constitute an intermediate life form that fills the transitional space between fish and land vertebrates. Unfortunately, Professor Miller has failed to provide a step-by-step account for how that intermediate form came into being ... especially in conjunction with the origins of the lung component of that so-called intermediary, but, in addition, lurking in the background, there are some unanswered questions concerning the origins of a gill system that is capable of extracting oxygen from water, and, then incorporating that oxygen into metabolic pathways that generate energy or eliminate waste materials during oxidation reduction cycles.

Professor Miller seeks to bolster the evolutionary account that purports to link rhipidistians and tetrapods by citing a 1998 discovery that occurred along a Pennsylvania roadside. Two paleontologists – Neil Shubin and Edward Daeschler – discovered a well-preserved, fossilized fish fin from the Devonian period (which, approximately, covers a period of time between 409 and 363 million years ago).

The foregoing fossil clearly revealed that the fish, from which the fin came, possessed eight, finger-like digits that are very similar to the digits that are found in land vertebrates or tetrapods. According to Dr. Miller, the 1998 discovery arises in precisely the right place (i.e., the region now known as Pennsylvania), at precisely the right time in evolutionary history (the Devonian period), and, as well, exhibits precisely the right set of characteristics (eight finger-like digits) that are needed to empirically document an evolutionary account capable of linking the group of fish known as rhipidistians with the first land vertebrates or tetrapods.

However, what Professor Miller does not provide is an explanation concerning how the aforementioned finger-like digits first came into existence, just as he did not explain how lungs (and, before that, gills) first came into existence. Instead, one is presented with so-called intermediary forms that lack a complete contextual history concerning the origins of all of their characteristics.

| Evolution Unredacted |

Dr. Miller seems to feel that the evolutionary history linking rhipidistians and land vertebrates is very evident. He believes he has demonstrated the existence of an unmistakable evolutionary pathway that is capable of marking, in a plausible manner, the transition from fish to fish-like amphibians to land vertebrates.

Nonetheless, there is a great deal that is missing from his overview. Among other things, as indicated earlier, he has not provided a step-by-step account that documents the origins of gills, lungs, and finger-like digits.

In the National Football League, there often are a variety of occasions during any given game in which a fair amount of time is spent examining a plethora of camera angles that can, if necessary, provide a frame by frame breakdown of a particular play in order to determine whether a player actually: Scored a touchdown, caught a pass, fumbled the ball, and so on. What Professor Miller does is comparable to handing referees a couple of snap shots and asking them to speculate about what happened before, between, and following those photos while simultaneously insisting that the referees evaluate the available evidence (i.e., the two snapshots) in accordance with the biases of evolutionary theory.

Questionable Facticity

Dr. Miller maintains that the most appropriate way to evaluate any objection that is directed toward evolutionary theory is to try to establish the extent to which those objections are rooted in scientific facts. The problem here is that the hen house of facts is being guarded by a group of very clever foxes (evolutionary scientists and researchers) who not only determine what hens (i.e., facts) are to be allowed into the chicken coop, but, as well, those individuals often spend a fair amount of energy trying to distance themselves from the mystery of the many hens that appear to be missing from the hen house (evolutionary theory) they claim to be guarding.

Science is missing a considerable number of facts concerning, among other things, the origins of: The genetic code, membranes, life, chemotropic organisms, photosynthesis, cyanobacteria, archaea, eukaryotes, metazoan life forms, aerobic organisms, the many kinds of glial, neuronal and other kinds of specialized cells that underwrite the activities of neuronal communication, immune defenses, circulatory capabilities, and respiratory functions ... not to mention (which I am mentioning) the origins of consciousness, intelligence, logic, language, emotion, talent, creativity, and so on.

Yet, one often hears from different biologists (e.g., Richard Dawkins and Kenneth Miller) that the evidence in favor of evolution is overwhelming. Apparently, one is supposed to take the evidence from discoveries such as those that took place in, say, 1991 and 1998 and, then, just concede that all of the many factual elements that are absent from those discoveries should be decided in favor of, and in conformity with, evolutionary theory.

The essential principle of evolutionary theory – namely, that <u>all</u> life forms are a function of the accumulation of small, random changes over time that are endorsed (selected) by a set of natural forces and conditions that give rise to differential rates of reproductive success – has never actually been proven. To be sure, there is a great deal of evidence to indicate that variation does occur in every population of organisms and that such variation does lead to differences in reproductive success as a result of the way that an array of natural forces and conditions interacts with the variants in those populations, but conceding the latter point does not force one to accept the parallel proposal that all manner of: Domains, kingdoms, phyla, classes, orders, families, genera, and species arose as a result of the accumulation of small, random changes over time that affected reproductive success rates.

There is considerable evidence capable of lending support to the idea that speciation might occur in a variety of circumstances, and some of that evidence has been reviewed in the previous pages of this chapter. Nonetheless, the aforementioned evidence, as considerable as it might be, is miniscule – i.e., it constitutes a very, very small sampling – relative to the millions of species that currently exist or have existed in the past.

At the present time, there are countless questions concerning the origin of various life forms – as well as questions concerning the origins of the metabolic pathways, proteins, organelles, behavioral characteristics, and specialized functioning to which those life forms give expression – that evolutionary theory cannot answer in concrete, definitive, step-by-step terms (as opposed to theoretical, speculative, and overly-general terms). Evolutionary biologists and most scientists seem to expect that everyone should be willing to give evolutionary theory the benefit of the doubt when it comes to all of the many unanswered questions concerning the origin puzzles that swirl about life and about its many, diverse capabilities, but there is nothing very scientific or reasonable about that sort of expectation.

If someone rejects evolutionary theory, Professor Miller would like those individuals to present an alternative scientific theory that is capable of fitting the available data more effectively than evolutionary theory is capable of accomplishing. Putting aside, for the moment, the fact that science is a game that Professor Miller likes to play and, therefore, one might note in passing how demanding – as Professor seems to be inclined to do -- that alternatives to evolution must be formatted in a manner that is acceptable to science seems rather self-serving and biased, nevertheless, one actually can provide Professor Miller with a scientific alternative that fits the facts better than evolutionary theory does while, simultaneously, avoiding its many problem.

More specifically, population biology is capable of accounting for all of the material that evolutionary biologists claim to be factual even as it refrains from potentially overstepping the available data by trying to claim that all new biological capabilities, functions, structures, metabolic pathways, and organizational wherewithal are a function of the accumulation of small, random modifications over time. When one combines the principles of population biology with, on the one hand, the idea of whatever variation arises through mutation (which need not be random and need not be all-inclusive), and, on the other hand, the forces and conditions that push and pull variation in directions that have differential reproductive success (i.e., natural selection), then, one can accommodate pretty much all of the facts that have been established through genetics and molecular biology without necessarily having to invoke the idea of evolution and all of the latter's philosophical baggage and problems.

The hermeneutical speculations (whether religious or evolutionary in nature) that attempt to account for the origin of life forms or the origins of the capabilities exhibited by life forms are merely unsupported hypotheses concerning those issues and, therefore, are not really all that compelling, rational, or scientific. Those sorts of speculations might take place within the context of a methodological framework that is rooted in the sciences that subsidize population biology, but they clearly are of a quality that is considerably less than factual in nature, and, therefore, at best, those sorts of speculations reside on the periphery – if not the fringes -- of science rather than at its center.

A hypothesis that is unsupported or factually challenged might be called a scientific hypothesis simply because it is spoken by a scientist or arises within the context of a scientific discussion. Nonetheless, such a hypothesis is substantially inferior to proposals that are rigorously supported by, and rooted in, an array of evidence.

For more than a 155 years, the hypothesis of evolution has been unable to establish in a step-by-step fashion that life or the origins of all domains, kingdoms, phyla, classes, orders, families, genera, and species (together with their many capabilities) are a function of the accumulation of small, random changes over time that are endorsed (selected) by a set of natural forces and conditions that give rise to differential rates of reproductive success. Consequently, the idea of evolution – as an account of how all life forms are a function of the accumulation of small, random changes over time that are endorsed (selected) by a set of natural forces and conditions that give rise to differential rates of reproductive success – should be recognized for the factually-challenged hypothesis that it is.

Professor Miller claims that: "If evolution is genuinely wrong, then we should not be able to find any examples of evolutionary change anywhere in the fossil record. (page 125 of: *Finding Darwin's God*)" I disagree.

One can find examples (e.g., the previously discussed material concerning: Lin in conjunction with ProNADO, Reznick's Trinidad guppies, the shells of Bahamian land snails analyzed by Gould and Goodfriend) that can be interpreted to constitute evidence (an interpretation that might, or might not, be correct) that evolutionary change might occur in certain circumstances … in other words, change that is a function of the accumulation of small, random changes over time that give rise to variants exhibiting differential forms of reproductive success. Nonetheless, the foregoing examples do not demonstrate that <u>all</u> changes taking place in life, domains, kingdoms, phyla, classes, orders, families, genera, and species are necessarily a function of the accumulation of small, random changes over time that give rise to variants exhibiting differential forms of reproductive success.

Evolution is an inductive argument. What occurs in some cases might not happen in other cases, and, as a result, one must resist the temptation to try to take the foregoing sorts of specific examples (i.e., Lin, Reznick, Gould) and (assuming them to be true) generalize their results to the millions of life forms that lived in the past or that populate the Earth now.

If one takes Professor Miller's foregoing words and alters them a little, one comes to the crux of the matter. Thus, one might paraphrase him and argue: If evolution is genuinely <u>right</u>, then we should <u>not</u> be able to find any examples of evolutionary change in the fossil record that might be inconsistent with the requirements of that theory.

However, as noted previously, there are numerous questions involving issues of origins that arise in conjunction with various kinds of life forms or their capabilities that cannot definitively be resolved by available evidence and, in fact, might never be capable of being properly resolved. Therefore, not only is Professor Miller incorrect when he claims that "if evolutionary theory is genuinely wrong, then we should not be able to find any examples of evolutionary change anywhere in the fossil record" (and he is wrong because there are examples that might constitute limited instances of evolutionary change without necessarily being able to take the principles underlying those examples and use them to account for all manner of change involving life forms), but, as well, Professor Miller might also be incorrect if he were to try to argue that if evolutionary theory is genuinely right, then we should not be able to find any examples of evolutionary change anywhere in the fossil record that do not comply with the requirements of evolutionary theory since, clearly, there are many instances of biological change in conjunction with the origins of life forms or their capacities that – at best -- remain uncertain as to whether evolution in a Darwinian sense is capable of demonstrating the precise nature of those changes.

Evolutionary theory does have the capacity to be able to offer a theoretical explanation that, purportedly, accounts for the foregoing sorts of events. However, the question that the evolutionary model cannot necessarily answer is whether, or not, its proffered explanations correctly describe what has taken place with respect to the emergence of new life forms or their capabilities.

Evolutionary theory is suggestive. Nonetheless, evolutionary theory is far from being definitive because it often lacks the step-by-step details that alone are capable of establishing the truth in relation to what is being suggested.

The Quality of Intelligence

According to Professor Miller, those who support a version of creationism known as "intelligent design" do God a great disservice by casting the Lord of the Universe in a role that seems akin to a clown prince who comes up with some idea for an organism, brings that life form to realization, and, then, relatively quickly, seems to feel that the original idea was not all that good and, as a result, decides to move in a different direction. On a number of occasions during *Finding Darwin's God*, Dr. Miller asks questions along the following lines: Namely, if the theory of intelligent design is correct, then why can't God – a supposedly intelligent designer -- get things right the first time? Why does God seem to be continually tinkering with creation and doing makeovers?

For instance, consider the case of proboscideans or elephant-like creatures. Approximately 35 million years ago, toward the beginning of the Oligocene epoch (during the Tertiary period), a life form known as *Paleomastodon* emerged on Earth that came equipped with an elephant-like trunk.

There also are some other differences that show up in the foregoing sort of elephant-like creatures. For example, those kinds of life forms exhibit skulls, jaws, and teeth that are different from other animals.

Some ten million years later, an organism with a similar form of trunk appeared when *Gomphotherium* began roaming about the African landscape. The latter creature might have been just a larger edition of the earlier Paleomastodon model, but it possessed, as well, a set of tusks that were not present in the earlier life form.

The foregoing trunk property also was present in two North American animals – namely, *Platybelodon* and *Deinotherium*. These fossils date from around the beginning of the Miocene epoch (which lasted from about 23.3 million years to 5.2 million years ago) that followed the Oligocene epoch.

Toward the latter part of the Miocene epoch, *Primelephas* emerges in Africa. This life form exhibits a trunk, as well as several tusks, that are very similar to what is possessed by larger species of modern elephants.

Beginning somewhere around 5.5 million years ago, Primelephas disappeared and was replaced by three other variations on the elephant-like theme. One of those three replacement life forms was the *Mammuthus*, or wooly mammoth, which became extinct, and the other two lineages of elephant-like creatures survived into the modern era and, eventually, assumed the form of *Loxodonta africana* and *Elephas maximus*.

Primelephas, *Mammuthus*, *Loxodonta africana*, and *Elephas maximus* constitute four genera. During the last six million years, 22 species have arisen in conjunction with those four genera.

Professor Miller maintains that proponents of intelligent design are required to argue that none of the foregoing 22 species have ancestral relationships with one another. Moreover, the existence of those 22 species induces Dr. Miller to wonder why an intelligent designer would try to create the impression that ancestral succession was taking place if – as, apparently, many proponents of intelligent design wish to argue – all 22 species constitute independent creations.

I have some familiarity with the arguments of intelligent design. At this point, I have no desire to either endorse their ideas or to reject them, but, instead, I would like to point out that neither I, nor anyone else, is obligated to follow their game plan.

For example, whatever the considerations might be (whether correct or incorrect) that underlie Professor Miller's claim that a system of intelligent design is <u>required</u> to maintain that, for instance, the 22 species of elephant-like creatures are necessarily ancestrally unrelated and must each constitute independent instances of creation, his argument is flawed because it fails to take into consideration various possibilities. For starters, there is really nothing preventing life from being able to exhibit a mixture of properties combining features of both ancestral dynamics as well as creational activity.

Existence does not have to be cast in the form of a zero-sum game in which either evolution is right or creation is right. Conceivably, there might be elements of each that – when properly understood – could be understood to be present in the life forms that have populated the Earth ... both currently as well as in the past.

Since elephant-like creatures emerged some 35 million years ago in the form of *Paleomastodon*, there have been a number of features associated with proboscideans that have been unlike properties found in other animals. Thus, previously noted characteristics such as: Skull, jaw, teeth, and trunk constitute novel features whose origins must be explained.

Evolutionary theory tends to gloss over this issue by alluding to the idea that mutations are responsible for those kinds of changes without ever actually providing evidence capable of demonstrating that the indicated modifications were, indeed, the result of a specific sequence of random mutations. Conceivably – that is, this is a logical possibility -- a set of random mutations brought about changes in DNA that gave rise to, among other things, a form of skull, jaw, teeth, and a trunk that had not appeared in any other previous form of animal, but until one can present clear evidence that this is what happened, then, the idea that random mutations generated those changes is nothing more than speculation.

Furthermore, even if one were to agree to the idea that proboscidean skulls, jaws, teeth, and trunks were due to a sequence of mutations, one cannot prove that such a sequence of mutations was random in nature. All one justifiably could say is that one does not know the precise nature of the mutational dynamics that led to the emergence of the aforementioned properties.

Professor Miller responds to the notion of a God that might have created various genera and species of elephant-like creatures by making a series of rhetorical-like remarks in different parts of *Finding Darwin's God* that seek to ridicule the sort of God Who, apparently, can't get things right the first time and, therefore seems to need to do things over and over again before either arriving at a satisfactory solution or giving up in frustration (i.e., with whatever species are extant), or Who keeps creating life forms that go extinct, or Whose notion of perfection seems to consist in a set of shifting standards, priorities and purposes, or Who seems intent on trying to test or deceive human beings by 'planting' evidence which appears to indicate that similar species are connected by heritable links when – at least, supposedly, according to Professor Miller's understanding of the idea of intelligent design -- those species represent independent creations.

To borrow from Professor Miller's way of phrasing things, the kindest thing that can be said about the foregoing sort of comments is that someone's ignorance seems to be fueling considerable arrogance.

Dr. Miller cannot cite any facts – as opposed to rhetoric – capable of demonstrating that God was unable to accomplish precisely what God wanted to do at each and every turn of biological history. Instead, despite being unable to put forth any evidence that he possesses the requisite sorts of insight to make the foregoing kinds of pronouncements, nonetheless, Professor Miller boldly pushes on into the unknown and, without justification, appears to assume that if he can't manage to grasp what God is up to, then, obviously, no such purpose could possibly exist.

Moreover, among other things, Dr. Miller appears to assume that each and every instance of creation necessarily must give expression to perfection in a way that he understands and, therefore, when researchers point out the nature of the imperfections that they believe are present in this or that aspect of creation, then, according to Dr. Miller, one is forced to choose between, on the one hand, the idea that evolution is correct or, on the other hand, the possibility that God made a lot of mistakes when creating different forms of life. Professor Miller never seems to consider, among other things, the possibility that, like the carpet weavers of Isfahan (Persia/Iran), God might have left imperfections, of one kind or another, in created entities as clues that the nature of perfection – whatever that might entail -- should be sought somewhere other than in created beings.

Another idea that Professor Miller does not seem to have considered is the possibility that different species of life forms might have some sort of significance that transcends their physical properties even as those same physical characteristics give expression to a symbolic way of alluding to some sort of greater meaning. Perhaps, for those who possess the right sorts of insight, then even with respect to those species that are similar to one another (as in the case of elephant-like creatures), those species could reflect different facets of metaphysical reality.

For example, the fact that God could have created four genera of elephant-like organisms that gave rise to 22 different species over a period of six million years might not make a whole lot of sense when considered in purely physical terms. However, when that same evidence is considered from the right kind of metaphysical perspective, then, it might make a great deal of sense.

Perhaps there is no metaphysical realm beyond the purely physical properties of the world. However, given that Professor Miller believes in God, he might want to consider beginning to explore the teachings of the mystics from a variety of spiritual traditions who, for thousands of years, have been claiming that the physical world is but an entry gateway or portal for the many dimensions of existence that transcend physical reality.

Alternatively, perhaps God, as some artists are wont to do, just threw various creative inspirations against the canvas of existence as acts of artistry that had no purpose other than to generate phenomenal manifestation of various kinds. In other words, God might have brought various created entities into existence because God had the capacity to do so and that is the end of the story, and as such, those creations were not intended to serve some notion of perfection or give expression to any sort of purpose other than to display God's desire to do things in one way rather than another and quite independently of whether, or not, someone like Dr. Miller understood what was taking place.

Whatever the qualities of truth – or falsity – that might be associated with any of the foregoing possibilities, what I do know is that Professor Miller hasn't provided a step-by-step account – nor do I believe he can produce such an account – that details how *Paleomastodon, Gomphotherium, Platybelodon, Deinotherium, Primelephas, Mammuthus, Loxodonta africana,* and *Elephas maximus – or any of the differences that distinguish those elephant-like beings from one another* -- came into being. Consequently, I consider all of the previously mentioned rhetorical flourishes of Professor Miller -- that appear to be intended to be little more than an attempt to denigrate the idea of creation in order to bolster the idea of evolution -- to be nothing more than the murmurings of someone who purports to know that which he provides considerable evidence to indicate he does not seem to understand.

One should not interpret the foregoing comments to indicate that I know what Professor Miller does not know. I am just as ignorant as

Professor Miller is with respect to many of the issues that have been raised throughout this chapter, but, nevertheless, I am prepared to entertain a variety of possibilities that Professor Miller seems not to even have considered, and, apparently, he fails to notice them not because those possibilities are irrelevant to questions concerning the nature of Being or irrelevant to the problem of trying to discover the nature of one's relationship with Being but because Dr. Miller's biases seem to prevent him from recognizing that those possibilities might well constitute an integral part of the woof and warp of the human condition.

Irreducible Complexity

Professor Miller takes exception with the notion of "irreducible complexity" that is put forth in Michael Behe's book: *Darwin's Black Box* as well as takes exception with the associated argument that life forms are filled with processes whose complexity is such that they could not have evolved in piecemeal fashion because those cellular systems require all of their component parts to function properly, and, therefore, the likelihood that there could have been the requisite number of 'just so' random mutations that were devoid of interim value, yet, accumulated over millions of years and were able to code for the foregoing sort of functional complexity is extremely remote ... to the point of being vanishingly infinitesimal. Dr. Miller points out that the foregoing ideas of Dr. Behe are really just updated versions of the 'argument from design' that was put forth by William Paley during the pages of the latter individual's 1802 book: *Natural Theology*.

Paley focused on the intricacies of human anatomy and the natural world. He considered the activities of nature to be comparable to the workings of a well-fashioned watch and, as such, constituted evidence that they had been made possible through the presence of a mind capable of intelligent design. He argued that since the eye could not operate properly unless its component parts – such as the optic nerve, retina, lens, iris – were all present and functioning correctly, then the eye would seem to constitute an unsolvable problem for evolutionary theory.

Darwin responded to Paley's argument by pointing out that one could conceive of a series of gradual modifications taking place in conjunction with, for example, the eye that would each have some kind of adaptive value. Over time, the accumulation of those kinds of interim, value-laden changes would enable organisms to make the transition from simple processes involving light-sensitivity to complex, integrated visual systems involving components such as the optic nerve, retina, lens, iris, and so on by means of the forces of natural selection that would automatically identify which set of modifications would be permitted to continue on.

Professor Miller augments the foregoing Darwinian perspective by noting that any capacity to sense light would possess adaptive value. In this regard, Dr. Miller refers to the eyespots of bacteria and algae that consist of a collection of proteins and pigments that are capable of orienting an organism with respect to the presence of light despite the fact that those eyespots do not possess any sort of lens system or links to a nervous system, and, in addition, Professor Miller alludes to the many varieties of "semi-eyes" and "pseudo-eyes" that occur in nature that might be considered to constitute intermediate forms that bridge the divide between organisms that have some capacity for sensing light and organisms that possess complex visual systems.

The general thrust of Dr. Miller's (and Darwin's) perspective is fairly clear. One is less clear about the nature of the specific dynamics that, over time, supposedly take life forms from simple to complex systems.

For example, how did the genetic coding arise that underwrites not only the collection of pigments and proteins that make up an 'eyespot' but, as well, also underwrites the organizational processes that bring together, and maintain, that group of pigments and proteins as a functioning collective? Or, more broadly, how did the genetic coding arise that underwrites the organizational dynamics that are responsible for the formation, structure and properties of various kinds of "semi-eyes" or "pseudo-eyes".

If, as Professor Miller suggests, "semi-eyes" and "pseudo-eyes" constitute intermediate forms between organisms that have some sensitivity to light and those life forms that possess complex visual capabilities, then, one should be able to trace the individual steps that form the ancestral ties that link different organisms. For instance, one should be able to provide a step-by-step account of the transitions that, first, enable organisms to form a 'eyespot', and, then, on the one hand, accumulate the necessary modifications that will give rise to "semi-eyes" and "pseudo-eyes, and, on the other hand, accumulate the changes that will be capable of bridging the evolutionary distance between organisms with "semi-eyes" and "pseudo-eyes" and those life forms with complex visual systems.

Dr. Miller does not provide the degree of detail that is needed to lend credibility to the foregoing sort of account. I suspect that at the present time neither he nor any other biologist can do so, and, instead -- like Darwin -- all they tend to offer are vague allusions to the possibility of such an account by establishing a few data points, drawing a line through those points, and, then, <u>assuming</u> that all the space before, between, and following those data points will fall somewhere along a line that has been drawn in accordance with the principles that underlie the Darwinian hermeneutical perspective.

They do not provide a step-by-step account with respect to the origins of 'eyespots', 'semi-eyes', or 'pseudo-eyes'. Nor do they offer a step-by-step account concerning the origins of the optic nerve, retina, lens, iris, and other facets of complex visual systems.

Like so many stage magicians, Professor Miller, Darwin, and other like-minded biologists seem to engage (and I'm uncertain whether this is done intentionally or unintentionally) in a form of misdirection that creates the illusion that evolutionary processes are taking place when this might not be the case. More specifically, the foregoing individuals allude to the possibility of a series of modifications that – if those changes actually occurred -- <u>might be capable</u> of accounting for the rise of complex visual systems from simple beginnings, but, then, those individuals engage in an endless amount of verbiage concerning a variety of issues that often cause audiences to lose focus concerning the fact that no detailed account is ever actually given.

Furthermore, a form of syllogistic-like reasoning is offered to lead people to believe that something has been demonstrated when this is not necessarily the case. In other words, advocates of evolution – such as Darwin – refer to a series of modifications, each one of which, supposedly, is useful, and, therefore, has adaptive value.

We are asked to imagine such a possibility. Unfortunately, as pointed out previously, the sequence of steps that takes one from simple systems of light-sensitivity to complex visual systems is not given, and just as importantly, no one actually demonstrates that every, intervening step along the foregoing set of sequential changes actually possesses adaptive value.

If the organism survives that serves as host for those changes, then, obviously, adaptive value of one kind, or another, is present. Nevertheless, the aforementioned changes in DNA coding might not entail either adaptive or maladaptive properties, and, therefore, tend to give expression to random-like events because, in many respects, they occur independently of considerations that are functionally dependent on adaptive value even as they occur in conjunction with an organism that exhibits properties and capacities that do have adaptive value.

One could agree with Darwin that if there were a sequence of modifications -- each of which had adaptive value -- that took one through a series of transitions that went from simple to complex systems for interacting with light, then, one would have shown that, over time, natural selection would be fully capable of accounting for the emergence of complex visual systems from simple beginnings. The problem is that neither Darwin, nor anyone else, has ever shown that such a set of continuously adaptive sequence of modifications actually exists in the ancestral lineage that supposedly links organisms that possess simple systems of light sensitivity to life forms that exhibit complex systems of vision.

During the *Origins of Species,* Darwin stipulates: "If it could be demonstrated that any complex organ existed, which could not possibly have been formed by numerous, successive, slight modifications, my theory would absolutely break down." (This quote appears in Chapter 6, which is entitled: 'Difficulties of the Theory'). The foregoing statement is not as critical as some might think.

Darwin left himself plenty of wiggle room. How does one demonstrate that a complex organ exists that "could not possibly have been formed by numerous, successive, slight modifications"?

Surely, one could conceive of a long series of successive, slight modifications that might have led to the formation of some given organ. Whether that sort of a series of events ever actually occurred is irrelevant to the issue of whether things could have happened that way and, therefore, that kind of an imagined sequence tends to refute the idea that such a set of events could not possibly have taken place even though acknowledging that possibility does nothing to prove whether, or not, the foregoing kind of sequence ever actually occurred.

Not content with imagined possibilities, Professor Miller explores the structure, properties, and dynamics of the cilium … a microscopic, tubular-like structure that is capable of vibrating, and, in some cases is capable of propulsion. Dr. Miler – a cell biologist – finds it amusing that Michael Behe is venturing beyond the boundaries of his own area of expertise – biochemistry – and, as a result, treats the cilium as an example of a biologically irreducible system when Professor Miller knows of many counter-examples ... as any competent cell biologist would also be able to do.

The particular form of cilium that Dr. Behe discusses in *Darwin's Black Box* is one that is typical among forms of eukaryotic organisms. That kind of cilium involves an arrangement of two central microtubules surround by nine other pairs of microtubules.

A microtubule is a very small, tubular structure involving the polymerization of alpha and beta forms of tubulin. As Professor Miller makes clear in his critical analysis of Dr. Behe's cilium example, there are many kinds of microtubule arrangements that can be found in eukaryotic and prokaryotic organisms, but Professor Behe only describes one of those possibilities – the 9 plus 2 structures.

Professor Miller runs through a list of organisms that employ alternative arrangements of microtubules. Thus, in passing, he mentions: (1) the protozoan *Diplauxis hatti* that has a flagellum (a threadlike, whip-like microscopic structure) with a 3 + 0 complex of microtubules (the '0' indicates that, unlike the cilium described by Dr. Behe, there is no central pair of microtubules in this flagellum); (2) the protozoan *Lecudina tuzetae* that sports a 6 + 0 set of microtubules; (3) the sperm of the *Anguilla* eel that exhibits a 9 + 0 arrangement of microtubules; (4) mosquitoes of the genus *Culex* have systems that do not possess a central pair of microtubules but, instead, have a single microtubule at the center of an arrangement of 9 + 9 microtubules (thus, it has a 9 + 9 + 1 structure), and (5) a variety of organisms that do not employ the radial arrangement of microtubules present in the foregoing kinds of organisms but, instead, exhibit non-radial microtubule systems that are capable of generating motion of one kind or another.

In *Darwin's Black Box*, Dr. Behe argues that the 9 + 2 microtubule system of the cilium is irreducibly complex. By citing the variety of alternative microtubule arrangements that were noted in the last paragraph, together with several other alternative arrangements of microtubules, Professor Miller believes he has shown that Dr. Behe is incorrect when the latter individual tries to claim that the arrangement of microtubules in eukaryotic cilia are irreducibly complex since, after all, if systems of microtubules exist that do not have to have either a series of 9 pairs of microtubules surrounding a central pair of microtubules, or systems exist that do not have to have a central pair of microtubules at the heart of a radial arrangement of nine pairs of microtubules, then, obviously, one seems to be forced to consider the possibility that simpler systems of microtubules could have arisen and, over the years, might have been gradually modified to establish more complex arrangements of microtubules.

Professor Miller does not discuss how the capacity to produce microtubules arose. Moreover, he does not provide an account of how the DNA coding that underwrites the systems of microtubules that he does discuss became capable of generating functioning arrangements of microtubules involving various kinds of structural properties.

There might not be irreducible complexity in the cilium system described by Dr. Behe. Nonetheless, Professor Miller's account also seems problematic since it fails to provide an account of: (a) How the capacity to produce different forms of tubulin (e.g., alpha and beta formats) initially arose, or (b) how the capacity to polymerize different strains of tubulin into microtubules came into being, or (c) how organisms acquired the organizational wherewithal that would enable those organisms to bring microtubules together into functional units involving different structural properties, and, finally, Professor Miller offers no step-by-step account for (d) how, over time, genetic transitions took place that went from organisms that had the capacity to construct microtubules to a succession of ensuing organisms that deployed more complex or different arrangements of microtubules.

Following his dismantling of the claims in *Darwin's Black Box* concerning the irreducible complexity of a cilium's 9 + 2 microtubule structure (and notwithstanding the problems that remain despite that alleged dismantling), Professor Miller provides an overview for a number of discoveries that he believes demonstrate how a Darwinian perspective is fully capable of accounting for the evolutionary development of complex cellular mechanisms. For instance, he discusses a series of experiments that took place in California during 1997 that explored the relationship between the human growth hormone and certain protein receptors.

Normally speaking, in order for a protein – say, human growth hormone – to affect what takes place within a given cell, the activating protein (or hormone) needs to bind to certain protein receptors that are present in the membranes that enclose those cells. The binding spot is three-dimensional in character, and, consequently, the hormone and the membrane receptor fit together like a lock and key that enables certain kinds of cellular activity to take place.

The California researchers, led by Dr. Atwell, began tinkering with the foregoing system. First, they removed the amino acid tryptophan from the binding site of the membrane protein, thereby changing the shape of the binding location and, as a result, interfering with the ability of human growth hormone to bind to the membrane protein.

Next, the scientists employed a variety of techniques that have been made possible by modern, genetic technology to alter five of the amino acids that form part of the human growth hormone. This set of procedures permitted approximately 10 million different combinations of human grown hormone to be generated.

The foregoing editions of human growth hormone were, then, filtered to determine which mutant form would best fit the binding area of the membrane protein that had been altered through the removal of the amino acid tryptophan. In this way, the researchers were able to discover the existence of a form of mutant human growth hormone that was capable of fitting the altered membrane receptor 100 times more tightly than nonmutant editions of the human growth hormone were able to accomplish.

The researchers concluded that their experiments demonstrated how it was possible to affect substantial changes in the way that proteins bind to one another by inducing mutations that affected just a few locations. In addition, the foregoing experimental work also indicates how different components of a system might be capable of evolving together.

Professor Miller keeps referring to the foregoing changes as being instances of random modifications that were introduced into the binding system that links human growth hormone with certain membrane protein receptors. However, use of the term "random" seems, at best, rather strained in nature.

There was nothing random about the removal of a tryptophan molecule from the binding site of the membrane receptor. This removal process might have been arbitrary in the sense that other parts of the binding site might have been targeted instead of tryptophan, but the act of degrading the receptor protein was anything but random.

To be sure, the removal of an amino acid from the binding site of the receptor protein could be understood as an example of what might have happened through truly random events in the wilds of nature. Nonetheless, since what did occur during the experiment did not take place in the wilds of nature but in a laboratory, the process is hardly random ... merely arbitrary.

Furthermore, the fact techniques of modern genetic technology had to be used to target five particular amino acid sites on the human growth hormone molecule in order to generate roughly 10 million mutant copies of that protein that were subsequently run through a set of filtering procedures to determine which mutant forms of human growth hormone might best fit the altered receptor protein tends to describe a situation that is about as non-random as one can get. Although the underlying idea might have been to consider the experiment to constitute some kind of simulation for what might have taken place or could have taken place in the wild, the fact that so many facets of the experiment were not random in character makes one question how credible such a simulation actually is.

Is it possible that something akin to the foregoing set of laboratory arrangements might have taken place in the wild? Yes, the idea that such a set of events might have occurred in the wild is <u>theoretically</u> possible, but this is not necessarily the same thing as saying that such a set of events did or would occur.

Among other things, one might note in passing that the various components of the hormone-receptor system that were being experimentally probed existed in conditions that were relatively devoid of the kinds of antagonistic forces that might be capable of compromising the integrity of those components as they "sought" to solve the binding problem. On the one hand, one might wonder how long a 'wild' population of organisms would have been able to last long enough to be able to generate 10 million combinations if the hormone-receptor binding issue had been critical to survival, and, on the other hand, if such an issue were not critical to survival, one wonders whether, or not, an organism in the wild would be inclined to put forth

10 million different genetic combinations in order to "solve" a non-essential problem.

Similarly, one cannot necessarily conclude that the foregoing experiment demonstrates how different components of a system might randomly evolve together in the wild. To begin with, there is little about the experiment that is random in nature and, therefore, that experiment really doesn't have much to offer with respect to providing insight into what might, or might not, be possible under non-experimental, allegedly random conditions.

The receptor protein in the foregoing experiment didn't evolve. That protein was modified, and whether such a modified receptor would come to have any actual adaptive value in the wild remains to be seen.

Moreover, the sense of adaptive value that was introduced into the Atwell experiment is artificial in nature since the researchers invented a dimension of adaptive value by creating mutant versions of the human growth hormone to see if any of those molecules might be capable of binding with the previously altered receptor protein. Neither the altered receptor molecule nor the mutant human growth hormone would necessarily have any adaptive value at all if it were not for the fact that the researchers used that receptor molecule to serve as the standard against which to measure success in their experiment – namely, whether, or not, mutant forms of human growth hormone would emerge that were capable of binding to modified versions of the receptor proteins.

Much more pertinent to the issue of evolution are questions concerning, for example, how cells came to be able to code for the 191 amino acids that make up human growth hormone, together with all of the ancillary coding that regulates when and where human growth hormone is to be produced in order to bring about cell reproduction, regeneration, and growth. One might also like to know how cells came to code for the membrane receptor proteins with which human growth hormone binds, and whether, or not, such coding came independently of the emergence of coding for the human growth hormone.

Professor Miller seeks to strengthen the argument that biological systems are capable of randomly generating solutions to existing problems when he briefly discusses the biology of the lactose system. Lactose is a sugar that serves as a source of energy for bacteria when a released enzyme splits the sugar molecule into galactose and glucose subunits that are, subsequently, further metabolized to create energy for the cell.

Galactosidase is the released enzyme being alluded to above that is used to cut lactose into two smaller molecules. When lactose is not available as a food source, the bacterial system turns off the gene that contains the coding that produces galactosidase.

The foregoing lactose system consists of both a structural and a regulatory component. The structural component codes for the series of amino acids that is necessary for the construction or synthesis of galactosidase, whereas the regulatory facet of the system controls the on/off switch for the production of that structural gene.

Obviously, the regulatory side of things must have some way of "knowing" or detecting whether, or not, lactose is available. In addition, the regulatory coding must have some way of determining whether, or not, to turn the gene on that is responsible for the production of galactosidase.

In 1982, Barry Hall devised an experiment to see what would happen if one blocked the ability of bacteria to make galactosidase. He did this by removing the structural gene that codes for that enzyme.

Once the foregoing step was instituted, the researchers didn't have to wait too long before mutant strains began to appear that – despite the absence of galactosidase -- possessed the enzymatic ability to split lactose into galactose and glucose. The foregoing enzyme was not due to the emergence of an entirely novel sequence of DNA coding (i.e., brand new gene), but, instead, was the result of some tinkering with the existing DNA coding.

On the one hand, the underlying coding for a second, structural enzyme that existed in bacteria was modified by means of a relatively simple mutation and, in the process retrofitted the second protein with the capacity to split lactose in the requisite way. On the other hand, the regulatory side of things also underwent mutational modification in a manner that enabled the "new" enzyme to become expressed at the appropriate time (i.e., when lactose was present).

During a follow-up experiment, Hall induced his bacterial cultures to grow in conjunction with, yet, another sugar – namely, lactulose. When he did this, he discovered that a "new," retrofitted enzyme had arisen that was able to produce allolactose.

Allolactose is the chemical signal whose presence tends to turn on the lactose gene. The presence of the foregoing enzyme enabled bacterial cells to activate the coding for lac permease ... a cell membrane receptor that helps regulate the rate at which lactose is able to enter the cell.

Professor Miller states that the 1982 Barry Hall experiments demonstrate how biochemical systems are capable of evolving before our very eyes. Without in any way wishing to deny that what happened in the Hall experiments actually happened, nevertheless, I am not sure that those experiments necessarily prove what Dr. Miller believes they do – namely, that Darwinian principles involving random mutations and principles of natural selection are capable of providing a complete explanation for what took place.

First, let's consider a bit of context. Beta-Galactosidase is a tetramer consisting of four identical chains of polypeptides made up of 1023 amino acids.

Neither Professor Miller, nor any other biologist can provide a step-by-step account that specifies how the coding for that enzyme initially came into being. Nor can they explain how the coding that governs the regulation of galactosidase production first came into being.

In addition, Dr. Miller cannot provide a step-by-step account for how the second enzyme – the one that is modified or retrofitted after the coding for the production of galactosidase has been eliminated – originally came into existence. Nor, can Professor Miller establish how the regulatory coding associated with that second gene first came into existence.

Consequently, one cannot necessarily claim that the Hall experiment offers a demonstration that completely accounts – in Darwinian terms – for how organisms are capable of coming up with new ways of engaging the environment. One cannot make the foregoing claim because the newly discovered ability is entirely dependent on a bacterial system that -- while remaining open to being modified in certain ways – is fully functional and cannot be proven – at least at the present time -- to have arisen through a set of dynamics that are Darwinian in nature.

As a result, one wonders about the precise character of the dynamics that led to the changes that occurred in conjunction with the adaptive DNA coding that occurred in relation to both a second enzyme, as well as with respect to the coding that regulated the turning on and off of the "new" structural gene. Did those changes constitute a purely random set of modifications, or were the aforementioned changes the result of a form of "intelligence" that is operating within bacteria and that, within certain degrees of freedom, gives expression to an experimental trial and error, tinkering process involving different facets of the DNA coding for the structural and regulatory facets of the bacterial system?

The bacteria in the Hall experiments exploited an already existing set of structural and regulatory genes. A few modifications arose in that system (which, depending on what is actually taking place, might, or might not, have been random in nature), and those bacteria transitioned from bacteria that had lost the ability to synthesize galactosidase to organisms that, after a few modifications, had become capable of generating and regulating a semi-new form of enzyme.

Professor Miller considers the foregoing modifications to give expression to a set of random events. However, he can't prove that what is taking place is random in nature.

Rather, Dr. Miller can only <u>assume</u> that what is going on in the Hall experiment is random in character. More importantly, he assumes that what is transpiring in the Hall experiment is random despite the fact that the Hall bacteria seem to be exhibiting qualities – although this might only be a matter of coincidence -- that are very reminiscent of intelligent behavior since it seems to involve being able to come up with make-shift solutions to various problems.

Organisms that display problem solving-like behavior are exhibiting a property that often is considered to be an indication that intelligence of some kind is present. Why assume that the problem solving-like behavior of Hall's bacteria is purely random in nature?

Even if the bacteria in the Hall experiment are not exhibiting signs of their own intelligence in conjunction with the manner in which structural and regulatory coding systems are being modified, this doesn't preclude the possibility that another form of intelligence might be present. Perhaps, God has taken pity on Hall's abusive treatment involving some of his bacteria and, as a result, provided at least a few of those bacteria with an alternative means for producing and regulating an enzyme that is capable of splitting lactose in the absence of galactosidase and, then, using the resulting components to generate energy.

Professor Miller might be right that the Atwell and Hall experiments indicate that many aspects of biological life are not necessarily governed by considerations of "irreducible complexity" in Dr. Behe's limited sense of the term (that is, applied to sub-systems within various life forms.) However, at the same time, Dr. Miller has not successfully shown that what is taking place in those experiments is necessarily a function of purely Darwinian principles involving nothing more than random mutations, natural selection, or differential rates of reproductive success.

As a result, there is a sense in which there seem to be phenomena (such as occurred in the Atwell and Hall experiment) that are made possible by life's inherent complexity and might not necessarily occur in the absence of that sort of complexity. Moreover, perhaps neither Professor Miller nor other evolutionary biologists have been able to provide a step-by-step account for how so many dimensions of life (such as, for example, the events that first led to the coding for galactosidase) arose in purely random ways or in accordance with Darwinian principles is precisely because that is not what took place.

Evolutionary theory does offer a tentative way of allegedly resolving the foregoing problem (e.g., the origins issue) through the idea of random mutations. However, such a proposal is not only firmly ensconced in nothing more than presumption but, as well, that proposal lacks the degree of specificity that would be necessary to induce a reasonable person to become inclined to accept the presumption as true.

In other words, if neither Professor Miller nor his fellow advocates for evolutionary theory can offer a step-by-step account for how the

DNA coding that underwrites the production of human growth hormone (the Atwell experiment) or galactosidase (the Hall experiment) first came into existence through random means, then, why suppose that Professor Miller's interpretation (or that of his colleagues) with respect to, among other things, the Atwell and Hall experiments is correct? More specifically, If the process or set of events that initially gave rise to the coding for 191 amino acids in the case of human growth hormone, or, alternatively, in the case of galactosidase, if the events leading to the laying down of coding for four monomer units consisting of 1023 amino acids each that are to be assembled into a tetramer complex cannot be proven to be a random process, then why should one suppose that the minor modifications of coding that occurred during the Atwell and Hall experiments were necessarily random in character?

Nothing has been proven. Instead, pretty much everything of critical importance in Professor Miller's account of the foregoing experiments is, in one way or another, being assumed.

Professor Miller continues to elaborate on his belief that scientists are able to explain the evolution of various facets of biological functioning by mentioning a 1998 article by Anthony Dean in the journal *American Scientist*. The article contains a detailed exposition concerning how two enzymes (different forms of isocitrate dehydrogenase – ICDH – which play parts in the citric acid or Krebs cycle) might have evolved from just one ancestral template.

The foregoing dynamic becomes more complex since all known forms of ICDH require either one of two co-factors in order to function properly. The co-factors are NAD^+ and NADH (the oxidized and reduced forms, respectively of nicotinamide adenine dinucleotide that are responsible for moving electrons – through acquiring or donating them -- from one reaction to another).

During the course of his 1998 article, Dean relies on the Neutral Theory of molecular evolution developed by Motoo Kimura to show how just a few random modifications in the DNA coding sequences underlying the synthesis of several key amino acids that make up the structure of one form of ICDH could have led to a slightly altered form of DNA coding that is necessary to produce the other form of ICDH. Professor Miller considers the foregoing explanation to constitute a key piece of evidence in support of the idea that biological mechanisms could have evolved through Darwinian means.

To argue that something might have happened in a certain way does not necessarily mean that events occurred in the way being hypothesized. Possibility is not necessarily synonymous with actuality.

Moreover, Dean's hypothetical account takes place in the context of a fully functioning cell. Yet, no step-by-step account is given for how the DNA coding originally came into being that makes such a functioning cell possible, and, as a result, one really has no empirical basis for supposing that the sorts of mutations discussed by Dean in his article – few though they might have been – were necessarily the result of random processes.

One might be willing to accept Dean's hypothesis that the DNA coding for one, or another, form of the ICDH enzyme might have been modified in a certain way in order to make the synthesis of the other form of ICDH possible. Nonetheless, such an acknowledgement does not force one to simultaneously maintain that those modifications were necessarily random in nature.

Therefore, one can acknowledge some of the ideas associated with the Dean article as being legitimate possibilities without feeling compelled to accept them as being true. Indeed, given the absence of the kind of corroborating evidence that would be needed to demonstrate that things happened in the way Dean proposes, one has considerable grounds for concluding that Dean's hypothesis remains a possibility and nothing more.

Professor Miller seems to want to treat the Dean material as if it constituted evidence that verifies the truth of Darwinian principles. However, given (1) that no one has, yet, been able to show that what Dean proposes actually happened and given (2) that even if – in line with Dean's proposal -- the DNA coding for one form of ICDH had been modified to make another form of ICDH possible, this would not necessarily show that those transitions were random in nature, and, as a result, one really has no sound, rigorous basis for considering Dean's perspective to constitute evidence in support of evolutionary theory.

Francois Jacob, the Nobel Laureate from France, used the term: "evolution by molecular tinkering" to refer to the process through which slight changes in the DNA coding for an organism occurred in order to generate new genetic possibilities. This sort of tinkering begins with DNA coding for some existing function, structure, pathway, organ, or the like, and, then, through small changes or modifications to that coding, some sort of new function, structure, pathway, organ, and so on, emerges within the modified organism.

The idea is clear enough. Whether, or not, organisms actually engage in that sort of molecular tinkering is another matter, and even if organisms did (and do) engage in molecular tinkering, those modifications are not necessarily random in nature.

There is nothing inherently contradictory about the idea that God might have provided organisms with certain degrees of freedom through which to experimentally explore ontological possibilities. Indeed, if molecular tinkering does occur, then, one would like to know whether that tinkering is random or non-random in nature.

Furthermore, the notion of "evolution by molecular tinkering" does not necessarily address the issue of how life came to be in the first place. The idea of molecular tinkering tends to be dependent on the existence of organisms that already have the capacity to function and survive.

In the absence of the foregoing kind of functionality, whatever allegedly random sorts of tinkering that might have occurred prior to the emergence of life would appear to entail little more than a highly improbable sequence of lucky coincidences. Unless, of course, the process of tinkering was being pushed, or pulled, in some directions rather than others by a force or set of forces that is non-random in nature.

However, contrary to what some proponents of evolution might wish to argue, natural selection does not qualify as the foregoing sort of non-random force since it operates <u>after</u> such tinkering occurs, <u>not before</u> it takes place. Natural selection is not the cause of that tinkering, but rather, natural selection is the process that demonstrates how some kinds of tinkering give rise to possibilities that are more viable than are other modalities of tinkering.

Professor Miller advances additional examples that he considers to constitute evidence that Behe's notion of "irreducible complexity does not necessarily govern complex biochemical mechanisms. For example, Dr. Miller mentions the 1998 work of Sunney Chan and Siegfried Musser in conjunction with the cytochrome c oxidase protein pump.

The foregoing biological structure consists of a complex of proteins that have the capacity to reduce oxygen and, in the process, produce water while also releasing energy in the form of an electrochemical potential that assists in the translocation of certain proteins across cell membranes. Among other things, Chan and Musser constructed a detailed evolutionary tree that showed how two of the foregoing proteins in the cytochrome c oxidase pump were very similar to an enzyme in the cytochrome bo_3 complex that is found in bacteria.

As a result, Chan and Musser suggested that the two proteins in the cytochrome c oxidase pump likely were modified versions of the cytochrome bo_3 bacterial enzyme. Supposedly, here was another example of a living cell that had tinkered with its molecular machinery and brought about the evolution of a system (i.e., cytochrome c oxidase protein pump) that was different from, but ancestrally related to, a previous structure (i.e., the cytochrome bo_3 complex in bacteria).

Professor Miller does not have much in the way of specifics to offer in relation to how the DNA coding for the other proteins in the cytochrome c oxidase pump came into being. What Professor Miller does say is that all of the other proteins in the cytochrome c oxidase also can be shown to exhibit properties that are similar to qualities existing in various microorganisms and that all of the modifications that occurred over time were part of a process that progressively enhanced respiratory efficiency in a variety of life forms.

The foregoing scenario makes sense. However, determining whether, or not, that account is true would require a body of evidence that is far in excess of what Chan, Musser, or Professor Miller are able to provide, and, therefore, one is uncertain whether an understanding that makes sense is also true.

The Chan and Musser hypothesis concerning the ancestral relationship between the cytochrome bo_3 enzyme in bacteria and the cytochrome c oxidase complex in, say, mammals, constitutes evidence that living systems are not characterized by qualities of irreducible complexity (contrary to the claims of Dr. Behe) <u>only if</u> someone can demonstrate that: (a) Two of the proteins in the cytochrome c oxidase pump actually were derived from an enzyme in the cytochrome bo$_3$ complex of bacteria, but, as well, only if someone can demonstrate that (b) all of the other proteins in the cytochrome c oxidase pump also owe their existence to a similar sort of tinkering process that took place within some ancestral lineage. In addition, in order for the Chan/Musser proposal to have real credibility, then (c) one must be able to provide an account that explains how the regulatory dimension that governs the process of gene expression for the cytochrome c oxidase pump complex arose through a similar tinkering dynamic.

At the present time, none of the foregoing sorts of definitive evidence exists. Consequently, what Professor Miller has accomplished by mentioning the 1998 work of Chan and Musser is not so much a matter of putting forth evidence that refutes the claims of Dr. Behe concerning the idea of "irreducible complexity" as much as the cytochrome c oxidase pump material that appears in *Finding Darwin's God* merely constitutes an outline for a hypothesis that provides an alternative hermeneutic to the work of Dr. Behe.

Apparently, Professor Miller is counting his chickens before they hatch. He is assuming that the Chan/Musser hypothesis is true, but he does not offer any evidence capable of verifying the truth of that claim.

Instead, Dr. Miller just strings together a number of similar claims (e.g., Atwell, Hall, Dean, and Chan/Musser) and suggests that the mere possibility of something being true constitutes evidence for a Darwinian approach to evolution. One can't help but wonder if the foregoing sorts of conjectures are the supposedly overwhelming "evidence" to which people like Professor Miller and Richard Dawkins are alluding, when they claim that evolutionary theory has assumed ascendancy in the 20th and 21st centuries, and, if this is the case, then there is something quite problematic inherent in those sorts of claims.

Blood clotting is another, alleged piece of evidence that is put forth by Professor Miller which he considers to favor an evolutionary account concerning the origins of various life forms or the origins of different sub-systems and complexes within those organisms. In human beings, the process of blood clotting consists of more than a dozen proteins but contrary to the claims voiced by Dr. Behe in

Darwin's Black Box, Professor Miller does not believe the foregoing biological process constitutes an example of the kind of "irreducible complexity" that Behe maintains is capable of stumping evolutionary theory.

Professor Miller begins by pointing out that fibrinogen plays a fundamental role in blood clotting. Fibrinogen is a glycoprotein that consists of three, non-identical polypeptide chains that collectively contain well over 400 amino acids that possesses a potentially sticky portion near the center of the structure

Prior to being activated, the aforementioned sticky portion of fibrinogen is rendered inactive because the complex molecule has a configuration that enables a sequence of negatively charged amino acids to cover the sticky portion of the fibrinogen molecule, and, as a result, the sticky portions of fibrinogen are not able to clump together with one another. However, when a cut occurs, the clotting system is activated through the release of thrombin, a protein-cutting enzyme that consists in hundreds of amino acids.

Thrombin removes the negatively charged amino acid chains that normally cover the sticky segments of fibrinogen. Once uncovered, the sticky portions of fibrinogen are able to begin clinging to one another and, in the process, begin to form a clot.

However, in order for the thrombin protease (cutting enzyme) to be synthesized and released, it must be activated. This requires the presence of another protease known as Factor X that, in turn, must be turned on by several other proteases – namely, Factor VII and Factor IX – which also must be activated by still other proteins.

According to Professor Miller, all of the foregoing steps help amplify the clotting process. Nonetheless, Dr. Miller assures his readers that a clotting system involving fewer steps could still work even though it might take longer for the more simplified clotting system to be effective.

In this respect, Professor Miller refers to the work of Russell Doolittle, a molecular biologist, who has been engaged in more than three decades of exploratory research concerning possible evolutionary pathways for the blood clotting dynamic. Doolittle is convinced that from an evolutionary point of view the complexities of blood clotting become much more theoretically manageable when one realizes that almost of the regulatory genes involved in the blood clotting cascade code for a single class of protein cutting enzymes known as serine proteases, and, therefore, in his opinion, many of the apparent complexities associated with the process of blood clotting really are just variations on an underlying set of DNA coding that share a lot of similarities ... variations that might have arisen as a result of a relatively limited number of mutations.

Serine proteases are found in both prokaryotic and eukaryotic organisms. One reason for their relative ubiquity is a function of the capacity of serine protease enzymes to cleave peptide bonds involving the nucleophilic, or electron donating, tendencies of the amino acid serine, and this plays a valuable role in a lot of biological functions that have nothing to do with blood clotting.

Therefore, both Dr. Doolittle and Professor Miller believe that serine proteases represent good candidates for the kind of molecular tinkering that, in time, might have led to the formation of novel structures, functions, pathways, and the like. In other words, the DNA coding underlying some given serine protease might have become modified through various instances of mutation that permitted those enzymes to transition from proteases that served some kind of non-blood clotting process to proteins that were re-purposed in conjunction with one, or another, edition of a blood clotting system.

Dr. Doolittle is of the opinion that many millions of years ago, a series of gene duplications took place involving some given form of serine protease. Over time, this series of gene duplications led to the emergence of a set of serine proteases that, in various ways, were incorporated into a number of different kinds of biological functions, one of which had to do with the blood clotting process.

To lend further credibility to the foregoing idea, Professor Miller argues that just as the gene for some primitive form of serine protease that was unconnected to the process of blood clotting might have been duplicated and, then, subsequently modified by mutations that, over time, became adapted for use in a system of blood clotting, so too, one should be able to say the same thing in relation to fibrinogen, another component in the process of blood clotting. In other words, if the fibrinogen protein that plays such a fundamental role in blood clotting had been the result of some sort of gene duplication that occurred in conjunction with a fibrinogen-like protein that had nothing to do with blood clotting and, then, subsequently was adapted by means of mutations for use in a blood clotting system, then, one should be able to find instances of fibrinogen-like proteins that had nothing to do with blood clotting.

Sure enough, in 1990, Russell Doolittle and Xun Yu discovered a fibrinogen-like sequence of DNA coding that was unconnected to the process of blood clotting. They found the fibrinogen-like sequence in a sea cucumber.

Professor Miller also explores what he considers to be the implications of Russell Doolittle's research for the origin of fibrinogen in crustaceans such as crabs and lobsters. Before proceeding on with the foregoing origins story, however, one should note that crustacean "fibrinogen" is not the same molecule that exists in human beings.

Because many facets of clotting in crustaceans are, at least on the surface, very similar to what takes place with respect to particular aspects of the clotting process in human beings, some people have referred to the molecule that plays a central role in the crustacean clotting dynamic as a form of fibrinogen. Nonetheless, the clotting molecule in crustaceans is very different from the fibrinogen molecule that is found in human beings, and, as well, the nature of the clotting process that occurs in crustaceans is very different from what occurs in human beings.

Although the fibrinogen molecule that occurs in crustaceans is very different from the fibrinogen molecule that operates in vertebrates, the crustacean fibrinogen molecule is very similar to another kind of protein that is found in crustaceans. This latter protein is known as "vitellogenin," and it is a fairly large molecule that is synthesized in a variety of cells before being: Deposited into the blood strea, delivered to the ovary and, then, broken down when enzymes from the ovary are released that cut vitellogenin molecules into smaller segments that are further processed to become the yolk in an egg.

Doolittle believes that at some point the gene coding for the vitellogenin molecule underwent duplication. Over time, mutations occurred that modified the coding for the vitellogenin protein and, in the process, brought about the re-purposing process that transitioned the role of the molecule from being a source of nutrition in the yolk of a crustacean egg to becoming a molecule that played a central role in the clotting process in crustaceans.

The foregoing account does not include an explanation for the origins of the vitellogenin gene from which the fibrinogen gene is supposedly derived. Of course, one could claim that the vitellogenin gene arose in a manner that is similar to the way in which fibrinogen came into being – namely, through gene duplication of some other gene, followed by the right series of mutations – nevertheless, such a claim cannot be proven to have occurred any more than one can empirically demonstrate that the fibrinogen gene arose through a duplication of the vitellogenin gene, and, then, the latter underwent mutational modification over time

Doolittle's proposal does not offer a step-by-step account that provides the details of when, how, or if the mutations that would have been necessary to convert a vitellogenin gene to a fibrinogen gene actually took place. In addition, he doesn't know whether those modifications were random or non-random, nor does his account provide information about for how the DNA coding that is responsible for regulating the expression of genes with different functions made the transition from directing the creation and flow of nutrients to the yolk of an egg, to helping to regulate the clotting of blood.

Notwithstanding the foregoing lacunae, the vitellogenin example is <u>consistent</u> with the idea that certain genes are duplicated and, then, somewhere down the ancestral line, are repurposed, via mutational modifications, to serve some function other than the one that the gene served in ancestral organisms. At the same time, the foregoing cases involving DNA coding for fibrinogen molecules in sea cucumbers and the coding for vitellogenin molecules in crustaceans are also consistent with the possibility that some form of intelligence was present that chose to re-purpose a given molecule – say fibrinogen, a serine protease, or vitellogenin – for some function (say, clotting blood) other than the one that might have been present originally.

Furthermore, proponents of evolution believe that the technical capacity to be able to line up sequences of DNA coding from different species and identify homologous (similar in nature) portions of those sequences means those species share a common ancestry. While this might be true in some cases, one cannot automatically exclude the possibility that homologous sequences in DNA coding might also mean an intelligence is present in different life forms that is utilizing the same or similar components during the design and construction of an array of alternative structures, pathways, and functions.

Just as one cannot suppose that simply because the same components – perhaps with slight modifications -- are being used to construct various buildings, machines, or electrical devices that, therefore, this means that the components in those buildings, machines, or electrical devices necessarily have an ancestral relationship, so too, one cannot assume that because the same kinds of biological components – perhaps with slight modifications – are being use to construct "new" enzymes, metabolic pathways, and the like that, therefore, this means the components in those organisms necessarily have an ancestral relationship. In short, one cannot just assume one's way through the nooks and crannies of life's mysteries.

Instead, if one wishes to place evolutionary theory on sound ground, one must be able to demonstrate that the foregoing sorts of biological homologies could only have arisen through an ancestral process involving nothing more than random events and genetic inheritance. However, at the present time, evolutionary biologists are not able to provide the evidence that would be needed to bring the foregoing sort of demonstration to empirical realization.

Naturally, Professor Miller could ask what he considers to be rhetorical-like questions about why any God or intelligent designer would go about the process of creation in a manner that seems to elude Dr. Miller's ability to fathom what is transpiring. Nonetheless, Professor Miller's lack of understanding concerning such matters should not be permitted to assume the status of a metric that is used to determine what significance should be assigned to the presence of homologies among various life forms.

Some would argue that the idea of God is not scientific in nature. Depending on what one means by the notion of "science", such an argument might, or might not, be true, but why permit one's ignorance concerning an issue to limit what possibilities are to be considered?

Is science doing human beings a favor by removing the foregoing sorts of possibilities from consideration? Or, is science doing human beings a disservice by arbitrarily filtering out those aspects of experience that might be relevant to discovering the nature of one's relationship with Being?

Irrespective of how one answers the foregoing questions, the Doolittle model has its limits. For instance, as intimated previously, one might wish to ask about how the coding for the hundreds of amino acids that comprise the fibrinogen-like protein found in the sea cucumber first emerged?

One will not be able to continue to argue ad infinitum that those genes are always the result of some prior process of duplication that becomes adaptively modified through mutation to serve some alternative function. At some point, the explanatory buck is going to have to stop and be held empirically accountable.

The process of gene duplication not only requires an ability to duplicate genes but, as well, that capacity requires concomitant, complementary processes that are able to provide a biologically viable, functional context of protected, dynamic space through which genes can be duplicated and modified on a regular basis. Both of the aforementioned capacities depend on the existence of a set of functional, structural, and organizational genes that cannot merely be assumed into existence by a Doolittle-like mechanism of duplicated genes followed by a set of "just so" mutations.

Of course, someone might wish to argue that "in the beginning" different sequences of DNA coding randomly arose that just happened to have the sort of functionality that was capable of sustaining some form of primitive life, and, then, genes were duplicated in subsequent generations that became repurposed, by means of random mutations, for other kinds of functions. However, the foregoing is really nothing more than a series of assumed events that conjure up biological functionality whenever evolutionists wave their mysterious wand of randomness ... something that occurs with the kind of monotonous frequency that strains, if not warps, the limits of even a reasonable amount of credulity.

Evolutionary theory is unable to account for how the code arose that assigns amino acids to triplet sequences of DNA. If one cannot account for how such a fundamental, critical assignment process might have come into being through random events, then, why should one suppose that the kind of events that supposedly govern the possibility that genes serving one kind of function are duplicated and, over time, are re-purposed to serve other kinds of functions as a result of a series of mutations that are random but, again and again, produce felicitous results?

Professor Miller tries to shore up his argument concerning the viability of evolutionary theory by introducing the notion of "selective pressure" (for instance, consider his discussion on page 156 of *Finding Darwin's God*). According to Dr. Miller, any mutation in, say, white cells or some other biological component, that brought about an increase in the property of stickiness in the process of blood clotting would be favored by natural selection, and as such signifies the presence of a selective pressure that favors the emergence of that kind of an enhanced degree of stickiness.

The term "selective pressure" that is being employed by Professor Miller during the foregoing discussion is misleading. If mutations are random in nature, then, irrespective of what a given set of conditions might favor, those conditions will not influence how random events play out within an organism.

If events are truly random, then, there is no "pressure" that is present to induce those events to turn out in one manner rather than another. To be sure, if random mutations were able to enhance the degree of stickiness present in some blood clotting dynamic, then, yes, that kind of modification likely would be favored by natural selection.

Nevertheless, the dimension of favored status has no ability to affect whether, or not, certain kinds of modifications will actually take place. Yet, contrary to the foregoing logic of random events, Dr. Miller's use of the term "selective pressure" encourages readers to believe that, somehow (in a manner that resonates somewhat – but not entirely -- with the ideas of Jean Baptiste de Lamarck), an organism's need for some component, structure, or pathway, seems to help bring about the changes that will best accommodate existing conditions of natural selection

Professor Miller is also misleading when he states that: "By now it should be clear that any claim that evolution cannot produce complex, well-designed biochemical machines is just plain wrong (p.152 of *Finding Darwin's God*)." While it is true that evolutionary theory does offer explanations concerning the emergence of "complex, well-designed biochemical machines," (for example, the previously discussed research of Atwell, Hall, Dean, Chan/Musser, and Doolittle), nonetheless, as previously discussed in considerable detail, those explanations tend to give expression to theoretical possibilities rather than actual instances of proof that Darwinian evolution is capable of producing various "complex, well-designed biochemical machines."

Dr. Behe does overstate his case when he tries to argue in *Darwin's Black Box* (page 145) that evolutionary theorists have failed to put forth any accounts concerning what might have happened in the past because, clearly – as Professor Miller points out on a number of occasions – many researchers have reflected on, discussed, and written about what might have happened and how things might have happened in conjunction with quite a few facets of evolutionary history. Nonetheless, Professor Behe might be quite right when he indicates that proponents of evolution have no specific knowledge concerning what actually did take place – say, on a step-by-step basis -- in any given instance of proposed evolution.

Although there are times during *Finding Darwin's God* when Professor Miller tends to clearly acknowledge that many of the examples he discusses in his book are purely theoretical and that, as a result, currently, we do not have sufficient evidence to prove that those possibilities are correct (e.g., his discussions on page 147 and again on page 158), nonetheless, there are other occasions (such as the previously quoted sentence from page 152 of his book) when, like Professor Behe, Dr. Miller makes claims that tend to overstate the strength of his own case. Despite Professor Miller's belief that the material he put forth in his book has shredded Dr. Behe's notion of "irreducible complexities," nevertheless, much of that "shredding" seems to involve nothing more than putting forth theoretical possibilities that, if true, would weigh heavily against Dr. Behe's perspective, but that – at least to this point in time -- have not been shown to be true and, therefore, Professor Miller's belief that evolutionary theory is fully capable of accounting for the emergence of complex biological machinery is – in the absence of actual empirical evidence – rather premature.

Even if it turns out that, say, Doolittle's model of repurposed duplicated genes or Jacob's notion of evolution by molecular tinkering is correct, at the present time, this understanding involves little more than being able to say, in very general terms, <u>what</u> happened. Given the foregoing, one still would not know the step-by-step processes that account for how and why the foregoing 'what' happened and, as a result, one would be unable to demonstrate whether the 'what' that took place was due to random or non-random phenomena.

Guerilla Warfare

On page 96 of *Only a Theory*, Professor Miller asks a question about what he feels are the only two alternatives that are relevant to the discussion taking place in that book. Is the genome designed or did it evolve?

Dr. Miller doesn't appear to realize there is, at least, one further possibility to consider. Namely, what about the possibility that the genome was designed to be able to evolve or change within certain degrees of freedom?

For example, Professor Miller refers to the fact that the genome not only consists of a set of functioning genes, but, as well, it contains a variety of broken genes, coding errors, and what appears to be an array of useless information (although in the latter case the information might appear to be useless because we don't understand the nature of that information). While Professor Miller considers the presence of those sorts of imperfections to constitute evidence that evolution is a random affair rather than the result of design, there is nothing inherently contradictory about the possibility that God invested the genome with a potential that enabled organisms – within certain degrees of freedom -- to be able to tolerate various kinds of imperfections, errors, or breakdowns.

Such a potential might serve a variety of functions. For instance, on the one hand – and within limits -- that kind of a potential might help buffer organisms against the occurrence of coding errors and broken genes of one sort or another that are natural events during the course of life, and, on the other hand, such a potential also might offer organisms an opportunity to try to correct those sorts of errors when the latter problems arise – to whatever extent this is possible – by taking ameliorative steps through processes such as changing diet, behavior, and the like.

No one lives forever. The errors that emerge in the genome might be part of the price that one pays for being able to exist, and, as well, the errors that are manifested in the genome might be part of the price that any given individual pays for engaging existence in one way rather than another.

That is, such errors and problems could constitute a life-style issue that might, or might not, be capable of corrective treatment. Some of those corrective steps might be automatic and, as such, form part of an organism's capacity to make changes under various internal and external circumstances, but other corrective measures might not be automatic and require choices of some kind to be made (a topic to which I shall return).

We don't necessarily know what the significance is of the coding errors, broken genes, and seemingly useless information that is contained in the genome. Professor Miller might be right that those kinds of properties are merely random residues of life, but, then again, he might be wrong since I do not feel he has put forth any arguments that are capable of successfully defending his perspective.

For example, Dr. Miller indicates that many animals have the capacity to synthesize their own vitamin C, but human beings have either lost this capacity or never really had it. Although human beings possess the five liver proteins that are necessary to synthesize ascorbate, or vitamin C, there is a component known as gulonolactone oxidase that is coded for by a gene that has become inactive in human beings (as well as guinea pigs, certain bats, and a variety of primates).

According to Professor Miller, the reason why gulonolactone oxidase or GLO has become inactive is due to the mutational errors that have built up over time. As a result, human beings must derive vitamin C from external sources.

In the light of the foregoing information, Dr. Miller raises a question. If God wanted human beings to be dependent on external sources for our supply of ascorbate, then why include the GLO gene at all in the human genome?

The foregoing question is kind of dumb. It seems somewhat reminiscent of questions like: How many angels fit on the head of a pin, but, in passing, a few comments should be made.

The CIA, NSA, FBI, and a variety of other intelligence agencies – which are merely human organizations and, despite their self-serving sense of self-aggrandizement are not Divine -- all have reasons for doing things that they do not necessarily share with other people in society – including Professor Miller. Irrespective of whether, or not, one approves of those reasons, the foregoing organizations continue to go about doing what they do, and the fact that someone – for example, Dr. Miller -- doesn't know why those organizations do what they do has nothing to do with what does, or does not, occur or the reasons why those sorts of things happen.

By means of a rather arbitrary process of questionable republican pedigree (see Article IV, Section 4 of the United States Constitution), some people are granted various levels of classification that permit those individuals to be read in on this or that secret associated with one, or another, intelligence or government agency. Similarly, for thousands of years mystics have indicated that ontological secrets are veiled by a kind of metaphysical version of classified information involving Divine mysteries and, as a result, one has to go through a process (i.e., the mystical path) in order to qualify for the sort of security clearance that enables one to be read into those kinds of matters.

By raising some of the questions that he does in *Finding Darwin's God* as well as in *Only A Theory*, Professor Miller seems to be indicating that since he is not able to come up with an answer to his own question, then, somehow this means there couldn't possibly be some reason of which he is unaware that might be governing what is going on. Moreover, he seems to want to use his possible ignorance concerning those matters as a form of rhetorical evidence in favor of evolutionary theory, and, this seems like – as previously intimated -- a kind of dumb thing to do.

Life is a journey. This is true both for individuals as well as for populations.

Some species did not lose the capacity to synthesize ascorbate. Some species did lose that capacity.

The journeys of individuals and populations are marked by the events that occur along the path of life for those individuals and populations. Professor Miller considers the fact that human beings lost their capacity to synthesize ascorbate to be of significance because it parallels what happened to a variety of primates – including chimps, gorillas, and orangutans – that are considered to have a close evolutionary relationship with human beings, and, therefore, suggests to Dr. Miller and other like-minded individuals that human beings inherited the dysfunctional GLO gene from some primate ancestor or from some ancestor held in common by primates and human beings.

Since Dr. Miler believes in God, let's consider various possibilities with that perspective in mind. For example, one might suppose that what makes human beings human is not necessarily their physical form but, rather, the quality of humanness could be a function of the soul that, by God's Grace, becomes linked to certain kinds of body.

Whatever might, or might not, be true with respect to the evolutionary descent of the physical body that eventually assumed the form of hominid organisms, what makes something human might be the presence of a soul and not necessarily the nature of the body associated with that soul. Consequently, one could imagine a series of primates and hominid-like creatures arising over time that might have been ancestrally related to one another but that, at some point, diverged from the foregoing lineage when provided with a soul, and, as a result, human beings became manifest for the first time … as such, the acquiring of a soul would be at the heart of a speciation event involving human beings.

As was indicated toward the beginning of this chapter, I could accept a great deal of what constitutes evolutionary theory and still not feel compelled to change much of anything with respect to my belief concerning the existence of God. The facticity of evolution – if that is what the data actually indicates – might affect this or that point of theology, but it would not necessarily demonstrate that God did not exist and, instead, might only demonstrate that the process of evolution was the means through which God went about creating various biological life forms. Nonetheless, I am somewhat agnostic when it comes to issues such as knowing the precise nature of the dynamics to which creation gives expression (although I am not agnostic when it comes to the existence of God).

We know that chimps, gorillas, orangutans, and human beings all seem to possess an inactive gene for the production of gulonolactone oxidase that plays a critical role in the synthesis of ascorbate or vitamin C. Does the presence of that inactive GLO gene prove that human beings are ancestrally related to the aforementioned primates, or does the presence of an inactive GLO gene demonstrate that human beings and primates are all organisms that are similarly vulnerable to the loss of GLO functioning, and as a result, at some point during life on Earth, humans and various primates lost the functioning of that gene due to a series of mutations, but despite their similarities in relation to the loss of functionality of the GLO gene, they are not necessarily ancestrally linked to one another.

Are the homologies in DNA coding among, on the one hand, chimps (as well as various other primates), and, on the other hand, human beings an indication that all of those life forms are ancestrally related? Or, is the presence of those homologies being interpreted to indicate that ancestral relationships exist when this is not necessarily the case?

Homologies in DNA coding are considered to be evidence that genetically ties one species with another. Yet, what is the nature of the proof that demonstrates that this is actually true?

An array of data from population genetics indicates that genes get passed on from one generation to the next. However, what data indicates that the homologies among different species are necessarily the result of genes that, at one time or another, were passed on from one kind of life form to another?

To be sure, one can conceive of instances – and evolutionary biologists spend a great deal of time providing this kind of documentation – in which, for various reasons, a given species bifurcates and, in time, a new species emerges that is ancestrally linked with the original species and, therefore, shares many facets of DNA coding with the latter organism even as the new species exhibits characteristics that differentiate it from the other species and prevent it from successfully breeding with members of the population that constitute a separate species. Nevertheless, accepting the foregoing point does not simultaneously force one to concede that the presence of homologies in DNA coding among different species is necessarily due to ancestral relationships.

Logically speaking, if God exists, then, presumably, there is nothing preventing God from using similar design protocols in species that are unrelated to one another. Similarly, there, presumably, is nothing preventing God from using similar sequences of DNA coding in unrelated species for purposes of generating similar kinds of proteins, metabolic pathways, and various other biological features that, over time, might, or might not, assume certain differences due to the occurrence of mutations.

Why did chimps, gorillas, orangutans, and human beings lose the functionality of the GLO gene? Many other animals retained the capacity to synthesize their own ascorbate, or vitamin C, so why was the capacity conserved in some species but not others?

Do species that retain functionality in the GLO gene enjoy the presence of some sort of mechanism that helps preserve that functionality ... a mechanism that is not present in various primates and human beings because, for whatever reason, it is not part of their design. Perhaps, certain primates, together with human beings, all share a similar vulnerability to the loss of GLO functioning not because of inter-species ancestral relationships but because that kind of vulnerability constitutes part of the structural character of the potential for their respective populations ... just as various other biological systems that characterize the organisms that are encompassed by different populations might be vulnerable to certain kinds of difficulties (e.g., diseases, breakdown, injury) because those sorts of problems are inherent in the potential associated with the properties displayed by the organisms within those populations and not necessarily because those problems are inherited from some common founding ancestor.

For example, all organisms that have visual capacity of one kind or another are susceptible to an array of problems involving their eye or eyes. That susceptibility is not necessarily a function of some manner of inter-species ancestry – since visual systems have emerged that are ancestrally unrelated to one another – but, rather, the foregoing sort of vulnerability might be a function of the way the visual system is structured and operates within a particular population of organisms quite independently of what is the case with the potential vulnerability inherent in the visual systems of other populations of organisms.

One also could raise questions concerning the mutations that lead to the loss of GLO functioning in primates and human beings and inquire about whether those modifications are merely a matter of random events, or are there various other, unknown principles and forces that are present which lead to the loss of functioning in a non-random manner? How – i.e., on what basis -- does one decide between the foregoing two possibilities?

The universe – whether Divinely ordained or non-Divinely ordained -- might have an ontological stance toward life that in <u>some</u> instances could be similar to, as previously discussed, the notion of "defensive indifference" in baseball (when the team on defense is so far ahead of its opponent that the former individuals are indifferent about whether, or not, an opposing base runner steals second or third). In other words, perhaps, life forms are given (by design or through evolution or both) different packages containing an array of constraints and degrees of freedom that determine what those organisms can and can't do, and, within certain limits, the universe is indifferent about whether, or not – for example -- primates and human beings lose their capacity to synthesize ascorbate because there is at least one back-up system in place that enables those organisms to be able to continue to survive even if they should become unable to synthesize their own ascorbate.

Chimps, gorillas, orangutans, and human beings all possess the sorts of degrees of freedom in their genetic potential that permit them to seek external sources for their needed supply of ascorbate when, and if, GLO functioning is lost. Other organisms might have no need for those sorts of back-up systems because their capacity to synthesize ascorbate remains intact.

As previously indicated, Barry Hall, among others, has shown that certain forms of life (e.g., bacteria) exhibit the capacity to develop alternative ways for synthesizing, say, the sugar lactose despite having lost the capacity to manufacture the enzyme galactosidase that plays a major role in the production of lactose. One wonders why chimps, gorillas, orangutans, and humans have not been able to conjure up a similar trick to do a work around with respect to the dysfunctional GLO gene.

Apparently, bacteria have degrees of operational freedom that various primates and human beings do not have. Similarly, primates and human beings possess an array of degrees of freedom that are not enjoyed by bacteria.

The foregoing sorts of degrees of freedom are part of the package that helps makes different organisms what they are. Those kinds of degrees of freedom help differentiate one species from one another irrespective of whether those capacities are a function of ancestral relationships or they are a function of the design that characterizes a given population of organisms.

To hear advocates of evolution say words to the effect that different organisms have different characteristics because that is the way the random generator of existence dealt out the hands of fate for those organisms is as much a non-answer – and just as annoying to the other side -- as when the proponents of creation state words to the effect that things are the way they are because that is the way God wanted it. Evolutionary theory and religion both rely on assumptions concerning the nature of the universe -- as well as rely on assumptions concerning the nature of one's relationship to Being -- but neither framework has the capacity to demonstrate – beyond a reasonable doubt to one and all -- that their underlying hermeneutic of the universe is true even though one, or the other perspective (and, perhaps, both) are correct with respect to, at least, issues such as, on the one hand, whether, or not, the universe is, in some ultimate sense, random or non-random, and, on the other hand, whether, or not, God exists.

Professor Miller believes that the dysfunctional nature of the GLO gene constitutes proof that human beings inherited that gene from primate ancestors. He might be correct, but he has not been able to provide the sort of definitive proof which shows that notwithstanding the fact that genes are passed down to subsequent generations within a given population -- and, perhaps such genes might even have been passed on from the members of certain other related populations that underwent speciation and, as a result, passed on those genes to members of what became a different species -- nevertheless, one still remains uncertain about whether the status of the GLO gene in various primates and human beings is a function of inheritance or due to the existence of similar vulnerabilities to which each of those species of organisms have succumbed during their respective – but ancestrally unrelated -- existential journeys through life.

Let's turn to another set of facts that Dr. Miller feels is even more persuasive than he believes the case of the GLO gene is when considering evidence about whether human beings are ancestrally related to various primates. This new set of facts involves hemoglobin.

Hemoglobin is a protein that has the capacity to transport oxygen – via the bloodstream -- to various biological destinations that have need for the latter molecule. Hemoglobin consists of a number of components, two of which are referred to as alpha-globin while two other constituents are known as beta-globin.

On chromosome 16 of the human genome there are five genes coding for beta-globin. One of those genes is given expression during the embryonic stage of development, while two other genes are expressed during the fetal stage of development, and another two of those genes are expressed when the individual becomes an adult.

A pseudogene is located in the interstitial space between, on the one hand, the two beta-globin genes that are expressed during fetal development and, on the other hand, the two beta-globin genes that are given expression during adulthood. The pseudogene consists of DNA sequences that are very similar to the other beta-globin genes that bookend the pseudogene, but the latter gene also contains a set of DNA sequences that render the gene inactive.

One of the dysfunctional sequences in the pseudogene prevents the exon regions in the gene that code for a protein from being transcribed into RNA sequences. A second dysfunctional region undermines the capacity of any RNA sequence that might somehow become transcribed from that gene to be able to bring about the synthesis of a protein, while an additional number of dysfunctional sequences in the pseudogene ensure that whatever protein might somehow might have been able to become synthesized would not be able to function properly.

Professor Miller does not indicate how the pseudogene came to be dysfunctional. Moreover, one wonders why five other beta-globin genes retained their functional capacities (i.e., were conserved) but the pseudogene became broken (i.e., was not conserved).

Obviously, human beings seem to be able to do quite well despite the presence of the pseudogene. One wonders what, if anything, the function of that gene might have been prior to becoming broken.

In addition, I find it somewhat strange that there seem to be at least six layers of constraint to ensure that the exon regions of the gene (the sequences that code for a protein) will never come to fruition or expression. The foregoing six layers of constraint could be just a coincidental function of the way a set of random mutations played out over time, or that arrangement might indicate that forces other than purely random mutation are ensuring that the gene does not get expressed.

Whatever the truth concerning the foregoing matter might be, Professor Miller considers the presence of the pseudogene in the human genome to be quite important because precisely the same sort of pseudogene is present in both chimps and gorillas. Not only does the same pseudogene exist in chimps, gorillas, and human beings, but, as well, all three species exhibit the same sort of sequential pattern of a dysfunctional nature.

Why would the same pattern of dysfunction be conserved across millions of years of evolution? If the process of mutation is random, and if there is nothing associated with the pseudogene that is dedicated to conserving its original functionality, then, surely, one might anticipate that the pseudogene in chimps, gorillas, and human beings would exhibit signs of further mutational modifications rather than having been conserved to retain exactly the same set and pattern of dysfunctional sequences.

The foregoing question is related to, but different from, the earlier question that asked why five beta-globin genes would be conserved while the pseudogene had been permitted to break and remain broken. Now, one wonders why the sequential properties of the pseudogene have been conserved instead of showing evidence of having been pushed in different directions as a result of the sorts of mutations – which might, or might not, be random in nature) that are likely to have happened in each of the three species over millions of years.

Professor Miller concludes that there is only one interpretation of the available data that makes sense. More specifically, he believes that some common ancestor of chimps, gorillas, and human beings must have undergone the degradation that led to the emergence of the pseudogene, and, then, that species passed the pseudogene on to its ancestors and, as a result, the pseudogene eventually would have been transferred further down the ancestral lineage to the line of species that became chimps, gorillas, and human beings.

There is at least one other alternative explanation for why the beta-globin pseudogene that appears in the genome of chimps, gorillas, and human beings contains the same set of DNA sequences. Neither the scientific side nor the religious side of Professor Miller is likely to find the following alternative to be very palatable.

I do not offer the following alternative possibility as proof of anything. Rather, the possibility is being mentioned in order to show that Professor Miller has failed to consider at least one possibility that might, or might not, be relevant to the pseudogene issue.

More specifically, on three separate occasions, the Qur'an -- which Muslims consider to be the final revelation that was communicated from God to human beings -- states that God caused certain human beings to become apes or made them into apes. The three instances occur in Surah 2, The Cow (2:65), Surah 5, The Dinner Table (5: 60), and Surah 7, Elevated Places (7:166). Given such a perspective, one could argue that the pseudogene that is found in chimps, gorillas, and human beings might have been the result of a transfer of physical properties (such as the hemoglobin/pseudogene arrangement) from human beings to chimps and gorillas, rather than being due to a transfer of physical properties (such as the hemoglobin/pseudogene) from chimps and gorillas to human beings.

Now, I realize that the foregoing possibility probably offends Professor Miller's scientific and religious sensibilities. Be that as it might, I do not believe that he – or anyone else -- is able to put forth either scientific or religious evidence to prove that the foregoing alternative is not possible or did not happen.

Whatever the strengths of science might be, that process is blind to whatever cannot be accommodated by its methodological and mathematical forms of engagement. As a result, although one might be willing to acknowledge that the claims of the Qur'an tend to fall beyond the parameters and purview of modern science, this concession is neither here nor there as far as the issue of whether, or not, the statements in the Qur'an are true.

From the perspective of modern science, what the Qur'an has to say about anything is considered to be irrelevant to the practice of science. Nonetheless, despite whatever wisdom might, or might not, be present in such an orientation, the existence of that sense of irrelevancy in science concerning the Qur'an and other spiritual issues might only indicate that science entails various kinds of lacunae that are affecting that method's capacity to establish the full truth about the nature of one's relationship with Being.

Moreover, even if it were the case that the pseudogene in chimps and gorillas did not find its way into those organisms as a result of certain human beings being made apes by the command of God, nonetheless, the existence of the pseudogene might have been placed in chimps, gorillas, and human beings independently of one another. For instance, based on my knowledge I have acquired concerning the Sufi mystical path during more than four decades of research – and limited though that knowledge might be – I would not be at all surprised to discover that the pseudogene could have some sort of metaphysical or cosmic significance that had nothing to do with ancestral relationships but was present in all three species as a sign or symbol concerning some deeper dimension of reality.

Professor Miller might be right, or wrong, with respect to his beliefs concerning the significance of the pseudogene in chimps, gorillas, and human beings. Whatever the epistemological status of that belief might be, he, apparently, failed to consider several issues (i.e., whether, or not, certain human beings became apes and the possible metaphysical significance of the pseudogene), and, as a result, his analysis of the pseudogene topic is somewhat arbitrary.

Scientifically, Dr. Miller might have been justified to proceed as he did in the matter of the pseudogene. Epistemologically, Professor Miller's perspective is fairly incomplete because, as noted above, there are a number of possibilities that carry implications for his position that remain unexplored by him.

Disputing the Indisputable

Professor Miller claims that: "In the real world of science, in the hard-bitten realities of the lab bench and field station, the intellectual triumph of Darwin's great idea is total. The paradigm of evolution succeeds every day as a hardworking theory that explains new data and new ideas from scores of fields." (*Finding Darwin's God*, page 165). The foregoing claims are both trivially true and, possibly, quite misleading.

The reason that the intellectual triumph of Darwin's ideas might be total in the world of science as well as in the labs and field stations where such science is carried out is not necessarily because Darwin's ideas have been proven to be true but, rather, because the individuals who tend to participate in the process of science might have chosen to buy into, or been induced into, or been persuaded into, or unduly influenced into accepting the delusional mind set to which evolutionary theory might give expression. Furthermore, while it could be true that the paradigm of evolution has an explanation for a great many things, one cannot necessarily also justifiably say (and the contents of the present book are the warrant for what is being claimed here) that the sorts of explanations that are being offered by proponents of evolution are necessarily true, and, if this turns out to be the case, then, the success enjoyed by the paradigm of evolution will be purely illusory in nature.

In 1998, the National Academy of Sciences released a document that was put together by a stellar group of experts that sought to bring the public up to date concerning the nature of science and biology and to overcome the public's resistance to the idea of evolution. Among other things, the foregoing report indicated that no one, on scientific grounds, could justifiably hope to viably sustain any sort of opposition to the claim that not only did all living organisms evolve from earlier life forms, but, as well, human beings are also subject to the same kind of evolutionary mechanisms as are all other life forms.

In terms of the principles of population biology, one could agree that later generations that arise from a given population derive, and, to varying degrees, alter a variety of genetic properties that were present in foregoing population of organisms. One also could agree that like other organisms, human beings also derive, and to varying degrees, alter a variety of genetic properties that were present in a prior population of human beings, and, therefore, human beings are subject to many of the same principles of genetics that characterize other forms of life.

Notwithstanding the foregoing concessions, what has not been proven – except, perhaps, at best, in a very limited sense -- is that the characteristics that are present in any given population of organisms necessarily owes the existence of those features to some common ancestor that, supposedly, was able to establish a lineage through which various genetic features were transmitted to a variety of different specie populations over time. New species – in a limited sense – might arise when a combination of mutations, together with different modalities of ecological separation and isolation, differentially affect the members of a given population and, as a result, bring about a bifurcation of the population and, over time, the emergence of a new species.

Nonetheless, the foregoing limited sense of speciation does not necessarily demonstrate that all species owe their existence and properties to such a process of speciation. To claim that speciation in the foregoing sense might occur, is one thing (and relatively non-controversial), but to try to argue that all speciation emerges through such a process is quite another matter and is riddled with a variety of problems, many of which previously have been explored during the course of this chapter.

Proponents of evolution are unable to provide a step-by-step set of specific transitions that lead from: Abiotic systems to living protocells; or, Archaea organisms to bacteria (or vice versa, or neither); or, anaerobic life forms to aerobic organisms; or, chemotropic systems to photosynthetic pathways; or, single-cell life forms to multicellular organisms; or prokaryotic cells to eukaryotic life forms, and so on. Evolutionary theory does offer a variety of tentative accounts concerning how – at least in general terms – the foregoing sorts of transitions might have occurred, but there is very little in the way of concrete proof that any of those tentative explanations are, indeed, true.

Evolutionary theory holds that the principles of physics and chemistry are adequate to account for all of the complexities to which various organisms have given expression since life first emerged on Earth. Nonetheless, as the foregoing paragraph suggests, evolutionary theory is unable to provide a set of specific, step-by-step events capable of demonstrating that the foregoing claim is true.

In all too many cases, evolutionary theory is held together by assumptions rather than hard facts. Assumptions might have a role to play in the process of science, but not everything that is being assumed is necessarily true, and, therefore, one has to exercise considerable caution when trying to assess the significance of any given statement by proponents of evolution.

Kenneth Miller is an expert in many aspects of biology and evolutionary theory, and he has served as an expert witness in a landmark legal case involving the on-going dispute between proponents of evolution and creationism. Among other things, he teaches courses in cell biology at one of America's great institutions of higher learning, and, as well, he writes biology textbooks that are used in high schools all across the United States.

Yet, as the previous hundred-plus pages of detailed analysis demonstrate in conjunction with various ideas that are contained in two of Dr. Miller's best-selling books – namely, *Finding Darwin's God* and *Only A Theory* – his position is not as unassailable as he seems to believe. Although Professor Miller feels that he has been able to show in the aforementioned books that evolution and Darwinian theory stand triumphant, I find (and throughout this book I have tried to provide the reader with a detailed sense of why I feel the way I do) that many of his arguments are unconvincing if not flawed.

Quantum Uncertainty

The last third, or so, of Professor Miller's book: *Finding Darwin's God*, explores a variety of issues that he believes will help complement his evolutionary perspective. An important component of the material presented in that final hundred pages of the aforementioned book concerns Dr. Miller's reflections on, and application of, a variety of principles drawn from quantum theory.

For instance, after reviewing some preliminary background material (toward the beginning of Chapter 7 – *Beyond Materialism*) concerning the foundations of quantum theory (e.g., the work of Max Planck in conjunction with the quantitative role of quanta and Albert Einstein's ground-breaking re-conceptualization of the photo-electric effect) Professor Miller proceeds to state two major points that will help frame the discussion that will ensue throughout the rest of the aforementioned chapter and carry over into subsequent chapters of his book. More specifically, Dr. Miller indicates that: (a) the uncertainties that are inherent in quantum theory are not a function of gaps in our knowledge that could be improved upon by better, more precise, modes of measurement, and (b) the probabilities that describe quantum phenomena exhibit patterns that indicate how large-scale physical and chemical events tend to be ordered even as single events remain indeterminate.

The foregoing points entail a number of problems. Since Professor Miller intends to use those two points to "inform" the foundations that shape the conceptual perspective to which he wishes to introduce his reading audience, critically reflecting on those points might be well advised.

Quantum theory reflects quantum methodology and vice versa. Stated in another way, quantum theory and quantum methodology exhibit a certain degree of resonance with the process of performing a tox-screen in order to detect the presence of certain kinds of drugs or molecules in a given sample in the sense that just as a tox-screen only permits one to identify the presence of items for which the test is screening, so too, quantum theory and quantum methodology permit one to detect the presence of only what the structural character of that theory and methodology permit one to identify or recognize.

Quantum theory and quantum methodology maintain that the nature of ontology is uncertain precisely because this feature of uncertainty is a function of the way in which quantum theory and quantum methodology engage reality. The conceptual baggage of uncertainty is present in the theoretical and methodological side of things rather than being present in the nature of reality.

Consequently, contrary to what Professor Miller claims, the probabilities that quantum theory and quantum methodology permit one to calculate don't "give order to the physical and chemical world" and they are not the reasons why "quantum indeterminacy does not produce universal chaos" (*Finding Darwin's God*, page 291). Rather, those probabilities constitute the metric through which quantum theory and quantum methodology measures the world ... a world that is quite independent of that metric but which has a character or nature that is capable of interacting with the quantum metric to the limit of the latter's capacity to engage reality.

Trying to measure the position of a particle will affect one's capacity to simultaneously measure that particle's velocity, and vice versa. Similarly, trying to measure the energy of a particle will be affected by the length of scale one uses to measure temporal facets involving that property of energy, and vice versa.

The methods being used to measure conjugate qualities such as mass and velocity or energy and time are self-limiting because they interfere with one another. They are capable of capturing or describing what is taking place only up to the limits that are inherent in the theory and methodology that underlies the metrics of quantum mechanics.

Thus, a certain kind of particle can be located within a certain distance of the nucleus of an atom, say, 88% of the time, but quantum theory and quantum methodology not only have no idea where that subatomic particle is during the other 12% of the time, but, as well, neither theory nor methodology can tell you how the particle comes to become manifest in the location that it does. Again, the dimension of uncertainty is inherent in the theory and methodology and not necessarily inherent in ontology.

Alternatively, one can use quantum theory and quantum methodology to describe what the odds are that a certain kind of atom will be likely to be in a reactive state at a given level of energy. However, neither quantum theory nor quantum methodology will be able to tell one what is going on during the time when that atom is not reactive or why, at a given energy level, the atom is reactive at certain times and not others.

Or, consider another example that is discussed by Dr. Miller. This example involves the phenomenon of radioactivity.

He begins by imagining some source of radioactivity that has been selected because it emits a beta particle (an energized electron) every second, on average, and will be able to continue doing so for some given period of time. According to Professor Miller, because each of the atoms is identical, one cannot know which atom will emit a beta particle, nor will one be able to predict in which direction the beta particle will be released.

If the radioactive substance really was composed of identical atoms, one might suppose that all of the atoms should release a beta particle at roughly the same time. The fact that such a mass release doesn't take place suggests there is something about the atoms that are releasing beta particles that differentiates them from the atoms that do not emit that kind of a particle.

At the present time, we do not know whether, or not, atoms that emit beta particles are, somehow, inherently different from atoms that do not emit beta particles. In addition, we do not know whether, or not, some sort of unknown process occurs, for unknown reasons, within certain atoms and renders them vulnerable to emitting beta particles, nor do we understand how a substance that exhibits the property of radioactive decay is able to keep track of its half-life properties.

We don't know that atoms are going to release a beta particle or in which direction those beta particles are going to be released because we don't fully understand the nature of radioactivity. Both quantum theory and quantum methodology cast a cloud of unknowing about the process of radioactivity that prevents us – at least up to this point in time – from understanding what is taking place, and, therefore, once again, the indeterminacy that surrounds radioactivity is a reflection of the problems inherent in quantum theory and quantum methodology and is not necessarily a reflection of the nature of reality.

In a way, quantum physicists remind me of Sergeant Schultz in the old television series: *Hogan's Heroes* who often would claim that he saw nothing, heard nothing, and knew nothing even though he clearly did see, hear, and know more that he professed. Quantum physicists have seen, heard, and know a great deal about how to calculate the probability distributions that describe different facets of the sub-atomic world but when they begin to talk about the relationship of sub-atomic realms to the classical world of everyday experience, they tend to hide behind the vagaries of the Copenhagen interpretation of quantum physics that contends, among other things, that the inherent nature of reality is indeterminate in nature.

If the inherent nature of reality is indeterminate in nature, then why do the probability distributions that arise from quantum calculations have the properties that they do. What are the actual ontological dynamics that underlie the so-called collapse of the wave function?

Quantum physicists never see the noumenal, ding an sich (the thing as it is in itself). They only apprehend a hermeneutical rendering of the phenomenal that is being framed and filtered by quantum theory and quantum methodology.

As a result, they are unable to provide an in-depth account concerning the fundamental nature of reality. Instead, they (at least those individuals, such as Dr. Miller, who make pronouncements about such matters) become lost in the inexplicable mysteries of indeterminacy.

Contrary to what Professor Miller claims in *Finding Darwin's God*, the foregoing probabilities do not give order to quantum indeterminacy. Instead, those probabilities give expression to the limits of what can be known through the filters being imposed on reality by the metrics of quantum theory and quantum methodology, just as a tox-screen will not necessarily inform one about all of the molecular components that are present in a given sample but will, instead, reveal only what the screening method has been set up to identify.

Quantum theory and quantum methodology are set up to identify probabilities of various kinds. Those probabilities reflect the nature of the metric being used and are oblivious to whatever else might be going on during the process when any given individual or group of individuals is engaging reality through the filters of quantum theory and quantum methodology.

Professor Miller claims that the inherent indeterminacy of nature indicates that the assumption underlying materialism is incorrect. More specifically, if materialism is predicated on the belief that when one understands the rules, laws, principles, mechanisms, and forces of the past, this will enable one to predict how the dynamics of chemistry and physics will unfold in the future, then, given – according to Dr. Miller – that quantum theory has shown how nature is inherently indeterminate, then materialism can't possibly be correct because the indeterminate nature of reality would prevent one from ever being able to know the past with sufficient precision to be able to determine how the dynamics of chemistry and physics will unfold in the future.

Before discussing Professor Miller's take on the subject of materialism and the challenge of indeterminacy, there are a few preliminary issues that should be clarified. For instance, quite apart from the issue of indeterminacy, materialism assumes that reality is nothing more than a function of physical and chemical dynamics

If the physical/material world is subject to forces other than the laws of physics and chemistry, then, irrespective of whether, or not, one can completely know the rules, laws, and principles of physics and chemistry that supposedly govern the material world, one will not be able to predict how the future will unfold because – if the foregoing possibility is true – the unfolding of the universe depends on forces other than purely physical and chemical ones. In other words, if the nature of reality cannot be shown to be reducible to principles of chemistry and physics, then irrespective of whatever the nature of the indeterminacy might be that permeates our understanding of reality (whether methodological or ontological in nature), one will not necessarily be able to calculate how the principles of physics and chemistry will unfold in the future.

Quantum theory is a form of materialism that, at the very least, places constraints on what can be known about the dynamics of material events. Consequently, quantum theory is not necessarily a repudiation of materialism as much as it seems to constitute a refinement of that idea's essential character, and, therefore, even if our relationship with Being is such that there is an indeterminacy – methodological or ontological – that limits our understanding of reality, this does not necessarily mean that the nature of reality is indeterminate, but, rather, it might only mean that our understanding of how physical and chemical systems unfold is subject to the sort of epistemological limitations that clothe our understanding of reality in veils of indeterminacy.

In addition, one might wish to consider the possibility that there might be other methods for engaging reality that permit one to by-pass the limitations of quantum theory – and a variety of mystical paths (rightly or wrongly) make such claims. If the foregoing sorts of claims were true (and I am not asking the reader to be inclined one way or the other on this issue but merely making a point of logic), then, irrespective of whatever kinds of indeterminacy might affect our material understanding of reality according to the principles of quantum physics, nevertheless, one still might be able to gain access to insights and understandings through non-quantum methods that disclosed – to varying degrees -- how chemical and physical systems unfold over time.

Having noted the foregoing points, let's return to Professor Miller's discussion concerning the relationship between quantum mechanics and materialism. Dr. Miller's belief that the inherent indeterminacy of nature is incompatible with a materialist notion of predictability appears to conflate methodology and ontology because it is based on the premise that quantum theory reveals something about the inherent nature of reality when, in fact, quantum theory and quantum methodology only reveal the nature of their own limitations for engaging reality.

Reality might, or might not, be inherently indeterminate. However, the nature of quantum methodology is such that one will never able to decide that issue by means of such a methodology because it is the very nature of that methodology which prevents one from being able to grab hold of reality in a more definitive fashion and, thereby, be in a position to actually know whether, or not, the nature of reality is determinate or indeterminate in character.

Professor Miller believes that quantum theory is capable of resolving a longstanding conundrum for those who are interested in pursuing a religious way of life. In other words, he feels that quantum theory permits one to be able to describe the world in an orderly fashion (via the probabilities that are provided through quantum theory and methodology) without forcing a person to feel compelled to conclude – as a strict materialism tends to require -- that the future is completely determined.

In essence, Dr. Miller has projected the methodological properties of quantum theory onto the nature of reality or ontology. Therefore, as a result of confusing or conflating methodology with ontology, Professor Miller believes that human beings have a way to avoid the implications of determinism that he considers to be entailed by a strict form of materialism, and although I feel that the foregoing belief of Dr. Miller is unjustified, this is not because I consider human beings to be completely determined (more on this later), but because I believe it is important to introduce a point of logic concerning the structural character of Professor Miller's argument at this point.

I do not know the ultimate nature of reality. What I do know is that conflating or confusing methodology with ontology does not entitle one to claim that quantum theory enables a person to do an end-around the issue of determinism since acknowledging that one's understanding of reality is indeterminate in nature cannot be used to demonstrate that reality is actually also indeterminate in character.

Because of the allegedly inherent indeterminate nature of reality that is a function of quantum dynamics, Professor Miller believes that mutations are every bit as unpredictable as are the dynamics of individual particles. As a result, he considers the future course of evolution to be unpredictable in nature.

Professor Miller often speaks in terms of random events. For instance, mutations, of one kind or another, are considered to be random in nature.

However, if the nature of reality is inherently indeterminate, then, one cannot possibly know whether those indeterminate events are, or are not, random in nature. Neither the notion of randomness nor the idea of indeterminacy can account for why certain kinds of probabilities rather than other possibilities characterize the way in which reality is manifested under various conditions.

If the nature of reality is inherently random, then, why do the same sorts of probability distributions keep occurring? If the nature of reality is inherently indeterminate, then, why do the same kinds of probability distributions continue to bubble to the surface with respect to the phenomena that are being described by quantum theory?

Whatever the answer to the foregoing question might be, randomness and indeterminacy do not seem to have synonymous meanings. They appear to constitute two different ways of describing phenomena that tend to conceptually push or pull one in the same sort of epistemological or hermeneutical direction – namely, one steeped in uncertainty and the unknown – even as they each (in their individual ways) – try to give the impression that something of a fundamental ontological nature is known about reality when, in fact, this might not be the case.

If the universe is inherently indeterminate, then, one cannot possibly know if events are random in nature. In other words, if the ontological character of the universe were inherently indeterminate, then, presumably, this would seem to preclude the possibility that the nature of that indeterminacy is necessarily random in character since if this were not the case, then the nature of the universe would no longer be indeterminate but random in character ... that is, one would know that the fundamental character of reality is random rather than indeterminate.

On the other hand, if reality is a function of purely random processes, then, the nature of reality is not indeterminate but random in character. Random phenomena might well have an element of indeterminacy about them, but this is not because reality is inherently indeterminate but, rather, is because reality is inherently random and, as a result, imposes a quality of epistemological indeterminacy on what can be known.

Putting aside, for the moment, the issue of whether, or not, evolution in Dr. Miller's sense actually occurs (i.e., an account that claims <u>all</u> forms of life arise through a Darwinian dynamic of speciation that is a function of a process involving variations – often random in nature -- that are differentially endorsed by forces of natural selection), one cannot necessarily claim that the future of evolution is indeterminate even though one might be willing to acknowledge that one cannot predict how evolution will unfold. Once again, one needs to distinguish between, on the one hand, being unable to determine the outcome of events because of epistemological problems inherent in a given way of understanding things (e.g., through the filters of indeterminacy or randomness) and, on the other hand, claims concerning whether, or not, the actual character of ontology is, or is, not determinate.

If reality is determinate in character, then what is the nature of that determinacy? Are there forms of determinacy that permit the sort of possibilities that are simultaneously ordered even as they contain degrees of freedom that permit variations in how some dimension of reality are manifested (which would be an important consideration with respect to any account of how choice might be possible in an universe that might be largely determinate in character).

Replaying Life's Tape

In the book *Wonderful Life*, Stephen Jay Gould explores different facets of the fossil discoveries that were made during the late 1980s in conjunction with the Burgess shale formations (finely layered sediments consisting of compacted clay and/or mud) that are located in British Columbia. Analysis of the Burgess fossils, together with more recent discoveries in China, has led to a transformation in the way that many paleontologists and other researchers think about life during a geological period known as the Cambrian that occurred between 510 and 570 million years ago.

Prior to the foregoing discoveries, many – if not most -- scientists thought that the life forms displayed in Cambrian fossils consisted of body plans that were, more or less, consistent with the body plans of many modern phyla. However, the analytical work of a number of researchers (e.g., Derek Briggs and Henry Whittington) indicated that the foregoing fossils encompassed at least nine extinct life forms that possessed characteristics that were not at all like the body plans found in modern phyla.

I will put aside questions about how either the extinct Cambrian life forms came into existence or how the body plans for modern life forms came into existence and concentrate on the fact that Professor Gould wondered about whether there was any way to determine why the foregoing life forms became extinct and, then, were replaced by organisms exhibiting a different kind of body plan. His answer to his own query was in the negative.

He maintained that the course of evolution is determined largely by chance, and consequently one cannot know how the process of evolution will unfold over time. In fact, Dr. Gould claimed – and Professor Miller seems to concur – that the role which chance played in the process of evolution was such that if one were able to rewind the tape of life and replay it, life likely would proceed down an evolutionary path that was different from the one that unfolded for life originally.

Dr. Miller expands on Professor Gould's foregoing position by mentioning how Einstein's often repeated quote concerning the idea that "God does not play dice" turned out to be incorrect because the physical evidence that supported quantum theory's principle of indeterminacy showed otherwise. Actually, quantum theory hadn't shown that reality is indeterminate in nature, but, instead, Einstein was unable to come up with a form of thought experiment that was capable of countering Bohr's ability to detect problems in conjunction with one, or another, structural feature of those thought experiments.

Bohr never demonstrated that the nature of reality is indeterminate. Rather, he pointed our various inadequacies entailed by the thought experiments offered to him by Einstein, and, as a result, Einstein was not able to prove – to Bohr's satisfaction (or that of a variety of other people) -- that God did not play dice with the dynamics of the universe.

Many people – including Bohr – believed he had won the argument by default. However, the only thing that had been clearly established was that Einstein had not been successful in his attempt to demonstrate that God did not play dice.

The absence of evidence does not necessarily constitute evidence of absence. Yet, in the matter of the Einstein-Bohr debate, many people seemed to believe that Einstein's failure to be able to put forth viable arguments against the idea of indeterminacy constituted a species of evidence against the idea of determinism.

Professor Gould's proposal about rewinding the tape of life is contrafactual in character. In other words, his idea is dependent on something happening – E.g., the rewinding of life's tape of occurrences -- that is contrary to what, in fact, has taken place. As such, Dr. Gould is not necessarily saying something about the nature of reality, but, rather, he is saying something about the character of his beliefs concerning reality.

Clearly, Dr. Miller endorses the element of chance that is present in Professor Gould's position. After all, the foregoing idea seems to fit in quite well with the notion of quantum indeterminacy.

However, Professor Miller distinguishes between random events and indeterminacy in a way that is different from the manner in which I previously proceeded with respect to those two issues. More specifically, he says that random events are those in which anything is possible and all possibilities have an equal likelihood of happening.

Dr. Miller, yet again, seems to be confusing or conflating ontology with methodology. A popular definition suggests that randomness involves a sequence or set of occurrences for which no algorithm can be established that would be capable of reproducing or generating that series or set of occurrences, and, as such, random events do not necessarily have anything to do with events that are equally probable.

Random events are processes for which no pattern seems to exist. Those kinds of events are not occurrences to which one can assign equal probability but, instead, they are an array of dynamics that seems to unfold in no identifiable order and, therefore, are unpredictable.

The notion of equal probability tends to crop up in conjunction with various kinds of methodology that seek to describe or model what might be taking place in a certain system of interest. The concept of equal probability is used to establish a base line against which to compare what actually is observed in a given system and to, thereby, be able to calculate whether, or not, the latter results are consistent with, or deviate from, the aforementioned baseline.

If throwing a die leads to results that are significantly different from an average of $1/6^{th}$ for any given face of the die (and there are various tests for determining what constitutes being significantly different), then, one could have reason to believe that the process of casting the die might not be random in nature. Similarly, if a coin is flipped and generates an outcome that deviates significantly from an average of ½ for either side of the coin during a long series of coin flips, then, one might have to consider questioning whether the coin flipping process is entirely fair (i.e., random).

Notwithstanding the foregoing sorts of considerations in which Dr. Miller appears to become entangled in his own skewed ideas concerning the nature of randomness, Professor Miller directs a certain amount of criticism toward those opponents of evolution who seem to confuse indeterminacy with randomness. The reason for doing so is because he believes that such individuals fail to grasp that indeterminacy is a central property in the mind of God.

Dr. Miller reminds his readers that there can be only one alternative to indeterminacy. According to Professor Miller, that alternative involves the sort of determinism that, among other things, would preclude God from being able to intervene in creation and change things.

Other than Dr. Miller's claims that the statements in the previous several paragraphs are warranted, I don't recall anything in either *Finding Darwin's God* or *Only a Theory* that demonstrates how the mind of God is necessarily indeterminate in nature. As previously indicated, Professor Miller has not actually shown how quantum theory demonstrates that the nature of reality is indeterminate, but even if quantum theory were capable of proving that the nature of physical reality is indeterminate, Dr. Miller has not shown that the mind of God is governed by those same principles of quantum physics.

In fact, earlier in this chapter, I indicated there is nothing inherently problematic with erasing the distinction between the natural and the supernatural and, thereby, permitting God to directly affect the way in which phenomena are being manifested through natural rather than supernatural means. At the same time, I also indicated that the foregoing possibility does not require one to be wedded to some form of pantheism in which the universe is God but, instead, allows for the universe to be akin to a virtual reality that has been programmed by God to run on the hardware of Divine capabilities.

If quantum phenomena were merely a modality of software that is used to help give expression to physical events in the universe, then Professor Miller would have to be able to demonstrate that God was incapable of programming the foregoing sort of software as a means of generating the probabilities that are able to underwrite the structural and dynamic properties of any phenomenon that God wished to bring about, just as a software programmer is capable of arranging coding to give expression to one kind of virtual reality rather than another. However, Dr. Miller has not put forth any arguments capable of demonstrating that God is, or would have been, incapable of the foregoing kind of programming.

Furthermore, as noted earlier, Professor Gould's idea about rewinding the tape of life is entirely contrafactual. Nonetheless, one could presume – without logical contradiction -- that God might be fully capable of rewinding the tape of life and bringing about precisely the same sequence of events as occurred the first time around.

Of course, as far as we know, the latter possibility might also be contrafactual in nature. Yet, other than our lack of knowledge concerning whether, or not, God has ever replayed, or would ever replay, the tape of life, there is nothing necessarily indeterminate about the possibility that if God wished to replay the tape of life, then God has the capacity to do so.

Professor Miller is concerned that if the universe is determinate in nature, then, one will deny human beings the sort of conceptual and moral space he believes is necessary for free-will choices to be possible in conjunction with issues involving the acceptance or rejection of God's existence, spiritual guidance, Divine love, and so on. Consequently, Dr. Miller believes that just as indeterminacy is inherent in the quantum properties of the universe, the same kind of indeterminacy is present in biological systems, and, by extension, the cognitive lives of human beings, and, as a result, choice represents the degrees of freedom inherent in the kind of indeterminacy that enables human beings to move in one direction rather than another by means of the tiny uncertainties that pervade a quantum universe.

However, what if one supposed that the nature of existence consisted of an indefinitely large series of interacting attractor-like basins (The term "attractor-like" is used because attractors are usually thought of as mathematical entities that give expression to a set of numerical values toward which a given dynamic tends to evolve, and the kind of attractor basin I have in mind is ontological and not just mathematical in character)? Since the structural properties of an attractor limit the way in which that sort of system might unfold over time, the dynamics of any of the aforementioned ontological attractors – taken individually or collectively – would be constrained by whatever the properties of those attractor basins might be, and their dynamics would be manifested within the sorts of parameters toward which those properties might tend despite beginning from a wide variety of possible starting conditions.

Attractors are determinate in nature but trying to capture the character of that determinacy often entails dimensions of methodological indeterminacy. In other words, attractor dynamics operate in accordance with determinate principles that, over time, unfold in ways that cannot always be predicted.

The realm of human choice might give expression to a dynamic that is attractor-like in its properties. For instance, choices are often determinate in nature but unfold in ways that become difficult to predict precisely because the dynamics of choice tend to be sensitive in a variety of subtle ways to the character of starting conditions as well as the presence of a wide variety of influences that are capable of affecting the trajectory of choice.

There could be degrees of freedom associated with the dynamics of choice that permit an individual to move in different directions within the limits afforded by the attractor-like basin that governs the parameters of behavior without necessarily causing those dynamics to break down and become something other than they are. Indeed, the foregoing degrees of freedom might be an example of the previously introduced notion of <u>defensive indifference</u> that, in certain circumstances, might be present in the universe.

Without wishing to claim that how one chooses is unimportant, nonetheless, it might be that irrespective of how one chooses, such decisions will not disrupt the dynamics of the universe because everything operates in accordance with the properties of the dynamics that are inherent in the attractor-like basins that governs each aspect of reality, and, yet, degrees of freedom might still be exercised within the context of that determinacy that help shape the character of that determinacy. In other words, God has the game of life well in hand, and, as a result, permits human beings to make whatever choices they like because the ramifications of those choices will not appreciably affect the overall outcome of the existential game even as those same choices carry considerable ramifications for how the lives of various individuals unfold over time.

Ontology is determinate in character. Nonetheless, there are certain degrees of freedom that are built in to that determinacy – such as choice -- that permit the dynamics of any given attractor-like basin to move in ways that cannot always be predicted even as the exercise of those degrees of freedom are determinate in nature and, therefore, never permits the dynamics to spill over into something that they are not and, consequently, behavior remains constrained by the determinate nature of whatever the particular properties of such a dynamic might be despite the presence of certain degrees of freedom that are present in those dynamics.

I have no idea how God would make any of the foregoing possible. Nonetheless, the aforementioned perspective is offered as a form of logic that runs contrary to the claims of Professor Miller that the only alternative to indeterminacy is a form of determinism that renders Divine intervention impossible as well as undermines the idea that human beings have the capacity to freely choose their destinies.

Moreover, even if one were to suppose that God were omniscient and, therefore, knew how any one, or all, of us would exercise the degrees of freedom that are inherent in the dynamics of the set of interacting attractor-like basins that we call life, nevertheless, that sort of knowledge would not have caused our choices to be what they are. Instead, that knowledge would merely be an awareness of what was going to take place ... just as an intelligence officer who know his, her or their foe or asset extremely well might know how a given person was going to respond without having caused that person to behave in one way rather than another.

Origins of Asymmetry

According to Professor Miller, molecular biology has discovered a material mechanism that is capable of providing details concerning the nature of inheritance. However, what molecular biology has not done successfully is to demonstrate that the material details of inheritance will necessarily lead to the process of evolution. The reality of the former (that is, molecular biology) might have demonstrated the possibility of the latter (i.e., evolution), but molecular biology has not shown that the mechanism of inheritance is responsible for the emergence of all life forms.

For example, Dr. Miller indicates that up until 1998, no one could explain how the embryonic cells of vertebrates were capable of establishing a spatial orienting axis that enabled the developmental process to differentiate between the right and left sides of an organism. As a result, some people wanted to claim that such a mode of orientation must be due to God, some sort of élan vital, or the like.

However, Professor Miller claims that when researchers in 1998 began to uncover the nature of the dynamics governing the biochemical and molecular biological processes that enabled organisms to establish an orienting axis that permitted developing cellular systems to give preference to one kind of orientation (e.g., right rather than left) instead of the other alternative (e.g., left rather than right), scientists were able to remove one more issue from the list of mysterious phenomena that might be due to something other than processes of biochemistry or genetics. As a result, Dr. Miller concludes: "I suppose you could say that God had lost another job ..." (page 214)

Although scientists gained insight into some of the details concerning the manner in which early on during the process of development cells undergo a process of asymmetric polarization (i.e., exhibiting or showing preference for a left-right, or right-left, orientation), researchers subsequently discovered that there were a number of ways that were employed by different life forms to accomplish the foregoing process of polarization. For instance, some organisms relied on the sterospecific character (left-handed versus right handed isomeric versions of a molecule) of a certain activating enzyme to orient various aspects of subsequent development, and there were other organisms that used molecular structures that were coded for by a maternal-effect gene to bring about or provide a signal for the asymmetrical polarization of subsequent development.

The foregoing considerations might have helped researchers understand how certain facets of development unfolded. Yet, those discoveries were not necessarily capable of satisfactorily answering questions concerning the origins of the organizational structure that regulated the expression of those different systems of asymmetric polarization.

For example, what particular sequence of steps brought about the DNA coding that underlies an organism's capacity to synthesize and release certain kinds of sterospecific enzymes at certain junctures during the developmental process and, thereby, bring about asymmetric polarization in subsequent growth? Or, one might inquire into the nature of the precise sequence of steps that were necessary to bring about the sort of DNA coding that enabled a maternal-effect gene to become active, or expressed, at a certain point during the developmental process and, thereby, releases a signal that the embryo recognizes as a sign to initiate asymmetric polarization during subsequent development?

Discovering that a process of asymmetric polarization occurs in accordance with principles of biochemistry and molecular biology is one thing. Knowing how different species first arranged DNA coding in certain ways to establish such a capacity is quite another matter.

The former sorts of discovery began in 1998. The latter kind of discoveries have not, yet, been made ... that is, currently, no one knows the identity of the set of step-by-step events that initially led to the accumulation of the sorts of DNA coding sequences that made different modalities of asymmetric polarization possible.

Do purely chance events govern the emergence of asymmetric polarization capabilities in various organisms? If so, what is the nature of the proof that would be able to demonstrate beyond a reasonable doubt (rather than the lesser standard of a preponderance of evidence) that, in fact, chance, random, or indeterminate events gave rise to the foregoing processes?

Or, alternatively, does accounting for the origins of the aforementioned sorts of organizational capabilities (in a step-by-step manner) transcend or exceed the explanatory potential inherent in the principles of physics and chemistry to be able to show, in a non-arbitrary manner, precisely how such a sequence of coding transformations took place? If one has scientific difficulty accounting for the origins of the capacity to bring about asymmetric polarization during the process of development, then, one cannot necessarily justify the removal of God from consideration – as Professor Miller appears to feel he is entitled to do -- when discussing certain aspects (e.g., issue of origins) -- involving the process of asymmetric polarization during development.

After all, there are many problems of an unresolved – and, perhaps, irresolvable -- nature entailed by the challenge of establishing the step-by-step, DNA coding events that enabled various organisms to acquire, for the first time, the capacity to push/pull embryonic development toward one form of asymmetric polarization rather than another. Perhaps, in time, some scientist or group of scientists might be able to determine how the organizational properties underlying the process of asymmetric polarization came about in any given organism, but that time does not appear to be now or in the near future.

Fundamentalism

The issue is not whether someone – e.g., Professor Miller – is able to offer a theory that can provide a plausible, meaningful, factually rich description concerning the natural history of life. I have no difficulty conceding that evolutionary theory satisfies the foregoing conditions – that is, evolutionary theory has many elements of plausibility, meaningfulness, and factual richness.

Nonetheless, one can have a great many facts at one's command and still not necessarily correctly understand the causal dynamics of the phenomena that are taking place in the context that is giving rise to so many different kinds of facts. In fact, evolutionary theory is a case in point since it has been able to generate a great many facts and, yet, that framework does not necessarily help one understand how all life forms came into existence.

The key challenge before us is whether, or not, the theory that evolution puts forth can be demonstrated to constitute an accurate description concerning the origins of <u>all</u> species as well as an accurate account concerning the origins of various potentials that are inherent in those species. My contention (based on the considerations that have been put forth during the previous 140 pages, or so) is that, at the present time, individuals such as Professor Miller cannot prove that the origins of <u>all</u> species are a function of the principles inherent in the theory of evolution even though that theory <u>might</u> be able to accurately explain – within certain degrees of freedom -- the origins of <u>some</u> species.

To claim – as Dr. Miller, Richard Dawkins, and many other proponents of evolution do – that the theory of evolution has won the debate concerning how best to account for the origins of species is incredibly premature and self-serving. Such a claim is a function of a hermeneutical orientation that in many respect seems to embrace a potentially delusional sense of confidence in the capacity of the basic principles of evolutionary theory to correctly account for the origins of all species despite a multiplicity of facts indicating that evolutionary theory cannot provide a step-by-step account for how many kinds of life forms first came into existence.

Professor Miller contends that the world that we observe is one that is governed by principles of physics and chemistry and, consequently, we are forced to conclude that life operates in accordance with the laws to which those principles of physics and chemistry give expression. There are several potential problems inherent in the foregoing perspective.

To begin with, one could easily argue – and with considerable success -- that what we observe in the world around us is often little more than what our modalities of hermeneutics permit us to see since a person's understanding of things often tends to frame and orient how experience is interpreted. The foregoing statement is not intended to relativize the truth but, rather, is meant to remind us that the relationship between what, on the one hand, is seen or observed and what, on the other hand, is actually present can be quite complex and is vulnerable to a variety of influences that filter and frame experience according to an array of expectations, interests, fears, beliefs, and biases that select, shape, and exclude various kinds of "facts".

Secondly, as physics has shown us again and again over the last 135 years, or so, whenever we think we have succeeded in reaching the very foundations of reality, some experiment or individual comes along to provide various kinds of evidence that indicate or suggest that, perhaps, there is a lot more to the nature of ontology than we originally believed to be the case. Currently, scientists might see the world in terms of the laws that give expression to the principles of chemistry and physics, but this does not necessarily preclude the possibility that those laws are themselves a function of an even more fundamental set of principles.

For example, some mathematicians believe the universe is inherently mathematical in nature. No one has been able to work out exactly how mathematical principles might bring the physical world into existence, but if it were true, then, ultimately, the laws of physics and chemistry are a function of dynamics that run deeper than the principles of chemistry and physics.

Furthermore, for thousands of years, mystics have been indicating that the world we see – Professor Miller's world of physics and chemistry – is illusory in nature. If true, then, the laws of physics and chemistry are not fundamental to the nature of reality but form something akin to a palimpsest in which the original mystical nature of the world has been effaced, or covered up, to varying degrees in order to be able to accommodate the laying down of another layer of parchment or manuscript that gives expression to the laws of chemistry and physics.

The fact of the matter is that in many ways we don't understand the fundamental nature of the world in which we are embedded. Physics, chemistry, biology, evolution, religion, mysticism, mathematics, psychology, sociology, history, and so on are all attempts to grapple with trying to discover what the fundamental nature of reality might be.

Professor Miller is entitled to believe that the world in which we exist is fully governed by the laws of chemistry and physics. Nevertheless, there are many questions swirling about the issue of origins in conjunction with the emergence of physics, the universe, life, various kinds of life forms, consciousness, intelligence, logic, or language, and, consequently, one can't necessarily be sure whether, on the one hand, God is required to operate in accordance with the physical and chemical laws of the universe, or, on the other hand, the laws of physics and chemistry are programmable principles that can be altered to accommodate God's intentions with respect to the nature of the virtual reality that God has created.

The form of scientific theology to which Dr. Miller seems to subscribe is somewhat like the religious theology to which all too many people subscribe. More specifically, Professor Miller seems to believe that just because something is written down in the book of nature, then, this means there is only one way to understand the significance of what has been written down.

As such, Dr. Miller appears to be something of an evolutionary fundamentalist. An evolutionary fundamentalist someone who seems to be unprepared, or unwilling, to consider the possibility that the laws of physics and chemistry might be all well and good as far as they go, but, nevertheless, allows his or her or their hermeneutical biases to prevent that person from being able to consider the significance of all of the data concerning, for example, the problem of origins (which has been a constant theme over the last 120-plus pages of this chapter) that might run counter to such a fundamentalist perspective.

God, Science and Evolution

Although Professor Miller tends to have specific human beings in mind when he criticizes certain beliefs concerning God's role in, or relationship to, the universe, I would like to offer some critical commentary of my own vis-à-vis Dr. Miller's position without necessarily conceptually aligning myself with any of the individuals toward whom Professor Miller is directing his thoughts. The foregoing form of firewall is being introduced to forestall the tendency of some people to try undermining the significance of what is being said by tagging me with one epithet, or another, that has nothing to do with my actual position.

For example, Dr. Miller suggests (see page 218 of *Finding Darwin's God*) that anyone who seeks to claim that the process of evolution could never account for the emergence of new species is advocating a position that places constraints upon God that not only prevents Divinity from acting in the present, but, as well, limits God to having been able to act or create only in the past.

Firstly, irrespective of whatever those individuals believe against whom Professor Miller might be directing his criticisms, one need not argue that the principles of evolution could <u>never</u> explain the origins of <u>some</u> species, but, instead, one might wish to maintain that the origins of a great many biological issues are shrouded in mystery (e.g., the genetic code, metabolic pathways, the first protocell, archaea, chemotropic life forms, cyanobacteria, eukaryotic life forms, metazoans, and so on), and, therefore, the principles of chemistry and physics are not necessarily capable of providing an accurate account in relation to how any of the aforementioned biological processes came into being for the first time. Evolutionary theory does, of course, have explanations to offer with respect to all of the foregoing issues, but whether any of those explanations are true, or not, is a separate matter, and, if they are not correct, then, while at some point in the future, evolutionary researchers might establish what the correct sequence of step-by-step events were that led to origins of this or that organism or biological process, at the present time, evolutionary theory does not know the answer to those questions and, conceivably, might never know the correct answer to such questions.

In addition, contrary to what Professor Miller claims, there is nothing in any of the foregoing considerations that necessarily would place limits on God's ability to act in the present or in the past. If it were true that random, chance, or indeterminate events were not able to bring about the origins of some given biological process (e.g., genetic code, circulatory system, immune system, endocrine system, nervous system, etc.) or life form and that, therefore, the creative talents of a Deity might be required to get something started, then a self-contained, God-independent, process of evolution would not be capable of correctly accounting for the origins of certain biological processes or organisms precisely because a set of organizing forces greater than those that are provided by physics and chemistry would be necessary to be able to underwrite the origins of the foregoing sorts of possibilities.

By claiming that the process of evolution might not be enough to enable one to correctly and fully account for the origins of various kinds of biological systems, one is not depriving God of the ability to act in either the present or the past. Instead, one is pointing out that a dimension of order, organization, and creative imagination needs to be injected into, or imposed upon, the processes of physics and chemistry in order to bring about the emergence of certain kinds of biological systems and organisms that the laws of physics and chemistry might not be able to accomplish on their own.

Presumably, God could employ the principles of physics and chemistry to regulate the dynamics of the universe. Nevertheless, whenever those principles need to be augmented, altered, or suspended for Divine purposes, then, God might change the nature of the material programming to accommodate those purposes.

In other words, physics and chemistry do not necessarily have the last word with respect to the manner in which the universe operates. Yet, God might well sit at the desk where the buck finally stops.

Consequently, one could agree that the everyday nature of reality often tend to be expressed in terms of the principles of physics and chemistry, Nevertheless, this doesn't mean that those principles are necessarily always capable of <u>initiating</u> the origins of, among other things, various kinds of biological processes and organisms even though, once initiated, the laws of physics and chemistry could be fully adequate to give expression to what had been initiated through God's creative programming presence.

According to Professor Miller, the quantum, indeterminate properties of the universe prevent human beings from being able to acquire a full understanding concerning the nature of reality. Furthermore, he contends that the ultimate, physical nature of reality is such that a full chain of causality is absent from the fabric of the universe.

Dr. Miller admits the foregoing considerations demonstrate there is a substantial form of epistemological inadequacy existing within science that, thereby, places limits on that kind of inquiry process to be able to provide a complete picture of the physical world. However, he also indicates that whatever the inadequacies of science might be, the indeterminate nature of the universe is not enough to prove that God exists.

While one might agree with Professor Miller that acknowledging the limits of science does not necessarily demonstrate that God exists, nevertheless, Dr. Miller is presenting issues in a problematic, if not distorted, fashion. For example, as previously noted earlier in this chapter, quantum physics has not shown that the ultimate nature of reality is inherently indeterminate.

One can as easily argue that the indeterminacy that is present in quantum physics is a reflection of the character of the theory and method that are used to engage reality. In fact, if quantum theory were really capable of demonstrating that the ultimate character of reality is inherently indeterminate in nature, then, quantum physics would be making determinate claims about something that, supposedly, was indeterminate in nature. Quantum theory and methodology do not so much grab hold of reality as they provide a set of frames and filters through which to engage reality and do so in manner that interferes with that process of engagement, thereby giving rise to a certain amount of indeterminacy.

In addition, although the nature of quantum physics tends to be entangled in various issues of indeterminacy, none of this precludes the possibility that, ultimately, the universe might operate in accordance with principles that are more fundamental than the laws of quantum dynamics. For instance, conceivably, the reason why various quantum events exhibit the probability distributions they do in different circumstances is a function of more fundamental principles or laws that regulate how quantum dynamics will unfold at any given time or in any given case.

Furthermore, quantum physics has not necessarily shown that lacunae exist in the chain of causality for the universe. Rather, quantum physics tends to entail the fact that there are lacunae present in that system's description of reality.

Even though -- as Professor Miller indicates in passing -- none of the foregoing considerations entitle one to say: "And, therefore, God exists", nonetheless, those considerations do tend to induce one to begin to wonder about, and, perhaps, try to find answers for, what makes different aspects of reality possible. However, the theory of evolution – as least as presently constituted – tends to place unnecessary limits on the direction and character of inquiry because, in effect, it filters reality by means of a number of conceptual barriers that do not permit, or acknowledge, the possibility that the theory of evolution might not be capable of providing an accurate account for the origins of <u>all</u> species, but, rather, might (but not necessarily) offer an accurate account for only <u>some</u> instances of speciation.

One should be willing to admit that the path of future scientific discovery is unknown, and, therefore, one ought not be too quick to conclude that because science cannot answer a particular question at the present time, then this means science will not be able to uncover the sorts of evidence in the future that will enable scientists to answer an array of questions that currently are not capable of being answered. Notwithstanding the foregoing sort of considerations, evolutionary biologists should be willing to admit that in the light of the many, many questions concerning the issue of origins that the theory of evolution is not presently capable of answering in any step-by-step fashion, then, at best, evolutionary theory, despite all its aspects of facticity, provides only a very limited system of understanding that, currently, leaves many issues unresolved.

One might be willing to concede that the conceptual edifice that gives expression to the theory of evolution tends to operate in accordance with the process of science. Nonetheless, when students are required to learn the theory of evolution as if it necessarily revealed the truth of things rather than being a lacunae-filled and, therefore, at best, only a partial account of certain aspects of natural history (and this seems to be the kind of attitude that appears to pervade the presentations of many people who teach and write about evolution ... or, at least, many of the ones that I have met), then, there is an aura of repugnance that is released through the indoctrinatory qualities that frequently tend to characterize the teaching of evolution -- whether in the classroom or in the many articles and books that are written by proponents of evolutionary theory – just as there is something repugnant present when indoctrinatory qualities pervade the teaching of any form of religion or spirituality.

Professor Miller claims he is interested in pursuing a traditional notion of God ... the sort of Deity that one prays to, marvels over, and wonders about. Dr. Miller believes the foregoing kind of God is the sort of Deity that is threatened by the theory of evolution.

If God exists – and I believe God does exist – then, I rather doubt that God feels threatened by the theory of evolution. Either (1) the theory of evolution is true, and God uses (or is required to use) the principles inherent in the process of evolution to bring about this or that form of speciation, or (2) the theory of evolution is not true – or true only in limited cases – and, as a result, is largely irrelevant to how God goes about arranging and engaging the universe.

Consequently, God would not appear to have anything to fear in conjunction with the theory of evolution. However, human beings who harbor certain kinds of theological beliefs might feel threatened by various aspects of the theory of evolution because if the principles that are given expression through evolutionary theory were true, then, one would be confronted with the challenge of how to proceed when one, or more, of one's basic beliefs appear to be contradicted by established facts.

According to Professor Miller, people pursue many kinds of religion, but there is only one form of science that exists. I disagree with the foregoing claims.

While people might develop an array of beliefs, understandings, orientations, interpretations, and practices in conjunction with the pursuit of religion, I believe there is only one kind of religion. More specifically, religion is the search for the truth concerning the nature of one's relationship with Being ... a search, truth, relationship, and Being that one considers to be sacred – that is, worthy of veneration and commitment (For much more on this approach to religion, see my work: *Final Jeopardy: Religion and the Reality Problem*).

Moreover, despite the fact that scientists have generated all manner of beliefs, theories, and models concerning how to interpret, understand, evaluate, and use various discoveries that arise through a process of observing, testing, and critically reflecting on that process, I believe that science – like religion (as defined by me previously) – consists of a process that involves seeking the truth concerning the nature of one's relationship with Being ... a seeking, truth, relationship, and Being that one considers to be sacred and, therefore, worthy of veneration and commitment.

The techniques, methods, and instruments that are used in science and religion might not be the same. Nonetheless, they both share a common purpose ... to discover the truth about the nature of one's relationship with Being.

People – whether theologians or scientists – have generated all manner of ideas about the nature of their relationship with Being. Despite those ideas, God remains what God is, and nature remains what nature is.

The cacophony of beliefs – whether religious or scientific – comes from the seeker's side of the foregoing equations. Reality is not other than what it is.

The challenge facing those who pursue either (or both) religion or science is to figure out what the nature of the truth is that accurately describes the nature of one's relationship with Being. The history of science and the history of religion, or the philosophy of science and the philosophy of religion -- considered both individually and collectively – are reflective and often critical commentaries concerning the character of the foregoing searchers.

Dr, Miller claims that science, unlike religion, shares in a common culture. He believes that one of the central principles governing that commonality involves being willing to participate in an open discussion of, and a critical reflection on, various issues concerning science.

Evolution Unredacted

The degree to which someone could be willing to be open to discussion and critical reflection on any given issue might not to be a function of whether the topic is science or religion. Rather, the key to being willing to engage in open discussion and critical reflection in conjunction with any given issue could be tied to how sincere, empathic, humble, courageous, honest, and fair one wishes to be when engaging another human being.

I have discussed issues with individuals who operate – at least on the surface of things -- out of a very different spiritual tradition (Buddhism, Hinduism, Native spirituality, Sikhism) than I do, and when those people (and I) exhibited the qualities of character mentioned in the last sentence of the previous paragraph, then, I found that we tended to share a common culture of inquiry regardless of topic. I also have discussed issues with those who claim to be scientists and when the foregoing qualities are not present, I tend to find that not much of a common culture exists to encourage either open discussion or critical reflection.

Although there are quite a few examples of the latter sort of situation that might be cited -- including many that arose during my life as a graduate student -- one of the most egregious examples I have personally encountered that casts a shadow of doubt on Professor Miller's claim that science takes place within the value boundaries of a common culture involves the following incident. Approximately 40 years ago, I was a participant in a committee convened by the provincial government in Ontario to address the issue of bias in school textbooks.

During a break in committee proceedings, I became involved in an exchange of opinions concerning the issue of evolution with a professor of anthropology who also was a member of the same committee. When the professor discovered that I had some reservations about the theory of evolution, he couldn't contain himself.

He went into a lecture mode and began questioning my intelligence and integrity in a variety of ways ... none of which had much to do with the topic of evolution. Among other things, he wondered how a graduate of Harvard could be so backward in his understanding of the world.

There are many additional examples that could be drawn from history to indicate that science does not necessarily occur within a framework of values that operates out of a shared culture of free inquiry, open discussion, and fair critical reflection. For example, one could cite the way in which Dr. Stanislaw Burzynski has been treated by much of the medical and science community for the last fifty years, or the manner in which Halton Arp was largely ostracized by the so-called community of astronomers in the United States, or the obstacles encountered by Dr. Judy Wood in conjunction with her materials science analysis of the World Trade Building destruction on 9/11, or the difficulties (e.g., obtaining employment or getting published) that were encountered during the last 30-40 years in many universities by a variety of physicists who might have resisted jumping onto the string theory bandwagon or voiced reservations about various aspects of Big Bang cosmology.

Professor Miller states that one of the most impressive discoveries made by cosmological scientists concerns the idea that the universe had a beginning. Whether the foregoing possibility was really an ontological discovery or merely a speculation that, over time, acquired the status of orthodoxy remains to be seen.

In fact, just prior to making the foregoing sort of claim, Dr. Miller indicates that, currently, scientists have not been able to figure out how the singularity (a point or instant during which space-time supposedly assumes infinite or indefinitely great density) that supposedly gave rise to the Big Bang actually might have worked. Indeed, the many difficulties that seem to permeate the process of trying to work out the physics of the singularity is precisely what has led an increasing number of scientists to reconsider the possibility that the universe might not have had a beginning after all.

Dr. Miller indicates that part of the evidence underlying the idea of a cosmological Big Bang (a name which, ironically, was the result of a sarcastic jab uttered by Sir Fred Hoyle in relation to theories that were non-steady state in nature) can be tied to the work of Edwin Hubble. More specifically, in 1929 Hubble discovered that the Frauenhofer lines (absorption spectra) found in the light from more distant galaxies seemed to be shifted more toward the red end of the spectrum than occurred in conjunction with stellar objects and galaxies that were considered to be closer to Earth.

Initially – and there is some evidence to indicate that he might have modified his original impression -- Hubble interpreted the increase in red shift values of various cosmic objects to be an indication that galaxies and stellar objects might be moving away from Earth. In other words, perhaps the materials that composed various galaxies and stellar systems had been jettisoned outward from some initial point of reference (i.e., a Big Bang).

What Professor Miller did not mention in his book, *Finding Darwin's God*, is that there is a body of research conducted by Halton Arp, Wolfgang Pietsch, Margaret Burbidge, and others that tends to undermine Hubble's initial claims concerning the possible significance of the red shift that can be detected in the analysis of light coming from distant galactic and stellar sources. According to Arp, Burbidge, and others, the presence of a red shift in the spectral analysis of light from stellar objects and galaxies did not necessarily indicate that the universe was expanding because there were quasars that appeared to be nearer to Earth that exhibited a much higher red-shift value than stellar objects that were considered to be much further away, and, yet, what had led Hubble, and others, to the idea that the universe might be expanding was because, supposedly, the data indicated that the further away a given stellar or galactic system was from Earth, then, the higher the value of the red-shift tended to be.

Notwithstanding the absence of the foregoing considerations, Professor Miller goes on to indicate that -- to date at least – cosmology does has not been able to put forth any evidence that can be considered to be capable of either demonstrating the existence of God or disproving God's existence (and, of course, even if God's existence could be proven, this does not necessarily resolve the issue of what God's actual nature might be). Similar things could be said in conjunction with issues involving the origin of life or the origin of various kinds of biological systems (e.g., genetic code, circulatory, pulmonary, immune, endocrine, and nervous systems), or the origin of the prototype for different species involving, for instance, chemotropic life forms, archaea, cyanobacteria, eukaryotic organisms, and/or metazoans.

In other words, despite the fact that many, if not most, evolutionary theorists are inclined to describe the origins of species issue in terms that seek to redact God's presence from the discussion, the possibility – if not reality -- of God often remains amidst the interstitial spaces that are created by the many unanswered questions that pervade the framework of evolutionary theory. Even Dr. Miller – despite being someone who believes in God – seems to be somewhat favorably disposed toward removing God from the framework of evolutionary theory by maintaining that the principles of physics and chemistry that underlie the process of speciation take place in a manner that does not require God's presence.

However well the foregoing kind of conceptual firewall might protect the purity and sanctity of science from being corrupted by the actual, or possible, presence of God, there seems to be something epistemologically dishonest about that kind of a stance. Talk of God might not be very scientific, but redacting God from the conversation appears to compromise -- to varying degrees, depending on the topic -- the epistemological credibility to which evolutionary theory gives expression.

There are a variety of reasons (many of which have been stated previously in this chapter) indicating that the attempt to seek the truth concerning the nature of one's relationship with Being is not necessarily a quest that can be resolved by science. Consequently, one wonders about the wisdom of requiring science to be a function of a set of materialistic processes (quantum in nature or otherwise) that often do not seem to be capable of fully or adequately accounting for the nature of our relationship with Being.

Methodologically speaking, one understands – if not appreciates -- why someone might like to keep the practice of science free from problematic, non-scientific sorts of influences. However, epistemologically speaking, one should be prepared to engage life through whatever means might legitimately be able to bring one closer to the truth, but one cannot automatically assume that the most legitimate means of realizing that aspiration is necessarily through science.

Why should human beings be forced to engage the fundamental questions of life thorough the arbitrary, and often artificial, filters of evolutionary science? Why shouldn't human beings be permitted to undertake the quest to discover the truth concerning the nature of their relationship with Being through a rigorous process of epistemological inquiry that is not necessarily tied to, or ruled by, the biases of evolutionary science or the biases of this or that form of theological speculation?

Relatively recently I read a book edited by Robert T. Pennock and Michael Ruse entitled: *But, Is It Science?* The collection of articles, essays and excerpted judicial decisions explored a variety of scientific, philosophical, legal, and educational issues that are encompassed by the evolution-creation controversy.

The foregoing book's central premise – voiced by scientists, philosophers, and several judges – was that the creationist perspective, whatever else it might be, is not science and, therefore, should not be admitted into, or be allowed to influence, the contents or methods of the science curriculum. This seems like an eminently practical point to make until one wonders -- irrespective of the scientific status of evolution -- whether, or not, evolution is true?

Evolutionary theory might give expression to very good science. Whether, or not, that theory gives expression to good epistemology might be quite another matter, and one wonders why, for example, students are forced to take a course – say, biology – that might teach a student how science works but contains elements (e.g., based on, or giving expression to, various facets of the theory of evolution) that very well either might not be true or might be true to only a very limited degree.

Somehow, the claim that evolution is the best scientific theory that is available rings hollow when considered against broader epistemological questions that raise a multiplicity of reasonable concerns about the degree of truth that is present in that theory. Assuming human beings are interested in seeking the truth concerning the nature of their relationship with Being, then, why do so many educational programs (whether in secondary schools or institutions of so-called higher learning) require students to learn material that might constitute good science but will not necessarily help a person to get any closer to – and, in fact, might push an individual further away from -- being able to seek and discover the truth concerning the nature of one's relationship with Being?

What value should be assigned to the idea that evolution is a scientific theory if the possible price for gaining access to such a system of thought is to be induced to lose contact with important facets of the truth? Instead of possibly being a Faustian-like bargain in which one acquires certain kinds of "knowledge" at the potential cost of one's soul, evolutionary theory might merely constitute a sort of false-positive dynamic that makes claims concerning the truth that turn out to be incorrect.

Being able to filter and frame existence through the lenses of the best available scientific theory concerning the origins of species might have little value. This is especially the case if that kind of hermeneutical orientation prevents one from being able to pursue a form of understanding that could lead to the truth even though it might not satisfy the conditions of science.

Let's not kid ourselves. The reason why the evolution-creationist controversy is so heated is because individuals on all sides of the matter understand that the underlying issue involves a conflict over not only the nature of truth, but, as well, it involves a conflict concerning the nature of the methods that are to be used to engage experience in order to try to access whatever facets of truth might be accessible.

Requiring individuals to engage various facets of life through scientific lenses – and this is what a science curriculum does – sometimes seems as arbitrary, artificial, and problematic as a process would be that sought to force individuals to engage life through some given theological perspective. The essential concern in any process that seeks the truth concerning the nature of one's relationship with Being should involve a rigorous process of critical reflection that is much broader than either science or theology tend to be.

Even the way in which issues are framed can limit and distort how one thinks about the foregoing matters. For instance, without necessarily agreeing with the following point of view, Professor Miller indicates that, from an emotional perspective, one of the most forceful critiques in opposition to the idea of the existence of God has to do, apparently, with certain structural features of the universe.

More specifically, part and parcel of the Copernican revolution that supposedly helped to displace human beings from occupying a central place in the universe involved replacing a geocentric model of the known universe with a heliocentric perspective. According to Copernicus, the universe didn't revolve about the Earth (and, therefore, some people concluded the universe didn't revolve about human beings), but, instead, the Earth was merely one of many bodies that revolved about the Sun.

Since the time of Copernicus, science has discovered that not only is the Earth a relatively small object journeying about a fairly average star, but, as well, our solar system constitutes a miniscule portion of a galaxy that does not appear to be special in any way and is, instead, just one system -- containing billions of stars and untold numbers of planets -- that exists along with an indefinitely large number of other galaxies. Consequently, some people have argued – in a pejorative sort of manner -- that if God created the universe just for human beings, then, God did so in a manner that took billions of years to unfold and, then, stranded human beings in a remote part of a planetary archipelago that was located somewhere toward the outer edges of an obscure galactic ocean and, therefore, presumably, hardly seems to give expression to a reason for celebrating either the glory of human beings or the glory of God.

Copernicus might have helped initiate the revolution that displaced human beings from the physical center of the universe, but such a revolution did nothing to displace human beings from, possibly, being part of the metaphysical (or spiritual) center of the universe. Just as the quality of being human is not necessarily a function of a particular kind of physical/biological form but might be due to the nature of the soul that allegedly has been breathed into a given modality of embodied existence, so too, metaphysical significance might have nothing necessarily to do with one's physical location within the universe but could be a function of whatever metaphysical (spiritual) role human beings might play in the overall scheme of things.

Wherever human beings might have been placed – irrespective of how small, off the beaten path, nondescript, and insignificant that physical location might be – then, presumably, such a location contains what is necessary for human beings to have the opportunity to develop their metaphysical potential – whatever that might be -- that is inherent in the soul. Perhaps human beings – irrespective of their outer form -- are those beings that, regardless of physical location on Earth or elsewhere in the Universe, have the capacity, via the potential inherent in the soul, to become metaphysically or spiritually realized individuals and, in the process, discover the truth concerning the nature of their relationship with Being.

The planet Earth is approximately 4.5 billion years old. The known Universe is roughly 13.8 billion years old, and the unknown Universe might be far more ancient.

Human beings supposedly arose on Earth about 2 million years ago. Organisms with the sort of soul that constitutes humanness also might have emerged periodically over billions of years in other regions of the known and unknown universe.

Has it really taken 13.798 billion years for human beings to show up? Or, is it possible that the human beings that have appeared on Earth are only among the most recent of such beings to emerge in the Universe?

Furthermore, even if nearly 14 billion years had to pass before human beings were permitted to show up, this is hardly a crucial epistemological consideration. Presumably, God gets around to things according to a Divine sense of propriety rather than in accordance with the temporal expectations of individuals who might have no idea of why things are the way they are or why various events happen in the way they do, but, nonetheless, use their ignorance for a metric in order to assess the nature of things – something that I previously noted that Professor Miller seems to be inclined to do from time to time.

A variety of people who are enamored with science, and, as a result, might tend to harbor a degree of disdain toward the idea of God and related topics are often as foolish and shallow in their way of framing things as are many individuals that approach issues in accordance with the dictates of a particular theological doctrine or set of doctrines. Consequently, as long as someone attempts to cast the ideas of God and religion in terms that are meant to frame those issues in the most self-serving of ways (whether from the perspective of evolutionary theology or religious theology), or as long as a person is reluctant to consider alternative ways of understanding a situation (whether scientific or religious), then those people might end up with notions – such as the alleged displacement of human beings from the center of the universe – that are more likely to obfuscate understanding rather than permit one to be able to seek the truth concerning the nature of one's relationship with Being in a manner that is as free as possible from all manner of problematic assumptions and conceptual biases.

Chance, Choice, and Determinism

Toward the latter part of *Finding Darwin's God*, Professor Miller raises some questions concerning the nature of chance and the role that the idea of chance plays in life, the world, and evolution. He views those issue through a lens that is ground from materials made of uncertainty.

More specifically, Dr. Miller believes there are outcomes for a variety of events that are genuinely unknown. Moreover, he seems to believe that the outcomes for those events are not only unknown to human beings but, apparently, those outcomes are unknown to God as well.

Professor Miller indicates that some people might wish to conceive of God as some sort of universal tyrant who controls every aspect of life. However, Dr. Miller rejects that sort of a possibility because, among other things, it appears to imply that God is responsible for every bad thing that happens.

Instead of pausing to dig more deeply into the nature of God's responsibilities as seen from the vantage point of limited understanding, Professor Miller moves on and asserts that the physical/material world has an existence that is independent of God, and, therefore, chance events involving uncertain outcomes are a genuine feature of the universe. For example, according to Dr. Miller, if someone flips a coin in order to decide which of two individuals will get to eat the last slice of pizza, then the outcome will be uncertain because God is indifferent toward such matters.

Professor Miller might, or might not, be correct with respect to his belief about God's indifference about whom gets the last piece of pizza. On the other hand, one wonders about the criteria that might be used to distinguish between the things toward which God is indifferent and those issues that are of interest to God.

In any event, indifference toward an outcome is not necessarily equivalent to uncertainty concerning that outcome. Conceivably, God could remain indifferent to the outcome of the coin flip and still know how the coin flip turns out without necessarily causing the coin flip to end up one way rather than another.

For example, one might suppose that Divinity's knowledge is so extensive that God would be able to grasp every facet of the coin flip [from: The precise nature of the force (when a person's intention is translated into action) that will be applied to the flip, to: The physics of the coin as it sails through air of a certain density, lands and bounces about on a surface characterized by some degree of elasticity, along with other properties, before settling down] and, therefore, be able to "see" how all those factors would come together to cause the coin to land on one side rather than the other without necessarily interfering with any facet of the coin flip or interfering with the conditions within which the coin flip took place. Alternatively, one might suppose that God could have known what the outcome of the coin flip would be before even before the coin was flipped because God has access to the entire multi-track tape of history and knows what is about to transpire even though those captured images depict a human being exercising free will to generate a coin toss … a coin toss that already has been faithfully stored on the tape of history and gives expression to an instance of free will or choice.

Can free will exist in a context that is completely known to God? As long as God doesn't cause the choice to assume one form rather than another, then, having knowledge concerning the character of that choice would not seem to negate the dimension of freedom that gives expression to a given choice since one might suppose that God's insight into how any person goes about exercising choice is so complete that God would never be surprised by how a person made his, her, or their choices.

Where does the human capacity to choose freely leave off and God's Will to be able to determine what will occur begin. Surely, God's Will already is entangled in the human capacity to choose because God made that kind of a capacity possible in the sense that certain degrees of freedom were built into the structural character of human beings that, under various circumstances, would enable free will to be exercised.

However, one might also suppose that various divinely generated constraints are likely to impinge on, and modulate, the human capacity to choose. As a result, the Will of God not only enables degrees of human freedom but, as well, constrains that same freedom in various ways ... some of which are known and many of which are likely to be unknown.

Whether God is indifferent to any of the choices that human beings make is not known. However, as pointed out previously, God might have an interest in whatever occurs but could operate in accordance with a modality of defensive indifference in which no steps would be taken to prevent human beings from undertaking certain kinds of action because those behaviors, whatever they might be, would not affect the outcome of the game that God had set in motion, but defensive indifference is not necessarily the same thing as harboring an indifference to whether, or not, human beings act in one manner rather than another.

God could assume responsibility for the outcome of the game, and, as a result, arrange general, and, sometimes, particular, features of that game to be in compliance with God's sense of responsibility toward the Universe. Nevertheless, human beings still would have the capacity to act freely – at least to some degree – despite the nature of the game being determined by a process that transcends the decisions of any given human being.

As noted earlier, Professor Miller claims that the events of the physical world are independent from the Will of God. He believes that the realm of chance, or uncertainty, is not only what permits the world to have a dimension of independence from God but, as well, it is only through such uncertainty that an independent, physical reality is possible.

Even if one were to suppose there were some dimension of uncertainty inherent in the nature of reality, one still has difficulty understanding what the issue of uncertainty has to do with the kind of capacity that would be needed to be able to make choices freely. If decisions are the causal result of events that are laced with uncertainty, then how does the property of choice fit into such a scenario?

There are two broad questions arising out of the foregoing considerations. (1) Are decisions (or, at least, some of them) the result of free exercises of choice ... that is, choices over which one has control, or (2) are decisions shaped by, and give expression to, a set of uncertainties over which one has no control?

To whatever extent the latter possibility is the case, then, one will be unable to exercise free choice. Uncertainty does not so much enable a person to be free but, instead, seems to siphon off freedom.

Professor Miller assumes that uncertainty is the key to the issue of freedom. However, while such uncertainty might be able to free human beings, to some degree, <u>from</u> Divine interference, that uncertainty does not necessarily provide individuals with the wherewithal to be able to <u>actively choose</u> -- and, thereby, impact the process of determining -- how <u>to</u> proceed.

Just as the issue of uncertainty in physics precludes issues of causality from being addressed, so too, the issue of uncertainty in the realm of human choice leaves unaddressed issues concerning the nature of the mechanism or means that makes the dynamics of actual choice possible. The presence of uncertainty might serve as a buffer against Divine interference and control, but that sort of function or service comes with a price that removes from human beings the capacity to be able to actually have control over what is decided, and this consideration would seem to go to the heart of the issue of real free will.

God is free to will one thing rather than another. Presumably, the One Who possesses such a capacity knows best how to create conditions that would enable human beings to make decisions and choices that are capable of being freely exercised.

Contrary to what is claimed by Professor Miller, what is central to the capacity to be able to choose freely or will freely is neither uncertainty nor indifference. Rather, at the heart of choice is a causal process that places ultimate and determinate control in the hands (or minds or hearts or souls) of the one who is making the choice

If people like, they can argue about whether or not God exists and whether, or not, human beings – with or without God's assistance – have the capacity to make free choices. None of those arguments, however, are capable of affecting the character of the logical points being put forth in the foregoing paragraphs – namely, that: (a) the capacity to choose has nothing to do with the quality of uncertainty or the possibility of Divine indifference; and, (b) the issue of choice depends on whether, or not, human beings have full control over the character of those choices.

Part of the reason why Dr. Miller wishes to embed the issue of free will within the clouds of uncertainty is to absolve God and, thereby, prevent Divinity from being held responsible for the many tragedies and horrors that occur in the world. Apparently, if God were responsible for such events, then, one could no longer consider God to be a good, loving entity.

Notwithstanding the possibility that many human ideas about the nature of love or goodness are highly arbitrary, self-serving, and deeply entrenched in ignorance concerning the nature of reality, one might also keep in mind a major theme in a 1964 book by Joanne Greenberg (using the pen name Hannah Green) that is captured in its title – namely, *I Never Promised You A Rose Garden*. The book is a semi-autobiographical account of a 16-year old individual who journeys into madness and struggles to regain control over her life and who, at a certain point, is counseled by the doctor who is helping her to remember that the doctor never promised that traveling the path back to sanity would be an easy process.

To be sure, life is filled with the "slings and arrows of outrageous fortune." Nonetheless, life also provides an opportunity to engage those slings and arrows and – through them -- learn about oneself, others, and the nature of one's relationship with Being.

Are death, pain, suffering, tragedy, and so on good things or bad things? Surely, one is not in any epistemological position to generate an informed opinion concerning the foregoing kinds of questions until one grasps – if one is ever able to do so – the actual nature of the universe.

Prior to that point, all judgments are premature. Consequently, while one understands Professor Miller's concerns with the possibility that if God is responsible for the universe, then, God would seem to have culpability for events that many people find abhorrent, nevertheless, perhaps God knows something that we don't and, therefore, human beings might need to temper their surface impressions concerning the nature of reality with a more nuanced manner of engaging and evaluating life.

Of course, irrespective of whether, or not, one grasps the nature of reality in all of its dimensional aspects one is often required to make choices about how to proceed. Whatever else life might be, it is a cauldron of bubbling possibilities that confront one with a variety of choices that exist not because of ontological uncertainty or Divine indifference but because the structural character of life is such that, somewhat ironically, one cannot escape the need to make choices irrespective of whether, or not, one wishes to do so.

Professor Miller claims that the physical world is, to some degree, independent of God's will. Among other things, he makes the foregoing claim in order to be able to create the sort of conceptual space within which the process of evolution might have the freedom afforded by uncertainty to bring about the sort of changes in the variability of any given population that are capable of leading, eventually, to the emergence of all the species that have ever existed, exist now, or will exist in the future (see the discussion on pages 234-236 of *Seeking Darwin's God*).

Unfortunately, whether intended or not, the foregoing scenario appears to be an exercise that is forged more by a process of conceptual slight-of-hand and misdirection than anything else. Dr. Miller is assuming that the kind of uncertainty that, supposedly, is inherent in the nature of reality is capable of generating all species that have been observed to appear on, and in, Earth, but Professor Miller never actually proves that the foregoing assumption is warranted in all cases ... instead, he tends to suggest that this is the case.

Therefore, evolutionary theory becomes a narrative-like template that is to be superimposed on the data of science to permit a person to make sense of that information in a manner that can be reconciled with the principles of science rather than constituting an account that can be shown, in a step-by-step fashion, to be an accurate reflection of what has transpired during the natural history of life on Earth. In compliance with the foregoing narrative, Professor Miller suggests that the ancestors of vertebrates didn't have to survive the Cambrian period, and mammals didn't have to give rise to primates, and primates didn't have to lead to the emergence of *Homo sapiens*.

According to Professor Miller, all of the foregoing sequence of occurrences could have turned out other than they did. However, chance, uncertain events pushed evolutionary history in certain directions and not others.

The realm of uncertainty becomes a cornucopia for the theory of evolution. Whatever has happened during the natural history of life on Earth must have been possible because the conditions inherent in uncertainty made it so.

Such flexibility is very convenient both theoretically and practically. Without ever having to prove anything, one merely has to claim that all species owe their existence to that which uncertainty makes possible together with that which is subsequently endorsed by the dynamics of natural selection.

Although Dr. Miller believes in God, he does not believe that God is in control of the physical universe. Rather, he feels that uncertainty and chance events are in control of what transpires in the world!

Seemingly, Professor Miller – like Darwin before him – wants God to bow in submission to the theory of evolution despite the fact that chance, uncertain events have never been proven to be the cause that underlies the origin of all manner of species. In addition, Dr. Miller – like Darwin before him – feels that science forces God to use a pathway constructed from a series of assumptions concerning the presence of certain kinds of chance, uncertain events that are needed to bridge a variety of evidential chasms that separate different kinds of evolutionary proposals and various questionable conclusions concerning those proposals.

Dr. Miller – like Darwin before him – wants to be able to reconcile his belief in evolution with the existence of God. That modality of reconciliation seems rather shaky because the whole ontology of chance, indeterminate events has not really been established in a convincing fashion.

As a cell biologist Professor Miller intimates that his profession permits him to gain insight into the nature of the dynamic through which chromosomes are formed during the process of meiosis as different contributions of the mother and father are brought together in a random, chance, indeterminate manner. How does one have insight into a random, chance, indeterminate dynamic and understand that such an event is, indeed, random, chance, and indeterminate?

Dr. Miller doesn't <u>know</u> that meiosis is a random, chance, and indeterminate set of events. Rather, he <u>assumes</u> those events are random, chance and indeterminate because that assumption saves the appearances of the kind of evolutionary theory that he wishes to promulgate.

Supposedly, according to Professor Miller, the "fact" that inheritance is a random, chance, indeterminate event does not affect God's plan for the universe. Yet, aside from presuming that chance, indeterminate events have the capacity to equip human beings with the sort of conceptual maneuvering room through which human beings allegedly enjoy a certain modicum of freedom, Professor Miller never really indicates how – or proves that -- God's plan requires, or makes use of, random, chance, indeterminate events.

Could God have been able to use random, chance, indeterminate events as part of, or to carry out, a Divine plan? Presumably, God could have done so if that is how God wished to proceed.

However, the issue is not what might have been or could have been. The issue is whether, or not, God – if God exists (and I believe God does exist) – actually did, or was required to, work (as Dr. Miller claims) with random, chance, and indeterminate events during the process of implementing whatever Divine plan was guiding the way in which the universe unfolded?

Professor Miller doesn't actually know the truth concerning the foregoing issue. He just assumes that his understanding concerning the matter is correct, and he forced to make such an assumption because, otherwise, his theory doesn't really explain what he claims it does.

The idea that the inherent nature of history is unpredictable and, therefore, might have unfolded in a different fashion than it did is, for Dr. Miller, "almost self-evident" (page 237 of *Finding Darwin's God*). One wonders about what reservations he might have had that induced him to qualify "self-evident" with the word "almost".

Of course, Professor Miller could be correct. History might be shaped by chance, random, indeterminate forces and events.

Nevertheless, the notion that history could have been other than it turned out to be is neither necessarily self-evident nor 'almost self-evident'. Furthermore, even though Dr. Miller disagrees with anyone who feels that God could not have – or would not have -- used chance, random, indeterminate events to help shape the world we see around us, Professor Miller is more than a little vague on the details concerning the nature of such a dynamic.

According to Dr. Miller, no one expects a religious person in the West to believe that God organized human events in a manner that was intended to bring about, say, the Civil War, the Holocaust, or any other historical event. Therefore, he believes no one should expect a religious person in the West to believe that God organized nature in a manner that was intended to bring about the success of one species rather than another or in order to cause human beings to emerge at a certain point in evolutionary history.

Somewhat inexplicably, Professor Miller likens the idea that God might determine how history will unfold as being akin to a process of rigging a game (see page 238 of *Finding Darwin's God*). With respect to the foregoing perspective, one might ask whether a computer programmer who creates the rules and conditions for a game is engaging in a process of rigging the outcome of that game or is the programmer merely providing a set of parameters that will offer both degrees of freedom as well as various constraints to those who participate in the game.

The game is organized so that the choices of the participants will determine how the game will unfold in any particular case. Nevertheless, the conditions and criteria that determine what constitutes winning are set by the programmer?

While God might not have arranged human affairs in order to bring about, say, the Civil War, God very well could have created human beings despite possessing an intimate, deep understanding about what human beings are capable of doing, and, therefore, might have been able to foresee – in considerable detail perhaps – the nature of the tragedies and horrors that human choices could bring about individually and collectively within a certain framework of time and space. Nevertheless, I'm not sure how any of the foregoing possibilities constitutes a process of rigging the game.

God created human beings with various potentials. Human beings have the responsibility for determining which dimensions of their potentials are going to be pursued.

As indicated during a discussion earlier in this chapter, one might consider the possibility that irrespective of whatever dimensions of life human beings – individually or collectively – decide to pursue, none of those choices will necessarily be capable of undermining or preventing the game of life from being what it was intended to be from the beginning ... for example, an opportunity for human beings to seek, and, perhaps, discover the nature of their relationship with Being. In other words, life constitutes a set of interacting attractor basins that operate within the Divinely established set of degrees of freedom and constraints that generate the dynamics that give expression to those attractor basins, together with their interactions with one another, but in accordance with the rules and principles of the game, none of those basins of potential will ever be induced to spill over into something that is not consistent with that potential.

Furthermore, one might suppose that like any good programmer, God would have been able to build features into the existential game that might be, on the one hand, frustrating, difficult, challenging, seductive, and/or painful, while, on the other hand, some of the features of the game might give expression to various kinds of fortuitous twists and turns of – possibly -- good fortune.

Similarly, one might suppose that God could arrange to bring into existence any manner of species that Divinity decided lent interest, variety, complexity, challenge, or possibility to the nature of the virtual world that had been programmed. Once again, I'm not sure why any process that involved organizing the manner in which natural history unfolds should be considered to be an exercise in trying to fix the game rather than merely serving as a way to establish various parameters that help define various properties of the game.

According to Professor Miller, the freedom to act forces God to permit the future of created beings to be left open or indeterminate. Nevertheless, as pointed out earlier, the freedom to act seems to depend on possessing the capacity to act freely rather than on whether, or not, the future remains indeterminate.

In other words, one can conceive of a possibility in which human beings are able to freely choose that which God knows they will choose. Moreover, even though the capacity to choose might have been established by God, God's ability to be able to understand why and how human beings make the choices they do is not necessarily the same thing as causing the choices that are made

Just as players in certain kinds of bridge tournaments are free to play pre-determined hands in any way they choose, so too, human beings are free to play the hands they are dealt in life in any way those individuals choose despite the fact that there is nothing indeterminate about the nature of the hands they have been dealt or the rules of the game that need to be observed during the process of playing those hands. The freedom to act does not necessarily depend on the process being open-ended but, instead, depends on the way in which the determinate features of life are engaged through human choices.

One could even suppose that the outcome of certain choices might already be pre-determined. Yet, a person's decision to emotionally or conceptually accept or reject those results could serve as the metric through which those decisions are evaluated rather than making the metric to be a function of the actual outcome ... e.g., it is not whether one wins or loses but how one plays the game that matters.

The J Wellington Wimpy Factor

A primary theme of evolutionary theory concerns the process of competition. Just as being hired is often the first step an athletic coach takes towards becoming fired, so too, being born is the first step organisms make toward becoming fired by life.

However, along the path that links birth with death, some organisms succeed in leaving numerous offspring while other organisms are less successful in that regard. Success is often a function of a battle for finite resources, including mates.

Yet, within the context of a highly competitive nature that Lord Tennyson described as being "red in tooth and claw," Darwin and others noted that there were many examples in which the quality of competitiveness seemed to be absent from the behavior of organisms. For example, worker ants share the hard-earned food they have secured with all the members of the colony rather than hoarding that resource for their exclusive use, and worker bees are prepared to give up their own lives in order to protect the rest of the hive, while in many herds of animals (including elephants and caribou), the strongest members of those herds can be observed spending time and effort protecting the herd's most vulnerable members rather than pursuing the narrow survival interests of the strong.

Since the groundbreaking work of Edward Wilson, sociobiologists (who explore the structural and dynamic properties of those aspects of genetic inheritance that are believed to underlie social behavior) have documented that cooperative, altruistic, and self-sacrificing kinds of behavior have selective value. In other words, although initially the foregoing kinds of behavior appear to be somewhat counterintuitive when considered against the backdrop of a world that seems to be governed by a rigorous, seemingly relentless spirit of competitiveness, nevertheless, cooperative and altruistic behavior can be understood to serve the long-term interests of the species, and as a result, help increase the likelihood that not only will the genes underlying cooperation and altruism survive but, as well, will help enhance the chances for other properties of that species to also be able to survive.

In effect, cooperative and altruistic behaviors help to perpetuate many of the genes that are present in the organisms that are engaged in that kind of behavior since the organisms that are being assisted carry many of the same genes that are present in the organisms that are assisting other organisms through some form of cooperative or altruistic behavior. As such, the foregoing kinds of behavior enhance species competitiveness and, as a result, the genes underlying those behaviors are likely to be favored by the process of natural selection.

While sociobiologists might have an explanation for the nature of the evolutionary function that is served through cooperative and altruistic behavior, that explanation does nothing to account for how such a capacity came into being in the first place. If one only asks why that sort of behavior exists, then evolution can provide an account that makes sense (i.e., it exists because it has selective value), but if one asks how that sort of behavior came to be possible – that is, what were the set of step-by-step transitions in genetic sequences that gave rise to the capacity to be co-operative or altruistic, then, the idea of evolution tends to make a great deal less sense.

Introducing the notion of natural selection into the mix does not resolve the uncertainty that permeates the question of origins. For, although natural selection might account for why certain qualities and properties are selected (e.g., they increase the likelihood of species success), natural selection cannot necessarily account for how that which is selected came into being in the first place … at least not without requiring someone to have to swallow a great deal of arbitrary speculation concerning such a possibility.

What was the series of transitions in DNA sequencing that caused the gene or genes underlying the capacity for cooperativeness or altruism? In order for the theory of evolution to be a good scientific account, one needs to do more than offer a plausible narrative that is capable of explicating the role that various kinds of properties, qualities, or behaviors play in helping a species survive … one also needs to provide a demonstrable step-by-step account that shows how different genes came into being that make the aforementioned properties and behaviors possible.

Evolution theory is quite accomplished in providing the former sort of account – that is, accounts that show how certain properties, once they arise, are likely to be favored by a given set of conditions (i.e., the process of natural selection). However, evolutionary theory is virtually devoid of evidence (as opposed to hypotheses and speculation) capable of showing how different genes came into existence.

Yet, Professor Miller repeatedly returns to the idea that the origins of all organisms come from the physical, material properties of chemistry and physics. Moreover, he even goes so far as to claim that even though human beings might find comfort in the idea that there is some sort of divine spark within us, nonetheless, Dr. Miller claims that human beings – along with the other organisms that currently occupy the Earth at occupied this planet at some point in the past -- can be reduced to being nothing more than chemistry and physics (For example, see page 250 of *Finding Darwin's God*).

Until Professor Miller can prove, among other things, that life, consciousness, intelligence, language, memory, reason, creativity, and spirituality are nothing more than a function of chemistry and physics, then, his foregoing comments seem more than a little premature. Furthermore, irrespective of whatever role chemistry and physics might play in the process of life, consciousness, intelligence, and so on, Dr. Miller really has little, or no, evidence to indicate that a variety of non-physical and non-material elements might be involved in the foregoing processes.

Despite the inclination of many scientists to suppose that life, consciousness, intelligence, and so on are processes of automatic self-assembly that occur when undergoing conditions that are characterized by forces that are far from equilibrium (Ilya Prigogine), or are a function of systems that are governed by non-linear chaotic dynamics, or that give rise to emergent properties as a result of the conditions of complexity that govern those sorts of phenomena, nonetheless, currently, there is very little evidence to indicate that life, consciousness, intelligence, reason, and so on can be reduced to a set of purely physical and chemical principles or laws.

Moreover, one can acknowledge that Professor Miller might be right when he indicates that we should not ignore the possibility that science will, in the not to distant future, be able to provide the kind of evidence that might be capable of demonstrating how life, consciousness, and a variety of other qualities are nothing but physical, chemical processes, but, one should also keep in mind the character of J Wellington Wimpy from the comic strip Popeye who is frequently promising to pay someone next Tuesday for a hamburger today. There is a certain quality of Wimpyness associated with the refrain that the promissory notes issued on behalf of science today in the form of speculations and various hypotheses will be paid for with scientific legal tender (proof) at some point in the future.

Just as currently evolutionary theory cannot account for the emergence of the set of step-by-step molecular transitions that led to genes capable of underwriting cooperative and altruistic behavior, so too, at the present time, evolutionary theory cannot account for the set of step-by-step transitions that brought about the construction of the genes that supposedly underwrite properties such as consciousness, intelligence, language, reason, and so on. What is consistently missing from evolutionary theory is a viable, detailed step-by-step account concerning the origins of the kinds of organized order that makes life, the genetic code, metabolic pathways, various kinds of biological systems (digestive, circulatory, pulmonary, immune, endocrine, or nervous), consciousness, intelligence, reason, language, creativity, and so on possible.

One can easily grasp the selective value that the foregoing capabilities might have once they arise, but one has much more difficulty trying to figure out how – or if – chance events molded by natural selection would necessarily be capable of generating the foregoing kinds of qualities that exhibit the intricacies of organized order.

Professor Miller claims (see toward the bottom of page 253 in *Finding Darwin's God*) that evolution is the means through which God freed human beings from the determinacy of strict materialism. Unfortunately, rather than provide the sort of step-by-step account that one might expect to be provided by an allegedly scientific explanation, Dr. Miller retreats into the speculative mysteries of indeterminacy theory and, as previously pointed out earlier in this chapter, he never really demonstrates (despite putting forth various philosophical, theological, and quasi-scientific arguments) that evolution, indeed, was God's way of freeing human beings from the limitations of materialistic determinism.

The Making of Flowers

Early on during the last chapter of *Finding Darwin's God*, Professor Miller provides an overview of research conducted by Elliot M. Meyerowitz concerning the process that causes plants to flower. Dr. Miller notes that there are four main components in a flower – namely, pistils, petals, stamens and sepals – and each of those four elements actually give expression to a modified form of leafiness.

In order for a flower to emerge, leaves need to be signaled that they should stop expressing the genes that manifest one, or another, kind of leaf and start expressing genes that will lead to the formation of flowers rather than leaves. The research team led by Professor Meyerowitz spent several years engaged in various kinds of genetic experiments before they were able to establish that there were four genes in a plant that had to be turned on and off in a particular order before a flower became possible.

After providing a summary of the flowering process, Professor Miller thinks back to his catechism classes that took place in the 1950s under the watchful tutelage of Father Murphy and proclaims that Father Murphy was wrong. More specifically, Professor Miller stipulates that, contrary to the claims of Father Murphy, God doesn't make flowers, but, rather, the process that induces genes to switch on and off in a specific order is what makes flowers.

Dr. Miller goes on to indicate that Father Murphy had made a mistake ... one that Professor Miller feels occurs fairly often amidst individuals who are religiously inclined. The mistake to which Dr. Miller is alluding concerns the tendency of some people to suppose that because, at a given point in time, science might not be able to answer why things are the way they are or be able to provide an explanation for what makes various kinds of phenomena possible, then, science will never know the answer to those kinds of questions.

In the early 1950s scientists did not understand how leaves led to the emergence of flowers. According to Professor Miller, Father Murphy used this state of knowledge (or lack thereof) to slip in the idea -- somewhat illicitly apparently -- to his catechism class that God was the One Who was responsible for the flowering of plants.

I don't know what Father Murphy had in mind back in the 1950s. Furthermore, I am uncertain whether, or not, he knew that scientists had no idea how plants created flowers and, as well, I am unsure about whether, or not, he believed that science would never discover how flowers are manifested in plants.

What I am fairly certain of, however, is that Father Murphy wasn't necessarily wrong when he proclaimed that God makes flowers. Despite the fact that Professor Miller seems to feel that he (i.e., Dr. Miller) has provided a full and adequate account for how flowers come forth from plants by discussing the induction process that leads to flowers, in reality, Professor Miller has not demonstrated what he claims to have shown when he provided an overview of the aforementioned work of Professor Meyerowitz.

More specifically, Dr. Miller does not provide a step-by-step account concerning how the four genes that underwrite the flowering induction process in plants first came into existence. In addition, Dr. Miller has not provided an account for how the processes that control the timing sequence for turning various genes on and off in a certain order came into being.

To be sure, in the absence of the foregoing kind of information, one cannot justifiably conclude that, therefore, God must have created flowers. On the other hand, knowing that four genes have to be turned on and off in a particular sequence does not preclude the possibility that God actually did make flowers unless Professor Miller can show that God is not the One Who, at some point, arranged for genes consisting of certain sequences of DNA molecules to be embedded in a regulatory system that turned those genes on and off in a certain order.

Consequently, Dr. Miller is not in the sort of evidential position that would permit him to say that Father Murphy was necessarily wrong about whom or what made flowers possible. As a result, at best his claim is premature and, at worse, his claim might even be false.

God does not need to have directly intervened in the process of induction through which various genes are sequentially expressed in a way that transforms leaves into flowers in order to be considered to be the One Who has made the flower. If God fashioned the laws of physics, chemistry, and biology and organized them to come together in a certain manner in order to give expression to a set of genes with various structural dynamic, and regulatory properties that will, at the appropriate time, induce flowers to blossom, then in what sense did God not make flowers?

Professor Miller's perspective is like saying that the manufacturer of a complex electronic component did not make that component because she, he, or they permitted the mechanism to work as it was intended to do rather than fiddling with that mechanism during its operating process. If God – rather than random, chance, indeterminate events molded by the forces of natural selection -- is the One Who fashioned flowering plants to have the sort of properties that were likely to be able to survive and leave progeny in a given set of environmental conditions, then God -- not chance, random, indeterminate events -- is the primary, proximate cause of flowering plants even if the process of induction occurs without further assistance from God.

Professor Miller has failed to demonstrate that God is not responsible for the origins of species because Dr. Miller has not succeeded in showing that the step-by-step processes through which all species arose were due to random, chance, indeterminate events that produced outcomes that were, subsequently, selected by the forces of natural selection. To be sure, Dr. Miller provides – as did Darwin – a possible account of, or explanation for, how some species might have originated at certain points, but what neither Professor Miller nor Darwin -- nor any other proponent of evolution -- has demonstrated is that the entire course of the natural life on Earth – both currently and in the past – came into being through a set of processes that constitute the central dynamic to which the theory of evolution gives expression and did so without the help of God.

Finding Darwin's God

The title of Professor Miller's book – the one with which I primarily been focusing on for the last 160 pages, or so, is: *Finding Darwin's God: A Scientist's Search for Common Ground Between God and Evolution.* I'm not exactly sure why trying to find "Darwin's God" should be the focus for any attempt to find common ground between God and evolution.

My uncertainty concerning the foregoing issue arises in several ways. Firstly, there is considerable ambiguity surrounding whether, or not, Darwin, once he became enthralled with the idea of evolution, actually retained any substantial faith in conjunction with the notion of God.

To whatever extent Darwin did hold on to some remnant of faith in the Divine, there can be little doubt that the naturalist expected God to bow down to Darwin's ideas about evolution rather than requiring the idea of evolution to adapt itself in some way to the possibility of God's existence and causal presence in relation to the natural history of life. Before the advent of evolution, the notion of God might have been a primary consideration in the life of Darwin, but once the idea of evolution became established in his mind, heart, and soul, then, God seemed to become, at best, a secondary consideration that was required to fit, as best it could, to the requirements of evolutionary theory.

Secondly, one wonders why 'Darwin's God' should serve as the standard for which one should strive during the search to find common ground between God and evolution. To whatever extent Darwin believed in God, his ideas might have been good ones or problematic ones, but until one establishes the quality of his ideas concerning the issue of God, then, one can't help but wonder why Darwin's ideas about God should form the metric for what is, and is not, acceptable with respect to finding common ground between God and evolution.

There are two broad kinds of common ground that might exist between the notions of 'God' and 'evolution'. One kind of common ground is purely conceptual in nature and really constitutes nothing more than beliefs that share certain, overlapping features with one another, while another possible species of common ground is a function of the manner in which the reality of God (to whatever extent this is empirically the case) and the reality of evolution (to whatever extent this is empirically the case) resonate with, or are congruent with, one another.

While Darwin's theological musings concerning his beliefs about the nature of God might, or might not, have their points of interest, surely, one is looking for more substantial forms of common ground than are entailed by whether, or not, there is a certain internal consistency to Darwin's thoughts with respect to God and evolution. One also would like to know whether, or not, there are reasons beyond the horizons of Darwin's beliefs about such matters that might establish a common ground for the notions of God and evolution.

Dr. Miller's aforementioned book might, or might not, enable a person to find Darwin's God. To whatever extent Professor Miller's quest has been successful, I feel the common ground he believes he has discovered is beset with a variety of highly unstable fault lines that raise many questions about whether, or not, Darwin's version of God is worth finding.

Section III: Anatomy of a Theory

| Evolution Unredacted |

Introduction

What do you know about evolutionary theory? Or, maybe there are two questions here: (1) What do you think you know; (2) What do you actually know?

In reality, if people were honest about the matter -- and quite irrespective of whether they believe in evolution or they are opposed to it -- most individuals probably would have to acknowledge that they know almost nothing at all about the actual nuts and bolts of the technical issues at the heart of evolutionary theory. Their belief concerning this matter -- whatever the character of that belief might be -- is, for the most part, rooted in two sources: (a) a largely unexamined acceptance of the opinion of others; (b) the extent to which evolutionary theory makes carrying on with the rest of their philosophical or religious perspective either easier or more difficult to continue to do.

In addition, the controversy surrounding evolutionary theory with respect to origin of life issues has been plagued by the fact that many of the advocates for various sides of this issue have been conducting the discussion on the wrong level. More specifically, people have been arguing mostly in terms of the evidence entailed by paleobiology ... that is, the anatomic/fossilized data that has been drawn from zoological and botanical studies. Unfortunately, the origin of life issue cannot be settled, one way or the other with any degree of certitude, when approached in this manner.

On the aforementioned level of discussion, one, at best, can obtain data that is either consistent with, or raises problems in, evolutionary theory as an explanation for the origin of life. However, there is no smoking gun (either for or against) to be found in such material -- just self-serving and heated rhetoric that tends to be cast in the garments of apparent rigor.

Furthermore, contrary to what many people believe, with the exception of a brief allusion to the possibilities that might exist in a 'warm little pond' somewhere ... a pond with just the right set of magical conditions ... Darwin has virtually nothing to say about the origin of life issue. The entire argument in his universally known but largely unread book is not about the origin of life but about the plausibility of a form of argument that alludes to, and presupposes, such a possibility without ever spelling out the mechanism.

The first part of the title of Darwin's historic work is: *On the Origin of Species by Means of Natural Selection*. There is a potential problem inherent in this title because the words tend to suggest that a species comes into being by a mechanism known as "natural selection". However, natural selection gives expression to a set of forces that operates after-the-fact of something having originated, and, therefore, at best, natural selection does not so much generate a species as much as natural selection operates on such a species once the latter has originated.

Natural selection acts on what is. It presupposes what is.

Natural selection does not cause what is, but, rather, it is an expression of those aspects of what is that might help determine which features of what is might continue to be. Natural selection introduces nothing new into the evolutionary picture, but, rather, the idea of natural selection only says something about the facets of that picture which might be most consonant with the dynamic of interacting natural forces existing at a given time and in a given location.

Therefore, the cause of that (whether a prebiotic collection of organic molecules or some primitive form of protocell) which natural selection comes to act upon still stands in need of an explanation. One cannot use natural selection as an explanation for that which natural explanation clearly presupposes without becoming entangled in completely circular thinking, and this sort of jaunt around the conceptual barn does not constitute an explanation of any kind.

Another problem with the previously noted title of Darwin's book is that it gives the impression that something is being selected ... as a person might make a selection among an array of choices. In truth, nothing is being selected since what exists in the way of a set of organic chemicals, or a set of protocells, or a set of species is either compatible (across a range of being more, or less, compatible) with the existing conditions of nature, or such chemicals, protocells, or species are not compatible. If random, such natural events do not select or choose.

What is compatible with the prevailing forces and conditions, survives. What is not so compatible tends not to survive. Nothing has been selected.

Another key idea in Darwinian theory is the notion of 'the accumulation of small variations'. The idea of the accumulation of small variations does not really account for either the origin of life, in general, or for the origin of different, particular biological blueprints, so to speak, on which the notion of species difference is based.

Variation presupposes that which is capable of such variation. Consequently, what needs to be explained is the origin of the capacity for variation.

Genetics is not the science that provides an account of the story of the origin of that capacity. Rather, genetics is merely a science that delineates how that kind of capacity operates once it has arisen.

Neither the ideas of natural selection nor variation help explain the origin of life. Only with the advent of modern molecular and cellular biology have we finally come into contact with the sort of information that allows one to make insightful judgments about the plausibility of evolutionary theory as an adequate account for the origins of life on Earth. When one integrates the disciplines of molecular and cellular biology with data derived from geology, hydrology, meteorology, and cosmology -- along with what has been learned about organic and inorganic chemistry -- then, one is in a position to work toward an informed understanding concerning the questions that surround and permeate the possibility of whether the modern neo-Darwinian theory of evolution offers an acceptable paradigm with which to approach origin of life issues.

In contradistinction to the original Scopes "Monkey" trial – when John Scopes, a high school science teacher, was put on trial for teaching material at odds with the Biblical account of the origins of man -- in *Anatomy of a Theory*, Robert Corrigan, a fictional character, has been put on trial for teaching material that is considered by the book's prosecutor to be inconsistent with evolutionary theory. However, the defendant in this case is <u>not a creationist</u> nor is his argument an expression of what has come to be known as "Creationist Science".

The current overview is <u>not about trying to prove the truth of this or that religious account</u> of the origins of either human beings, in particular, or life, in general. *Anatomy of a Theory* is about the process of interpreting empirical evidence and subjecting that data to various methods of critical reflection.

Unlike works such as *Inherit the Wind* (which is largely the account of a clever lawyer's legalistic and philosophical dismantling of the simplistic arguments of a rather flawed personality who desired to be regarded as a defender of the faith), *Anatomy of a Theory* addresses the issue of whether, or not, science, as presently understood, can be said to demonstrate the validity of evolutionary theory as an account about the origin of all life. As such, the present overview focuses on the issue of evolutionary theory itself and does not get sidetracked with irrelevant considerations ... however interesting these later twists and turns might be in purely human terms.

At this juncture, some people might wish to make the critical comment that the foregoing really has little to do with modern evolutionary theory. The latter is an elaboration upon the seminal ideas of Charles Darwin and, as a result, is sometimes referred to as neo-Darwinian thought. If one would like to critically explore modern evolutionary theory, then one must stay within the confines of the neo-Darwinian paradigm as it is.

If someone made this sort of a comment, I might say something along the following lines. If such an individual is saying that modern evolutionary thought has no explanation for the origin of life on Earth, then let this fact be known far and wide so that everyone will clearly understand that the theory of evolution has absolutely nothing to say about how life came to exist on the planet Earth, and I will accept that perspective. Moreover, with the exception of changing a little terminology here and there in the discussion that follows, the following critical exploration concerning origins of life still poses a challenge to modern, scientific understandings concerning the issue of the origin of life.

More often than not, however, when people speak about the origin of life from a scientific point of view, they tend to use the term "evolution" in a broader sense than did Darwin. More specifically, such people tend to convey the idea that however life came into being (on Earth ... or arose elsewhere and, then, was somehow transported to Earth -- perhaps through meteors), it did so through purely "natural" evolutionary processes that generated increasing complexity involving prebiotic/inorganic chemistry that was, then somehow, 'naturally' transformed, in some evolutionary manner, into biotic chemistry, out of which the first protocells emerged – that is, the first species of life, and, at this point, neo-Darwinian theory would become relevant.

Anatomy of a Theory is primarily a critical exploration of this broader, more inclusive sense of 'evolution'. However, there are a variety of ideas entailed by such a discussion that carry implications for neo-Darwinian thought concerning evolution as well.

There are things about *Anatomy of a Theory* that are true. First, it contains a lot of technical material. Secondly, everything that is necessary for understanding this material has been included within the context of the direct and cross- examinations that take place during the trial and, as such, it is a largely self-contained work.

However, this work is not the sort of discussion that one can rush through. As with anything else worth the effort -- and I believe this book is worth the effort – *Anatomy of a Theory* takes time to digest and appreciate.

If you are ready to make the commitment to attempt to come to grips with the essential issues of evolutionary theory, then *Anatomy of a Theory* is waiting to be read. Be the first kid on your block to actually know what one is talking about when the conversation turns to evolutionary theory in relation to the origin of life problem ... and the foregoing point actually brings us to a third thing, alluded to previously, about *Anatomy of a Theory* that is true.

More specifically, if an individual cannot grasp the point-counterpoint of the discussion in this book, then, one is not in a conceptual position to argue intelligibly or honestly either for, or against, evolutionary theory. Whatever one might have to say on such issues will be entirely derived from the opinions of others -- opinions that might, or might not, be correct but with respect to which one will have no direct, personal understanding, knowledge or insight.

Opening Remarks

Upon arrival in Chicago, I took one of the shuttle buses from the airport that made the rounds of different hotels in the downtown area. After getting off the bus at my destination, the Balmer House, I confirmed my reservation at the main desk, picked up my key card and proceeded to the assigned room on the twenty-first floor.

I spent about ten or fifteen minutes in the room unpacking. Once this task had been completed, I went downstairs in search of the symposium registration desk.

After the signing in requirements had been met, I picked up a brochure that listed the various lectures, panels, discussions and so on that had been scheduled for the symposium. I quickly perused the day's listings.

The only event that struck my fancy was a moot court session on evolutionary theory to be held on the fourth floor, beginning at 3:00 p.m., about twenty minutes from now. I decided to go and see what it was like.

I fully expected the worst. At the same time, I held out a certain amount of hope that there might be some degree of entertaining diversion to be derived from the trial.

The whole thing would be very trying, indeed, if the participants took themselves too seriously and lacked a sense of humor. Equally daunting was the prospect that few, if any, of the individuals taking part in the moot court might know anything about modern evolutionary theory.

Images of Spencer Tracy and Frederick March came to mind from *Inherit the Wind*. There had been a remake of the movie in which Jason Robards played a Clarence Darrow-like character to Kirk Douglas's version of William Jennings Bryan.

I had enjoyed both movies but always felt the cards had been stacked rather unfairly in the debate. The crux of the drama had not really focused on evolutionary theory per se but on a clever lawyer's dismantling of a simplistic presentation of a narrowly conceived religious position held by a somewhat flawed personality. Hopefully, the moot court session was not going to repeat the same mistake, except in reverse -- that is, to use a clever lawyer's debating tactics to defeat a simplistic presentation of evolutionary theory.

If done properly, the trial setting could provide a valuable opportunity for a good educational experience. I preferred not to think about what the result would be if things were done improperly.

I eventually found my way to the indicated room. When I walked through the doors, two things surprised me.

For some reason, I was expecting a relatively small venue ... perhaps from having seen too much of the stage settings for the old, pre-revival, Perry Mason television series. The room selected for the trial was quite large and had been set up like an actual court complete with a jury box, witness stand, lawyers' tables, a raised desk-like affair for the presiding magistrate, and a large area at the back of the courtroom for the audience.

The other feature that I found interesting was the size of the crowd. Nearly every seat was taken. I was lucky to find a vacant chair.

The members of the jury already were assembled in their seats. Individuals that were performing as lawyers were at their respective tables.

A door to the left and behind the judge's bench opened, and a diminutive, attractive, forty-something, black-robed, brown haired woman entered the hall. As she did, a court officer stood up and said: "Hear ye! Hear ye! Hear ye! All rise, Moot Court is now in session, the Honorable Justice Karen Arnsberger presiding over the matter of the people versus Wayne Robert Corrigan in the City of Chicago, in, and for, the County of Cook, on June 26, in the year of our Lord, 2009. Draw nigh, and ye shall be heard."

The court officer watched the judge settle into her chair. When he was satisfied, the man announced: "Please be seated."

As the Judge waited for the noise of the audience's seating dynamics to subside, she shuffled and re-arranged some of the papers before her. When relative quiet had returned to the room, she scanned the court and, then, said: "In accordance with agreements reached in chambers between the prosecution and defense concerning pre-trial motions filed on various aspects of the procedural format to be observed during the course of this trial, the following principles will be in effect:

"(1) Due to considerations of time, the prosecution and defense each will be entitled, if so desired, to call a maximum of two witnesses;

"(2) With the exception of certain provisions ... provisions that have been agreed to by all parties concerned -- standard rules of evidence will be in effect throughout these proceedings;

"(3) Prospective jurors has been polled by both the defense and prosecution prior to the start of this moot court session and jurors have been selected and impaneled on the basis of their perceived capacity to judge the matter before the court in a fair and impartial manner. During the selection process, both sides were given the right to challenge seven of the candidates without the need to show cause for dismissal;

"(4) Again, out of consideration for the time constraints under which we are operating, neither the defense nor the prosecution will be permitted the opportunity for redirect examination;

"(5) The decision of the jury shall be read in open session on the last day of the symposium."

Putting the paper down from which she had been reading, she addressed each of the lawyers: "Are these the conditions to which you have agreed?"

Both responded, almost simultaneously, but slightly out of synchronization: "So stipulated, Your Honor."

"Very well," she replied.

She shuffled through a few more papers and stopped when she found the desired document. "Mr. Corrigan, will you please stand."

After the defendant -- a curly-haired, freckled youngster who looked to be in his mid-twenties -- had arisen, Judge Arnsberger said: "Wayne Robert Corrigan, you are being accused of teaching material to students that is in direct conflict with the facts of evolution as well as with the principles and methods of science. How do you plead?"

"Not guilty, Your Honor," came the response.

"All right, Mr. Corrigan, you may sit down," she indicated. Turning to the lawyer for the prosecution, she asked: "Are the people ready to proceed, Mr. Mayfield?'

"The people are prepared, Your Honor," he informed her. Looking in the direction of the table for the defense, she asked: "Is the defense ready to proceed Mr. Tappin?"

"We are, Your Honor," he stated.

"Good," she asserted, "then, let us proceed with opening statements. Mr. Mayfield, you are up first, and, gentlemen, please remember the meter is ticking."

Pushing his chair back as he arose, the lawyer for the prosecution -- who looked, sounded, and acted like he came from a family of moneyed- gentry ... walked to a point in front of the jury box, about midway between the two ends. He placed his hands momentarily on the railing atop the three-foot partition that enclosed the jury area and briefly made eye contact with various jurors as he looked first to his right and then to his left, as he surveyed the members of the jury.

Removing his hands from the railing, he began to address the jury as he slowly walked back and forth along the front length of the boxed area. Every so often, he would stop and face the jurors in front of him and speak as if he were talking just to them.

"Ladies and gentlemen of the jury, some seventy-five years ago, a man by the name of John Scopes was placed on trial for teaching evolution to his students. He was accused of promulgating theories and ideas that ran contrary to established religious doctrines concerning the origins of human beings.

"Today, you are being asked to pass judgment on a case that, in many ways, is quite similar to the Scopes case, but with a major difference. The defendant, Mr. Corrigan, has been accused of teaching material that is contrary to the facts of evolution and in opposition to established principles, practices and methods of science.

"Personally, I find it very disheartening that just as we begin our journey into a new millennium, and some hundred and forty -plus years after the publication of Charles Darwin's classic study: *The Origin of Species by Natural Selection*, we find ourselves unable, apparently, to put this matter behind us. I consider this situation to be unsatisfactory because for nearly one hundred and forty years, there has been an exponential growth of data from many different fields of scientific endeavor, all of which points in one direction -- namely, that evolutionary theory has been demonstrated to be a valid, consistent, empirically grounded, rigorously examined and scientifically satisfying account of the origins not only of species but of life itself.

"To be sure, as is true in any area of scientific research, there are differences of opinion concerning the value and use of various kinds of methods, techniques, and interpretations in evolutionary theory. However, none of these differences has anything to do with bringing into fundamental question, nor are they capable of undermining or refuting, the shared understanding and agreement of scientists concerning the essential character of evolution.

"At the heart of evolutionary theory is one simple truth. The origin-of-life, the origin of species, the transition from one species to another, -- these all are completely explicable in terms of known natural principles and processes.

"In other words, the principles of physics, chemistry, cosmology, geology, meteorology, and climatology, when combined with a few simple ideas such as natural selection and variation, provide a definitive, exacting and sufficient framework through which to understand the origins of life along with the biological phenomena that such origins set in motion. In short, the dynamic interaction that results from the interfacing of the forces operating through these various principles and processes is all that is necessary to be able to provide an adequate account of why certain phenomena and forms, rather than other phenomena and forms, were selected to play crucial roles in the emergence and perpetuation of different life forms.

"To employ principles and forces beyond the natural realm is to violate what is known as Ockham's razor. This long-venerated tenet of scientific methodology advises us not to multiply assumptions or concepts beyond what is needed to adequately account for any given phenomenon.

"Translated into more modern language, Ockham's razor is really the law of parsimony.

"Keep things simple. Do not complicate matters unnecessarily.

"Evolutionary theory operates entirely within the purview of this law of parsimony. Indeed, as far as the issues surrounding the origins of life are concerned, evolutionary theory is the only account that operates in accordance with this fundamental principle of rigorous methodology.

"The Scopes trial was caught up in emotion, dogma, and cultural biases. These influences settled like a dense fog around the minds and hearts of the jury and made reaching a fair and impartial verdict on the issues of that case very difficult.

"As a result, John Scopes lost the case. He lost the case despite the fact that the overwhelming character of the trial evidence revealed through testimony as well as cross-examination demonstrated that the charges against the defendant were entirely without merit.

"You, the members of the jury, have been selected because of your stated willingness to rise above issues of emotion, dogma and cultural bias. You have been selected because of your commitment to render a free and impartial judgment in the matter before us based solely on considerations of facts, logic and reasonableness of deliberations.

"The prosecution intends to demonstrate, within the limits being imposed on this trial, that evolutionary theory has been established beyond any reasonable doubt. Consequently, anyone, in this day and age, who would teach material that stands in opposition to a theoretical framework that has been developed and agreed upon during the last one hundred and forty-plus years can only do so by denying the facts of the matter and by refusing to observe sound scientific practice and principles.

"This is precisely the violation of which Mr. Corrigan is being accused. If the prosecution is successful in the presentation of our case, as I believe we will be, then you, the women and men of this jury, will, beyond any reasonable doubt, find Mr. Corrigan guilty as charged."

Once again, the prosecutor briefly ran his eyes down the two rows of impaneled jurors, stopping here and there to engage the eyes of this or that juror. When he had finished, he said: "Ladies and gentlemen of the jury, I want to thank you for the careful attention that you have given to my opening remarks. I am confident you will give the same considered attention to the evidence governing the case before you."

Mr. Mayfield turned and went back to his table. As he sat down, one of his assistants whispered something in his ear.

Judge Arnsberger turned to the lawyer for the defense. "Surf's up, Mr. Tappin," she informed him.

Before getting up, he picked up one of the sheets from the tabletop, looked at it for a few seconds, and, then, put the paper back down. He continued to sit for another five or ten seconds, as if in thought, and, finally, quickly rose and made his way to the jury area.

Evolution Unredacted

In speech and manner, Mr. Tappin appeared to be the opposite of Mr. Mayfield. With the exception of his thinking processes, everything about the defense lawyer was casual, informal and laid back.

Like the lawyer for the prosecution, Mr. Tappin appeared to be in his early thirties. Like Mr. Mayfield, the defense lawyer was moderately handsome but in a rough and ready manner and, therefore, somewhat at odds with the prosecution lawyer's aura of urbane sophistication.

"Good afternoon, ladies and gentlemen of the jury," the defense lawyer began.

"Good afternoon" was the collective, somewhat mumbled response from the jurors.

"I would like to thank my learned adversary for the wisdom of his comments," Mr. Tappin stated. "With his well-known and respected capacity for conciseness, Mr. Mayfield's introductory statement has focused on the most important elements of this case.

"The legal matter before you is not about ... or at least, it should not be about ... emotion, dogma and cultural biases. On the other hand, this case is about facts, logic and reasonable deliberations.

"These proceedings will not be about evolutionary science versus what some adversaries of evolution refer to as 'creation science'. This is so because my client is not an advocate of creation science, nor is this what he teaches in his classroom.

"My client, Mr. Corrigan, does not find any philosophical, or even religious, inconsistency between the vast majority of the tenets of evolutionary biology and a belief in a Divine Being Who creates the material and physical world. Mr. Corrigan is willing to admit the plausibility, if not tenability, of a position that says that evolution is merely the manifest form of the means through which God creates physical/material reality.

"The nature of Mr. Corrigan's faith is not so feeble that it depends on presupposing a particular conception of creation that precludes the possibility of evolution. He doesn't have a vested interest or axe to grind in this respect.

"Mr. Corrigan's concerns lay elsewhere. He is worried about issues such as truth, proof, logical argument, understanding, explanation, interpretation, and the integrity of the exploratory process.

"The case of the defense will not be about whether the second law of thermodynamics is inconsistent with the theory of evolution. We are quite prepared to live with the entirety of thermodynamic theory, including the relatively recent work on the phenomenon of dissipative structures that, sometimes, arise under conditions in which a system is far from equilibrium.

"The defense will not involve any arguments about whether the fossil record does, or does not, create problems for evolutionary theory. In addition, we will not try to exploit the controversies surrounding punctuated equilibrium theories as a means of undermining the framework of evolutionary biology.

"The position of the defense does not depend on the raising of questions about the reliability of dating methods based on radioisotopes. Furthermore, we have no intention of trying to use to our advantage differences of opinion concerning the role that, say, lunar samples play in pinning down the time of events on Earth, or the way in which, for example, high temperatures can affect the significance and interpretation of Carbon^{12} and Carbon^{13} ratios as an indirect procedure for helping to establish the possible presence of life at a given period of time in the early history of the Earth.

"There will be no attempt by the defense to take quotes of noted evolutionary scientists out of context and try to use these quotes as evidence against evolutionary theory. We are only interested in taking a look at what the best science of our day has to say in support of the case for evolutionary theory with respect to origin of life issues.

"Ladies and gentlemen of the jury, so far, I have told you what the case for the defense will not be. I have not, yet, indicated what our case will be, so let me take this opportunity to rectify that omission.

"The contention of the defense is as follows. When closely examined, evolutionary theories concerning the origins of life consist of little more than a rather argumentative mixture of: questionable assumptions, speculative conjectures, problematic inferences, arbitrary interpolations or extrapolations, ambiguous evidence, and a wonderfully serendipitous confluence of events quite beyond the ability of science to demonstrate with any degree of plausibility except, perhaps, to the true believers among evolutionary theorists who are more in need of faith to prop up their theories concerning the origins of life than are many followers of religious traditions.

"The defense will be asking you, the members of the jury, not to be dazzled by the technical virtuosity of modern science. We will be asking you not to be intimidated by the use of technical terms.

"However, the defense will be asking you to keep in mind the importance of such basic, fundamental questions as: How? Where? When? What? and Why? In addition, the defense will be asking you not to shunt aside or marginalize the number of questions that go unanswered within the evolutionary perspective.

"The defense believes that if the members of jury are prepared to persist in asking simple questions along the lines we have indicated, and if you are willing to keep a running total of the questions that, after all is said and done, lack a satisfactory answer, you will arrive at one conclusion beyond any reasonable doubt. This conclusion is that my client, Wayne Corrigan, is not guilty of teaching material in conflict with either the facts of the matter at hand or with the methodological tenets and principles of scientific investigation.

"Ladies and gentlemen of the jury, I would like to thank you for your kind attention to my opening statement. I also would like to leave you with one suggestion.

"Pause for a few seconds, sit back, relax and take a few deep breaths. For, in approximately ten to twenty seconds, you might not get the opportunity to do so again until these proceedings have concluded.

"Thank you, again," Mr. Tappin stated and returned to his seat. A few jurors seemed to be following his suggestion.

Approximately fifteen seconds later, Judge Arnsberger announced: "The prosecution may call its first witness."

What on Earth is Happening?

At this time, the prosecution calls upon Professor Alan Yardley," proclaimed Mr. Mayfield. As he uttered the name, he looked back toward the audience.

A tall, thin, bearded man -- who appeared to be in his late thirties or early forties -- stood up in the area where Mr. Mayfield was looking. The man strode to the witness stand and remained standing while the oath was administered by a court officer: "Do you promise to tell the truth, the whole truth and nothing but the truth, so help you God?"

The witness answered: "I do."

The court officer then informed him: "You might be seated." Once the witness was settled in his chair, the court officer said: "Will you state your name and address for the record, please."

"My name is Alan Ross Yardley," he replied. "I presently reside at One Finch Beak Road, Daphne Major, the Galapagos Islands."

The court officer returned to his seat. Mr. Mayfield approached the witness.

"Dr. Yardley," he requested, "will you state your current occupation and title."

"I hold the Charles Darwin Chair for Biological Sciences at the University of Galapagos," he responded. "I am a full professor and teach a variety of courses dealing with different facets of evolutionary biology."

"How long have you held your present position, Dr. Yardley?" Mr. Mayfield inquired.

"For seven years," Dr. Yardley answered.

"Professor," Mr. Mayfield said, "would you be kind enough to list your major publications."

Dr. Yardley was about to begin when the defense lawyer arose. "If it pleases the court, Your Honor," Mr. Tappin indicated, "in the interests of saving time, the defense is quite prepared to acknowledge the expertise of Professor Yardley in the field of evolutionary biology. His reputation as a first-rate scholar is recognized internationally, and we feel there is no need to go through the usual procedures for establishing expertise with respect to this witness."

"So noted," acknowledged Judge Arnsberger. "Thank you for expediting matters, Mr. Tappin."

The defense lawyer nodded his head and sat down. He began writing something on a piece of paper and, when finished, showed it to his assistant.

"Your Honor," the prosecutor said, "before beginning examination of my witness, I would like to introduce into evidence, at this time, the People's Exhibit, marked 'A'." While saying this, he had returned to his table, picked up a collection of material, checked its identity, and delivered the bundle of papers to Judge Arnsberger.

The Judge examined the papers briefly and made a few notations, presumably, in her own log of the trial. Having done so, she said: "You may proceed Mr. Mayfield."

Returning to his table, he picked up another, similar bundle and walked back to the witness. Handing the papers over to Dr. Yardley, the prosecutor said: "These papers, Professor, which have just been introduced into evidence as People's Exhibit 'A', constitute a detailed curriculum for the courses on evolutionary biology that are being taught by the defendant, Mr. Corrigan. Dr. Yardley, have you had a chance to study these papers prior to the beginning of this trial?"

The professor quickly worked his way through the pile of documents. "Yes, prior to the beginning of these proceedings, I have looked through this set of documents," confirmed the professor.

"What is your opinion, Professor Yardley, of the educational merit of these curriculum materials as far as the teaching of evolutionary biology is concerned?" the prosecutor inquired.

"Well, in certain ways," he asserted, "they appear to be reminiscent of the kind of material that is taught under the misleading title of creation science. And ..."

"Objection, Your Honor," Mr. Tappin blurted out.

"On what grounds?" Judge Arnsberger asked.

"Your Honor, as has been clearly stated in the defense's opening statement, Mr. Corrigan's position is not that of the so-called 'creation scientists'. Unless the prosecution demonstrates in what way the position of Mr. Corrigan is 'reminiscent' of the position of the creation scientists and unless the relevance of that reminiscence to the present case can be established, then, all references to creation science are really immaterial and irrelevant to these proceedings, as well as being quite prejudicial to the interests of my client."

"Mr. Mayfield," inquired Judge Arnsberger, "does the prosecution intend to provide the court with the sort of demonstrations and connections about which Mr. Tappin is concerned?"

"No, Your Honor," indicated the prosecutor.

"Very well," she said. "The objection of the counsel for defense is sustained, and the statement of the witness will be stricken from the records. You'll have to begin again, Mr. Mayfield."

Nodding his head in compliance with the directive, the prosecutor turned backed to the witness. "Professor Yardley, in the light of what has just transpired, how would you sum up your objections to the curriculum materials of Mr. Corrigan?"

"Perhaps," the professor began, "the most diplomatic way to state what is problematic about the content of Mr. Corrigan's course material is that it is consistently antagonistic toward the precepts, findings, conclusions, principles, orientation and general framework of the modern theory of evolutionary biology. In other words, Mr. Corrigan seems to want to debate and question issues and themes that, for the most part, have long been accepted as settled among the vast majority of scientists all over the world."

"Dr. Yardley, is this 'antagonistic' flavor of Mr. Corrigan's teaching material, only directed at specific aspects of evolutionary theory or is the tenor of his attitude more general in character?" the prosecutor asked.

"Quite general, I would say, but it is manifested in specific ways at virtually every level of evolutionary inquiry. For instance, Mr. Corrigan seems unwilling to accept much of what has been agreed upon with respect to issues involving prebiotic chemistry, or the origins of the first proto cells, or the emergence of prokaryotic and eukaryotic forms of life, as well as ..."

"Professor Yardley, I'm sorry for interrupting you," Mr. Mayfield apologized, "but three or four terms, in quick succession, have occurred in the testimony, and I feel they should be explained by you ... in a brief fashion if possible ... for the benefit of the jurors. Perhaps you could start with the term 'prebiotic'."

"Certainly," the professor said, "I would be most happy to do so. 'Prebiotic' chemistry refers to the study of all chemical processes, whether inorganic or organic, that are thought to have occurred prior to the appearance of biological or living systems on Earth.

"These prebiotic chemical systems are believed to have evolved over the course of millions of years, into, first, quite primitive cellular forms of life known as 'protocells'. Such protocells were, however, sufficiently developed to exhibit three properties.

"First, they contained some kind of membrane mechanism that provided a certain amount of protection for, as well as enclosed an area involving, a variety of chemical reactions necessary to sustain life on some minimal basis. Secondly, there would have had to be a method of metabolism that would permit the coupling of certain sources of energy with the building up and tearing down of chemical substances that result in the regulation of cell functioning and structure. Thirdly, such a protocell would need a means of storing and replicating information concerning the capabilities of the protocell that would enable the entity to reproduce itself and generate other protocells of a similar enough nature to be able to perpetuate the life cycle in future generations.

"Before proceeding, however, I should point out something. Among evolutionary biologists, as far as the issue of protocells is concerned, the aforementioned three m's ... that is, membranes, metabolism and memory ... might operate in ways that are quite different from what goes on in the current, modern life forms with which we are familiar, such as prokaryotes and eukaryotes.

"Prokaryotic forms of life consist of single-celled organisms in which the genetic material of such an organism is not enclosed by a true nucleus within the cell but, instead, floats freely in an area known as the nucleoid. By and large, most prokaryotes are one species or another of either bacteria or blue-green algae.

"Eukaryotic forms of life, on the other hand, include all those organisms whose cells contain a true nucleus, consisting of a bilayered or double membrane, which ropes off, so to speak, a roughly circular area within the cell that stores the genetic blueprints for the cell. These eukaryotic organisms might be either single-celled or multiple-celled in character and, for the most part, involve all forms of life other than the aforementioned bacteria or blue-green algae prokaryotic life forms."

"Thank you, Professor Yardley, for your very concise definitions of the technical terms," said Mr. Mayfield. "I'm sure we will be relying on this ability of yours quite a lot in the testimony that lies ahead of us.

"Dr. Yardley, you have indicated in your previous testimony that Mr. Corrigan's curriculum materials take exception with well-established and generally agreed upon issues and themes at virtually every level of evolutionary theory. Maybe the most effective way in which to proceed is to spend some time providing an overview for the members of the jury concerning the theoretical framework for modern evolutionary biology.

"In this manner we will be able to develop, hopefully, a much clearer understanding of that to which Mr. Corrigan stands in opposition. Moreover, in the process of coming to this understanding, you can provide evidence that, when contrasted with the material in Exhibit 'A', will demonstrate the truth of the allegations contained in the People's charges against Mr. Corrigan.

"Let's start, Professor Yardley, with first principles. Could you provide us with an outline of the currently accepted understanding of the formation of the Earth and what ensued from that as far as the conditions which are believed to have arisen to give expression to the prebiotic environment out of which life is said to have originated."

"Objection, Your Honor," Mr. Tappin asserted. "While the defense is willing to concede Dr. Yardley has expertise in the specific area of evolutionary biology, we are not prepared to concede his expertise in areas of cosmology, meteorology, climatology or geophysics."

"Under other circumstances, Mr. Tappin," the judge indicated, "I might be inclined to agree with you. On the other hand, earlier on, you waived your right to establish the precise nature of the parameters within which the expertise on evolutionary biology falls.

"Furthermore, unless I am mistaken, Mr. Tappin, in your opening statement you seemed to indicate that in order to set the stage for the case of the defense, you wished to concentrate on what the science of our day claims to be the best version of the evidence in support of evolutionary theory. Why don't we give them a chance to stick their head into the lion's mouth before trying to lop it off?

"I'm going to allow the witness to answer this line of questioning. Objection overruled.

"However, Mr. Mayfield, let's understand what is being said here. I don't want you taking undue advantage of the latitude that is being extended to you by the court, or else I will step in and revoke your privileges in this regard. Have I made myself clear?"

"Like a Norwegian fiord, Your Honor," he acknowledged.

"Dr. Yardley," the prosecutor said, "let me rephrase, somewhat, my previous question to you. Among evolutionary biologists, what is the generally agreed upon understanding concerning the conditions prevailing on Earth during prebiotic times?"

"To properly answer your question in even a cursory manner," stated the professor, "one must understand that prebiotic times entail a number of different stages and kinds of interacting evolutionary forces. These include: the evolution of the solar system as it relates to planetary formation; geological evolution; atmospheric evolution; hydrological evolution of the physical character, distribution and effects of the waters of the Earth; together with chemical evolution, especially as this development relates to the generation of increasingly complex forms of hydrocarbons that are the bread and butter of organic chemistry.

"I'll try to give a brief overview of all but the last of these areas. The topic of chemical evolution will require considerably more time.

"Obviously, my brief account of the issues beyond the horizons of chemical evolution will be leaving out a great deal of detail. Nonetheless, I believe people will be able to grasp the character of the general picture that is being constructed.

"To begin with ... ahh! Mr. Mayfield ... it is all right that I proceed in this way isn't it?" he asked.

"Of course, Professor Yardley," the prosecutor confirmed. "If I feel any clarification is necessary, I'll be sure to intervene.

"Moreover, Professor, I realize some minimum degree of technical language and explanation will be necessary. However, while avoiding as much distortion and oversimplification as possible, if you could try to make your account as clear and succinct as possible, this would be greatly appreciated.

"This is probably asking the impossible of you. Nonetheless, I believe the more you are able to approach the 'impossible' as a limit, the more easily will the jurors understand the validity of the allegations being made against Mr. Corrigan."

Professor Yardley paused briefly, seemed to gather his thoughts, and began to speak. "At one point in the development of cosmological theory," he said, "scientists believed planets were formed by a very rapid gravitational collapse of interstellar dust clouds once, depending on circumstances, certain critical densities within those clouds had been achieved.

"Today, based in large measure on the findings of the Apollo space program's crater studies of the moon, most scientists have abandoned the foregoing theory and, now, believe in an accretion theory of planet formation. In other words, they believe planets come into being, not through gravitational collapse of dust clouds, but by gradually growing in size by means of a series of collisions with other objects of varying sizes.

"For example, one begins with specks of cosmic dust that collide with one another to form tiny particulates. Particulates collide with other particulates as well as cosmic dust to form larger, gravel-sized objects.

"This cosmic gravel, in turn, collides with cosmic dust, particulates and other gravel-sized objects to generate larger and larger objects. Eventually, something the size of a small planet, called a planetesimal, is produced, and, then, later --through continued collisions -- objects the size of the moon, and, finally, the Earth, emerge.

"The process of planet formation might have required a hundred million years give or take a few hours. This period of primary formation and evolution of the Earth has been determined, on the basis of radioisotope studies of the rate of conversion of uranium to lead, to have been completed approximately 4.55 billion years ago.

"As the objects grow larger, then, relatively speaking, there are fewer and fewer large size objects running around in space with which to collide. Collisions, of course, do continue to occur. Nonetheless, the number of years between large-scale, or even moderate-scale, collisions begins to increase.

"At first, after the formation of a planet the size of Earth has taken place, the occurrence of collisions will be separated by periods of time lasting hundreds, followed by thousands, of years. Later, the interval between collisions will become hundreds of thousands of years and, then, millions, if not tens of millions of years.

"The last great collision on Earth was believed to have occurred some sixty-five million years ago at the Chicxulub crater, some 300 kilometers in diameter, near the northern tip of the Yucatan Peninsula. This collision is thought to have led, both directly and indirectly, to the extermination of many, if not most, species of life, including the dinosaurs, living on Earth at the end of the Cretaceous era.

"In any event, most evolutionary biologists are agreed that life on Earth probably could not reasonably have been thought to have had the opportunity to establish a firm foothold until the frequency of these collisions had declined to, at least, less than once every ten or twenty million years. The reason behind this thinking is that whenever objects big enough to create craters of diameters equal to, or greater than, say, 265 kilometers, collide with the Earth, they cause, among other things, a one hundred-degree Celsius, transient rise in the temperature of the Earth's atmosphere.

"This would cause obvious, destructive havoc with the vast majority of origin-of-life processes that might have been going on in a prebiotic environment on Earth. There must be, consequently, enough undisturbed breathing room, so to speak, within which biological organisms would have a plausible opportunity for emerging spontaneously through purely natural chemical and physical processes.

"Most of my colleagues set the lower limit of the relatively undisturbed breathing space time that is considered to be necessary to account, reasonably, for the origins of, say, the first protocells, to be around ten to twenty million years. Such intervals of cosmic quietude are not likely to have taken place on Earth prior to about 4.44 - 4.41 billion years ago.

"These kinds of calculation are based on statistical projections derived from radioactive dating of the cratered surfaces of the moon. For instance, if one assumes there will be a proportionate increase in the number and size of large impacts as one goes from the smaller surface area of the moon to the larger surface area of the Earth, then, scientists have concluded there were about 15-16 collisions on Earth which were larger than the ones that caused the largest of the moon craters, Imbrium. These collisions would have taken place at some point after 4.3 billion years ago.

"Since collisions do not take place in accordance with a fixed schedule, they are a stochastic or probability phenomenon. Therefore, if we take the 15 or 16, previously mentioned, large-sized collisions with Earth and average them out over a period of time, we would have to wait for all of these collisions to take place before we could begin to talk about conditions on Earth that were minimally conducive, as far as collision activity is concerned, to the origins of life in a prebiotic environment.

"The time at which the last of these large-scale collisions is believed to have occurred is somewhere between 4.3 and 3.8 billion years ago. We should begin to find traces of life somewhere in this time-frame, and, in fact, we do, but I'll come back to this."

Professor Yardley picked up a jug on a table near the witness stand and poured water into a small drinking glass. He took a long drink, finished the glass, replaced it on the table, and continued on.

"When, as a result of the gradual process of accretion, the Earth grew to roughly its present size, our planet was not considered by scientists to be a static, dead entity. In fact, there were several theories about, for example, the formation of the core of the Earth that have ramifications for theories concerning the origins of life.

"One theory, the older one, maintained that the Earth started out as a cold body. Its interior layers did not begin to heat up until hundreds of millions years later when there had been a sufficient amount of heat generated by the radioactive decay of various elements in the Earth.

"Consequently, rather than sinking to the core early on in the formation of the planet, heavier elements, like iron, remained fairly close to the surface for many millions of years. Moreover, since iron tends to react with oxygen, this reaction would have severely restricted the amount of oxygen that could have combined with carbon to form an atmosphere consisting of large amounts of carbon dioxide.

"According to this theory, the volcanoes created by the thermal activity of the Earth's interior layers would have caused the spewing forth, or out-gassing, into the exterior regions of the planet, of large amounts of nitrogen and carbon that would combine with hydrogen. These reactions would have led to an atmosphere consisting, predominantly, of methane and ammonia.

"If, on the other hand, one subscribes to the collision or accretion theory of planet formation, as most modern researchers do, then, one comes up with a very different sequence of events than is painted by the older theory that started off with a cold Earth. According to the up - dated theory, the many violent collisions that were typical of the Earth's early years would have generated thermal conditions sufficient both to melt the interior regions of the Earth, as well as the heavy elements, like iron, which were on the surface.

"As a result, the interior of the Earth, some two to four hundred kilometers below the surface, would have formed what is known as a 'magma ocean'. Among other things, this 'ocean' would have underwritten the activity of volcanoes for millions of years and would have served as the 'sea' by means of which the plate tectonics of landmasses would have manifested themselves.

"In addition, the heavy metals, such as iron, would have sunk, in the form of a dense liquid, thereby differentiating the Earth, through the formation of a magnetic core, at a very early stage of the planet's evolution. Iron, consequently, would not have been available to react with oxygen as the old theory hypothesized, and, consequently, this would have cleared the way for oxygen and carbon to combine to form an atmosphere consisting, to a considerable degree, of carbon dioxide instead of the methane and ammonia called for by the previous model.

"Calculations involving the atmospheric-mantle ratios of two isotopes, argon^{40} and xenon^{129}, suggest that as much as 80-85 percent of the Earth's atmosphere probably was out-gassed in the initial million years of the existence of Earth as a planet-sized body. The remainder of the atmosphere was slowly out-gassed during the following 4.4 billion years.

"In addition to large quantities of carbon dioxide gas, there is believed to have been considerable amounts of nitrogen gas in the prebiotic atmosphere. Furthermore, although trace amounts of sulfur dioxide, methane and ammonia also are considered to have formed part of the early atmosphere of the Earth, no oxygen was believed to be present in the Archean era atmosphere that lasted from about 4.54 until roughly 2.5 billion years ago.

"This assertion concerning the relative absence of any oxygen content in the Archean era atmosphere has been backed up by a variety of studies. For instance, research has been done in relation to the stability of certain compounds such as uranium oxide and iron oxide, and these studies strongly suggest that the oxygen content of the Archean era atmosphere prior to two billion years ago appears to have been extremely low."

"Excuse me for interrupting, Dr. Yardley," the prosecuting attorney interjected, "could you, perhaps, explain the significance of the relative lack of free oxygen in the Archean era atmosphere?"

The professor nodded in acknowledgement of the request and said: "Essentially, free oxygen is highly reactive and tends to remove hydrogen atoms from any compounds it encounters. If free oxygen were present in the Archean era atmosphere with anywhere near the concentration of roughly 20 percent of our current atmosphere, the tendency of oxygen to oxidize or to take hydrogen from other compounds would interfere, in a fundamental way, with many important chemical reactions in a prebiotic environment.

"If one were attempting, as evolutionary biologists are, to account for the transition from simple hydrocarbons to the more complex forms of hydrocarbons that are necessary to the emergence of biological organisms through natural processes, the presence of substantial amounts of free oxygen would undermine one's efforts. If the Archean era had an oxidizing atmosphere, this would constitute a major theoretical problem for evolutionary biology.

"Fortunately, we are not faced with such a difficulty. As I suggested earlier, the available evidence indicates oxygen was not present during the Archean era except, at best, in minimal, trace amounts."

"Thank you, professor," Mr. Mayfield stated. "Please continue with your overview."

Dr. Yardley seemed to be searching in the air for where he had left off in the previous discussion. Apparently finding it, he said: "The process of core formation through the downward displacement of dense liquids consisting largely of molten iron is believed to have generated enough heat to raise Earth's temperature by as much as 1500 degrees Celsius. Such temperatures, in turn, could have helped create a set of conditions on the surface of the planet that might have culminated in a runaway greenhouse effect that, for a period of time, would have resulted in a melting of the surface of the Earth, creating a magma ocean of truly global proportions.

"This forms part of a theoretical scenario that is referred to as the 'hot world hypothesis'. A number of scientists have conjectured that, among other things, the Earth's crust would have been extremely thin during this period of geological evolution.

"These researchers believe that such a thin crust would have been very prone to cracking, and, one of the results of this would be the prevalence of a great many more hydrothermal vents than exist currently. These hydrothermal vents were channels to subterranean rivers and oceans of molten rock.

"Such hydrothermal vents would have helped create conditions for such phenomena as underwater geysers. In addition, they could have played an important role in providing a set of conditions out of which life might have first arisen.

"Modern researchers, however, also link the origin of the oceans and their concomitant hydrogen cycle with the previously mentioned process of out-gassing. Voluminous quantities of water would have been released by the heating of the Earth's mantle.

"This water vapor would have condensed, subsequently, into the extensive precipitation that formed the oceans. In addition, this process of condensation would have created a cooling trend that, eventually, would have helped to cool the atmosphere and surface of the planet down to the range of 40 degrees to 80 degrees Celsius that is believed to have prevailed at the time of the emergence of life from the prebiotic environment.

"In any event, most scientists agree this sequence of steps involving: A, the formation of the Earth's core, B, the gradual evolution and retaining of an atmosphere consisting of large amounts of carbon dioxide, C, the formation of oceans, as well as, D, the cooling down of the surface to temperatures in the range of, say, 40 to 80 degrees Celsius, was not likely to have been completed before 4.44 to 4.41 billion years ago, some eleven to fifteen million years after the emergence of the Earth as a planet-sized body. This figure coincides roughly with the evidence mentioned earlier concerning the gradual lessening of collisions with Earth of objects sufficiently large to interfere with, or frustrate, the prebiotic processes that eventually resulted in the formation of either protocells or biological organisms.

"There is further, independent data that helps confirm the foregoing time frame. These studies concern the mineral zircon.

"Zircon does not dissolve during the process of erosion. This mineral becomes deposited in sediment in the form of particles.

"Zircon particles are capable of lasting for billions of years. As such, zircon can provide evidence concerning the time of formation of a relatively stable surface crust.

"Ancient particles of this mineral have been found in Western Australia. These specimens were dated as having been in existence from around 4.1 to 4.3 billion years ago.

"The discovery and dating of these zircon particles is said to demonstrate there was a differentiated crust, consisting largely of silicon-derivatives, already in existence by that time. With the exception of various volcanic islands that had risen above sea levels, the aforementioned crust was believed to have been covered by a global ocean whose pH value is commonly set at 8.0, plus or minus 1 ... that is, this massive ocean was considered to have a pH that was either slightly basic in character or was relatively neutral.

"Among the oldest fossils discovered by scientists are structures known as stromatolites. Communities of marine microorganisms consisting mostly of cyanobacteria have produced those structures.

"Stromatolites are a combination of sedimentary material of various kinds that have been trapped in an inorganic secretion generated by these organisms. The ones that were produced at least 3.55 billion years ago are homologous with, or very similar in structure, character and appearance to the ones that are produced today.

"The oldest known stromatolite structures have been found in the lower strata of the Warrawoona Group of rock formations in Western Australia. This Group is the second oldest well-characterized rock formation that is known to scientists.

"The oldest such rock formation that, so far, has been encountered is the Isua Supracrustal Belt in Southwestern Greenland. This has been dated at about 3.77 billion years ago.

"The Isua formation consists of high-grade metamorphic rocks that have gone through a process of reformation under conditions of extremely high temperature and pressure. Consequently, any direct fossil evidence that might have been contained in this rock formation would have been destroyed.

"However, there is some indirect evidence that has been discovered at Isua to suggest a bacteria-like organism might have existed in Greenland some 3.85 billion years ago. This evidence is based on an analysis of the ratios of two isotopes of carbon, C^{12} and C^{13}, that were found in a hydrocarbon specimen taken from the rock formation.

"Since C^{12} tends to be used preferentially in biological processes rather than C^{13}, and since the ratio of C^{12} to C^{13} found in the sample of hydrocarbon was high, some scientists have been quite excited by the implications of the findings. They have concluded, despite possible methodological contraindications, that these findings on carbon isotope ratios might mean the hydrocarbons being examined were produced some 3.8 billion years ago during a process of photosynthesis in which an organism converted carbon dioxide into oxygen along with various hydrocarbon compounds.

"Interestingly, the term 'Isua' is translated from the Inuit language as being equivalent to the English phrase: 'the farthest we can go'. Whether this is true as far as the earliest evidence for life is concerned remains to be seen.

"Be this as it may, if the scientific interpretation of the significance of this analysis of the Isua hydrocarbon is correct, then, the earliest evidence for life has been placed just some 750 million years from the time the Earth reached planetary size. Furthermore, if the interpretation of the carbon isotope ratios is correct, living organisms have been located only 200 - 400 million years from the time when the prebiotic conditions on Earth are thought to have begun to stabilize with respect to a broad set of planetary, geological, atmospheric and hydrological parameters considered to have an important bearing on the issue of the origin-of-life.

"This period of 200-400 million years establishes the temporal framework within which modern evolutionary biology has attempted to delineate a plausible sequence of steps in chemical evolution. This sequence would provide an account of the dynamic of factors considered necessary to produce a working prototype of a living organism capable, minimally speaking, of processes of photosynthesis similar to what is suggested by the Isua hydrocarbon.

"Conceivably, there might have been some primitive form of life, a protocell, which existed prior to the emergence of the first modern prokaryotic-like microorganism. On the other hand, its manner of cellular functioning probably would have been very different from, and, therefore, a matter of speculation relative to, the kind of DNA-based organism that is indicated by the earliest evidence we possess either with respect to the indirect evidence of the Isua rock formation in Greenland or the direct evidence of the Warrawoona Group in Australia.

"On the basis of the available evidence, the Isua hydrocarbon and the Warrawoona prokaryotes constitute remnants of the last ancestor that is shared or held in common by all existing life forms. More distant or ancient ancestors, in the form of various kinds of primitive protocells, do not necessarily form part of the biological lineage of all current life forms.

"As such, these kinds of protocells would be regarded as spontaneously arising experiments in life that, for whatever conditions of natural selection, fizzled out at some point. These experimental failures, if you will, are to be distinguished from the appearance of the first, sustained, experimental biological success story to emerge from the prebiotic environment and that represents the last common ancestor of all subsequent life forms."

"All right, Dr. Yardley," the prosecutor said, "you have established a general framework within which, and through which, a person can engage the more difficult issues surrounding chemical evolution. For the benefit of the jurors, let's try to break up the themes of chemical evolution into units that, to the degree this can be accomplished, will become a little bit more user friendly for those of us who are relatively uninitiated in such matters.

"Professor, if you had to list four or five areas of discussion that you consider to be crucial to developing some minimal appreciation of how evolutionary biologists go about explaining the transition from prebiotic chemistry to the first life forms, what areas would you cite?"

Hesitating only slightly, Dr. Yardley replied: "First, one should address the ways in which more complex hydrocarbons either evolved out of chemical reactions amongst simple hydrocarbons or became available to the prebiotic environment through means other than chemical reactions. Secondly, there would have to be some discussion of the systems of energy that were helping to drive the chemical reactions in the prebiotic environment.

"At some point one would have to talk about the formation of proteins by the linking together of amino acids through peptide bonds. This would be of great importance because of the many different roles that proteins have in biological organisms, including: hormonal functioning; muscular contractions; the variability of morphology or structural form among species; electron transport in both photosynthesis and respiration; antibody activity in the immune system; and, the transport of nutrients, ions and so on across the membrane barrier.

"Quite obviously, one also would have to explore the processes surrounding the formation of nucleic acids, especially, of course, deoxyribonucleic acid or DNA and ribonucleic acid, RNA. Both of these molecules have fundamental roles to play in the processes of replication, transcription, translation and energy-coupling reactions that are central to the continued existence of both individual organisms as well as a given species.

"Finally, one would have to discuss the role that lipid formation plays in, for example, the structure and function of cell membranes. Biological membranes help regulate the passage of compounds into and out of the cell, and, in doing so, provide a relatively protected, enclosed environment in which various vital chemical reactions can take place under much more favorable conditions than might be prevailing in the medium that is surrounding the cell's exterior".

"In view of the limited time available to us, Dr. Yardley, I am hoping you will be able to summarize some of the research evidence concerning the different areas you have just mentioned that scientists believe helps establish a compelling case in support of the modern theory of evolution. In fact, Professor, maybe the easiest way to proceed is to allow our discussion to unfold in accordance with the sequence of topics you have listed.

"Consequently, if you will, Dr. Yardley, begin with the first theme you cited as being important to the foundations of modern evolutionary theory. This concerned, I believe, the generation and availability of complex hydrocarbons in the prebiotic environment."

"There are," the professor said, "two broad approaches to explaining the existence of complex hydrocarbons in the prebiotic environment. One approach focuses on the chemical reactions and dynamics that are likely to have occurred on the Earth in prebiotic times.

"The other approach, which is not necessarily in conflict with, or in opposition to, the first approach, gives emphasis to the possibility that various hydrocarbons, both simple and complex, might have been transported to Earth through carbonaceous chondrite meteors, comets and interplanetary dust particles. I'll start with this second approach.

"The term chondrite is derived from the millimeter-sized structures -- known as chondrules -- that can be found distributed throughout the interior matrix of a meteor along with other kinds of stony minerals. The origin of these chondrules still has not been determined although they are believed to come from the aggregates of silica minerals that were generated through the melting and fusion occurring in the solar nebula during the early stages of the evolution of our solar system.

"Approximately 5-6% of these stony, chondrite meteorites also contain different amounts of carbon compounds. For obvious reasons, this subset of stony meteorites is referred to as carbonaceous chondrites.

"Usually speaking, carbonaceous chondrite meteorites contain up to several percent, by mass, of carbon materials, of one sort or another. Moreover, some of these compounds include complex hydrocarbons.

"For example, the Murchison meteorite that fell in Australia in 1969 has been studied quite extensively. Six of the basic twenty amino acids found in Earth organisms were discovered in that meteorite.

"There also were at least twelve other kinds of amino acid compounds found in the meteorite. Although, as far as we know, these other varieties of amino acid do not occur in biological organisms on Earth, their presence is considered significant because it suggests, under the right prebiotic conditions, many different species of complex amino acids are capable of being formed.

"Some people have disputed the Murchison findings, claiming that the amino acids discovered in the meteorite were there as a result of contamination by organic matter from Earth. While most researchers do not accept such claims, there is a small aura of controversy lingering about the Murchison meteorite.

"This charge of contamination cannot be leveled at the findings of another study involving two meteorites that have been discovered in Antarctica. These meteorites had been buried in the frozen depths of Antarctica's ice for some 200,000 years.

"Many varieties of amino acid were found in those two meteorites. A little less than half of these amino acid compounds were quite different from the ones that are found in living organisms on Earth.

"The definitive proof concerning the extraterrestrial origin of these amino acids has to do with their optical properties. More specifically, by optical properties, I mean the direction in which a solution of such amino acids can rotate the plane of polarization of polarized light that is passed through such a solution.

"On Earth, when one shines polarized light through a solution of amino acids taken from a biological or living source, then, in such a solution, all twenty of the amino acids that form the proteins in Earth organisms will rotate, to the left, the plane of polarized light shining through the solution. This is a distinctive signature of the amino acids of Earth organisms.

"On the other hand, if one throws together a batch of amino acids in the laboratory, one will end up with what is called a racemic mixture. In other words, there will be equal numbers of what are called, in accordance with an agreed upon convention, left- and right-handed amino acids.

"This means that if one were to shine polarized light through solutions made up of this racemic mixture, one would find the direction of rotation of the plane of polarization shifting in different ways. Sometimes the direction of rotation would be to the left, and sometimes the shift in the plane of rotation would be to the right.

"When, however, amino acids from these meteorites were placed in solution, they shifted the plane of polarization exclusively to the right. This was entirely unlike what happens with either the racemic mixtures of amino acids in the laboratory or the amino acid solutions drawn from organisms on Earth.

"At least two conclusions follow from this. First, the only explanation we have for the origins of the amino acids in the Antarctic meteorites involves sources that are extraterrestrial in nature. Secondly, the existence of such complex hydrocarbons suggests that when conditions are right, whether on Earth or elsewhere, amino acids will arise through natural processes.

"In addition to amino acids, other kinds of complex compounds have been found in some carbonaceous chondrites. One researcher, for instance, discovered hydrocarbon compounds that appeared to have properties that could have played a role in membrane formation.

"This same researcher also found a yellowish pigment-like material that was able to absorb energy when light was shone on it. This pigment might have been some sort of precursor to, or an early competitor of, the chlorophyll pigment system that eventually emerged in some Earth organisms."

Professor Yardley paused in his presentation to pour another glass of water. Once he filled the glass, however, he did not drain the glass as he had done previously.

He held the glass in his hand and took only occasional sips. After one of the samplings, he said: "The material strength of carbonaceous chondrite meteorites often is so low many of them are unable to traverse the Earth's atmosphere without undergoing an airburst phenomenon in which they break up, and there is a release of many megatons of energy. Nonetheless, this sort of disintegration results in an increased surface-area-to-volume ratio of the remaining fragments that might allow some of the remnants to reach the ground with their organic payloads still intact.

"Researchers, in fact, have recovered fragments from catastrophic airbursts that are about a millimeter in size. Those who have examined such fragments have observed no signs of heating in their interiors and, therefore, any organic compounds that could have been there would have been protected from the effects of both the explosion as well as the heat of friction from passage through the Earth's atmosphere.

"Comets have been hypothesized, by some researchers, to be another potential means of transporting various kinds of hydrocarbons to Earth. These individuals have estimated -- on the basis of different methodological considerations -- that the composition of comets might have a hydrocarbon content that constitutes up to 14% of the mass of the comet.

"However, certain kinds of disparities between, on the one hand, the cratering records of the satellites of some of the outer-most planets, and, on the other hand, the cratering records of the so-called terrestrial planets that are closer to the sun, have led some scientists to maintain that very few comets are likely to have collided with Earth. Considerable uncertainty surrounds the role, or lack of it, which comets might have played in delivering organic molecules to Earth.

"There are some scientists who have argued that a far more important method of bringing organic compounds to the Earth might involve what are known as interplanetary dust particles. These particles, which might be the remnants of comets or asteroid-asteroid collisions, are about a micron in size, ... about one -thousandth of a millimeter.

"This might seem excessively small, but one should keep in mind, many bacteria are no more than one micron in diameter. Moreover, bacteria contain many, many, very complex hydrocarbon molecules.

"Not only are interplanetary dust particles big enough to contain, potentially, a variety of complex hydrocarbons, some of these particles might have just the right kind of mass properties that would prevent them from being incinerated by the frictional heat that is generated during entry into the Earth's atmosphere. Some researchers have calculated that those dust particles that are between: 10^{-12} to 10^{-6} grams, would be decelerated sufficiently in our atmosphere to allow such particles, which have been radiation-hardened by their trip through interplanetary space, to reach the surface intact.

"If the dust particles were smaller than this, they probably would be destroyed by the photolysis that is brought about by the ultraviolet part of the spectrum of sunlight. If, on the other hand, the dust particles were to approach the size of, say, small pebbles, they would be destroyed by organic pyrolysis, or the decomposition brought about by the heat of friction when traversing the Earth's atmosphere.

"Approximately 10% of an interplanetary dust particle's composition is in the form of hydrocarbon molecules. In addition, some individuals have estimated that the collective mass of the particles that enter our atmosphere outweighs many of the smaller, grapefruit -sized, meteorites by a ratio of approximately 100,000 to 1.

"Some researchers have calculated that carbonaceous chondrite meteorites and comets, when considered together, could have transported as much as 10^{20} grams of organic carbon, or hydrocarbons, to Earth during the prebiotic period that led to the origin-of-life through natural chemical processes. If one adds this amount to that which is believed to have come through interplanetary dust particles, then one is talking about quite a lot of organic carbon materials.

"Irrespective of the precise extraterrestrial or exogenous source of the hydrocarbons, evolutionary biologists believe these organic contents would have been released over time. Heavier, water - soluble compounds, like amino acids, would have dissolved in the global ocean.

"Low-density hydrocarbons, on the other hand, are likely to have become concentrated on the surface of the ocean ... much as an oil-slick does today.

Eventually, these molecules, like so much flotsam, would surf on the tides to the shores of volcanic islands or continents that were in the process of formation.

"The same mechanism of tidal transportation, of course, also would occur in relation to the heavier water-soluble compounds that went into solution in the ocean. The process probably just would have taken longer."

Mr. Mayfield was about to ask another question when a man came through the door behind, and to the left, of the judge. The man approached the judge and seemed to be relaying some message to her in the form of a folded piece of paper.

Judge Arnsberger took the paper silently and nodded her head in acknowledgement or thanks to the man. She scanned the piece of paper briefly, and, then, put it down.

"Mr. Mayfield," she said, "before you continue with your direct examination of this witness, I'm afraid there is an urgent matter that awaits me in chambers. I ask for your indulgence and extend my apologies, but I need to call a short recess often to fifteen minutes."

Having made her announcement, she banged her gavel. She quickly got up from her chair and soon disappeared behind the door through which the messenger recently had come.

Beach Front Property on a Warm Little Pond

The door at the front of the courtroom opened and the judge entered. A court officer said: "All rise," and, then, a short time later: "Please, be seated."

"You may continue with your examination of the witness, Mr. Mayfield," Judge Arnsberger directed. "I should remind the witness that he is still under oath."

"Dr. Yardley, I believe," indicated the prosecutor, "you were talking about meteorites and carbonaceous chondrites before the recess. Would you continue on with your testimony please?"

"Actually," the professor stated, "I was just about to begin talking about something else when the recess was announced. As I suggested earlier in my testimony, meteorites, comets and interplanetary dust particles are only one approach to explaining the presence of various kinds of hydrocarbons, both simple and complex, in the prebiotic environment of early Earth history. The other approach, to which I will now turn, concerns the chemical processes that are believed by evolutionary biologists to have been operating prior to, but that eventually brought about, the advent of biological organisms.

"Serious experimental work in the area of prebiotic chemistry has been going on for nearly fifty-five years in laboratories all over the world. Symposia and conferences dedicated to this subject take place on a regular basis, and, in addition, there are academic journals that publish articles dealing with virtually every facet of the prebiotic chemistry in which life is believed to have had its origins.

"Obviously, I cannot possibly present all of that material at this time. What I can do, however, is to try to provide some of the broad-brush strokes of the picture being painted by researchers.

"Although a few scientists, such as Alexander Oparin in the Soviet Union and J.B.S. Haldane in England, had been doing work on this topic during the 1930s, many people cite the early-1950s work of Stanley Miller and Harold Urey at the University of Chicago as marking the real beginning of serious investigation of the conditions necessary for the chemical origins of life. They were the first to put things to the test under laboratory conditions.

"In an oft-cited, classic experiment, Miller gathered some gases, such as methane (CH_4) and ammonia (NH_3), believed to be present in the early

Archean era atmosphere, subjected these gases to a continuous spark discharge, which was intended to simulate the action of lightening, and examined the results after a number of days. The laboratory procedure had generated a variety of amino acids, some of which are found in living organisms and some that are not present in life on Earth.

"Amino acids are complex hydrocarbons. They consist of three parts.

"One part is a carboxyl group, having a formula of COOH. A second component is an amino group with a formula of NH_2.

"The third aspect of the amino acid is a side chain. This varies, in a unique way, with each, different amino acid.

"Some 16-17 years after Miller's experiment, the Murchison meteorite was found in 1969, and scientists were able to demonstrate a number of similarities between the products of Miller's experiment and the hydrocarbons found in the meteorite. For instance, they discovered that the kind and quantities of amino acids found in the Miller experiment were very, very similar to the kind and quantities of amino acids found in the meteorite.

"In any case, by 1953, Miller had produced the first experimental evidence that natural chemical processes could produce complex organic compounds that are fundamental to life on Earth. Over the next forty-odd years many other experimental results would be forthcoming from Miller and other researchers.

"In one series of experiments, Miller and Urey discovered that roughly 10% of the carbon molecules contained in the gases of their experimental set-up eventually were converted into known organic compounds. Furthermore, as much as 2% of this converted carbon was involved in the generation of amino acids within the experimental apparatus."

The prosecutor, Mr. Mayfield, who had been listening intently to the professor, suddenly came to life, so to speak, and said: "Dr. Yardley, earlier you had indicated that an oxidizing atmosphere ... in other words, an atmosphere composed of, say, oxygen, which strips other compounds of hydrogen ... tends to interfere with chemical processes that build complex hydrocarbons from simple hydrocarbons. Is there a name for the sort of atmosphere that is conducive to the generation of complex hydrocarbons from simple ones?"

"Yes," he replied, "the kind of atmosphere to which you are referring is known as a reducing atmosphere. Molecules that can donate hydrogen atoms, or, more precisely, electrons, to other substances tend to dominate that kind of atmosphere.

"Methane and ammonia, the gases used in Miller's experiment, are both considered to be relatively good reducing agents. This means they tend to be involved in chemical reactions involving, to simplify things somewhat, the donation of some of their hydrogen atoms or electrons, which then interact with other hydrocarbon compounds to help make possible, under the appropriate conditions, the formation of even more complex hydrocarbon molecules.

"In one sense, all organic compounds are actually different gradations of reduced forms of carbon. Generally speaking, this is due largely, although not necessarily always, to the presence of hydrogen in such compounds.

"Creating different kinds of reducing atmospheres under experimental conditions, investigators were able to produce a variety of amino acids. Glycine, valine, alanine, proline, glutamic acid and aspartic acid all have been generated through different kinds of electric discharge experiments.

"In another experiment, when sunlight was passed through a solution of paraformaldehyde $(CH_2O)_3$, ammonia (NH_3), and ferric chloride, the amino acids asparagine and serine were produced. On the other hand, when solid ammonium carbonate was exposed to high doses of gamma rays, small quantities of the amino acid, glycine, along with formic acid (HCOOH), resulted.

"In 1961, another scientist, Juan Oró, wondered if amino acids could be generated under laboratory conditions if one used chemical processes that were even simpler than those involved in Miller's earlier experiments. Previous experiments had proven that if one exposed a mixture of hydrogen, nitrogen and carbon monoxide gases to a spark discharge, the reaction would yield hydrogen cyanide (HCN), which is a very reactive intermediate compound.

" Oró combined hydrogen cyanide with ammonia (NH_3) and water (H_2O). This chemical reaction produced a number of different amino acids, just as the Miller experiment had.

"In addition, among the product residues of his experiment, Oró discovered something else. This molecule was a purine ... a nitrogen-containing base of considerable importance.

"The particular purine found by Oró is known as adenine. This molecule is one of two purine bases having a general formula of $C_5H_4N_4$, and three pyrimidine bases, each of which has a general formula of $C_4H_4N_2$. When any of these are combined with either of the pentose sugars, ribose or deoxyribose, together with a phosphate group, then, RNA or DNA is produced.

"Adenine is also one of the components of adenosine triphosphate. This latter molecule is one of the fundamental energy-providing compounds in most organisms.

"In addition to adenine, a number of other useful products could be produced by means of reactions involving hydrogen cyanide. These products included a variety of intermediate precursor molecules that constitute steps on the way to purine or pyrimidine formation, and the products of the reactions included, as well, a number of pyrimidine base molecules that are found in the nucleic acids of some, but not all, living organisms.

"Subsequent experiments demonstrated the possibility of generating, through natural chemical processes, the other nucleic acid bases ... namely, uracil, cytosine, guanine and thymine, which are found in the vast majority of organisms on Earth. These reactions also started with hydrogen cyanide (HCN), but they required, as well, the presence of two other simple carbon compounds: cyanogen (C_2N_2) and cyanoacetylene (HC_3N), which are believed to have been present on the prebiotic Earth.

"Still other experiments were able to demonstrate that the pentose sugar, ribose, an important component of RNA, could be produced quite easily. This chemical process merely involved a series of spontaneous reactions involving molecules of formaldehyde CH_2O.

"Again and again, scientists were showing, experimentally, the possibility of starting with simple compounds and combining them to produce complex hydrocarbons. More importantly, these products were not just arbitrary molecules, but, rather, they were fundamental building blocks of compounds, such as proteins and nucleic acids, that are crucial to the life process.

"Researchers felt their laboratory experiments were recreating the conditions of prebiotic Earth and demonstrating that chemical reactions important to the origins of life would occur spontaneously. The whole process was relatively simple and straightforward.

"Initially, for example, atmospheric gases, such as methane and ammonia, would react together to generate a variety of simple hydrocarbons, like hydrogen cyanide and molecules known as aldehydes, which are compounds that contain a CHO group ... such as formaldehyde (CH_2O). Next, the products of the first round of reactions ... namely, aldehydes, hydrogen cyanide and ammonia ... would enter into a second round of chemical interactions that would result in such intermediary products as amino nitriles. These products, in turn, would react with the water of the ocean to produce ammonia and amino acids, like glycine.

"People such as Sidney Fox were able to discover, experimentally, alternative methods for the prebiotic generation of various kinds of amino acids - methods that were different from the ones outlined by Miller and Oró. When Fox heated urea [$CO(NH_2)_2$] and malic acid ($C_4H_6O_5$) at temperatures of 150 degrees Celsius, he was able to obtain aspartic acid.

"Fox also was able to construct chains of amino acids through a process of thermal co-polymerization or cooking. He referred to these chains of amino acids as 'proteinoids' because they had certain structural similarities to the proteins found in living organisms.

"The recipe for thermal co-polymerization of amino acids is fairly simple. One starts with some quantity of a given amino acid, such as glutamic acid.

"One places this quantity of amino acids in an oil bath and heats it at 170 degrees Celsius for an hour. When the timer goes off after an hour, one blends in a finely ground mixture of other kinds of amino acids.

"One heats this new mixture for an additional three hours at the same temperature as before. In addition, one heats it in an atmosphere of carbon dioxide.

"When the mixture has cooked for the requisite period of time, one allows it to cool under controlled conditions. When it is ready, one can examine the residue of this process and find polymerized or chemically linked sequences of amino acids of varying length and composition.

"Many of the proteinoid polymer chains consisted of up to 100 amino acids. The nature of the bonds linking the amino acids varied in character, but some peptide linkages, the kind that occur in proteins in living organisms, were observed among these bonds.

"The thermal co-polymerization process is capable of providing yields, by weight, of up to fifteen percent of the total mixture. These portions are considered by evolutionary biologists to be quite ample yields, although most of the rest of us might feel them to be too small to share with friends for a late-night snack.

"There are variations on the foregoing recipe. Glutamine, another of the amino acids occurring in living organisms, is substituted for glutamic acid. Phosphoric acid is also added.

"In addition, one skips the step of pre-heating prior to the adding of other ground-up amino acids. Everything else stays, more or less, the same, yielding roughly similar results as before.

"One can play around with parameters such as the temperature and time, at which and for which, respectively, the mixture is cooked. One also can alter the ratios of the reactants and/or phosphoric acid to be used in the process.

"Experiments focusing on the manipulation of these variables have permitted proteinoids with different kinds of character to be produced. For example, one can increase the percentage of neutral and basic amino acids that were incorporated into the polymerized chain.

"In 1977, a scientist by the name of Usher demonstrated that when one used relatively low temperatures, one could generate phosphodiester bonds between the phosphate and ribose sugar portions of nucleic acids. This is an important step in generating fully functional DNA and RNA molecules.

"In 1978 Juan Oró showed, experimentally, that if one heated fatty acids, an important building block of lipids, and, then, dried them in the presence of phosphate and glycerol, one could synthesize simple phospholipids. Phospholipids are fundamental to the formation of cell membranes in most living organisms.

"Stanley Miller has synthesized a compound under prebiotic conditions that is known as pantetheine. This molecule has been observed to link amino acids in some organisms.

"Many of the compounds produced in these kinds of experiment are quite soluble in water. Researchers have hypothesized that these molecules, at one point or another, probably would have gone into solution in the ocean, and, later, they would have become part of more concentrated solutions when washed, by winds and tides, into the margins of marine lagoons, tidal pools and other intertidal regions, from which water was being evaporated.

"This process of enhanced concentrations through evaporation is thought to be important by many researchers since, quite frequently, the presence of water seems to inhibit the process of polymerization or chaining of, say, molecules. Sidney Fox, along with other scientists, has found, for example, that in order to bring about the polymerization of amino acids, the conditions within the experimental apparatus should be anhydrous ... that is, done in the absence of water.

"Similarly, experimenters have discovered that ribonucleotides will not form oligomers or small chains of up to ten units of nucleic acids unless done by means of anhydrous heating. Furthermore, such heating must occur in the presence of both: a nucleotide triphosphate and cyanamide (CH_2N_2), a condensing agent."

"Dr. Yardley," the prosecuting attorney intervened, "what is a condensing agent?"

"Condensation," answered Professor Yardley, "involves a rearrangement of atoms in order to produce a molecule of greater complexity, density or weight. Condensing agents assist this process.

"Some scientists have hypothesized that exposed mineral, lava or sand surfaces, where temperatures might have reached 100 degrees Celsius, could have served as crucibles on which films of organic compounds, that washed in from the ocean, might have formed covalent bonds. This would have taken place through condensation reactions.

"Furthermore, these researchers theorize that once more complex hydrocarbons formed, some of these molecules might have migrated, through one natural process or another, downward a few centimeters below the surface. Such a micro-environment would have helped to protect the newly-formed compounds from degradation reactions driven by light and heat."

Pausing for a moment, Dr. Yardley finished the remainder of the water in his glass. He replaced the glass on the table in a way that suggested he was thinking about something else.

When he had settled in his seat, his lips were pursed. Finally, he spoke again.

"Even if one were to suppose," he added, "that some of the starting ingredients cited in the previous experiments were not produced in abundance through chemical reactions on prebiotic Earth, one should remember that exogenous or extraterrestrial sources might have helped supplement the normal, earthly complement of these compounds. Water, hydrogen cyanide, ammonia, cyanoacetylene, and formaldehyde ... all of which I mentioned earlier, are found in interstellar dust clouds and might have found their way into meteors, comets or dust particles and, then, subsequently, been transported to Earth.

"Some scientists, in fact, have estimated that in the first 700 million years of the Earth's existence as a planet, the Earth is likely to have passed through 4-5 interstellar clouds, taking roughly 600,000 years to complete each such passage. For each year of passage, this would have resulted in, approximately, between 1-10 million kilograms of material being added to the Earth.

"This is thought to be one or two orders of magnitude, or powers often, less than what has come through interplanetary dust particles. Moreover, like these latter dust particles, only a small percentage of this total mass would be in the form of simple carbon or hydrocarbon molecules. Nonetheless, even a limited percentage of such astronomical figures still would constitute a substantial amount of carbonaceous material available to the prebiotic environment."

"Although," said the lawyer for the prosecution, "I'm quite certain, Professor Yardley, you could provide the jurors with a great deal more information on laboratory experiments that are intended to simulate the conditions on prebiotic Earth, I would like to shift gears slightly. Earlier in your testimony, you had alluded to the importance of having some degree of understanding of the systems of energy that, in a prebiotic environment, would have driven many of the chemical reactions you just have been describing.

"Would you please tell the court a little about this facet of evolutionary thought? Once again, Professor, and I apologize for being a one-note-Norman on this matter, to whatever extent possible, try to strike a balance between avoiding both oversimplification and too much technical complexity."

Dr. Yardley sighed slightly and, then, took a deep breath. He looked briefly at the table by the witness stand, noticed that the pitcher had not much water in it, and made a few motions to Mr. Mayfield indicating he would like the jug to be refilled.

As one of the officers of the court went about the business of getting more water, Professor Yardley started to speak. "There are," he began, "five or six energy possibilities that are likely to have been available to prebiotic Earth for the purposes of bringing about certain kinds of chemical evolution.

"The first possibility requires no external input of energy. These involve physio-chemical forces, such as hydrogen bonds, which, very likely, played a significant role in helping certain molecules in the prebiotic environment to organize or self-assemble into more complex, and biologically relevant, packages.

"For over forty years, thanks to the monumental work of, among others, Watson and Crick, scientists have known that the purine, nucleic base adenine in DNA and RNA pairs spontaneously with the pyrimidine, nucleic base uracil in RNA or the pyrimidine, nucleic base thymine in DNA. Similarly, the purine, nucleic base guanine pairs, in spontaneous fashion, with the pyrimidine, nucleic base in both DNA and RNA.

"These pairings are known as Watson-Crick bonds and are a specific example of hydrogen bonding. The complementary pairs of nucleic bases that are strung along two strands of DNA or RNA are brought together in stable configurations by these bonds and, in the process, help lend the double helical structure to the joining of these strands with which most of us are familiar from school and the media.

"Hydrogen bonds occur as a result of the positive and negative, or dipolar, characteristics that arise in compounds containing hydrogen, oxygen and nitrogen atoms arranged in the right kind of geometrical configuration. More specifically, nitrogen and oxygen are both relatively electronegative in nature, whereas hydrogen tends to be electropositive in character.

"This means oxygen and nitrogen are inclined, under certain circumstances, to draw toward their nuclei a few of the electrons of geometrically well-placed, neighboring hydrogen atoms or molecules. As a result, the affected hydrogen atoms of these neighboring molecules become electropositive and, therefore, have a tendency to establish bonds with other neighboring atoms or molecules that offer electrochemically compatible opportunities.

"These hydrogen bonds bring a certain amount of stability to the manner in which, under certain circumstances, atoms and molecules arrange or organize themselves. Consequently, they are thermodynamically favored arrangements because of their tendency to help stabilize the way energy is distributed in a molecular configuration.

"Hydrogen bonds are characteristic of what are referred to as polar molecules. The polar aspect of these molecules is rooted, as indicated previously, in the process of creating electrochemically-charged dipolar, or positive and negative, regions.

"Polar molecules, such as water and ribonucleic acids, have very different physical and chemical properties from non-polar molecules that do not possess such dipolar regions. Many hydrocarbons that do not contain nitrogen and/or oxygen tend to be non-polar in nature.

"The bottom line on all of this is that hydrogen bonding, of which Watson-Crick pairing in complementary bases of DNA and RNA is an extremely important example, is an instance of a spontaneous, thermodynamically favored generation of greater complexity. A chemical reaction is said to be spontaneous if it can take place without requiring any additional energy.

"The reason a reaction can take place without the need of additional energy is because the energy available to the system has a natural tendency to redistribute itself until no further redistribution of that energy is capable of occurring in a spontaneous fashion. This redistribution process leads to a stable configuration of energy distribution that is why a reaction is said to be thermodynamically favored since, under most circumstances, the thermodynamic nature of chemical reactions is to spontaneously follow whatever pathways are available that will lead to such stability.

"Spontaneous reactions yield energy. In other words, if one measures the potential energy of the final, stabilized state of this kind of reaction, one will find less energy than was present at the beginning of the reaction.

"One of the reasons why the final state of spontaneous reactions is stable is because not all of the energy that is being released remains in a chemically useable form. Some of the released energy is in the form of heat that is unavailable ... that is, it cannot be harnessed to run the reaction in a reverse direction, back to the original, initial state prior to the reaction's commencement.

"The term 'free energy' is often used to refer to the form of energy in a given chemical system that is available to be redistributed, if possible, in a way that allows the system to find, if not already realized, its most stable configuration of energy. This configuration is that point at which the available free energy reaches, through the spontaneous activity characteristic of the system in question, its lowest level consistent with such stability.

"As I indicated previously, in the process of yielding or releasing energy during the time required for a spontaneous reaction to run 'downhill' to its stable state, there is a portion of the released energy that is transformed to a form of energy, namely heat, other than free energy. Entropy is a measure of the amount of energy that has been converted from its free energy form to its non-free energy form.

"Spontaneous reactions always result in a decline of free energy. In other words, the total amount of free energy of the products of a chemical reaction always will be less than the total free energy of the initial reactants of the reaction.

"Consequently, in the process of spontaneously seeking out a stable state of redistributed energy ... that is, a state of lowest possible free energy ... free energy is lost. The entropy, the amount of energy in a non-free form, tends to increase.

"Spontaneous chemical reactions in which energy is released to the environment are known as 'exergonic' reactions. Chemical reactions in which energy needs to be acquired from the environment are known as 'endergonic' reactions.

"One can use the released energy of spontaneous, 'downhill', exergonic reactions to drive 'uphill', non-spontaneous, endergonic reactions. This is referred to as a 'coupled reaction'.

"Non-free forms of energy are generated during both the downhill and the uphill portions of these coupled reactions. Consequently, the total amount of entropy will be increased during the process.

"As long as one has downhill reactions to sponge off, then, uphill reactions are possible. However, in order to keep a sequence of coupled reactions going, one becomes engaged in a constant process of borrowing from Paul to pay Peter who has borrowed from Mary in order to pay George, and so on.

"Non-spontaneous reactions always are in need of arranging a loan of energy from the spontaneous energy generators of the world in order to be able to activate the free energy potential of the non-spontaneous system. When there are no downhill reactions available from which an uphill system can borrow, things come to a sort of dynamic halt known as equilibrium in which its uphill, non-spontaneous character does not change, despite the fact activity still is going on within the system.

"There is a minimum amount of free energy that has to be borrowed by, or introduced into, an uphill, non-spontaneous system in order to bring about a chemical reaction. This minimum amount of energy is known as the free energy of activation or the activation energy.

"One of the major issues of evolutionary theory is to provide plausible accounts of how spontaneous, downhill generations of energy were coupled with non-spontaneous, uphill systems of molecules to generate arrangements of hydrocarbons of increasing complexity. Spontaneous chemical reactions that are thermodynamically favored will take one only so far.

"Therefore, while phenomena such as hydrogen bonding and Watson-Crick pairing are important ways of introducing additional organization into a system without having to borrow additional energy, much more is needed to be able to account for the gradual transition, or evolution, from simple hydrocarbons to the emergence of living systems. Many, if not most, of the chemical reactions that are needed to account for how life arose from a prebiotic environment are of the uphill, non-spontaneous variety rather than the downhill, spontaneous kind, and this means, as suggested earlier, the need to find coupling mechanisms of one sort or another.

"There are a fair number of coupling candidates that would have been readily available under prebiotic conditions. I'll list the candidates first, and, then I'll explore a few of these possibilities.

"First, although not necessarily the most important, are electrical discharges. In a prebiotic environment, these would be manifested through lightening.

"A second candidate would be ultraviolet radiation. Various molecules are capable of absorbing different dimensions of the ultraviolet portion of the spectrum of electromagnetic radiation. When a molecule absorbs ultraviolet light of the right wavelength, the energy of the light can be utilized to help drive certain kinds of chemical reactions involving such a molecule.

"A third possibility for a source of energy capable of driving some non-spontaneous, uphill reactions would be ionizing radiation. Gamma radiation, together with so-called cosmic rays, would be examples of this kind of candidate.

Prebiotic heat would be a fourth coupling candidate. For instance, a surface that had been heated to high temperatures ... either by sunlight, or by a nearby volcano, or by a hydrothermal vent ... such a heated surface might have provided an environment that helped bring about condensation reactions and the forging of various kinds of covalent bonds among molecules lying about on that surface.

"Another possibility involves the energy associated with shock waves. Such waves, for instance, accompany lightening discharges but are distinct from the electrical energy of those discharges.

"In addition, shock waves occur when meteors traverse the Earth's atmosphere. Such waves also are generated when there is an airburst of, say, a carbonaceous chondrite in our atmosphere.

"Tremendous amounts of energy are released under these circumstances. This could be coupled with, and utilized by, various uphill systems.

"There is a further possibility that is not really a source of energy but that would have an important impact on whether or not the minimum energy of activation was achieved in a, heretofore, non-spontaneous, endergonic set of molecules. This additional candidate concerns the process of catalysis.

"A catalyst is capable of helping reactions to proceed by, among other things, helping to lower the normal, minimal level of energy that usually needs to be imported in order to activate a given chemical reaction. A wide variety of non-protein, non-enzymatic mechanisms -- ranging from clays, to metal ions, to RNA ... have been proposed as possible catalytic agents in a prebiotic environment.

"Since, previously, I already have given something of a taste for what is possible, experimentally, with the electrical discharges of Miller's experiment and the anhydrous, heat driven experiments of Fox, I would like to touch on a few of the other possibilities. Once again, this treatment won't be exhaustive, but it will provide members of the jury with a framework of sorts through which to understand this aspect of the evolutionary model.

"There have been a number of laboratory experiments that explored certain aspects of the phenomenon of shock waves. For instance, the heat, in the vicinity of 3000 degrees Kelvin or more that is generated by rapidly expanding gases in shock wave tubes has been used to produce such hydrocarbons as hydrogen cyanide and amino acids in different kinds of gas mixtures with reducing properties.

"A few researchers have hypothesized that organic compounds might have been synthesized in the atmosphere when meteors passed by and, in the process, created conditions similar to the shock heating experiments in the laboratory. After being synthesized, these compounds would have found their way, through one means or another, to the ocean.

"Once in the ocean, one of three things is likely to have occurred. The newly synthesized molecules would have reacted further with molecules in the ocean; or, these molecules would have been carried to tidal pools and other intertidal zones where they would become concentrated and readied for further reactions when the water in these pools and zones evaporated; or, some combination of the first two possibilities.

"Some scientists have calculated that meteors with a mass between: 10^{-14} to 10^2 grams, enter the Earth's atmosphere with sufficient frequency to deliver about 1.6×10^7 kilograms of mass to the Earth each year. If one were to assume these meteors traveled with a velocity of 15 kilometers per second, the meteors collectively generate about 1.8×10^{15} joules of energy per year, which is equivalent to many megatons of explosives.

"Ah ... Professor, before you continue," Mr. Mayfield interrupted, "could you explain what a joule is."

"A joule," Dr. Yardley explained, "is a unit of work or energy equivalent to the work that is done, or the heat generated, in one second, by an electric current of one ampere against a resistance of one ohm and ... " Stopping, Professor Yardley smiled sheepishly and raised his eyebrows somewhat. "Sorry," he said, "I don't think my answer is quite what you were looking for, Mr. Mayfield."

After thinking about the matter for a few seconds, the professor informed the lawyer: "The easiest, maybe most recognizable, thing to say" he offered, "is this. A watt of energy is equivalent to 1 joule per second. However, one should keep in mind that, strictly speaking, a watt is a measure of power, whereas a joule is a measure of energy. Power deals with the rate at which energy is expended."

Dr. Yardley looked at the prosecution lawyer with a more hopeful expression, seeking, apparently, acceptance for his new approach. When Mr. Mayfield motioned his head and made a face, both of which seemed to suggest: Why don't we move along before things get worse, Dr. Yardley returned to his testimony concerning the energy created by atmospheric shock waves.

"100 percent of the kinetic energy of meteorites of the previously indicated size is lost to the atmosphere. Researchers maintain that some fraction of this energy is converted into the generation of atmospheric shock waves. Estimates of the fraction of the energy being converted in this manner run from 30% downward.

"Working along similar lines, researchers have made calculations for the amount of energy that is converted to shock waves for other kinds of phenomena. For example, the airbursts of carbonaceous chondrites with a radius that is less than, or equal to, 300 meters, is believed to generate about 1.5×10^{14} joules of energy per year, which is the equivalent of a huge amount of high explosives.

"When a meteorite does not airburst and strikes the ground, if the meteorite is sufficiently big in size, it will generate a post-impact vapor plume. Some researchers have calculated that such post-impact vapor plumes could generate as much energy as 6×10^{17} joules per year in the form of shock waves that, once again, would be the equivalent of many megatons of high explosives.

"In addition to the energy being converted into shock waves capable of synthesizing certain organic molecules, researchers have estimated that a small percentage of the carbon in the meteorite will be incorporated into organic compounds when the meteorite vaporizes upon impact. This percentage is considered to be about 4%, which would have yielded approximately 4.6×10^6 kilograms of organic materials per year on prebiotic Earth.

"Most of this incorporated carbon shows up in the form of carbon dioxide and carbon monoxide. However, several percent of the carbon is incorporated into various kinds of hydrocarbons, and there is a still smaller percentage being converted into such compounds as hydrogen cyanide, as well as aldehydes, like formaldehyde.

"If one adds all of these different kinds of energy and mass values together, one can begin to develop a thermochemical model of shock synthesis under both reducing and relatively neutral atmospheric conditions. Scientists have discovered that the efficiency with which organic compounds can be synthesized through shock waves is very dependent on the compositional character of the atmosphere in which the shock wave occurs.

"For example, in a reducing atmosphere of methane, nitrogen and water vapor, for each joule of energy generated by shock waves, one can produce approximately $10^{17.5}$ molecules of hydrogen cyanide. Simultaneously, lesser amounts of simple hydrocarbons like C_2H_2, C_2H_4, and carbon soot also will be produced.

"After all the calculations are done, this works out to be a yield of 1.2×10^{-8} kilograms of organic material is generated for each joule of shock-created energy in a reducing atmosphere. In a neutral atmosphere, on the other hand, consisting of, for instance, carbon dioxide, nitrogen and water vapor, a yield of 2.5×10^{-16} kilograms of hydrogen cyanide is produced for each joule of energy, but yields of formaldehyde (H_2CO) remain roughly equivalent to what occurs in a reducing atmosphere.

"Similar calculations have been carried out in relation to both lightening and coronal discharges in the atmosphere. For example, in a reducing atmosphere, lightening is estimated to have been likely to generate 3×10^9 kilograms per year of organic material from the 1×10^{18} joules per year of energy created.

"In a neutral atmosphere, lightening is calculated to have been likely to produce 3×10^7 kilograms of organic material per year from the same amount of energy yields. However, as the atmospheric ratio of hydrogen gas relative to carbon dioxide drops from, say, 2 down to 0.1, the yield of hydrogen cyanide, formaldehyde and amino acids drops by a factor of several magnitudes or powers of ten.

"If one combines all the different ways of using energy that would have been available on prebiotic Earth to generate organic materials, scientists estimate that about 10^{11} kilograms of organic material would have been produced each year in a reducing atmosphere. However, in a relatively neutral atmosphere, consisting of mostly carbon dioxide and about 10% hydrogen gas, approximately 10^9 kilograms of organic materials would have been produced each year, but this yield will fall considerably as the relative percentage of hydrogen gas drops.

"In the light of these calculations, evolutionary scientists have come to the following conclusion. If all the organic materials produced by these various means were fully soluble in oceans comparable in extent and depth to our present oceans, and if these organic materials had a mean lifetime of approximately 10^7 years with respect to thermal degradation in relation to mid-ocean hydrothermal vents, then the steady-state equilibrium of organic materials in prebiotic times would have been about 10^{-6} grams of organic solute for each gram of ocean water in a neutral atmosphere, and approximately 10^{-3} grams of organic solute for each gram of ocean water in a reducing atmosphere.

"Modern researchers in evolutionary theory believe that if the early Archean era atmosphere were strongly reducing in character, the predominant method of generating organic materials might have been through shock waves. Lightening would have been considerably less predominant in its effects in this regard, and the roles of ionizing radiation and radioactive disintegration would have been quite negligible."

"So," said Mr. Mayfield, "if someone wanted to put all of this information into perspective in a relatively simple manner, what would be the bottom line?"

"I guess" replied Professor after a few seconds hesitation, "one should return to the scenario I outlined earlier. Moreover, for the sake of simplicity, let's concentrate on just one of the hydrocarbons -- namely hydrogen cyanide -- that is likely to have been produced by one, or more, of the energy sources about which I have been talking.

"First, energy from shock waves or lightening or ultraviolet radiation is coupled with atmospheric gases such as methane (CH_4), ammonia (NH_3), and hydrogen (H_2), all of which serve as reducing agents, giving up hydrogen atoms or electrons to other substances. This coupling leads to the production of hydrogen cyanide.

"Secondly, the HCN or hydrogen cyanide that is formed becomes dissolved in water vapor in the atmosphere. Eventually, this becomes precipitation or rain that falls into the ocean.

"Thirdly, once in the ocean, the hydrogen cyanide would oligomerize or gather together in small quantities here and there. These oligomers of HCN would then undergo hydrolysis in the ocean.

"Hydrolysis is a process in which water interacts with a substance and tends to separate out the atoms of a substance such as hydrogen cyanide (HCN) by hydrating them, that is, surrounding them with water molecules. Furthermore, since water is a polar molecule involving, as previously indicated, dipolar regions of electronegative and electropositive charge, the polar character of water combines with the atoms that are being separated out through the process of hydrolysis to recombine to form different kinds of molecules.

"For instance, just to give you some idea of what is being said here, suppose one had a one-liter solution of one-tenth molar concentration of hydrogen cyanide and left it for a year. As a result of hydrolysis, after one year, one would find quite tiny, but detectable, amounts of the purine nucleic base adenine as well as larger, but still very small, quantities of the amino acid glycine.

"If we project such liter-size processes into the context of the trillions and trillions of liters of the oceans of the world, and if we left things for millions of years rather than one year, we are very likely to discover substantial amounts of a wide variety of complex hydrocarbons, many of which probably will be of fundamental importance to issues concerning the origins of life."

"Dr. Yardley, in the context of the present discussion, what relevance would the process known as a 'Strecker synthesis' have?" asked Mr. Mayfield.

"In synthetic, organic chemistry," responded the professor, "a Strecker synthesis generally involves bringing about the hydrolysis of, say, an amino nitrile in the presence of a strong acid. An amino nitrile joins together some kind of amino group or radical with a cyanogen or compound containing the group CN.

"Many researchers have accepted the pH value of the early, prebiotic ocean to be around 8, plus or minus 1. This means that the ancient ocean was considered to be either slightly basic, if it had a pH of 8-9, or relatively neutral, if its pH was around 7.

"Under such conditions, Strecker synthesis, which usually is done in the presence of a strong acid, would require a long time to hydrolyze organic compounds in the early, prebiotic ocean. Some researchers have set this figure at around 10,000 years.

"However, relative to tens and hundreds of millions of years, 10,000 years is really just a drop in the ocean so to speak. This kind of synthesis would have had the opportunity to run to completion many times over during the course of the Archean era.

"I should note, Mr. Mayfield, that although the Strecker synthesis process is considered by evolutionary theorists to be an adequate means of producing amino acids in the ancient oceans, some sort of additional mechanism of concentration and condensation would be required to produce, say, the purine, nucleic base, adenine. This is where processes such as evaporation, freezing and dehydration, along with hot, anhydrous conditions, which are believed to have been present in certain intertidal zones, would play important roles in chemical evolution on early earth."

"At this point, Dr. Yardley," requested the lawyer, "would you say a little about current thinking in relation to the nature and possible origins of membranes? I believe such a discussion will bring us a little closer to providing the jurors with a proper, introductory overview of evolutionary theory, by means of which they will be able to reach an informed judgment on the matter before the court."

"I suppose," Professor Yardley mused, "that molecules known as amphiphiles are as good as place as any with which to begin talking about the origins of membranes. Amphiphiles have sort of an aura of split personality about them.

"One part of this kind of molecule has hydrophilic properties and, as a result, is inclined to enter into interactions with water. The other part of the molecule entails hydrophobic characteristics and, therefore, tends to avoid, whenever possible, interacting with water.

"When amphiphiles are immersed in an aqueous environment, the hydrophobic aspects of the molecule curl up into small spheres known as vesicles. These tiny spheres form a protected space within which, given the right conditions and chemical reactants, various chemical processes could take place.

"The hydrophilic-portion of amphiphiles -- that is, the parts of the molecule which have an affinity for water – surrounds the hydrophobic aspects of the molecule. Not only do these hydrophilic portions represent an additional layer of separation between water and the interior, hydrophobic aspects of the amphiphile molecule, the water- loving components of the molecule also are free to enter into reactions with water.

"As such, amphiphile molecules possess some of the basic features of biological membranes. More specifically, the membranes of living organisms tend to be bilayered or have two membranes that are separated from one another by a relatively short distance, and the whole bilayered structure surrounds the interior of the cell.

"To be sure, the layered arrangement in amphiphile molecules is not quite the same as the sandwich structure of biological membranes. Most notably, there is no separate, distinct region between the hydrophilic and hydrophobic components of the amphiphile molecule, as there is in true, biological membranes.

"Nonetheless, in both true membranes as well as amphiphile molecules, one does have a double- layered arrangement surrounding an interior space or spaces within which chemical reactions could take place. Furthermore, both true membranes as well as amphiphile molecules consist of hydrophilic and hydrophobic aspects.

"Consequently, amphiphile molecules could be considered to constitute a rather crude facsimile or early precursor, of later, more complexly evolved, biological membranes. Interestingly enough, in this respect, some researchers maintain that exogenous organic materials -- that is, organic materials from sources such as meteorites and interplanetary dust particles -- might be quite rich, perhaps even preferentially so, in amphiphilic vesicles or spheres.

"Lipids, which are one of the main components in biological membranes, come in different varieties. As far as biological membranes are concerned, some of the more important lipids are composed of, among other things, a hydrophobic hydrocarbon component linked to a hydrophilic phosphate group, together with certain alcohols and/or bases.

"Lipids do not form polymers or chains of monomer units as, say, amino acids and nucleic acids do. This is because lipids are stabilized through non-polar physical forces instead of the covalent chemical bonding that characterizes polymerized compounds.

"These non-polar physical forces are essentially thermodynamic in nature. Non-polar hydrocarbons, such as oil, do not enter into solution when placed in water ... water being a polar molecule.

"Hydrocarbons have a tendency to disrupt, at least in part, the array of hydrogen bonding present in water. The most stable thermodynamic arrangement ... that is, the arrangement in which all of the molecules of a system have achieved their lowest chemical potential for reactivity ... is one in which hydrocarbon molecules aggregate into a separate phase form, such as droplets, away from water molecules.

"This process of phase separation between non-polar hydrocarbon molecules and polar water molecules is known as the hydrophobic effect. This effect serves as a significant force helping to stabilize various kinds of macro molecular systems, including membranes, in biological organisms.

"The hydrophobic effect does not involve any chemical transformations. It only reflects the natural preference, or self-organizational drive, of molecules to arrange themselves in ways that distribute the energy of the molecular system in the least chemically reactive, and, therefore, most stable state.

"Some evolutionary scientists have suggested that the hydrophobic, hydrocarbon portion of lipids might have been synthesized or formed by a Fischer-Tropsch-like reaction. This process starts with carbon monoxide and hydrogen that are placed under pressure, ranging from one to fifty atmospheres, as well as heated to temperatures that might vary from 180 to 300 degrees Centigrade.

"Usually, this reaction is done in conjunction with a catalyst of some sort. Many catalytic possibilities exist, but, quite frequently, the ones that are used are either nickel or iron supported by a layer of silica.

"Phospholipids, which are one of the fundamental building blocks of biological membranes, come in several forms. They are polar molecules in which the phosphate group has a negative charge, the alcohol group has a positive charge, and the complex hydrocarbon tail is hydrophobic in nature.

"More importantly, phospholipids, once formed, have been observed to assemble, spontaneously, into stable lipid bilayers and vesicles within an environment of water because of the aforementioned thermodynamic forces that are at work. The hydrophilic components of the molecule form the portions of the bilayer that will be in close proximity to water molecules, whereas the hydrophobic portions of the molecule form the aspects of the bilayer that will be phase separated from water molecules.

"Some scientists have approached the issue of the first, primitive cellular prototype from a different direction than that of amphiphilic molecules composed of both hydrophobic and hydrophilic components. These researchers have focused on certain kinds of proteinoid micro spheres that have been observed to form under certain experimental conditions.

"Once again, this sort of protocell structure would form a phase separation between the outer, aqueous environment and the inner regions of the micro sphere formed by the proteinoids. These inner regions could serve as a location for various kinds of chemical reaction to take place under conditions that are, to some extent, protected and stable.

"All membranes of living organisms consist of a combination of phospholipids and proteins. Therefore, if one were to combine the idea of proteinoid micro spheres and amphiphilic molecules, one would be getting quite close, in some respects, to the structural character of modern biological membranes.

"A cell is really a microenvironment bounded by a membrane. The phospholipid portion of the membrane constitutes a permeability barrier that helps stabilize and protect the microenvironment of the cell's interior.

"However, the down side of a permeability barrier is that it can keep out various kinds of molecules that might be necessary to chemical reactions going on in the interior regions of the bounded micro-environment. In biological cells, this problem is solved by a variety of proteins, referred to as transmembrane proteins, which extend from one membrane layer to the other membrane layer of the bilayered structure.

"These transmembrane proteins might serve different functions. Some of them provide channel ways, linking the external aqueous world with the internal bounded microenvironment.

"Some of these membrane proteins might function as carriers, or active transports, for certain kinds of molecules. Still other forms of these membrane proteins might form part of an ion pump system that brings various ions into the cell or gets rid of such ions de pending on circumstances and needs.

"When, as evolutionary biologists believe to be the case, proteinoids, at some point, became incorporated into self-assembling, phospholipid membrane structures, a major step would have been taken toward the first protocell. Various experiments of nature might have ensued then, exploring different arrangements and kinds of proteinoids in the membrane, some of which were naturally selected because of their ability to serve, in some minimum fashion, as channels, or carriers or parts of an ion pump system.

"Researchers feel those proteinoids would have been favored that had particular kinds of primary structure. More specifically, the sequence of amino acids constituting the primary structure of the protein should be such that, under the influence of purely thermodynamic, self-organizing forces, the tertiary folding pattern brought about by these thermodynamic forces would need to have arranged hydrophilic and hydrophobic aspects of the proteinoid in a certain manner.

"On the one hand, hydrophilic portions of the transmembrane proteinoid would need to be at the opposite ends of the membrane where they would be exposed to water molecules surrounding the cell as well as within the cell. On the other hand, those portions of the proteinoid structure that were hydrophobic should be folded away in the region between the two bilayers -- a region that consists of hydrophobic lipid molecules.

"Prior to the appearance of such phospholipid-proteinoid micro spheres, there might have been transitional structures. Liposomes, for example, are small vesicles composed of fluid, lipid bilayers.

"Liposomes have the capacity for reversible breakage. In other words, under various conditions, they can break open and, then, spontaneously reseal.

"Thus, when liposome vesicles are agitated in an aqueous environment, they will break open at various points and, afterward, reseal.

This process of breaking and resealing enables the liposome to capture any solutes that might happen to be in the environment.

"Similarly, when liposomes are dried, they often form multi-layered structures. Solutes can become trapped within these structures. When the dried liposome becomes re-hydrated, the trapped solutes become sealed within the microenvironment of the liposome's interior.

"The property of being able to break and reseal could serve another function beyond providing a mechanism for admitting different kinds of solute materials into the liposome's interior region. Growth, division and multiplication, of a sort, also could be associated with this capacity to break and reseal.

"If one were to add some of the potential properties of a liposome to those of phospholipids and proteinoids, then the possibilities become even more intriguing. Such an amalgamation of properties is coming much closer to what we would recognize as a cell-like structure or protocell."

"I believe," indicated Mr. Mayfield, "we are almost to the end of our conceptual journey, Professor Yardley." The prosecuting lawyer went to his table and was handed some papers by his colleague.

As Mr. Mayfield slowly returned to the area of the witness stand, he was busy going through the papers. Apparently, he was either looking for something or briefly reviewing the material prior to launching into the next phase of direct examination.

When he was near the witness stand, he stopped and studied the papers for a few more seconds. When he had finished, he asked: "Professor Yardley, what is the so-called Central Dogma of molecular biology?"

"Essentially," Dr. Yardley replied, "it says that DNA makes RNA that makes proteins. In living organisms, the available evidence is overwhelmingly in support of this principle."

Pursuing the issue, Mr. Mayfield inquired: "Does this principle raise any problems in relation to accounts concerning the origins of life?"

"Yes, it does," the professor responded. "In everyday terms, it leads to the question: Which came first, the chicken or the egg?"

"Could you elaborate a little, Professor Yardley?" requested the prosecuting lawyer.

"Briefly stated," the professor summarized, "if DNA is necessary to make, first, RNA and, then, proteins, but the synthesis of DNA polymers depends on the presence of catalytic or enzymatic proteins, then, how can one start with DNA which is dependent on the very molecule that it is supposed to make? On the other hand, if proteins depend on the existence of DNA and RNA molecules, then how can proteins come into being prior to that on which they depend?

"If we have DNA reprise the role of the egg to protein's stirring rendition of the chicken, we, once again, are faced with an ancient paradox. In the present case, the problem becomes: which came first, the protein or the DNA?"

"Is there any plausible way out of this dilemma, Dr. Yardley?" asked the prosecuting attorney.

"Until relatively recently, this paradox constituted a major stumbling block to providing an overall plausible explanation for how life originated from prebiotic beginnings through purely naturally processes. The situation vis-à-vis this paradox began to change around 1983.

"In that year, two researchers, Sidney Altman and Thomas Cech, quite independently of one another, made a breakthrough. They discovered what has come to be known as a 'ribozyme'.

"A ribozyme is a polymerized or chained sequence of molecules that is drawn from RNA and exhibits some of the properties of a protein enzyme or catalyst. In the case of the Altman-Cech findings, the RNA sequence that had been discovered was able to cut and join pre - existing strands of RNA.

"This ability to cut into a given sequence of RNA and, then, to splice such sequences together is of considerable importance. Broadly speaking, not only do such capacities allow for the possibility of building longer sequences of RNA, but cutting and splicing, constitute tools that could play fundamental roles in processes of both replication and the rearranging of ribonucleic acid sequences to generate, or experiment with, alternative genetic characteristics.

"Most importantly, ribozymes do not presuppose anything else to accomplish these functions. In other words, the chicken/egg paradox evaporates since an RNA sequence that is capable of acting as an enzymatic molecule in relation to other RNA molecules, depends on neither proteins nor DNA in order to come into being. RNA molecules are serving as both hereditary blueprints as well as catalytic agents for the generation and development of such blueprints.

"RNA molecules have a further advantage, at least in relation to DNA molecules. The ribonucleotides in RNA ... that is, the bonded triads of ribose sugar, phosphate and nucleic base that are chained together to create sequences of RNA ... such ribonucleotides are more easily synthesized than are the deoxyribonucleotides of DNA.

"On the other hand, deoxyribonucleic acids are more stable than are ribonucleic acids. Consequently, whereas the easier path of synthesis would have conferred an evolutionary advantage on RNA molecules over DNA, the property of greater stability would have conferred, later on, an evolutionary advantage of DNA molecules over RNA.

"Many theorists cite this dimension of greater stability as probably one of the primary factors that led to a gradual evolutionary transition from RNA-based protocells or life to DNA-based protocells or life. At some point, DNA displaced RNA from the latter's role as keeper of the genetic memory.

"During the 1960s, some twenty years before the discovery of the first ribozyme, three scientists, Francis Crick, Carl Woese and Leslie Orgel, all working independently of one another, had each suggested that RNA might have had evolutionary priority over both DNA and proteins. Today, the original proposal of these three scientists has evolved, through the contributions of a variety of theorists and researchers, to become a theory known as 'the RNA world'.

"In the RNA world, as one might anticipate, RNA plays a central role. RNA, as the carrier of genetic information, as well as the agent responsible for catalyzing reactions, becomes responsible for generating all of the steps considered necessary to produce the first precursor of life capable of self-replication and evolutionary change.

"One can lend support to the idea of the RNA world theory with a number of recent findings. For example, consider the work of Harry Noller Jr..

"He was doing research on ribosomes that frequently are called the protein factories of a cell. Ribosomes consist of, on the one hand, ribosomal RNA, which differs in certain ways from non-ribosomal RNA, and, on the other hand they contain various kinds of protein. Both of these components are joined together to form what are known as ribonucleoprotein subunits.

"Each ribosome is assembled from two such subunits. Each of these subunits is slightly different in size and kind from the other.

"Furthermore, different kinds of ribosomal subunits can be found in prokaryotic, or non-nucleated organisms, and in eukaryotic, or nucleated organisms. However, just to complicate matters, some of the former, prokaryotic kinds of ribosomal units also can be found within certain eukaryotic intracellular organelles, or membraned centers, such as the energy-related factories known as mitochondria and chloroplast.

"In general terms, ribosomes travel along the length of various strands of messenger RNA. Messenger RNA is a single-stranded transcription of the triplet nucleic bases that are carried by DNA molecules. These triplet, nucleic base sequences, or codons, constitute the letters, so to speak, designating the specific word from the dictionary of twenty amino acids that is being called for by means of the messenger RNA.

"As a ribosome travels along the strand of messenger RNA, the ribosome helps forge a linkage, known as a peptide bond, between amino acids. The ribosome does this by taking the amino acid called for by one triplet nucleic base sequence of messenger RNA and connecting the indicated amino acid with another amino acid that is being called for by the subsequent RNA triplet nucleic base sequence of the same strand of messenger RNA.

"The ribosome accomplishes its task of fashioning polymers or chains of amino acids ... that is, proteins ... with the assistance of a further kind of RNA, known as transfer RNA. This form of RNA consists of between 70-80 nucleotides that are specially modified or adapted to be able to interact with the construction area formed by both the ribosome and the strand of messenger RNA.

"One portion of the transfer RNA carries the amino acid being called for. Another section, known as the anticodon, links up with the appropriate codon section of the messenger RNA, and the final sequence of the transfer RNA links up with the ribosome.

"Thus, transfer RNA delivers the required amino acid to the active site of the interaction between the ribosome and messenger RNA. This tri-partite co-operative effort continues, using a succession of different transfer RNA molecules, until the fully formed protein, which is being specified by the collective set of triplet codons of messenger RNA, has been completed.

"Harry Noller, the scientist I mentioned earlier doing research into the nature and functioning of these ribosomes about which I have been talking, discovered something of considerable importance to the RNA world theory. He found evidence suggesting that ribosomal RNA appears to play a major catalytic role leading to the formation of peptide bonds between amino acids being delivered by transfer RNA to the site of interaction between messenger RNA and the ribosome.

"The proteins present in ribosomes also have a catalytic role to play. Yet, this role appears to be limited to one of enhancing the degree of efficiency of the process already set in motion by the ribosomal component of the subunit.

"This molecule, consisting of ribosomal RNA and protein, is known as ribonuclease-P, and it is considered to be a true enzyme. Not only does it accelerate the rate of the formation of peptide bonds significantly over what would occur in the absence of such a molecule, but, as well, the molecule survives the chemical reaction and is capable of repeating the process with other transfer RNA molecules.

"On the one hand, the self-splicing ribozyme mentioned earlier is not considered a true enzyme. Although that molecule does have an enzyme-like function that involves capacities for cutting and splicing, nonetheless, at the end of the chemical reaction, the molecule does not get restored to its original, pre-reaction form.

"On the other hand, scientists, like Gerald Joyce, have been able to take this research a few steps further. Through a variety of procedures, he was able to generate ribozymes ... that is, RNA with catalytic properties that could cleave a number of different kinds of chemical bonds, including the peptide bonds that link amino acids together in biological organisms.

"In 1993, researchers at the Scripps Research Institute in California synthetically created a small sequence of RNA, sometimes referred to as the Scripps molecule, which had some amazing properties. First, the molecule began to make copies of itself within an hour after it had formed.

"Secondly, the copies of this molecule began to make copies of the copies. Finally, these copies began to evolve and display a variety of chemical properties that had not been anticipated.

"In another development, around 1994, Jack Szostak isolated a relatively short sequence of nucleotides, known as an oligonucleotide, which had catalytic-like properties. This catalyst could join together other, short sequences or oligonucleotides.

"In addition, this same catalytic agent could utilize energy from a triphosphate group in order to underwrite the polymerizing or chaining character of that molecule. This is important because triphosphates play fundamental roles as suppliers of energy for chemical reactions taking place in a living cell.

"Other researchers have proposed alternative routes for, say, the synthesis of RNA oligonucleotides that could be considered complementary to the previous findings. For example, James Ferris discovered that montmorillonite -- a relatively, common clay -- is capable of synthesizing RNA oligonucleotides."

"Dr. Yardley," said the prosecuting lawyer, "I believe we have covered enough information to provide the jurors with a good, though necessarily abbreviated, overview of the evolutionary perspective concerning the origins of life from prebiotic beginnings. If you were to sum up the general thrust of your testimony, what would you say?"

The professor stared off into the space near the ceiling at the back of the courtroom. After about ten seconds of deliberation, he stated: "If one runs through the available evidence in support of evolutionary theory, of which my testimony is but a very small sampling ... if one considers all the cosmological, geological, meteorological, hydrological and chemical data, then, I believe there is only one way to make consistent sense of the existing evidence.

"Biological organisms arose gradually, as the result of a series of steps, each of which was selected by prevailing circumstances that favored such a step over other possibilities existing at the time of selection. This fortuitous confluence of natural forces required tens of millions, if not hundreds of millions, of years to complete themselves.

"Among other things, this confluence of forces included various kinds of energy interacting with the gases in the atmosphere to generate simple hydrocarbons. These hydrocarbons subsequently precipitated out into a set of hydrological conditions that, perhaps through a Strecker synthesis process, were conducive to the formation of a sequence of progressively more complex hydrocarbons, such as amino acids, purines and pyrimidines.

"In addition, when these complex hydrocarbons were subjected to further processes of dehydration and condensation in various intertidal zonal regions, then, eventually, a variety of proteins, nucleic acids, lipids, and carbohydrates formed that were incorporated into bounded -- or membraned -- micro-environments from which arose the first protocells capable of self-replication. This capacity, very likely, was as a result of, initially, RNA catalytic activity that, at some point, became transformed into a DNA-based living organism.

"Throughout the sequence of gradual, evolutionary steps leading from the formation of the Earth, to the first cellular system capable of self-replication and genetic experimentation, spontaneous processes of self-organization played important roles. In other words, although chemical kinetics ... the study of the paths and rates of actual reactions ... constitutes an essential part of evolutionary thinking, nevertheless, thermodynamic forces also spontaneously led to arrangements of energy distribution that had important evolutionary ramifications for the forms and functions that different molecules came to have.

"Although there are certain details of the foregoing scenario that are presently eluding our grasp, we -- that is, evolutionary scientists -- believe all of the basic components are, in principle, now present for a rigorous, consistent, and plausible account of the origins of life through purely natural processes. Moreover, scientists and researchers, collectively, are quite confident, despite the fact there might be certain details that currently are missing from our account, that these same details will be forthcoming in the near future by virtue of the sort of scientific discoveries that are being made every day around the world."

"Thank you, very much, Dr. Yardley," said the prosecuting attorney, "for your illuminating, expert testimony." As the lawyer walked back to his table, he said: "Your witness, counselor."

The attorney for the defense was about to rise, when the judge said:

"Mr. Tappin, we are approaching -- if not encroaching on -- the dinner hour. Before you start your cross-examination, I think we will adjourn for meals.

"The jury is instructed not to discuss these proceedings either among themselves or with anyone else. Court will be in recess until 7:30 p.m. ." With that pronouncement, she banged her gavel.

Ah, Sweet Mysteries of Life

Judge Arnsberger entered the courtroom, and everyone had risen in concert with the command to do so that was given by one of the court officers. Again, in obedience to a directive, we all sat down.

"Mr. Tappin" stated the judge, "you may begin your cross-examination. Dr. Yardley, please remember, you still are testifying under oath."

Picking up a note pad from the table in front of him as he arose, Mr. Tappin approached the witness stand. Smiling at the professor, the defense lawyer said: "Dr. Yardley, I would like to commend you on an excellent presentation during direct examination."

The professor angled and dipped his head slightly in acknowledgment of the compliment. The smile on his lips was a tentative one, and the look in his eyes was wary in character.

The two looked like a cobra and a mongoose ready to do battle. Which was which was a toss-up.

Beginning the conceptual competition, Mr. Tappin briefly referred to the note pad he was carrying and stated: "In your discussion concerning meteorite impacts of the early Earth, Professor, you indicated that the scientific models dealing with what was happening on Earth, and when, were based on various studies conducted in relation to the lunar cratering data acquired through the Apollo space program. Is this correct, Dr. Yardley?" the lawyer asked.

"Yes, that's right," the professor answered.

"To the best of your knowledge," inquired Mr. Tappin, "what is the oldest time frame for which a radiometric date has been fixed in relation to the lunar samples?"

"That would be the Apollo 16 and 17 uplands data," Dr. Yardley responded. The radiometric dating process has established a time frame of between 3.85 and 4.25 billion years ago for the lunar samples taken from the craters in the areas of the two, aforementioned Apollo expeditions."

"Do the samples from the uplands represent the most heavily cratered areas of the lunar surface?" Mr. Tappin asked.

"No, they don't," the professor indicated.

"Therefore, Dr. Yardley, am I right in assuming that, at the present time, we don't have any radiometric data from these more heavily cratered areas of the moon?"

"Your assumption is correct," affirmed the professor.

"Then, this would seem to suggest," the lawyer stated, "that we don't know whether the more heavily cratered areas are older or younger than the lunar samples that have been brought back to Earth, or, perhaps, a bit of both ... that is, some craters might be older, and some might be younger."

"Yes, at present, the age or ages of the more heavily cratered areas of the moon only can be estimated," the professor acknowledged. "More precise dates must come from radiometric testing of samples from those areas."

"How would one go about estimating the age of areas of the lunar surface for which we have no direct data?" Mr. Tappin inquired.

"Well, this is really not my area of specialization," pointed out Dr. Yardley, "but, I suppose, a lot would depend on one's choice of decay rates and how one fitted this to the available lunar cratering data."

"Dr. Yardley, would the choice of decay rates substantially affect one's conclusions, both with respect to amounts and times, in relation to the models of extraterrestrial bombardment of early Earth?"

"Whether or not one's conclusions would be affected substantially, depends on what one means by the word 'substantially'," the professor replied. "In general, however, the use of different methodological or radiometric starting points obviously will have some kind of impact on one's conclusions."

"If I understand you, Professor," Mr. Tappin said, "the choice of decay rates with respect to lunar cratering data could increase or decrease estimates of such variables as: how many meteorites, what size and when such meteorites collided with the Earth. Is this, essentially, the case?"

"In broad terms, yes," Dr. Yardley confirmed. "As I indicated in my earlier testimony, the model concerning the influx of meteorites into the Earth's atmosphere is largely a stochastic or probabilistic one.

"Consequently, a range of values is possible," indicated the professor. "The ones I have given to Mr. Mayfield are best-estimate projections based on carefully worked out models of probability that are believed to have governed what transpired on early Earth as far as meteorite activity is concerned."

"Dr. Yardley, in your direct testimony, I believe you stated many evolutionary researchers are of the opinion that much of the heavy meteorite bombardment of early Earth probably began to taper off somewhere between 4.44 billion and 3.8 billion years ago. Is this true?"

"Yes," the professor affirmed.

"You also testified, did you not Dr. Yardley, that many scientists contend an extremely large meteoric impact occurred on Earth approximately 65 million years ago off the Yucatan peninsula, and there is evidence to indicate this collision might have destroyed most of the species in existence on Earth at the time?"

"I gave such testimony, yes," admitted the professor.

"Was the Yucatan crater the result of a statistical anomaly?" asked the defense lawyer. "In other words, can we assume that between, say, 3.8 billion years ago and 65 million years ago, there were probably few, if any, large-sized meteoric impacts on Earth?"

"Such an assumption would be a reasonable one," the professor said.

"What makes the assumption reasonable, Dr. Yardley?" inquired the lawyer.

"Well, for one thing," Dr. Yardley answered, "the very fact life continues to exist, and, on the basis of paleontological data, has existed for over 3.5 billion years, indicates there cannot have been too many large-sized meteorite collisions with Earth. If there had been, we probably wouldn't be having this conversation."

"In your opinion, Professor, would living organisms have a better chance of surviving such a catastrophic event than various prebiotic arrangements of complex hydrocarbons?" Mr. Tappin asked.

At this point, Mr. Mayfield jumped up and firmly stated: "Objection, Your Honor. The question is highly hypothetical and speculative."

"Mr. Tappin" probed Judge Arnsberger, "do you care to respond to the objection?"

"Yes, Your Honor, I do," replied the defense lawyer. "On the basis of both direct testimony, as well as on the basis of evidence derived from cross-examination to this point, the nature of science has been shown to involve, among other things, the use of assumptions, hypothesis, conjecture, probability, projections, estimates, interpolations and extrapolations. Therefore, I fail to see on what plausible grounds the prosecution could object to the defense's desire to explore certain hypothetical and speculative issues concerning the origin-of-life problem from a scientific perspective."

"Mr. Tappin has a point, Mr. Mayfield," the judge indicated. "I'm inclined to cut him some slack on this line of questioning provided the attorney for the defense doesn't roam too far astray.

"Objection overruled. The witness should answer the question," she stated.

Turning his attention from the judge to the lawyer for the defense, Dr. Yardley replied: "In my opinion, the answer to your question would depend on quite a few variables. For example, one factor would concern whether the size of the meteor impact was sufficiently large to vaporize the ocean, or merely big enough to boil, to the point of evaporation, the 200-meter layer beneath the ocean's surface known as the photic zone."

"Excuse me, Professor," interrupted the defense lawyer, "what is the photic zone?"

"The 200-meter photic zone represents the depth to which light penetrates with sufficient energy to be able to sustain photosynthetic autotrophs. Photosynthetic autotrophs are organisms that synthesize their organic requirements by using sunlight as a source of energy to convert inorganic materials, such as carbon dioxide, to molecular forms capable of being used by the organism to sustain itself."

"Thank you," said Mr. Tappin, "please continue."

"The first kind of impact mentioned previously ... that is, one capable of vaporizing the ocean, would involve, roughly speaking, about 5×10^{27} joules of energy. This amount of energy would be delivered by an object that was around 440 kilometers in diameter and/or had a mass of 1.3×10^{20} kilograms, traveling at approximately 17 kilometers per second.

"The second kind of impact ... that is, one capable of boiling away the photic zone, would require about 4×10^{26} joules of energy. The object would have a mass of approximately 1.1×10^{19} kilograms and a diameter of about 190 kilometers.

"The Chicxulub, Yucatan crater, by way of comparison, is calculated to have been created from an object that is some 300 kilometers in diameter. Thus, it is intermediate in size between meteorites capable of evaporating the ocean and meteorites able to boil away the 200-meter photic zone near the ocean's surface.

"If the size of the impact were of the ocean-evaporating kind, then, neither living organisms nor various complex arrangements of hydrocarbons would have been likely to survive to any appreciable degree. To understand why this is so, one needs a few facts about the nature of the collision being discussed.

"With an impact of this magnitude, roughly a quarter of the energy arising from the collision would have been directed toward vaporizing the water of the ocean. Another quarter of the impact energy would have been radiated upward toward the atmosphere, and the remaining fifty percent of the energy would be buried in the vicinity of the impact.

"The heat generated at the point of impact would be sufficiently great to melt, if not vaporize, most of the crustal material ejected from the crater being formed by the force of the collision. The temperature of these materials probably would reach around 2000 degrees Kelvin or 1727 degrees Celsius.

"Furthermore, the heat released through these melting and vaporizing materials would have been radiated in at least two directions. There is a thermal wave of some 2000 degrees Kelvin that would have been generated upward toward the atmosphere, as well a thermal wave that would have been radiated downward.

"The rock vapor that radiated upward would have surrounded the globe for a period of time, raising the atmospheric temperature considerably. By the time the rock vapor had rained out, so to speak, from the atmosphere, half of the ocean would have existed in the form of a hot steam that would have added about 140 times of our present sea level pressure to the atmosphere.

"A short while after the rain out of the rock vapor, which would take several months, the uppermost portions of the steam atmosphere would have cooled enough to generate a relatively thick, moist zone capable of convectively reflecting substantial amounts of heat back to Earth. A number of researchers believe this would have led to the runaway greenhouse threshold, or beyond, at which time the rest of the ocean would boil away.

"There are a number of factors that could affect the character of the foregoing sequence of events. The amount of carbon dioxide in the atmosphere would be one consideration, especially given that the manner in which CO_2 is distributed among earth, atmosphere and the ocean is quite complex, with different greenhouse and temperature scenarios following from different modalities of distribution.

"In addition, the amount of cloud cover -- as well as whether the cloud cover was at higher or lower altitudes -- could affect the amount of infrared radiation that is absorbed and radiated back to Earth. On the other hand, cloud cover also could affect the amount of sunlight that might be reflected away from the Earth.

"Eventually, depending on the actual atmospheric temperature, pressure, and so on, the water content of the atmosphere would begin to precipitate out and fall back to Earth and, in this way, reform the ocean. This period of cooling and ocean re-formation would probably take between 2,000 and 3,000 years to be completed.

"The impact of a meteorite sufficiently large to boil away the 200- meter photic zone of the ocean also would have catastrophic results, although, obviously, not quite as pronounced as those that I have just described. For one thing, after an impact of the lesser kind now being addressed, the atmospheric disturbances and restoration of the ocean to relatively 'normal' conditions would take merely 300 years, rather than 2-3000 years as previously indicated for the larger kind of impact.

"If the nature of an existing ecosystem is such that it is dependent, ultimately, on photosynthetic autotrophs, then, the sterilization of the photic-zone would wipe out the ecosystem. In other words, when the bottom link of the food chain in a given ecosystem disappears, then all of the heterotrophs higher up the chain that depend on that link also will disappear."

Professor Yardley noticed the expression on the face of the defense attorney. The professor seemed to reflect for a second on what he had just said.

Upon, apparently, intuiting the question about to be asked, he started to speak again. "Heterotrophs," he added, are organisms that depend on other life forms, usually photosynthetic or chemosynthetic autotrophs, to provide them with the organic materials that can be used to derive energy by which to synthesize their organic needs."

Mr. Tappin smiled in acknowledgement of Dr. Yardley's correct intuition. The defense attorney gave a slight motion of his hand indicating for the witness to proceed.

"However," pointed out the professor, "not all life forms live within the photic zone, and not all life forms necessarily are dependent on photosynthetic autotrophs in order to survive. There are chemosynthetic autotrophs, involving a few species of bacteria, which derive their source of energy for organic synthesis completely independently of light energy.

"These organisms accomplish this by means of the oxidation of various reduced inorganic compounds. For instance, some of these chemosynthetic autotrophs, like the colorless sulfur bacteria, have the capacity to generate energy by oxidizing hydrogen sulfide to sulfur, while other organisms, like certain nitrifying bacteria, possess the ability to produce energy through oxidizing ammonia to nitrite.

"If these chemosynthetic autotrophs lived far enough below the photic zone, or lived sufficiently deep beneath the earth's surface, so as not to be affected by an impact large enough to vaporize the photic zone of the ocean, then such organisms might stand a very good chance of surviving this sort of catastrophic event. Similarly, complex, prebiotic hydrocarbons located out of harm's way in the same fashion as these chemoautotrophs also would be likely to survive a collision of this lesser kind.

"Thank you, Dr. Yardley," said the defense lawyer, "I believe you have answered my question quite adequately. Now, let's see if I understand the overall character of this part of your position as stated in direct testimony.

"Earlier, you informed Mr. Mayfield and the court that researchers have concluded, based on lunar radiometric analysis, there were as many as 15-16 meteorite collisions on Earth that were greater than the impact creating the largest crater on the moon and, therefore, might have been sufficiently big to evaporate the oceans of early Earth. You further testified, Dr. Yardley, that researchers contend the last of these ocean-vaporizing events probably took place somewhere between 4.44 billion and

3.8 billion years ago. Is my understanding correct on both of these points, Professor?" inquired Mr. Tappin.

"Yes, it is," Dr. Yardley agreed.

"On the other hand," the lawyer proceeded to say "none of this would preclude any number of lesser collisions capable of sterilizing the photic-zone from having occurred. Presumably, the Yucatan crater serves as indirect evidence for such a statement since it was considerably larger than what is minimally necessary to boil away the photic-zone and, yet, here we are talking about it. Would I, more or less, be correct in asserting this, Dr. Yardley?"

"In general," replied the professor, "I would be prepared to go along with you except I would add one proviso to what you have said."

"Yes, Professor, what would this proviso be?" inquired the lawyer.

"If one had too many impacts capable of sterilizing the photic - zone," suggested the professor, "then, this could prove to be as problematic, in its own way, to the development o f life or to the development of prebiotic systems as were impacts of the ocean - vaporizing variety. Such impacts do occur ... as the Chicxulub, Yucatan crater demonstrates ... but we believe the available evidence indicates these kinds of collisions, probably, were relatively rare events after 3.8 billion years ago, the time when the last of the ocean-vaporizing events is believed to have occurred."

"Yet, Dr. Yardley," the defense lawyer said, "the fact of the matter is there really is very little, if any, available evidence to indicate how many impacts there might have been, from, say, 3.8 billion to 3.5 billion years ago, that were capable of boiling away the photic zone. Is this not correct, Professor? Yes or no?"

"You would have ..." Dr. Yardley began to say. The defense attorney interrupted.

"Your Honor, I find the witness' answer non-responsive," Mr. Tappin stated.

"Dr. Yardley," Judge Arnsberger explained, "you must answer the queries of the defense counsel in accordance with the form in which the questions are being asked. In this particular case, your only options are 'yes' or 'no'"

"Thank you, Your Honor," acknowledged Mr. Tappin. "Would you like me to repeat the question, Dr. Yardley" he asked.

Shaking his head in a negative fashion, the professor sighed and said: "Yes."

"So, to restate the matter, Professor," the lawyer paraphrased, "statistically speaking, there might have been: no impacts, or one impact, or a few impacts, or more than a few impacts, of a size sufficient to boil away the photic zone of the ocean during the indicated period between 3.85 billion and 3.5 billion years ago. Is this correct?"

"Yes, that is correct," Dr. Yardley replied.

Flipping the page on his note pad, Mr. Tappin scanned the contents of the page for a few seconds and said: "Professor, in your earlier testimony concerning indirect, isotopic evidence for the existence of life 3.85 billion years ago that has been discovered at the Isua rock formation in Greenland, you mentioned, in passing, certain kinds of methodological contraindications with respect to the previously stated interpretation of that evidence. Would you explain" requested Mr. Tappin, "at this time, a bit more about the nature of these possible counter-indications?"

"As I said earlier," noted Dr. Yardley, "during the fixation of atmospheric carbon dioxide, living organisms tend to discriminate against the Carbon^{13} isotope and prefer its Carbon^{12} counterpart. This is due to the kinetic character of the enzyme responsible for the fixation of carbon in so-called C_3 plants ... that is, plants in which a three-carbon acid is the first product of photosynthesis.

"Consequently, one will find organic sediments exhibiting depleted amounts of Carbon^{13} relative to atmospheric CO_2. On the other hand, inorganic carbonate sediments, such as limestone, will tend to display elevated levels of Carbon^{13} relative to atmospheric CO_2.

"If one encounters a sample that fits the depleted Carbon^{13} profile, such evidence can be interpreted to mean that the profile was produced by a C_3-like plant that has a carbon-fixing enzyme with this tendency. The issue, unfortunately, is not always straightforward.

"This is especially true in cases where the sample is drawn from a rock formation, such as Isua, where the rocks have, at some time, been subjected to temperatures in the range of 450 to 700 degrees Celsius. Such high temperatures might bring about what is referred to as a partial re-equilibrium of any carbon isotopes that are present in the rock formation.

"This partial re-equilibrium of carbon isotopes tends to elevate the Carbon^{13} values for organic samples. At the same time, this process causes a lowering of the Carbon^{13} value for the inorganic carbonate sample.

"When this happens, the results are skewed. Under such circumstances, one might not know if one is dealing with an inorganic carbonate with a lowered Carbon^{13} value, or if one is dealing with an organic material with an elevated Carbon^{13} value.

"Some people have interpreted the Isua carbon isotope evidence to mean that the samples in question were produced by a carbon-fixing enzyme similar in character to the enzyme existing in C_3 plants of today. Other investigators are not so sure if this interpretation is correct."

"What ramifications follow from these different interpretations, Dr. Yardley?" inquired the lawyer for the defense.

"If the first interpretation I mentioned is true -- that is, if the Isua sample is actually organic in origin -- then, evidence would have been established that pushes back the earliest known life form to at least 3.85 billion years ago, several hundred million years, and change, prior to our previous oldest, fossil evidence drawn from the Warrawoona Group in Western Australia. If, on the other hand, the Isua sample turned out to be an inorganic carbonate with thermally skewed low Carbon^{13} values, giving a false positive for organic matter, then, the oldest known evidence for the existence of life would stand at around 3.55 billion years ago, give or take thirty million years, or so."

"If," hypothesized Mr. Tappin, "the organic interpretation of the Isua isotope evidence is correct, then, presumably, this would suggest an upper boundary had been established for ocean-vaporizing meteorite impacts. In other words, given the catastrophic character of this kind of collision as outlined by you earlier. then one might be hard-pressed to account for the continued existence of photosynthetic life forms like the proposed Isua organism. Would you agree with this, Dr. Yardley?"

"Yes, I concur," the professor indicated.

"On the other hand," offered the lawyer, "depending on circumstances, the location, the hardiness, and the luck of our hypothesized Isua organism, this photosynthetic autotroph might or might not survive an impact capable of vaporizing the photic zone. Is this correct?"

"Yes, I think so," stated Dr. Yardley.

"Now, Professor," continued the lawyer, "this approximate date of 3.85 billion years ago puts us at the upper, or later, limit of the period between 4.2 billion and 3.8 billion years ago that you cited earlier as the time during which the last of the 15-16 ocean-vaporizing meteorite collisions with earth is projected to have occurred. If one were to claim the final ocean-vaporizing impact were to have occurred some 4.2 billion years ago, then one has, approximately, 425 million years to play with in order to account for the origin-of-life. Is this right, Dr. Yardley?"

"Right," replied the professor.

"However," remarked Mr. Tappin, "on the one hand, there is no compelling evidence to suggest one would be justified in adopting the earlier 4.2 billion year bench mark as one's starting point. On the other hand, there is some evidence ... namely, projected photic-zone vaporizing and ocean-vaporizing meteorite collision like the one near the Yucatan Peninsula some 65 million years ago ... suggesting the parameter of 4.2 billion years might be a tad premature. Do you feel my characterization of the situation, Dr. Yardley, is unfair?"

"Not really," admitted the professor. "The starting point for origin - of-life scenarios has considerable theoretical and empirical looseness to it."

"If," Mr. Tappin conjectured, "scientists suddenly were to discover evidence indicating the incorrectness of the organic interpretation of the Isua sample, then, in your opinion Dr. Yardley, would the arbitrary nature of this starting point issue change much?"

"Yes and no," the professor responded.

"Would you please elaborate," requested Mr. Tappin.

"The fixing of a time frame that establishes a non-catastrophic period of time having conditions conducive to a prebiotic account of the origin-of-life always will have an element of arbitrariness about it. Nevertheless, using the later 3.55 billion-year Warrawoona date as the time when life initially had become firmly established is friendlier to evolutionary models than is the Isua date of 3.85 billion years ago.

"The later, Warrawoona dating of life fits in more comfortably with the available data than does the earlier, Isua dating. By this, I mean the earlier dating of life has more problems to overcome in a shorter period of time than does the later dating of life.

"Among other things, the earlier, Isua dating of life is overlapping with the meteorite impact data, which we have discussed, much more than is the later, Warrawoona dating of life. There are more likely to have been both ocean-vaporizing and photic-zone vaporizing impacts associated with the earlier, Isua-dating than with the later, Warrawoona dating of life."

"Still, Professor Yardley, wouldn't you agree," inquired the defense attorney, "that one of the bottom lines in all of this is the following? In the light of the meteorite impact data, do we really have any non-arbitrary way to theoretically determine the amount of time with which we have to play around, so to speak, as far as providing a plausible evolutionary account of the origin-of-life is concerned?

"In other words, are we not merely guessing in relation to the basic question? Do we have any empirical means of pinning down how much historical or Archean time we actually have to work with in order to provide an account of the transition from prebiotic conditions to the first protocell or full-fledged organism that is plausible?

"Isn't one as justified in saying there were only 4,000 years, or less, say, between the last catastrophic meteorite impact and the laying down of the physical evidence, whether direct or indirect, for the first appearance of life on Earth, as one is claiming there was some 425 million years between these two points in history? Aren't evolutionary scientists arbitrarily selecting the latter time interval, during which life allegedly arose simply because it proves to be less embarrassing and problematic for their theory than the 4,000 year scenario would be?"

"I believe," responded the professor, "there is a difference between: making educated, empirically based conjectures about the origin-of-life and creating myths concerning those origins. I maintain there is a difference between, on the one hand, making conjectures with respect to which one can seek out evidence both for or against, and, on the other hand, developing systems of beliefs that are removed from empirical data as well as from rigorous demonstration."

"Dr. Yardley," interjected the defense lawyer, "one can agree entirely with what you just have said, but you haven't addressed the essential thrust of my previous line of questioning. Let me restate the issue in another manner.

| Evolution Unredacted |

"Fact one: in your testimony, Professor Yardley, you indicated researchers have maintained there probably were 15-16 meteorite collisions with the Earth occurring sometime after 4.3 billion years ago. Furthermore, these collisions were projected to possess more force than the ones causing the largest lunar crater Imbrium.

"Fact two: the magnitude of these events would be sufficient, at the higher level, to vaporize the ocean, or, at the lower level, to vaporize the photic zone.

"Fact three: these collisions were believed to have occurred somewhere between 4.3 and 3.8 billion years ago.

"Fact four: these events were stochastically distributed across a 500 million-year interval.

"Fact five: the first indirect, potential evidence for the existence of life is dated around 3.85 billion years ago.

"Fact six: the first, direct fossil evidence for the existence of life is dated from about 3.55 billion years ago.

"Fact seven: an event intermediate between a collision that would have vaporized the ocean and one that would have vaporized the photic zone occurred approximately 65 million years ago.

"My questions to you Dr. Yardley are these: One, given the foregoing facts, when precisely, during the interval between, say, 4.3 billion years ago and 3.55 billion years ago, did the 15-16 projected collisions with Earth occur?

"Two, given the foregoing facts, is one justified in treating the event that took place 65 million years ago, as part of the stochastic distribution of the original 15-16 events?"

Dr. Yardley looked at Mr. Tappin, apparently considering the questions. The professor started to speak and, then, stopped.

Finally, he said: "There really is no way, at the present time, to answer your first question with any precision. As far as the second question is concerned, I'm not sure the Yucatan crater should be considered as part of the original stochastic distribution profile.

"I suppose, nonetheless, a case might be made by some individuals to include, on justifiable grounds, the Yucatan event in the original stochastic distribution. The object that collided with Earth some 65 million years ago might well have been a remnant of the original debris that had been bombarding the Earth during the Archean era and on which the projected 15-16 collisions is based."

"Would you agree, then, Dr. Yardley," inquired the defense lawyer, "that, on the basis of the available evidence, someone who claimed the last ocean-vaporizing collision took place 3.73 billion years ago would be as justified in her or his claim as the person who claimed the last ocean-vaporizing collision took place 3.54 billion years ago?"

"Yes and no," replied the professor. When Dr. Yardley realized Mr. Tappin was waiting for the answer to be expanded on, the professor said: "I agree, reluctantly, with your basic point about the unknown nature of the historical time that was actually available to be able to go from prebiotic conditions to biological organisms through natural processes.

"On the other hand," the professor added, "if the Isua sample does have organic origins, then, the person who claimed the last ocean - vaporizing event took place 3.54 billion years ago is somewhat in conflict with the facts because of the evidence for the existence of life at both 3.85 billion years ago, as well as 3.55 billion years ago. Seemingly, a continuity of some sort has been established through the two kinds of dated evidence for the existence of life at Isua and Warrawoona."

"Isn't it conceivable," asked Mr. Tappin, "that life might have originated more than once? After all, Professor, in your direct testimony you spoke about the possibility of protocells and organisms existing in the early Archean era that were not part of the lineage that is linked, in any way, with the last common ancestor of all modern forms of life. Were you not suggesting during your testimony that life could have arisen, in various forms, more than once?"

"Yes," Dr. Yardley acknowledged, "I was suggesting this. However, the fossil evidence discovered at the 3.55 billion-year old Warrawoona Group contains the imprints of eleven different kinds of microorganisms. One would be asking a lot to suppose this much diversity could arise so quickly after an ocean-vaporizing event of the sort you have hypothesized."

"I agree with you," confirmed Mr. Tappin. "Such a scenario might be stretching things to the point of snapping, but this is not my problem, Professor, it is yours.

"You are the one who says he has a plausible account of, or explanation for, the origin-of-life from prebiotic beginnings. The viability of that claim is what is being probed through this cross-examination."

Without pausing, Mr. Tappin pressed on. "Dr. Yardley," he asked, "are you familiar with the so-called 'faint young sun paradox'?" "Yes, I am," responded the professor.

"Would you explain to the court the nature of this paradox?" Mr. Tappin requested.

"On the basis of various calculations performed by astronomers, many scientists accept as likely that 4 billion years ago, the sun actually was some 25-30 percent dimmer than today. If this is so, then a possible paradox emerges.

"More specifically, considered in terms of the current atmospheric conditions of the world, if the sun were 25-30 percent dimmer than is presently the case, then, the upper 300 meters of the ocean would freeze, along with rivers, lakes and inland seas. In addition, under these circumstances, the ice sheet covering the Earth would reflect much of the rest of the sun's incoming light, thereby preventing any thawing from taking place.

"Evidence, on the other hand, derived from a variety of sedimentary rocks indicates liquid water was in existence around 3.8 billion years ago. Furthermore, direct fossil evidence demonstrates the existence of biological organisms as early as 3.55 billion years ago.

"The paradox is as follows. How could liquid water and biological organisms exist in environmental conditions that should have been frozen due to the presence of a faint young sun?"

"Is it not possible," inquired Mr. Tappin, "that various combinations of hydrothermal vents, volcanic islands, and so on, in different parts of the Earth, could have generated a set of relatively localized conditions capable of, over time, producing both sedimentary rocks as well as sustaining life forms?"

Professor Yardley shrugged his shoulders. His face had an expression that seemed to be a blend both of skepticism as well as a considering of possibilities in relation to the defense attorney's suggestion.

The professor's head bobbed back and forth slightly, and he appeared to be weighing things in his mind. Finally, he said: "Maybe, but researchers have come up with a number of other possibilities."

"Would you outline a few of these possibilities?" requested the defense attorney.

"Since astronomers calculate the early sun probably would not have overcome its faintness until around 2.5 billion years ago," Dr Yardley began, "the challenge is to devise ways capable of permitting the Earth to compensate for the sun's relative dimness during the Archean era. The ways that have been devised concern conjectures about the compositional character of the paleoatmosphere -- that is, the Earth's early atmosphere.

"For example, during the 1970s, there were several attempts to resolve the faint early sun paradox. The first proposal focused on methane and ammonia, while a second suggestion concerned carbon dioxide.

"Ammonia and methane both absorb, and, therefore, trap, certain portions of the infrared spectrum that is being produced by the Earth as the planet is heated by solar radiation. The absorbed infrared energy heats up the atmosphere, and the atmosphere, in turn, begins to radiate infrared wave lengths, some of which return to the Earth's surface in the form of what many people have referred to as the 'greenhouse effect'.

"If there were enough methane and ammonia in the atmosphere, then considerable amounts of infrared energy would be absorbed and, eventually, radiated back to the Earth. In fact, some researchers believe this process might have been able to generate and radiate sufficient heat back to the Earth's surface to compensate for the faint early sun.

"There are, however, several problems with the methane/ammonia compensation hypothesis. To begin with, both methane and ammonia are susceptible, in varying degrees, to photolytic dissociation, or breakdown, as a result of the effect of ultraviolet radiation.

"Moreover, both methane and ammonia tend to enter into reactions with the hydroxyl radical [OH] which arises as a result of the photolysis -- or breakdown by ultraviolet radiation -- of H_2O. While some of these hydroxyl radicals would combine with the hydrogen gas coming from volcanic emissions, enough free hydroxyl radicals still might have been available for chemical reaction with a great deal of methane and ammonia, and, consequently, removed these molecules from the atmosphere.

"In addition, ammonia is quite soluble in water. Therefore, NH_3 tends to be lost from the atmosphere through rainout.

"There have been some studies indicating that the presence of protective buffers, such as water vapor in the case of methane, and hydrogen sulfide in the case of ammonia, can affect the rates and extent of photo destruction of methane and ammonia. Furthermore, another study suggested the photolysis of methane could produce several hydrocarbons, such as hydrogen gas and methylene (CH_2), which are efficient absorbers of infrared radiation.

"Despite this sort of data, the overall effect of photolysis, chemical reactions and rainout, likely would have resulted in the removal of most of the methane and ammonia molecules that might have been present, at some point, in the Archean atmosphere. Therefore, an atmosphere composed largely of methane and ammonia would not have had a very long lifetime unless there was some continuous source of production for these molecules.

"Today's atmosphere consists of a mixing ratio of about 1 part per billion of ammonia as well as 1.6 parts per million of methane. The presence of these molecules in our atmosphere is entirely the result of biogenic production.

"Once the Earth had differentiated, through the formation of the magnetic core, and, in the process, removed much of the Earth's iron from the surface, there would have been no chemical mechanism on prebiotic Earth, of which I am aware, capable of producing, on a continuous basis, either ammonia or methane.

Thus, these molecules wouldn't be able to solve the faint early sun paradox.

"An alternative theory to the methane/ammonia hypothesis, which also arose during the 1970s, focused on the possible role of carbon dioxide as a means of compensating for the dimness of the faint early sun. Carbon dioxide, like methane and ammonia, is capable of absorbing infrared energy being radiated from the surface of the Earth and, as such, is a greenhouse gas.

"For reasons closely related to the elimination of methane from the theoretical picture, carbon dioxide became a strong candidate for providing a means of compensating for the coolness of the faint early sun. More specifically, when methane is oxidized by the presence of [OH] radicals created through the ultraviolet photolysis of water vapor, carbon dioxide is a product.

"Thus, the oxidation of much of the methane in the early Archean era is considered by many researchers to be a good candidate for helping to generate a considerable amount of carbon dioxide. To this, one can add the substantial portions of volcanic emissions that consist of carbon dioxide.

"In modern times, currently active volcanoes have been estimated to release some 4×10^{10} kilograms of carbon per year. Most of this is in the form of CO_2.

"One reasonably could assume that the amount of carbon dioxide released through volcanic activity during the Archean era was, undoubtedly, far greater than is the case today. Nevertheless, almost any estimates one came up with in this regard would be both speculative and arbitrary to a large extent.

"Furthermore, how much of this out-gassed carbon dioxide would have remained in the atmosphere during the Archean era depends on the amount of this material that would have entered into solution with the ocean, as well as on the amount of carbon dioxide which became incorporated into inorganic carbonate formations such as limestone. Unfortunately, knowing the amounts of carbon dioxide that are in any given form ... gas, solid or liquid ... at any given time is fairly difficult to pin down with any precision in the best of times, let alone some 4 billion years ago.

"Estimates of the amount of carbon dioxide in the atmosphere during the Archean era vary over a wide set of possibilities. Some people believe the amount of carbon dioxide in the prebiotic atmosphere rapidly decreased during the Archean era and remained at relatively low levels thereafter. Other researchers maintain the amount of carbon dioxide at the beginning of the Archean era was high and continued to remain relatively high for some time.

"Among those theorists who contend the amount of carbon dioxide in the atmosphere was fairly substantial, there are again differences in projected amounts. There are researchers who indicate there might have been as much as 100 bars, or 100 standard atmospheres, worth of carbon dioxide gas in the Archean atmosphere. Others suggest the amount of carbon dioxide in the ancient atmosphere might have been between 10 and 20 bars or standard atmospheres.

"A 100-bar atmosphere of carbon dioxide would result in surface temperatures of about 230 degrees Celsius. With a more modest 10 to 20 bars of atmosphere, the Earth's surface temperature is likely to have ranged between, say, 85 and 110 degrees Celsius.

"Both of these scenarios would create surface conditions capable of compensating for the coolness of the faint early sun, thereby eliminating the paradox created by the existence of sedimentary rocks and fossil evidence. Furthermore, even the 100-bar carbon dioxide atmosphere would not necessarily generate temperatures that automatically lead to a runaway greenhouse effect in which all of the surface waters would boil away and be present in the form of clouds or steam.

"The saturation water vapor pressure under such circumstances would be about 30 bars, or so. Consequently, when one adds this to the existing 100 bars of pressure of carbon dioxide, the temperature would have to be raised another 100 degrees before the ocean would start to boil under that kind of pressure."

"Didn't you indicate, Dr. Yardley, that the impact of a meteorite somewhat larger than one capable of vaporizing the photic zone of the ocean would generate a transient rise in temperature of 100 degrees?" asked Mr. Tappin.

"Yes," the professor confirmed.

"So," the defense lawyer suggested, "in the context of a 100-bar carbon dioxide atmosphere, the impact of a meteorite smaller than the one that created the Yucatan crater might be capable of triggering a runaway greenhouse effect?"

"Possibly," stated the professor. "The actual outcome might depend on a lot of different factors."

"All right, Dr. Yardley, let's see if I have this right," Mr. Tappin said.

"Firstly, the early sun is thought to have generated 25-30 percent less luminosity than the sun of today. Under current circumstances, a sun this dim would have resulted in the freezing, among other things, of the oceans to a depth of 300 meters.

"Secondly, the 'big freeze' could have been avoided by an atmosphere with the right kind of compositional character. In other words, the faint early sun paradox could be avoided if the Earth's atmosphere contained enough greenhouse gases to be able to, first, absorb from the

Earth, and, then, radiate back to the planet, sufficient levels of infrared energy to compensate for the 25-30 percent dimmer luminosity of the early sun.

"Thirdly, there are, in broad terms, two competing theories concerning the compositional make-up of the Archean era atmosphere. One theory champions methane and ammonia as the greenhouse gases of choice, while the alternative theory advocates carbon dioxide.

"Fourthly, in neither theory do we know, except in very broad terms, what the precise character of the composition, temperature or pressure of the Archean era atmosphere was. On the other hand, in both cases, there would have been enough infrared radiation absorbed and re-emitted by the respective gases of each theory to counter the cooling effects of the faint early sun.

"Finally, there are substantial arguments for, and against, each of the competing theories. Dr. Yardley, does my brief summary capture the gist of the matter vis-à-vis the faint early sun paradox?" inquired Mr. Tappin.

"Yes," acknowledged the professor, "I would say you have captured all of the highlights."

"Is there any preference among researchers between either of the two theories outlined by you, Dr. Yardley?" asked the defense lawyer.

"The early preference," noted the professor "had been for the methane/ammonia hypothesis. Relatively, recently, however, the preference scales have been tipping rather heavily in the direction of the carbon dioxide perspective."

"Does anything rest on these preferences beyond resolving the faint early sun paradox issue?" wondered Mr. Tappin.

"Quite a bit, actually," stated the professor. "The methane/ammonia hypothesis is far more conducive to providing plausible accounts for the evolution of prebiotic systems than is the carbon dioxide hypothesis.

"The methane/ammonia atmosphere constitutes a reducing environment. Due to the way this kind of atmosphere provides an environment that is conducive to chemical reactions believed to be capable of leading to increasingly complex organic molecular forms, a methane/ammonia atmosphere lends fundamental support to the emergence, eventually, of a variety of biologically important complex hydrocarbons such as amino acids, purines, pyrimidines, ribose sugars and so on.

"On the other hand, a carbon dioxide atmosphere is, at best -- and depending on what other molecules are considered to be present in such an atmosphere ... only slightly reducing, and, therefore, much less conducive, and, perhaps, even antagonistic, to the gradual buildup of the increasingly complex molecular forms required by evolutionary theory. In general, the more hydrogen gas there is postulated to be in a carbon dioxide dominated atmosphere, the greater will be, up to a point, the reducing character of that atmosphere. Alternatively, the more the ratio of $[H_2]$ to $[CO_2]$ falls away from 1, the less reducing will such an atmosphere be.

"Many researchers believe nitrogen, not hydrogen, was the most common gas next to carbon dioxide in the Archean era atmosphere. A nitrogen/carbon dioxide dominated atmosphere would have been either neutral or, possibly, according to some researchers, quite reactive with a propensity to breakdown, rather than build up, more complex hydrocarbons such as amino acids.

"As a matter of fact," pointed out Dr. Yardley, "this problematic dimension of a carbon dioxide dominated atmosphere inspired a couple of theorists ... around 1994, I think ... to develop another approach to the faint early sun paradox. In effect, these researchers seemed to feel there was no need to try to find ways of compensating for the cooling effect of a faint early sun.

"The starting point for their theory is to assume the Earth froze as a result of the early sun's 25-30 percent lower luminosity. The freezing would have created a 300 meter thick layer of ice near the ocean's surface.

"According to the architects of this theory, the layer of frozen ice would have served to protect chemical activity going on in the water below the frozen zone. In addition, the cold, but unfrozen, ocean water would have helped to preserve whatever organic molecules were formed since the decomposition of organic molecules is slower at these lower temperatures.

"Furthermore, these theorists allowed for the influx of large meteorites every million years or so. These large-scale impacts would have melted the ice and helped stir things up, so to speak, in a variety of ways involving shock-synthesis of various hydrocarbons, mixing of organic materials, energy distribution and so on."

"How can one be sure," queried Mr. Tappin, "there would have been any ocean at all beneath the frozen zone, or if there were liquid ocean water below such a zone, how would one know how deep the water would be? If the chill caused by the faint early sun was present from the very beginning of the Archean era, then how would this affect the formation of the ocean?"

"The answer to your question," remarked Dr. Yardley "would depend on a lot of different factors. For instance, scientists believe the process of core formation is likely to have raised the overall temperature of the planet to some 1500 degrees Celsius.

"Obviously, things would have to cool down considerably before lasting bodies of water could have begun to form on the surface. Before this point had been reached, there probably would be a time when the water being released into the atmosphere as a by-product of the core formation process would exceed the saturation level for water vapor in the atmosphere.

"The precise character of this saturation level would depend on things such as atmospheric temperature, pressure and composition. Once such a level was exceeded, then, for a time, there probably would have been a rapid precipitation and evaporation cycle in which water would not have collected on the surface, but humidity would have been quite pronounced.

"At some juncture, surface temperature, as well as atmospheric composition, temperature and pressure, along with water formation and precipitation would have collaborated to create conditions conducive to the generation of relatively stable bodies of water. How one factors a faint early sun into this process of ocean formation is rather difficult to say because so many of the variables being considered are uncertain.

"I'm sure a number of computer models concerning the nature of ocean formation in the Archean era have been developed. Depending on starting assumptions, different models likely would designate different depths of water as the point that would have to be reached before a frozen layer starts to form.

"Hydrothermal vents and volcanic activity also would have to be thrown into the mix since they both are capable of affecting water temperature, locally and, possibly, even globally. With each new variable that is added, the model becomes more complicated and, consequently, providing an answer to your question is less and less straightforward."

"Given," began Mr. Tappin, "what you have been saying, Dr. Yardley, in response to my question, is one being unfair to the facts if one were to argue that the manner in which one pieces together those facts is very much dependent on, or driven by, the assumptions one makes concerning the nature of conditions in which one believes those facts are embedded?"

"No," indicated the professor, "someone arguing in the fashion in which you have suggested would not be treating the facts unfairly. Indeed, in science, one constantly should be examining the relationship between established facts and the assumptions surrounding one's use of, or interpretation of, those facts."

"Could one," asked the defense lawyer, "not also say the following? When the facts of a matter have not been established clearly, then, the relation among assumptions, interpretation and 'facts' becomes, potentially, quite problematic?"

"Yes, I would agree with that," Dr. Yardley replied.

"Therefore, in the matter at hand, Professor... namely, the question of whether or not a 300-meter frozen zone would have formed near the surface of the Archean-ocean as a result of the dimness of the faint early sun ... we appear to be faced with a rather problematic situation. This is so, because given, as seems to be the case, that we don't know such things as the composition, temperature and pressure of the Archean - atmosphere; or, the rate of Archean-ocean formation; or, the water vapor saturation levels of the Archean-atmosphere; or, the degree to which hydrothermal vents or volcanic activity are present, and so on; then, in a very real sense, except in extremely broad terms, we don't know the facts of the matter, do we?"

"No, we don't," confirmed Dr. Yardley. "This is one of the reasons theoretical models are constructed.

"Scientists take what is known about the laws of nature, together with whatever data might be available concerning the conditions surrounding a particular problem, such as the present issue of a frozen zone above the Archean era ocean. Next, certain assumptions are made about how natural laws might be manifesting themselves under certain conditions.

"The implications of these assumptions are worked out in the form of a model. Essentially, the model says that if certain assumptions are true, then, under specified conditions, natural laws will generate certain kinds of predictable activity in the context of those given conditions and assumptions.

"At this point, if possible, controlled experiments are performed that focus on, or isolate, different variables shaping the problem being considered. By comparing the results of these experiments with the character of one's model, one has an opportunity, over time, to correct, eliminate, refine and/or confirm different facets of the model."

"We have before us, Dr. Yardley, three different -- models, I guess -- or theories concerning the faint early sun paradox," noted the defense lawyer. "Is there an experimental way," the lawyer asked, " of deciding which, if any of these models, are an accurate reflection of what happened on Earth during the Archean era?"

"Not really," observed the professor. "Certain experiments might carry various kinds of implications and ramifications for such models that will have to be taken into consideration.

"Experimental results might raise questions about, or pose problems and challenges for, a particular model. Generally speaking, however, what happens is that researchers will merely modify their models in the light of the experimental data.

"Since we, to some extent, are working in the dark concerning what the precise nature of the conditions were during the Archean era, we frequently are limited to saying that different kinds of models are consistent with, rather than proved by, the known facts. Yet, the known facts might be, more or less, equally consistent with quite different models, depending on the assumptions one makes and how one chooses to interpret, and piece together, the available facts in the context of one's model.

"All models are conditional in nature. In other words, the accuracy or reflective capacity of a model, vis-à-vis 'reality', or the facts, or one's field and laboratory experiences, is dependent on the rigor with which, and degree to which, one's assumptions can be shown to be plausible, or justified representations, of the prevailing conditions surrounding some issue or phenomenon.

"Modern scientists cannot recreate the Archean era conditions. At best, we can try to simulate certain facets of what, on the basis of the available data, we believe those Archean era conditions to have been.

"In the light of these simulations, we extrapolate and interpolate backwards to the Archean era. In this fashion, we try to link, as well as we can, our simulations, whether computer or experimental, to the available empirical data and known physical/chemical laws of nature.

"Many models will work in upper and lower boundaries as part of their conditional statements concerning the nature of reality. In other words, if certain variables operate at the upper boundary limits of the model, then, certain things are said to follow. If, on the other hand, these same variables operate at the lower boundary limits of the model, other kinds of things maybe said to follow.

"For example, quite a few simulation experiments in evolutionary theory concerning the Archean era are now, and have been for a number of years, examining the issue of organic synthesis under a variety of prebiotic conditions. The same experiment or simulation will be run a number of times under, say, a variety of conditions involving different atmospheric compositional packages.

"On one experimental run, a particular organic synthesis will be attempted with a methane/ammonia atmosphere. Other runs of the experiment will be done in the presence of, perhaps, different ratios of hydrogen and carbon dioxide gases.

"On the basis of these experimental results, a researcher will reach certain tentative conclusions. For instance, she or he might say: when the composition of the atmosphere consisted of a particular mixture of methane and ammonia, the synthesis went forward at such a rate and with such-and-such an efficiency yield. However, when the same synthesis was attempted with a certain ratio of hydrogen and carbon dioxide, the synthesis either did not occur, or it occurred at a reduced rate and with reduced efficiency yields of such-and-such a nature."

"Dr. Yardley, why don't we," suggested the defense counsel, "run some data by you and see how you handle it in the context of an evolutionary model? Perhaps, this exercise will help the court and the jurors to get a better feel for some of the issues that are, I believe, at the heart of the present trial."

The professor gestured a willingness to go along with such an exercise. He poured himself a glass of water and waited for the defense lawyer to begin.

"Let's return, for a moment," Mr. Tappin directed, "to the theory which assumes that the world froze, at some point, in response to the faint early sun. You indicated previously that one of the inspirations behind the construction of such a theory was to avoid the potential problems associated with a carbon dioxide dominated atmosphere in the Archean era.

"Presumably, one of these difficulties is that carbon dioxide has the potential to be highly reactive with complex hydrocarbons. As a result, CO_2 will help break the more complex molecules down into less complex and less interesting organic materials as far as the origin-of-life issue is concerned.

"Another difficulty posed by a carbon dioxide dominated atmosphere is the following. Experiments have shown that many kinds of organic synthesis are less likely to proceed or do so in very limited fashion, in such an atmosphere.

"If the Archean era atmosphere were dominated by carbon dioxide -- with very little, or no methane and ammonia -- how would the 'let the world freeze' assumption avoid the ramifications of this kind of atmosphere? In other words, where would the simple hydrocarbons come from out of which more complex hydrocarbons are to be synthesized, and what sources of energy would underwrite this underwater synthesis?"

"If one were to assume," Dr. Yardley responded, "there were little, or no, atmospheric production of hydrocarbons, like hydrogen cyanide (HCN) or formaldehyde CH_2O, then one would have to look to other sources such as carbonaceous chondrites, interplanetary dust particles or interstellar dust clouds, for either these more complex kinds of hydrocarbons or their simpler precursors such as methane and ammonia. Another possibility might be through hydrothermal vents from which hot water, rich in dissolved materials, spills out into the ocean."

"If," posited Mr. Tappin, "the surface of the Earth is frozen over, how do extraterrestrial materials get to the underlying ocean?"

"One possibility," replied the professor "is that these materials might have gone into solution during the earliest stages of ocean formation, prior to the establishing of a frozen zone near the surface of the ocean. Another possibility arises in conjunction with the asteroid impacts that, conceivably, could have provided a mechanism for mixing exogenous/extraterrestrial organic materials lying frozen on the surface with the oceans lying 300 meters below the frozen zone."

"Dr. Yardley, isn't the asteroid impact possibility a bit like dropping a hydrogen bomb on the Antarctic regions and seeing if anything interesting happens?"

"Objection, Your Honor," proclaimed Mr. Mayfield. The question is argumentative."

"Sustained," ruled Judge Arnsberger. "Rephrase the question, Mr. Tappin."

Starting again, the defense counsel asked: "Do we have any good reason to believe the impact of an asteroid sufficiently large to melt 300 meters of ice encircling the globe would have anything but destructive consequences for whatever residual exogenous organic materials might be lying frozen on the surface of the Earth?"

"No, I suppose not," answered Dr. Yardley.

"Is it fair to say, Professor," inquired Mr. Tappin, "that any conjectures concerning what might or might not have survived such a catastrophic event are quite presumptive and arbitrary in character?"

"Yes, I think that would be fair to say," Dr. Yardley acknowledged.

"Would you also agree, Professor," pressed Mr. Tappin, "that in view of the many uncertainties surrounding both the issue of the formation of the Archean era ocean, as well as the uncertain nature of the circumstances and conditions connected to the emergence of the 300-meter frozen zone, any conjectures concerning what had or had not entered into solution prior to the appearance of the frozen zone are equally presumptive and arbitrary?"

"I would have to offer a provisional yes to your question," Dr. Yardley stated."

"What is the nature of your qualifying provision?" inquired the lawyer.

"If," the professor hypothesized, "exogenous or extraterrestrial organic materials were reaching Earth through interplanetary dust particles or by the Earth's passage through interstellar clouds or by means of carbonaceous chondrites, then one would have to consider the possibility that these organic materials might be available to enter into solution should the opportunity arise."

Mr. Tappin briefly left the area of the witness stand and returned to the defense table. He whispered something to his colleague who rifled through some material on the table and pulled out a sheet of paper that he handed to the defense lawyer.

As the lawyer came back toward the witness, he started to speak. "Professor Yardley," he inquired "are you familiar with a 1993 report by a NASA experimental team concerning the composition of interstellar dust?"

"In general, yes, I am familiar with that report," answered the professor, "but some of its details are rather fuzzy in my mind."

"Let me refresh your memory," offered Mr. Tappin. "The NASA scientists examined a number of star-forming clouds in the Milky Way galaxy."

"In every star-forming cloud, without exception, examined by the NASA team, they discovered that carbon in the form of microscopic diamonds dominated these clouds. In fact, these microscopic diamonds were found in huge numbers and at planetary masses.

"The findings of the research team have been described as a challenge to existing theories of both galactic and star formation. These prevailing theories assumed that interstellar clouds were composed of softer hydrocarbons, somewhat similar to gasoline or candle wax.

"Dr. Yardley, in the light of your previous answer about the availability of different exogenous sources for entering into solution should the opportunity arise, what are the implications of the largely hard-carbon, or microscopic diamond, composition of interstellar clouds?"

"Probably," surmised the professor, "one would have to revise downward one's estimates of the quantities and the kinds of soft hydrocarbons that might have come to Earth by means of its passage through such interstellar clouds. How much these estimates would have to be revised in a downward direction would depend on the extent to which the hard-carbons dominated these interstellar clouds."

Once again, Mr. Tappin returned to his table. On this occasion, his colleague was waiting for him, giving the defense counsel some new material in exchange for the paper in the lawyer's hand.

Approaching Dr. Yardley, the lawyer for the defense stated: "Professor, not too long ago, there was a study that examined the character and composition of a substantial number of extraterrestrial dust grains, which you have referred to as interplanetary dust particles. In more than 50 ice samples taken from a core drilled in the ice of Greenland, and, therefore, representing thousands of years of elapsed time, these researchers found only an extremely tiny amount of amino acids.

"The scientists conducting this experimental analysis concluded that amino acids couldn't have arrived in interplanetary dust particles in amounts that would have any significant bearing on issues concerning the origin-of-life. How would you respond to this finding, Dr. Yardley," asked the defense counsel, "in the light of your previous qualifying provision concerning the availability of exogenous or extraterrestrial organic materials for entering into solution prior to the formation of a 300-meter ice layer caused by a faint early sun?"

"Obviously," the professor noted, "one's estimates again would have to be revised downward. How much, and in what way, would depend on what other kinds of organic materials were found in the analyzed samples."

Shuffling through the papers in his hand, the lawyer selected another document. "Undoubtedly, Dr. Yardley," the lawyer said, "you are aware of the fact that a great deal of the organic materials found in interplanetary dust particles exists in the form of polycyclic aromatic hydrocarbons and amorphous carbon, both of which offer far less promise for the origin-of-life question than do amino acids or purine and pyrimidine nucleic bases. Is my assumption concerning your knowledge correct, Professor?"

"Your assumption is correct," Dr. Yardley replied. "However," he added, "some pathways of synthesis have been proposed that permit one to go from amorphous carbon and polycyclic aromatic carbons to amino acids."

"Yet," countered Mr. Tappin "those pathways are not without their controversial dimensions. Is it not the case Dr. Yardley that other researchers have disputed the proposed pathways of synthesis to which you are referring?" queried the lawyer.

"That's right," the professor admitted.

"Dr. Yardley," inquired Mr. Tappin, "you have testified previously that no one knows, for sure, about the origins of interplanetary dust particles. Is this correct?"

"Essentially, yes," the professor confirmed, "although, as I indicated earlier, some researchers have conjectured these dust particles might have arisen as a result of asteroid-asteroid collisions."

"However, Professor," challenged the defense counsel, "would one be justified in saying there is no proof or evidence in support of such a conjecture?"

"Yes," Dr. Yardley agreed, "one would be justified in saying no hard proof or evidence exists with respect to that conjecture."

"Furthermore," inquired Mr. Tappin, "would one also be justified in pointing out that the conjecture which you have described does not really explain how these dust particles came to contain different kinds of organic materials?"

"Yes," the professor admitted.

"In point of fact, Dr. Yardley," pressed the defense counsel, "given our ignorance about the origins of interplanetary dust particles, we really have no reliable and valid way of projecting backward from current data involving interplanetary dust particles to what might have been going on during the Archean- era?"

"That's correct," replied Dr. Yardley.

"In other words, Professor," continued Mr. Tappin, "we have little or no evidence concerning either the rates of production of interplanetary dust particles or whether the levels of mass influx of such particles that are currently observed would have remained constant across more than 4 billion years, and, therefore, be indicative of the influx of interplanetary dust particles that might have occurred during the Archean era. Is this correct, Dr. Yardley?" inquired Mr. Tappin.

"Yes, I would say so," responded the professor.

"Is it not also true, Dr. Yardley," probed the defense counsel, "that we have no hard, rigorous, reliable data on the amount, or kinds, of extraterrestrial organic material that would have been lost in the Archean era due to: pyrolysis, while in transit through the atmosphere; or, meteorological and geological ablation after air bursting; or, destruction as a result of the effects of shock waves or impact with the Earth; or, ultraviolet decomposition?"

"What you say is true," Dr. Yardley acknowledged.

"Consequently, Professor," concluded the defense lawyer, "the mass influx figures you cited during your direct examination testimony are pure conjecture based on, among other things, the assumption that everything we observe today with respect to interplanetary dust particles has remained essentially unchanged for four billion years. Is this correct?"

"Yes, it is," the professor agreed, "but I would point out that continuity plays a fundamental role in many aspects of the natural laws that govern physical and chemical phenomena."

"Dr. Yardley would you say there are qualitative differences among inorganic chemistry, organic chemistry and biochemistry?" Mr. Tappin asked.

"I would say," answered the professor," there are principles and properties that are shared in common by these disciplines, as well as areas of qualitative difference in which properties and principles that are unique, in a sense, to each of these disciplines do manifest themselves."

"Do you feel one would be justified," inquired the counsel for the defense, "to say the problem of accounting for the emergence of life from prebiotic beginnings is, in part, a reflection of the fact that the transition from organic chemistry to biochemistry involves, at least at the present time, more unresolved problems of a qualitatively different kind than one would encounter in making the conceptual transition from inorganic chemistry to organic chemistry?"

"Yes," confirmed the professor, "at the present time, what you have said is the case. Nonetheless, evolutionary scientists firmly believe the current situation will not last forever.

"We all feel," Dr. Yardley added, "that one of these days a researcher or scientist will demonstrate or discover how the last, unknown steps in the transition from organic chemistry to biochemistry took place. When this happens, the transition from organic chemistry will be no more mysterious than is the transition from inorganic chemistry to organic chemistry."

"Be that as it may, Professor," responded Mr. Tappin, "let me point out the obvious. Evolutionary biologists do not currently have such knowledge.

"More importantly, as far as the present aspect of the cross - examination is concerned, even if evolutionary biologists did possess such knowledge, certain facts still cannot be denied. For instance, let us assume there is some continuous set of chemical principles that allows one to make the transition from organic chemistry to biochemistry through purely natural processes.

"Nevertheless, there still are phenomena which occur in biochemical systems that do not take place in the reactions of systems which are organic but non-biochemical in character. Is this not so, Professor?" inquired Mr. Tappin.

"Yes," confirmed the biologist.

"In brief, Dr. Yardley," the lawyer summarized, "things do not always remain the same over time. If they did, we wouldn't be having this debate about why post-prebiotic times exhibited properties that were not present in prebiotic times.

"What happens now is not necessarily what was happening in the past. Moreover, what happened in the past is not necessarily what is happening now.

"This general principle, if you will, is demonstrated by the qualitative differences between biochemical processes compared to purely organic ones. This principle also might be demonstrated by possible differences in the rates of mass influx of interplanetary dust particles between today and 4 billion years ago.

"Would you agree, therefore, Dr. Yardley" asked Mr. Tappin, "that although we would expect the same conditions to exhibit the same properties over time, we cannot expect different circumstances automatically to lead to the same manifested properties? In fact, isn't the problem with which we are confronted in this matter of the mass influx rates of interplanetary dust particles, a variation on this theme?

"We need to determine the precise nature of the conditions under which an individual is justified in concluding that the things which are observed today are the same as what would have been occurring in the Archean era. Is this not part of the problem before us, Professor?"

"Yes, I think I could live with your characterization of things," Dr. Yardley stipulated.

Mr. Tappin began to speak and was interrupted by Judge Arnsberger. "Mr. Tappin, I'm sorry, but in view of the lateness of the hour, I feel we would be well advised to adjourn these proceedings for the day.

"I hope you will agree that the present time seems to offer a natural point of transition in your cross-examination. In any case, you will be able to pick things up again at 10:00 a.m. tomorrow morning."

Turning her attention to the jury, she said: "Please remember, ladies and gentlemen, my previous instructions to you. You are prohibited from discussing this case either with fellow jurors or with others whom you might come into contact.

"Court is adjourned until 10:00 a.m., Thursday morning," announced the judge. Her gavel fell in confirmation of her words.

An Ocean of Difficulty

Going through the papers in his hand, the defense counsel removed several sheets. Walking over to his table, he returned the unwanted sheets to his colleague.

Standing in front of the defense table, Mr. Tappin said: "Professor, in your direct examination testimony, you indicated, I believe, that the Murchison meteorite contained 6 amino acids similar, in most respects, to amino acids occurring in living organisms. In addition, 12 other kinds of amino acids not found, as far as is known, in living organisms on Earth also were discovered in the Murchison meteorite. Is my recollection of this testimony correct?" asked the lawyer.

"Yes," Dr. Yardley confirmed.

"To the best of your knowledge, Professor," Mr. Tappin inquired, "has any recovered meteorite ever contained all twenty of the amino acids found in living organisms on Earth?"

"Not to my knowledge," the professor answered.

"Furthermore," continued the defense counsel, "you testified that the amino acids found in the 200,000-year old meteorite in Antarctica had optical properties that were opposite to the ones displayed by amino acids found in Earth organisms. Is this correct, Dr. Yardley?"

"Yes, it is," the professor responded.

"In addition, Dr. Yardley, I believe you stated earlier that in most cases outside of biological systems, amino acids tend to form racemic mixtures in that there are roughly equal numbers of left- and right-handed optical isomers. Is my understanding correct in this respect?" Mr. Tappin inquired.

"Yes," said the professor.

"Moreover, previously, you testified that only 5-6 percent of meteorites consist of carbonaceous materials, and organic materials constitute only a small part of this carbonaceous subset of meteorites. Is this right?"

"Correct," affirmed the professor.

"Finally, Dr. Yardley, isn't it the case that most of the organic material found in meteorites such as Murchison exists in the form of a complex kerogen-like polymer that is poorly defined and consists of a variety of aromatic groups, monocarboxylic acids and aliphatic hydrocarbons? In fact, isn't it true, Professor, that only a very small fraction ... measured in parts per million ... of the organic material found in meteorites contains molecules, such as purines and amino acids, which have any potential relevancy to issues concerning the origin-of-life?"

"That is right," the professor indicated.

"Well, Dr. Yardley," the attorney stated "if we factor in all of the foregoing possibilities, we seem to be left with very uncertain, and possibly negligible, amounts of usable organic compounds from exogenous sources. In other words, given that organic materials form only a tiny portion of an already small subset of meteorites, and given that many of these exogenous organic materials exist in forms, or as kinds, which are not used by Earth organisms, and given that a considerable amount of this organic material might be destroyed through pyrolysis, hydrolysis, photolysis or impact, and given that we really don't know the rate or mass of carbonaceous chondrite influx during the Archean era, are not any statements about the amount and kinds of useable exogenous organic materials that arrive, and survive, very speculative and arbitrary?"

"Yes, I suppose so," Dr. Yardley admitted.

"Earlier," Mr. Tappin noted, "you mentioned, briefly, the possibility that hydrothermal vents might have played a role in the 'let the Earth freeze' model that arose in response to, among other things, the faint early sun paradox. Would you expand on this a little?" the lawyer requested.

"Some people," the professor said, "began to look seriously at hydrothermal vents as a possible locus for the origin-of-life when, a few years ago, rather extensive ecosystems were discovered to have developed around some of these vents. These ecosystems consisted of many exotic sorts of organism, including blind shrimp and giant tube worms.

"The food chains of these ecosystems were rooted in various kinds of microorganisms. These microorganisms were sulfur-eating life forms.

"Thermophilic, or heat-loving, microbial organisms also have been found living in the steam bath-like conditions of the hot springs at Yellow Stone National Park. In general, however, no one has discovered life forms on Earth capable of surviving in temperatures above 112 degrees Celsius."

"My understanding, Professor," indicated the lawyer "is that these organisms are capable of living under such conditions because they possess specialized proteins that allow them, among other things, to dissipate heat. Apparently, there also are proteins in various species of cold water fish capable of binding to, and controlling the growth of, ice within the organism, and, as a result, helping the organism adapt to cold water conditions. Is this correct?"

"Yes," replied the professor. When he saw the defense lawyer signaling him to continue on with his discussion, he said: "Some researchers hypothesized that life might have originated with thermophilic organisms.

"Other scientists have hypothesized that life originated elsewhere. In time, however, these organisms might have migrated to the hydrothermal vents in order to seek resources exuded by the vents or as a protection from the extraterrestrial bombardment of the Archean era Earth, or, maybe, both.

"Presumably," reflected the defense counsel, "if organisms migrated to the vents, then, regardless of whatever forces drove organisms to, or induced them to seek out, these hydrothermal vents, nonetheless, in order to survive these organisms would have to be adapted, in some minimally feasible fashion, to the thermal conditions of the vents. Is this not so, Dr. Yardley?"

"That's right," acknowledged the professor.

"But, the process of migration presupposes the existence of such organisms and assumes the existence of such adaptive capabilities. So, we are getting ahead of ourselves.

"Has anyone," Mr. Tappin asked, "devised a plausible theory of how life would have originated in the vicinity of the hydrothermal vents?"

"Not really," replied the professor.

"Dr. Yardley, in your direct examination testimony concerning the period of core differentiation of the Earth, you indicated some scientists believed the Earth's crust would have been relatively fragile at that time, and, therefore, conducive to the formation of these hydrothermal vents. Is this right?"

"Yes," responded the professor.

"Does the water in the ocean remain relatively static, or does it circulate?" Mr. Tappin asked.

"The water in the oceans of our day circulates extensively," the professor reported. "In fact, we believe any given volume of water eventually will circulate through every portion of the ocean."

"What about the Archean era ocean?" inquired the lawyer?

"I think the same scenario probably was the case," offered Dr. Yardley. "Between tidal forces and convection currents, of one sort or another, a circulatory system of some kind likely would have been present."

"If my information is correct," Mr. Tappin stated, "the temperatures associated with hydrothermal vents are in the vicinity of 350 degrees Celsius. What would be the effect," queried Mr. Tappin, "of hydrothermal vents on complex hydrocarbons that had dissolved in ocean waters and were brought into contact with these vents through the process of circulation?"

"A lot would depend on the extent, length and character of the contact," replied Dr. Yardley. "In general, the more direct, the longer, and the more extensive such contact, the more likely would be the tendency of any given complex hydrocarbon to denature or decompose."

"Would one be justified in arguing," asked the defense counsel, "that given some unknown number of hydrothermal vents on the bottom of the Archean era ocean, then the formation of a 300-meter ice layer above the ocean, due to the effects of a faint early sun, would not necessarily offer long-term stability to complex hydrocarbons that had, in one way or another, arisen?"

"As long as the molecules were able to stay in cold or cooler waters," Dr. Yardley pointed out, "then, their average life times probably would be enhanced to some degree. On the other hand, to whatever extent such molecules could not stay in cold or cooler conditions, then the average length of life for such molecules would be decreased as a function of the different kinds of forces of decomposition, including temperature, to which these molecules were subjected.

"For example, one scientist has studied the effects of heat energy on the amino acid alanine. This molecule is one of the more stable amino acids.

"The researcher found that at a temperature of twenty-five degrees Celsius, the mean life of alanine is estimated to be 10^{11} years. Yet, the mean life of this molecule is calculated to be just thirty years in length when the temperature is raised to 150 degrees Celsius.

"Less stable amino acids will break down more readily at such temperatures, and, therefore, they will have even shorter mean life times than alanine. In fact, less stable amino acids might begin to break down at temperatures somewhat lower than 150 degrees Celsius.

"Generally speaking, the more complex a hydrocarbon, the more unstable it tends to be in the presence of heat. For instance, proteins, DNA, and RNA all tend to denature and decompose when exposed to sufficient amounts of heat much more readily than might be the case with their component parts."

"Dr. Yardley," said the defense counsel, "I presume the aforementioned effect of heat on complex organic molecules would remain the same whether one is talking about hydrothermal vents or elevated surface temperatures caused by a super greenhouse effect. Is this presumption correct?"

"Yes, of course," remarked the professor.

"Therefore," Mr. Tappin observed, "all three theories that have been proposed as possible ways of resolving the faint early sun paradox, face, each in its own way, a potential problem with respect to decomposition of complex hydrocarbons as a result of potentially prolonged exposure to heat energy, either in relation to hydrothermal vents or to enhanced greenhouse effects. Would you agree with this assessment of the situation, Dr. Yardley?"

"In broad terms, I suppose so," answered the professor."

Looking briefly at the papers in his hand, Mr. Tappin walked toward the witness stand. When he was a few feet away, he came to a standstill.

"Professor, earlier you testified that scientists believe there was little or no free oxygen in the early Archean era atmosphere. Given," postulated the defense counsel, "all the talk these days about holes in the ozone layer and how ozone absorbs ultraviolet radiation, and, in the process, protecting living organisms from the destructive effects of such radiation, I was wondering what the situation would be in the Archean era. More specifically, would the faint early sun lessen the presence of ultraviolet radiation?"

"Oddly enough," Dr. Yardley began, "although the overall, net luminosity of the early sun was lower than today's sun, nonetheless, on the basis of astronomical observations of young stars comparable to our early sun, the ultraviolet radiation of the early sun is considered to have been greater than is the case with our present sun. Consequently, in the absence of oxygen, the ultraviolet effect would be more pronounced than it is today, even in those areas, such as the Antarctic, where the ozone hole has grown to such a disturbing size."

"Where does ozone come from?" Mr. Tappin asked.

"When free oxygen is available," Dr. Yardley explained, "ultraviolet radiation tends to split oxygen molecules into separate atoms of oxygen that are quite unstable. These unstable atoms of oxygen will combine with oxygen molecules to produce O_3 or ozone.

"Studies have indicated there was no appreciable presence of atmospheric oxygen until sometime between 2.1 and 2.03 billion years ago. As a result, between 4.55 billion years ago, and 2.1 billion years ago, there would have been no way for ozone to be manufactured in the Archean era atmosphere."

"What are the ramifications," Mr. Tappin inquired, "of this combination of enhanced ultraviolet luminosity, as a result of the faint early sun, and the absence of ozone, due to the absence of oxygen, as far as the development of increasingly complex hydrocarbons is concerned?"

"Ultraviolet light," replied the professor "is like most forms of energy. They are all two-edged swords.

"In the right amounts and for the right length of time, energy is capable of bringing about many kinds of chemical reactions among organic molecules. In the wrong amounts and for the wrong length of time, energy can be quite destructive in its effects upon hydrocarbon compounds.

"In limited doses, ultraviolet radiation can help underwrite, among other things, the synthesis of a wide variety of organic molecules. Beyond a certain limit, however, such radiation begins to have an adverse effect, even on those compounds that, originally, it might have had a hand in helping to synthesize.

"Photolysis refers to the breakdown or decomposition of materials by the action of light. Prolonged exposure to ultraviolet radiation brings about photolysis.

"These remarks notwithstanding, the results of photolysis sometimes can bring about reactions that have a potential, under the right circumstances, for building more complex hydrocarbons. In other words, the products of photolysis might recombine with other organic materials.

"For example, one team of researchers observed that when methane gas is subjected to photolysis, methyl (CH_3) and methylene (CH_2) radicals are produced. Subsequently, these two radicals were observed to enter into reactions that resulted in heavier hydrocarbons.

"These researchers calculated that the equivalent of one bar or atmosphere of methane gas could have been polymerized by means of ultraviolet radiation over a period of some 10^6 to 10^7 years ... in other words, between one and ten million years. They further proposed that such heavier hydrocarbons would have precipitated out of the atmosphere and formed a layer of hydrocarbons on the surface of the Earth measuring anywhere from one to ten meters in thickness."

"Dr. Yardley," interjected the defense council, "in the light of our previous discussion about the nature of the atmospheric composition of the Archean era, couldn't one respond to the findings of this methane photolysis research in several ways? For example, if the Archean atmosphere were methane-dominated, this finding might have some value in origin-of-life scenarios, but if the Archean -atmosphere consisted of little or no methane, their finding is meaningless as far as the origin-of-life issue is concerned. Would you agree with this assessment of the situation, Dr. Yardley?"

"Not entirely," the professor indicated. "Even if there were little methane in the atmosphere, the synthesis of important precursors ... such as hydrogen cyanide, formaldehyde, and, maybe, a few amino acids, still is possible.

"A great deal would depend on the ratio of hydrogen (H_2) to carbon dioxide gas (CO_2) that existed in the Archean-atmosphere.

"As I testified previously, if the ratio were about 2, then, some researchers feel this kind of atmosphere would have reducing properties comparable to a methane-dominated atmosphere.

"As the ratio of hydrogen gas to carbon dioxide drops, the production efficiency by ultraviolet light also will drop. As one approaches a ratio of, say, one-tenth of hydrogen to carbon dioxide, then production efficiency by ultraviolet light is calculated to drop by at least two magnitudes or by a factor of around 100.

"Researchers suggest hydrogen might have arisen throughout ... gassing from Archean era volcanoes. Hydrogen also might have been generated through the photo-stimulated reduction of ferrous iron in the photic zone of the ocean."

"Doesn't this photo stimulated reduction of ferrous iron assume," observed the defense counsel," that the surface of the Earth has not been frozen over due to the effect of the faint early sun?"

"Obviously," the professor responded.

"In addition," continued the lawyer, "doesn't the temperature of the exosphere, some 400 miles above the Earth, have to be factored into the equation concerning hydrogen? Doesn't the rate at which hydrogen escapes from the Earth's atmosphere increase as the temperature of the exosphere rises?"

"Yes, that is correct," acknowledged the professor.

Turning over one of the papers in his hand, the defense counsel ran the fingers of his right hand down the page. At a point near the bottom of the page, he stopped and inquired: "Are you familiar with Shimizu's study on exospheric temperatures in a methane dominated Archean era atmosphere?"

"Vaguely, yes," Dr. Yardley answered.

"Shimizu had concluded," reported the lawyer, "that a methane dominated Archean era atmosphere would have had an exosphere whose temperature exceeded 1300 degrees Kelvin or more than 1000 degrees Celsius. The study suggested these temperatures would have made an atmosphere of such composition very short-lived.

"If one were to assume," Mr. Tappin postulated, "that a super greenhouse effect in a carbon dioxide-dominated atmosphere also were capable of generating comparable kinds of exospheric temperatures, then might one conclude, with some degree of justification, that there could be a relatively high rate of exodus of hydrogen from such an atmosphere?"

"Possibly," Dr. Yardley offered.

"Moreover," Mr. Tappin countered, "irrespective of the kind of atmosphere in which organic materials might have arisen by means of ultraviolet synthesis, if such organic materials were to continue to remain in the same exposed condition to ultraviolet radiation, then they will, after a time, begin to break down or decompose through the process of photolysis. Is this right?" inquired Mr. Tappin.

"That's pretty much the upside and the down side of things," answered the professor.

"Let us assume," proposed the defense counsel that a methane-dominated atmosphere, or its hydrogen/carbon dioxide equivalent, existed. Let us further assume that the equivalent of one atmosphere of methane gas, or its hydrogen/carbon dioxide equivalent, was polymerized to more complex hydrocarbons through ultraviolet photolysis over a period of some 1 to 10 million years.

"Despite allowing such assumptions as given, one still would have to consider the following possibility. The one to ten meters of organic material that we are assuming had precipitated out would now be subject to one to ten million years of further photolysis, not to mention possible hydrolysis, and, depending on surface temperature, pyrolysis. Is this about right, Professor?"

"More or less," Dr. Yardley said.

"In addition," continued the defense counsel, "if there were an extraterrestrial event of sufficient magnitude to vaporize the ocean, or vaporize the photic zone, or the size of the Yucatan meteorite, then, the one to ten meter layer of hydrocarbon material that has been postulated by some, would be, shall we say, history. Would you agree with this?"

"Given your premise, that conclusion follows," admitted Dr. Yardley.

Referring briefly to the paper in his hand, Mr. Tappin asked: "Professor, would one be correct in stating that only a small fraction of the light energy coming from the sun is in the form of ultraviolet wavelengths that are sufficiently small to be capable of being absorbed by molecules such as H_2O, CO_2, CH_4, and NH_3?"

"Yes," agreed the professor.

"Would one also be correct," inquired the lawyer, "if one said the following: when more complex molecules are formed, then, the absorption profile or spectrum of these molecules shifts in the direction of longer wavelengths where a great deal more energy is available from the light being radiated from the sun?"

"Again, yes," the professor affirmed.

"Dr. Yardley," continued the defense counsel, "do most of the relatively low wavelength ultraviolet photochemical reactions take place in the upper or lower atmosphere?"

"The upper atmosphere," responded the professor.

"Is it possible," queried the lawyer, "that the compounds that formed in the upper atmosphere through low wavelength ultraviolet photochemical reactions are now vulnerable to photolytic decomposition in relation to a broader range of energies as the absorption spectrum of these more complex compounds moves in the direction of longer wavelengths?"

"Yes, this is a possibility," the professor acknowledged.

"In other words, Dr. Yardley," the defense counsel summarized, "a variety of compounds could have been synthesized in the upper atmosphere by means of low-wavelength ultraviolet photochemical reactions, and, then, these newly formed compounds could have been decomposed through the photolysis brought about by longer wavelength ultraviolet radiation to which these compounds had become susceptible by virtue of their greater complexity, and, this all could take place before the organic materials ever reached the ocean or surface of the Earth. Isn't this a very real possibility, Professor?"

"Yes, it is," Dr. Yardley stipulated.

"Seemingly," Mr. Tappin suggested, "there is something of a race between two opposing forces here: photolytic production of compounds and photolytic decomposition of organic materials. Which of these two forces dominates in a given context will significantly shape what does and does not get to the ocean. Is this correct Dr. Yardley?"

"I would say so," the professor confirmed.

Once again, Mr. Tappin went to the table for the defense and exchanged the papers in his possession for ones being offered by his colleague. Turning back toward the witness, the lawyer said: "Dr. Yardley, in your direct examination testimony concerning the coupling of shock wave energy to hydrocarbon synthesis, you cited a number of figures."

Reading aloud from the papers in his hand, the lawyer summarized the material. "One, meteorites entering the atmosphere with a mass between 10^{-14} - 10^2 grams would generate, collectively, about 1.8×10^{15} joules per year. Two, carbonaceous chondrite airbursts of objects that had a radius less than, or equal to, 300 meters would generate, collectively, approximately 1.5 x

10^{14} joules per year. Three, the post-impact vapor plumes of meteorites striking the Earth's surface would produce – collectively -- about 6 x 10^{17} joules per year. Are these figures correct, Dr. Yardley?" asked the defense lawyer.

"Yes," the professor indicated.

"What sort of a conversion factor is used to come up with these figures?" Mr. Tappin inquired. "In other words, what percentage of the total impact energy actually is believed to be directed toward, or available for, shock synthesis?"

"The conversion factor," replied the professor, "would be a function of the kind of assumptions one made in developing the thermochemical model one used to calculate energies, efficiencies and so on. The amount of total energy that is capable, potentially, of being converted to synthesis reactions starts at about twenty to thirty percent and works its way downward from there depending on the factors being taken into consideration."

"Presumably then, Dr. Yardley," remarked the lawyer "the figures you have cited are not cast in stone. The actual energies that might be directed toward synthesis reactions might be less, perhaps even, considerably so, than the figures you have cited. Would you agree with this?"

"To a certain extent," the professor responded. "At the same time, these figures are not randomly pulled out of a hat. They are the end result of quite a bit of rigorous reflection and take into consideration a great deal of scientific knowledge."

"I'm sure," admitted the lawyer, "that what you say is true, Dr. Yardley. However, the same thing could be said with considerable justification at almost every stage of science for the past several hundred years, and, yet, despite this, models have changed and calculations have been revised. Isn't this so, Professor?"

"I suppose so," replied the professor.

"Dr. Yardley, if one varied the value of atmospheric pressure in one's model, how would this affect calculated energy values with respect to meteorite influx?" Mr. Tappin wondered.

"Within certain limits," the professor suggested, "increasing the atmospheric pressure would help to aerobrake incoming objects. This would decrease, to some extent, the velocity of these objects and, consequently, would tend to affect the total amount of impact energy manifested during the passage of the meteorite through different parts of the atmosphere."

"Would it be fair," Mr. Tappin inquired, "to say that we do not know, in any of the three sets of figures, how the energy is distributed over time or across space? In other words, Professor wouldn't some days, hours or minutes receive disproportionate amounts of these yearly allotments of energy relative to other days, hours and minutes? Similarly, wouldn't it be the case that the billions of cubic miles of atmosphere that surround our planet will not all receive an equal and even distribution of the yearly allotments of energy for any of the three ways of generating energy?"

"This is likely to be the case" Dr. Yardley answered.

"Is it not also true," asked Mr. Tappin, "that the efficiency of shock wave synthesis decreases in relation to increases of impact energy? In other words, isn't it true that the yield of organic materials per unit of impact energy decreases as the impact energy increases?"

"That's correct," responded the professor.

"Consequently," the lawyer reasoned, "there will be variable yields of organic materials due to shock synthesis as a function of the impact energy for any given set of spatial and temporal coordinates. Would you agree with this, professor?"

"I would," Dr. Yardley acknowledged.

"In the light of the foregoing considerations," stipulated the defense counsel, "could one argue in the following fashion? Is it conceivable that some of the shock wave energy created in the atmosphere by microscopic-sized meteorites might generate shock wave energy in such a way that either: (a) the pattern of energy distribution across space might not be sufficiently concentrated in any one area to bring about organic synthesis; or, (b) that the precise character of the pathway of the shock wave created by the passage of the microscopic-sized meteorite might not engage any molecules capable of being synthesized with the available energy?"

"Yes, I guess such a scenario is conceivable," acknowledged the professor, "but your description of the situation is very vague?"

"Yes, it is, Dr. Yardley. It is vague in precisely the same way as when one says that meteorites ranging in size from 10^{-14} to 10^2 grams enter the atmosphere and generate 1.8×10^{15} joules of energy per year, and no precise indications are given as to what is happening at any given moment in time and space.

"To speak in terms of yearly energy yields can be quite misleading. We have no way of knowing whether, at any given point in time and space, we have too much energy, which will tend to decrease shock processing yields, or too little energy and, therefore, not enough to generate sufficient energy of activation for a particular synthesis to occur. Would you agree with this, Dr. Yardley?"

"Yes," replied the professor. "Citing yearly energy production in isolation can be misleading. Knowing the particulars of this energy distribution across time and space would be much more important and helpful."

Glancing at the material in his hand, Mr. Tappin inquired: "Speaking of particulars, Dr. Yardley, isn't it true that the shock waves from the post-impact vapor plumes you mentioned don't match up well with the atmosphere as far as how their energy is distributed across space and time? In other words, don't these post-impact plumes rise considerably above the atmosphere and, as a result, release a great deal of their energy outside of the portions of the atmosphere where chemical synthesis is likely to take place?"

"That's right," the professor acknowledged. "I believe some researchers have suggested that, perhaps, only as little as one-sixtieth of the energy from these vapor plumes might be distributed in the atmosphere and, therefore, be available, potentially, for synthesis reactions."

"Therefore, would one be correct in assuming," Mr. Tappin asked, "that not all of the energy generated by shock waves will necessarily be coupled with certain molecules in the atmosphere to produce various synthesized hydrocarbons, and, therefore, some of the available energy will be lost?"

"Yes," replied the professor. "Generally speaking, however, researchers speak about the energy yield in such contexts. In other words, rather than talking about the amount of energy that might or might not be lost, researchers average the mass of the synthesized material across the available energy and, consequently, speak in terms of the amount of material yielded per unit of energy.

"Nevertheless, each unit of energy does not necessarily participate in the synthesis of some given amount of organic material. Energy yield constitutes a relational index of sorts that links the totality of materials synthesized with the totality of energy available for such synthesis.

"For instance, previously I had discussed certain laboratory experiments that investigated the amount of energy generated by rapidly-expanding gases in shock-wave tubes. Researchers involved with these studies found that 332 nanomoles of hydrogen cyanide (HCN) was the yield, on average, for each joule of shock wave energy present in the tube.

"A nanomole is one billionth of a mole, and a mole is the amount of a given substance that is equivalent to the molecular weight of that substance as expressed in grams. So, 332 nanomoles per joule constitutes, in and of itself, a very small quantity of synthesized material, but when multiplied by the total energy of the shock-wave, the overall quantity becomes much larger."

"Yet, isn't it true, "noted Mr. Tappin, "that energy yield figures will vary from one atmospheric composition to the next? For instance, given the same magnitude of shock-wave energy in a reducing and a neutral atmosphere, the energy yield of, say, HCN tends to be significantly higher in a reducing environment as opposed to a non - reducing atmospheric environment. Is this correct?"

"Yes," the professor agreed. "In general, the mass of organic materials capable of being shock-synthesized in a given kind of atmosphere - that is, the organic synthesis efficiency - is very dependent on the compositional character of the atmosphere being considered.

"For example, in one thermochemical model of shock-synthesis to which I alluded to earlier," the professor pointed out, "a methane dominated reducing atmosphere is calculated to give a production efficiency of about $10^{17.5}$ molecules of HCN, hydrogen cyanide, per joule of energy. In addition, this was accompanied by the production of a few other kinds of simple hydrocarbons such as C_2H_2 and C_2H_4.

"However, in a carbon dioxide/nitrogen-dominated atmosphere, the production efficiencies for HCN were calculated by this thermochemical model to be approximately $10^{7.5}$ smaller than in the reducing atmosphere. On the other hand, the production efficiencies for formaldehyde (H_2CO) in the neutral atmosphere were calculated to be roughly comparable to what would be obtained in a reducing atmosphere."

"Would you agree, Dr. Yardley," the defense counsel asked, "that thermodynamic calculations might tell one whether or not certain kinds of reactions are possible, but they can say nothing about whether such reactions will occur, nor anything about at what rate they will proceed, nor the path that will be taken by such a reaction?"

"That's right," the professor confirmed.

Mr. Tappin turned the papers in his hand over and began examining the other side. "Dr. Yardley, doesn't the energy yield index you were talking about previously vary with the nature of the energy source involved in any given case?"

"Yes," replied Dr. Yardley. "In experiments with artificial lightning, for example, researchers observed an energy yield of about 3 nanomoles of HCN per joule of energy in a methane-dominated atmosphere versus an energy yield of, approximately, 1000 times less than this in an atmosphere dominated by carbon dioxide."

"Professor, is the magnitude of the energy associated with artificial lightning the same as is associated with natural lightning?" asked the lawyer.

"No, there is a considerable difference between the two," the professor indicated. "In general, natural lightning is far more powerful than artificial lightning."

"Consequently, given what you have said previously," posited the defense counsel, "one would expect the shock-processing yield per unit of energy for natural lightning to be less than that of artificial lightning since the yield per unit of energy decreases as the energy of the impact increases. Is this correct, Professor?"

"This would be consistent with what has been said," Dr. Yardley confirmed.

"Would one be shaky or firm grounds" Mr. Tappin inquired, "if one were to argue that just as there might be various kinds of organic synthesis that occur in the shock-wave wake of meteorites and lightning, so too, meteorites and lightning also can cause the decomposition of materials through pyrolysis and so on?"

"Fairly firm grounds, I would imagine," answered the professor."

"Therefore," stated Mr. Tappin, "in somewhat analogous fashion with respect to ultraviolet radiation, in those circumstances when shock-wave energy is present, there are forces, both of synthesis as well as decomposition, which are taking place, so to speak, side by side. Are we not dealing here, Professor, with the fact that what is being given with the hand of synthesis, is, to some extent, being taken away by the hand of decomposition?"

"Yes," replied Dr. Yardley. "This seems to be the case."

"Apparently," observed the defense lawyer, "we require some kind of 'net energy yield' figure. We need to be able to determine whether the upside, or the down side, of photolysis, pyrolysis, hydrolysis and other factors is dominating any given feature of the Archean era world. Would you agree with this?"

"Yes, I do," the professor affirmed, "but this is easier said than done."

"Dr. Yardley, in the initial origin-of-life experiment performed by Stanley Miller, methane, ammonia, hydrogen and water vapor were used to simulate what was believed, at least at that time, to be the composition of the Archean era atmosphere. In addition, a continuous spark discharge was applied to the gaseous mixture in order to simulate the presence of lightning in a prebiotic world.

"After letting this experiment run for a number of days, the materials synthesized during the course of investigation were examined. Is this very general description of Miller's experiment accurate for the most part?"

"Yes," the professor indicated.

"We know," Mr. Tappin continued, "that questions have been raised by other scientists and researchers in relation to whether or not the Archean atmosphere actually was predominately methane/ammonia in character. I was wondering, however, about the spark discharge aspect of the experiment.

"What was the magnitude of the electrical discharge?" asked the defense counsel.

"Somewhere around two to four watts, I believe," the professor offered.

"Correct me if I am wrong, Dr. Yardley," requested the lawyer, "but I'm not familiar with any 2-4 watt lightning discharges that run continuously for several days. Are you?"

"No," smiled the professor.

"Dr. Yardley," stated the defense counsel, "one might assume that continuous spark discharges from a coil are different in character from lightning bolts and their associated shock waves. Would such an assumption be correct?"

"Well," the professor replied, "the two certainly involve different magnitudes of energy, but the underlying physics is essentially the same. Of course, lightning would not be continuous, but the sparking mechanisms used in the experiments are continuous in nature."

"Would one be unreasonable," Mr. Tappin queried, "to expect different sorts of outcome if one, first, were to expose a certain mixture of gases to a single bolt of lightning and, then exposed the same kind of gaseous mixture to a continuous spark of 2-4 watts for a number of days?"

"No," replied the professor, "probably not, but neither would one be unreasonable if one were to anticipate some degree of overlap in the product outcomes of the two experiments. For instance, both the 2-4 watt spark discharge as well as the lightning bolt might generate some amount of hydrogen cyanide (HCN) in the right kind of atmosphere."

"Can one assume," Mr. Tappin inquired, "that if lightning occurs in the Archean era atmosphere, one will observe amino acids being formed as occurred in the Miller experiment?"

"No, one couldn't assume this," the professor remarked. "In point of fact, the Miller experiment involved a continuous circulation of the gases through the chamber where the electrical spark was being discharged.

"Initially, molecules like formaldehyde and hydrogen cyanide would be synthesized. Then, as these molecules along with the original gases continued to be exposed to the electrical discharge of the spark chamber, slightly more complex molecules in the form of amino nitriles would have been formed.

"Amino nitriles plus water plus continued exposure to the electrical discharge yielded amino acids such as alanine or glycine plus ammonia. There also were a variety of amino acids synthesized that do not occur in any of the biological organisms with which we are familiar."

"Professor Yardley, you have previously testified," Mr. Tappin indicated, "that extremely tiny amounts of hydrogen cyanide were formed when artificial lightning was discharged in a methane-dominated gas mixture, and, you also have testified that hydrogen cyanide was generated during an early stage of Miller's original spark-discharge experiment. Is this correct?"

"Yes, it is," Dr. Yardley remarked.

"You also testified that formaldehyde (H_2CO) is generated during one of the early stages of the Miller experiment. Were there any findings concerning the production of formaldehyde in the artificial lightning studies of which you are aware?"

"In the limited studies that have been carried out," replied the professor, "no formaldehyde formation has been detected. Furthermore, as far as I know, even the figures that come from purely theoretical thermochemical calculations indicate no formaldehyde formation is to be expected in relation to lightning discharges, whether these are artificial or natural."

"Yet, Dr. Yardley, in the Miller experiment, the formaldehyde produced by spark discharge combined with the hydrogen cyanide produced by spark discharge and entered into reaction with ammonia, one of the gases in the supposedly simulated Archean-atmosphere of the experiment, and all of this resulted in the formation of amino nitriles. Is this correct, Professor?"

"That's right," Dr. Yardley agreed.

"Isn't it also the case, Professor," the lawyer inquired, "that researchers believe ion-molecular and free radical reactions, rather than lightning-like shock synthesis, are the essential processes involved in synthesis reactions in spark discharge experiments?"

"Yes," acknowledged the professor.

"In what sense, then, Professor," asked the defense counsel, "can one say the Miller experiment is a simulation experiment, given that it probably simulates neither the atmospheric composition of the Archean era nor the character of lightning discharges, nor the products of lightning discharges, and given that, previously, you have suggested amino acids were formed in the ocean through a Strecker-like synthesis process rather than in the atmosphere through electrical discharges?"

"As far as the features that you have pointed out," replied the professor, "the Miller experiment really isn't much of a simulation experiment. What it does show is this: if one continuously exposes a gaseous mixture of the right molecular composition to an electrical discharge of a certain magnitude, one can generate a series of chemical reactions that will culminate in the formation of complex hydrocarbons that have implications for origin-of-life issues.

"One would have had, perhaps, a closer simulation of certain aspects of actual Archean era prebiotic conditions if one had removed the products of each activation step so that the products of one set of reactions would not have been exposed to the energy source a second time. This process of removing synthesized reaction products at each step of the experiment would have simulated, to a degree, the passage of molecules, synthesized in the Archean era atmosphere, to the ocean, where they would have been protected from further exposure to various forms of energy impinging on the atmosphere."

"Dr. Yardley, wouldn't one have an even better kind of simulation," Mr. Tappin asked, "if one exposed the products of each reaction step to all of the conditions and forces that could have acted upon them in an Archean era context, including the ones that could decompose or destroy such products?"

"Yes, I guess so," agreed the professor, "but there is a practical limit to what can be accomplished in the laboratory."

"Yet," the lawyer countered, "wouldn't you agree that the more we will allow such limitations to distance us from the actual conditions of the world, then the more we will introduce distortions, biases and error into our experimental procedure? Moreover, wouldn't these kinds of distortions skew our capacity to interpret accurately the significance of what our experiments have to say about the nature of the physical world, whether in relation to the natural phenomena of our present day, or those of the Archean world?"

"I would agree," responded the professor, "that we must continuously seek to probe the limitations of our current experimental methods in order to devise, where possible, better experiments and procedures that will permit us either to overcome, or compensate for, such limitations."

"Professor, one could agree with every word you have just said," Mr. Tappin maintained, "but your words do not address or answer the problem before us. To what extent, do the simulations, calculations, estimates, experiments, conjectures, hypotheses and models of prebiotic, evolutionary theory reflect the conditions, forces, processes and dynamics of the Archean era Earth?

"On the basis of testimony that you have given, Dr. Yardley, Miller's experiment doesn't simulate, or emulate, the Archean era world in any way. What his experiment establishes is this: if you do certain things, certain things happen.

"Given that the things which the experiment has done are not necessarily what happened in the Archean era world, then, the fact certain things have been observed to happen might be interesting, intriguing or suggestive, but they don't necessarily shed any light on what actually took place during prebiotic times. Isn't this so, Dr. Yardley?"

"I would agree," the professor admitted, "that the Miller experiment, or others like it, don't prove what happened in the Archean era world. Nonetheless, such experiments generate data that can be incorporated into a process of theory construction that permits the scientific community, over time, to understand, in a consistent, rigorous fashion, a wider and wider body of technical information about an array of interconnected physical and chemical phenomena."

"Yes, Dr. Yardley," Mr. Tappin said, "but the question is this: to what extent does this condition of understanding a wider and wider body of technical information about an array of interconnected physical and chemical phenomena in a consistent, rigorous fashion provide one with a correct understanding of what actually did happen during the Archean era... rather than with just an understanding of what might have happened or what could have happened if all of the conditions, assumptions, and conjectures on which that scientific model is founded were really true? You see, Dr. Yardley, I'm far from convinced evolutionary theorists know, or have any way of proving, whether or not their belief system is capable of getting outside of itself and reflecting anything of the actual nature of reality."

"Objection Your Honor," announced Mr. Mayfield. "My learned colleague is making speeches."

"Yes, sustained," Judge Arnsberger indicated. "Let's move along Mr. Tappin. You'll have time enough for this sort of thing in your closing remarks."

As the counsel for the defense looked over the papers in his hands, he said: "Very well, Your Honor. I apologize to the court for my outburst."

Turning toward the witness, Mr. Tappin asked: "In the Strecker - like, amino acid synthesis scenario that you outlined during direct examination testimony, on what chemical reactants does this kind of synthesis depend?"

"As long as the concentrations of hydrogen cyanide and aldehydes, such as formaldehyde, do not drop too low," pointed out the professor," then researchers believe the Strecker synthesis will be an effective means of converting the aforementioned reactants to amino acids over the course of some 10,000 years."

"What concentration levels," queried Mr. Tappin, "are considered to be minimally necessary for the Strecker synthesis process to be able to proceed?"

"These would be roughly of the order of a 10^{-6} molar solution," Dr. Yardley replied. "This means there should be at least 10^{-6}, or one- millionth, of a mole of solute for each liter of solvent."

"What kind of collective production rates," asked the lawyer, "have been estimated for, say, hydrogen cyanide as a result of ultraviolet radiation, lightning discharges, and shock-synthesis?"

"The figures that I have seen used most frequently," Dr. Yardley answered, "have an upper and lower boundary. These boundaries reflect whether one is talking about a reducing or a relatively neutral atmosphere.

"In the case of a reducing atmosphere such as methane and ammonia, researchers have worked out a production yield of about 100 nanomoles, or 100 billionths of a mole, per square centimeter, per year. This would have resulted in a 3.3×10^{-4} molar concentration of hydrogen cyanide in the Archean era ocean over a period of 10 million years.

"On the other hand, if one were dealing with a relatively neutral atmosphere, the production rate of hydrogen cyanide would have been as much as several orders of magnitude less than 100 nanomoles ... somewhere around 1 nanomole, give or take a few nanomoles ... per square centimeter, per year. Over a ten million year period, this would have resulted in a 10^{-6} molar concentration of hydrogen cyanide."

"Therefore," Mr. Tappin observed, "the estimated concentration of hydrogen cyanide arising from a relatively neutral atmosphere is right at the minimal limit of what is necessary for the Strecker synthesis to proceed in the Archean era ocean. Is this correct, Dr. Yardley?"

"Yes, that's right," confirmed the professor.

"What assumptions, if any, Dr. Yardley, are made with respect to the conditions in the Archean era ocean in which such a Strecker synthesis is alleged to have taken place?"

"Usually," responded the professor, "researchers assume an ocean pH of either 7 or 8, which is comparable to what we find in the oceans of our present day. Moreover, the temperature of the water is assumed to be about 0 degrees Celsius."

"How do these assumptions, or do these assumptions," inquired the defense counsel, "affect molar concentration estimates for hydrogen cyanide?"

"At a pH of 7 and 0 degrees Celsius, a 3.5×10^{-5} molar concentration of hydrogen cyanide has been calculated for the Archean era ocean. This is based on the reducing-atmosphere production figure of 100 nanomoles per square centimeter, per year."

"Assuming a pH of 8 and, once again, 0 degrees Celsius, one comes up with a 4×10^{-6} molar solution of hydrogen cyanide. This estimate also presupposes the reducing-atmosphere production yield figures cited previously."

"Am I correct in stating," Mr. Tappin asked, "that if one were to use the lower neutral atmosphere production yield rates, rather than the higher, reducing-atmosphere production rates for hydrogen cyanide, then at pH 7 and 0 degrees Celsius, one would have a molar concentration of about 3.5×10^{-7} since the neutral-atmosphere production-yield rates are several orders of magnitude lower than the reducing- atmosphere production rates for hydrogen cyanide?"

"Yes, you would be correct," the professor admitted.

"Professor Yardley," pressed the defense counsel, "is this 3.5×10^{-7} figure for neutral-atmosphere Archean era oceans greater than, or less than, what is minimally needed to be necessary for the Strecker synthesis to proceed in the Archean era ocean?"

"This would be less than what is minimally necessary for the Strecker synthesis to proceed," the professor indicated.

"Moreover, Professor Yardley," postulated the lawyer, "if one were to assume a pH of 8 and 0 degrees Celsius in the Archean era ocean, as well as presuppose the lesser production-yield figures of a relatively neutral-atmosphere, would one be correct to conclude that this would result in a molar concentration of approximately 4×10^{-8} for hydrogen cyanide?"

"Yes," agreed the professor.

"Is this molar concentration," the lawyer continued, "of 4×10^{-8} for hydrogen cyanide greater than, or less than, the minimal necessary concentration of hydrogen cyanide required for the Strecker synthesis to go forward in the Archean era ocean?"

"Again, this concentration is less than what is minimally required," the professor confirmed.

"Dr. Yardley, what would happen," Mr. Tappin queried, "to hydrogen cyanide concentration figures if one were to raise the temperature of the water to, say, 25 or 50 degrees Celsius, but keep the pH at either 7 or 8?"

"If," assumed the professor, "one were to work on the basis of the reducing-atmosphere production yield figure of 100 nanomoles per square centimeter, per year, then at pH 7 and 25 degrees Celsius, the molar concentration of hydrogen cyanide in the Archean era ocean would be about 2×10^{-8}. In addition, at pH 7 and 50 degrees Celsius, the molar concentration of hydrogen cyanide would be about 3×10^{-9}.

"If, on the other hand ..."

Before the professor could continue, Mr. Tappin interrupted and asked: "Dr. Yardley, don't the figures you are citing indicate that even when one assumes reducing-atmosphere production-yields, which are favorable to the prebiotic evolutionary model, the concentration levels of hydrogen cyanide in the Archean era ocean are insufficient for the Strecker synthesis to proceed?"

"That is correct," the professor admitted.

"Obviously, then," the defense counsel reasoned, "if one were to use the production-yield figures for a neutral-atmosphere, which are several magnitudes of order lower than the reducing-atmosphere production figures, the concentration estimates for hydrogen cyanide in the Archean era ocean would be about 2×10^{-10} and 3×10^{-11}, respectively, for 25-degree Celsius and 50-egree Celsius temperatures at pH 7. Is this right, professor?"

"Yes, it is," Dr. Yardley stated.

"So, these last concentration figures cited," indicated the lawyer, "both for reducing, as well as for neutral-atmosphere production rates of hydrogen cyanide, would be insufficient to sustain Strecker synthesis in the Archean era ocean. This is correct, isn't it?"

"Yes," said the professor.

"Furthermore," Mr. Tappin added, "if one were to work out the concentration figures at pH 8 or pH 9, for either 25 degrees Celsius or 50 degrees Celsius, then, quite irrespective of whether one were working on the assumption of a reducing-atmosphere or the assumption of a neutral-atmosphere, all of the concentration levels for hydrogen cyanide would be far below what is minimally necessary to sustain an amino acid Strecker synthesis in the Archean era ocean. Isn't this the case, Dr. Yardley?"

"Yes, it would be," the professor acknowledged.

"Moreover," the defense counsel continued, "you did previously testify, Professor Yardley, that evolutionary scientists believe the pH of the Archean era ocean was 8, plus or minus 1, did you not?"

"That's right," said the professor.

"Therefore," reasoned the lawyer, "to single out an Archean era ocean with a pH of 7, at 0 degrees Celsius, under conditions of a reducing-atmosphere, is to describe a situation in which everything is stated in terms that are favorable to the idea of a natural account of the origin-of-life from prebiotic conditions. Alternatively, such a way of describing things is to ignore the very real possibilities that the Archean era ocean did not have a pH of 7, or a temperature of 0 degrees Celsius, and might not have existed in conjunction with a reducing-atmosphere.

"All of these other environmental conditions that are possible in the Archean era world would, if true, bring into serious question the plausibility of an evolutionary theory account of the origins-of-life. Would you agree with this, Dr. Yardley?" queried the lawyer.

"If these other possibilities were the case, then, yes, questions of plausibility would begin to arise in relation to such an evolutionary account," admitted the professor.

"What, if any, other assumptions are made concerning the conditions under which the Strecker synthesis is believed to proceed in the Archean era ocean?" Mr. Tappin inquired.

"Well, for one thing," the professor replied, "the Archean ocean is assumed to be comparable in depth and extent to the oceans of today. If the Archean era ocean were shallower or less extensive than current oceans, then, this would serve to increase, somewhat, the concentration figures previously cited. How much this increase of concentration might be, would depend on how much smaller and shallower the Archean oceans were relative to modern day oceans."

"Yet," countered the defense counsel, "couldn't one logically assume that the Archean era ocean was larger, not smaller, than current oceans? After all, the continents had not necessarily established themselves at this period of the Archean era.

"Perhaps, there was more, not less, water during the Archean era, and, therefore, the concentration figures mentioned previously are all inflated somewhat. Isn't this a possibility, Dr. Yardley?"

"Yes, I suppose so," the professor said.

"Isn't it also the case," queried the lawyer "that researchers believe a permanent ice cover formed in the Antarctic only about 20 million years ago, and somewhat more recently in the case of the Arctic region? And, therefore, Professor, might one be correct in assuming that until 20 million years ago, there was quite a bit more water in the Archean era oceans, again diluting the previous concentration figures for hydrogen cyanide?"

"Possibly," Dr. Yardley offered.

"Furthermore, Professor," Mr. Tappin pressed, "doesn't water expand when it is warmer, and if this is correct, isn't there a lot of evidence to indicate that the Archean era atmosphere was sufficiently warm to heat the ocean waters quite a bit above the 0 degrees Celsius temperatures that are being assumed in the Strecker synthesis model, and, therefore, wouldn't this expanded water tend to increase the volume of the solvent, reducing the concentration levels of hydrogen cyanide?"

"Quite possibly," responded the professor.

"Are there any further assumptions," asked the defense counsel, "which frame the conditions under which the Strecker synthesis is believed to have proceeded in the Archean era?"

"There are two more assumptions that I can think of," Dr. Yardley stated. "First, researchers tend to assume all HCN that is produced, by whatever energy pathway, is fully dissolved in the Archean era ocean. Secondly, scientists, generally, assume neither hydrolytic nor thermal degradation will appreciably affect the amount of hydrogen cyanide solute in solution."

"If," postulated the lawyer, "the surface of the Earth were frozen over, as some theorists have proposed in conjunction with the faint early sun paradox, couldn't this affect the amount of hydrogen cyanide that would be able to enter into solution in the Archean era ocean that is alleged to exist below the 300 meter layer of ice?"

"Yes, I guess it could," the professor replied.

"Moreover," Mr. Tappin continued, "if, as some other researchers, alluded to by you, have maintained, there were a layer of between one and ten meters of hydrocarbons floating on top of the ocean, presumably as a result of their non-polar and, therefore, non-soluble nature, then, couldn't this scenario also affect the opportunity of all hydrogen cyanide to enter into solution with the Archean era ocean?"

"I suppose so, yes," indicated the professor.

"In addition," Mr. Tappin pressed, "if we leave aside issues of hydrolytic decomposition, isn't the assumption about the relatively negligible extent of the thermal degradation brought about by hydrothermal vents rather arbitrary and speculative?"

"I believe," the professor offered, "that this assumption about thermal degradation might be based on the roughly ten million years that is required for any given volume of water to circulate throughout the ocean and, presumably pass some given hydrothermal vent. When one compares this period of ten million years to the period of approximately 10,000 years required by the Strecker synthesis, thermal degradation probably would constitute a negligible factor."

"Doesn't this way of thinking," queried the lawyer, "seem to be assuming there is only one hydrothermal vent that is being used as a point of reference for calculating the figure of ten million years necessary for water to completely circulate throughout the ocean? If there were many hydrothermal vents, as might be expected from an early Archean era in which, according to your testimony, some researchers have claimed that the Earth's crust might have been especially vulnerable to such hydrothermal breakthroughs, then the ten million figure that signifies the amount of time required for a given volume of water to circulate through the ocean might be true, but it is irrelevant if many such vents exist at many different points along the bottom of the Archean era ocean. Isn't this so, Dr. Yardley?" asked the defense counsel.

"What you say is a possibility that would have to be taken into consideration, in some way, I suppose," the professor said.

"What about photolysis, Dr. Yardley?" inquired Mr. Tappin. "I noticed you didn't mention this as a possible source of degradation, but wouldn't it have to be factored in, at least with respect to the 200-meter photic zone of the ocean?

"In other words, Dr. Yardley, since all hydrogen cyanide going into solution would have to pass through this photic zone, couldn't photolysis play a major role in affecting the amount of hydrogen cyanide solute available, and, therefore, the molar concentration of this molecule? Moreover, wouldn't this especially be the case given, as you have indicated, that the ultraviolet luminosity of the faint early sun would have been substantially greater during the Archean era?"

"This is a possibility," the professor admitted, "but degradation losses due to things such as ultraviolet light or ionizing radiation are very difficult to measure and, therefore, one has some difficulty in establishing a basis for making estimates in relation to them."

"Whatever the nature of such difficulties, Professor, our ability or inability to measure something really doesn't stop that something from having an effect on us does it?"

"As a matter of fact," stated the professor, "there are interpretations of quantum mechanics which do suggest that reality only comes into being with the act of measurement."

"Dr. Yardley," Mr. Tappin responded, "I believe this is getting more into the realm of philosophy than hard science. However, if you want to begin to grapple with the paradox of how to explain the existence of a prebiotic world prior to the advent of the process of human measurement, I believe you will find evolutionary theory will be in even more difficulty than I, and my client, already believe to be the case."

Looking at the material in his hands, the defense counsel said: "Most of the discussion of the past little while has been about hydrogen cyanide. Very little has been said about formaldehyde, but you previously had stated the Strecker synthesis in the Archean era depends on certain minimal levels of molar concentration being maintained not only for hydrogen cyanide, but for aldehydes such as formaldehyde, as well.

"However," added the lawyer, "in earlier testimony and cross-examination, we established that formaldehyde is not generated during lightning shock-synthesis and also that most of the formaldehyde that might be generated through ultraviolet radiation synthesis is also vulnerable to ultraviolet photolytic degradation. My question, professor, is this: From whence do the necessary levels of formaldehyde, or other aldehydes, come that are supposed to maintain concentration rates capable of sustaining Strecker synthesis in the Archean era ocean? Even if one could establish requisite production-yield rates, wouldn't all the difficulties that beset the matter of hydrogen cyanide concentration levels also apply to formaldehyde levels of concentration in the Archean era ocean?"

"To the best of my knowledge," Dr. Yardley indicated, "the figures on formaldehyde are less well established than are those for hydrogen cyanide. Nevertheless, I would agree, in general terms, that all of the issues that you have raised in relation to hydrogen cyanide concentration levels would also have to be raised in conjunction with formaldehyde, or other aldehyde, concentration levels in the Archean era ocean."

"Professor Yardley, let's assume," posited the counsel for the defense, "that I were willing to forget all the problems that have been raised with respect to the concentration issue. Do we have any way of knowing what proportion of the amino acids formed in the Archean era ocean through Strecker synthesis would be the twenty varieties of amino acid occurring in living organisms rather than the many other kinds of amino acid that are possible- some of which have been discovered in meteorites?"

"I imagine," answered the professor, "there are individuals with the talent to be able to come up with some kind of thermochemical model that would provide a set of theoretically-driven distribution values for all the different kinds of amino acid that might be possible. However, such a model would be affected by so many variable considerations, conditions and forces, I'm not sure even our current supercomputers could keep track of the problems that would arise in this kind of model.

"One could assume less complex amino acids might tend to be somewhat disproportionately represented in relation to more complex amino acids. On the other hand, a wide array of localized thermodynamic conditions might arise that could run against these sorts of tendencies.

"If temperatures in the ocean were low, say, near 0 degrees Celsius, then one would expect thermal decomposition to be low. However, some amino acids, like alanine and glycine, have far greater stability than do other amino acids, like serine.

"Consequently, stability properties would have to be factored in even if the water temperature were to remain near 0 degrees Celsius, which is unlikely. This is unlikely because within the last twenty to thirty million years there is evidence that bottom water temperatures can vary as much as 10 to 15 degrees as the Earth goes through various climatic transitions.

"What variations in water temperature, top or bottom, might have been taking place across hundreds of millions of years in an Archean era ocean and atmosphere are anybody's guess. Furthermore, how the decomposition tendencies of the twenty amino acids that occur in living organisms would stack up to the decomposition tendencies of all the other amino acids that are possible is another issue that would have to be factored in.

"Then, of course, one would have to work in the decomposing effect that hydrothermal vents and active volcanoes would have on amino acids that had been formed. Since we really don't have any idea of how prevalent either of these processes was during the Archean era, this introduces a further unknown into any prospective model that is being constructed.

"The effects of ultraviolet radiation in the 200-meter photic zone would have to be considered. In addition, once hydrolysis had done its magic and helped amino acids to form, then, the newly -synthesized, more complex amino acids become even more vulnerable to the forces of hydrolysis than is the case for the molecules that reacted together to form them.

"Furthermore, one cannot assume the only sort of synthesis reactions going on in the Archean era ocean are ones that lead to the formation of amino acids. Other, non-amino acid kinds of hydrocarbon are likely to have arisen, and this means there would have been chemical competition for available reactants, with unknown ramifications for the rate and extent of amino acid formation, both in relation to the twenty amino acids that are important to life forms, as well as in relation to the other varieties of amino acid that are not important to life forms on Earth."

"Dr. Yardley, is there," Mr. Tappin inquired, "any reason or mechanism you know of which would have led to the specific selection of the twenty amino acids fundamental to life forms on Earth from among the myriad numbers and kinds of other amino acids that are likely to have arisen in the Archean era ocean through Strecker synthesis?"

"No," the professor answered, "I know of no plausible theory that would explain the selection process that we believe went on during the Archean era. It might well have been a stochastic process, and since we don't know enough about the factors shaping that process, we really cannot do anything but speculate why certain probability distributions might have been thermodynamically and/or kinetically favored over other probability distributions."

"Professor Yardley," continued the defense counsel, "with respect to the amino acids synthesized in the Archean era ocean through the Strecker process, would they have formed a racemic mixture ... that is, a mixture consisting of roughly equal numbers of both left-handed and right-handed optical isomers of the various kinds of amino acid?"

"If our laboratory experiments are any indication," the professor replied, "then, yes, the Archean era mixture is likely to have been racemic in character. Nevertheless, I previously have mentioned a meteorite found in the Antarctic that contained some exclusively right-handed amino acids, and this discovery does carry some potential implications for what might have occurred in the Archean era ocean."

"Are you aware, Dr. Yardley," asked the lawyer, "of any plausible account that might explain why one might end up with a set of same-handed optical isomers rather than a racemic mixture of amino acids?"

"Over the years," stated the professor, "there have been a number of proposals directed toward this problem of chirality or handedness. The only hypothesis that I have found to be plausible is one proposed back in the 1950s.

"Essentially, this hypothesis assumes that when sunlight passed through the atmosphere of the Archean era, light took on a small degree of polarization. As a result, the polarized ultraviolet component of sunlight during the Archean era might have had a preferential tendency to degrade right-handed optical isomer forms of amino acids, leaving intact the left-handed optical isomer forms that have been observed in the vast majority of Earth organisms."

"Dr. Yardley, don't most of the biologically important carbohydrate molecules tend to exhibit right-handed optical isomer preference?" Mr. Tappin inquired.

"Yes, that's right," the professor indicated.

"So, wouldn't one expect," postulated the lawyer, "that the same polarized ultraviolet component of Archean era sunlight that degraded right-handed amino acid isomers would also degrade right-handed carbohydrate isomers? Consequently, how does one account for the fact one finds right-handed carbohydrate isomers playing fundamental roles in living organisms?"

"This is a problem," Dr. Yardley admitted, "but there might have been other kinds of selection mechanisms at work in addition to the polarized ultraviolet component of Archean era sunlight."

"Does anyone," challenged the defense counsel, "know what these other selection mechanisms were that are assumed to have been operative during the Archean era?"

"Not at this point in time," answered the professor.

"Dr. Yardley, even if," the lawyer hypothesized, "one were to accept the polarized-light hypothesis as the reason why left-handed amino acids were selectively favored over right-handed amino acids as far as ultraviolet degradation is concerned, this still leaves at least two problems. First of all, the polarized light assumption doesn't explain why DNA would possess a tendency to call for exclusively left-handed amino acids to be synthesized in the cell. Secondly, one still hasn't explained how the twenty amino acids common to life forms on Earth came to be selectively favored over the other left-handed amino acid optical isomers that would have survived being degraded by slightly polarized ultraviolet radiation. Would you agree with my assessment of the situation, Dr. Yardley?"

"As far as the second problem is concerned," stated the professor, "I would agree no fully satisfactory account presently exists for explaining why the twenty left-handed amino acid isomers were selected over other possible left-handed amino acid isomer candidates. As far as the first problem described by you is concerned, something could be said.

"Selection forces would have favored the DNA and/or RNA system that would have arisen that relied on the optical isomer form of amino acid that was available ... in this case, the left-handed amino acid isomer. If a DNA and/or RNA system would have arisen that depended on the existence of a pool of right-handed amino acid isomers, then given that polarized ultraviolet light had selectively destroyed all, or most, of these kinds of isomer, such a DNA/RNA system would not have been favored by the prevailing conditions of the Archean era world. Prebiotic conditions would have favored the DNA/RNA system that called for, or needed, left-handed amino acid isomers."

"Excuse me, Dr. Yardley, perhaps, I don't understand the situation," said Mr. Tappin. "Although your account or explanation makes sense in the context of having assumed that a left-handed-amino acid-preferring DNA/RNA system already had arisen, your account doesn't really explain how such a left-handed-amino-acid-preferring DNA/RNA system arose in the first place ... does it?"

"No, it doesn't," the professor acknowledged.

"In fact," continued the lawyer, "wouldn't one be justified in arguing that the process of natural selection really is incapable of accounting for change over time except in a post-facto manner? By this, I mean that although natural selection can help explain why certain capabilities, once they arise, might have been selectively favored by existing conditions, nevertheless, natural selection cannot explain how such capabilities arose in the first place, can it, Professor?"

"Well," Dr. Yardley responded, "some theorists do speak in terms of the idea of 'evolutionary pressure'. In other words, they believe the collective character of any given set of conditions might, in a sense, generate a certain amount of pressure to induce the sort of changes that would be favorably selected by those conditions."

"How does this process of inducement work?" Mr. Tappin asked. "How does the physical/chemical world induce a given system to change both its structural character, as well as its way of operating, so that the system adopts a structure and set of processes that would be selectively favored by the prevailing conditions of that physical and chemical world?"

"It's a very complicated issue," replied the professor. "There is a great deal of work going on with the science of complexity, as well as chaos theory and the theory of dissipative structures that is directed toward trying to answer questions like this."

"Has anyone," inquired the lawyer, "come up with a model in any of these disciplines that has been accepted by the scientific community as a plausible account of how prevailing physical and chemical circumstances induce a system to generate structural and dynamic changes that are, capable of taking advantage of precisely the conditions that prevail in the world at a given time?"

"Not yet," responded the professor.

"Then, Dr. Yardley, would one be doing injustice to the available evidence," Mr. Tappin pressed, "if one were to say, at least at this point in time, that the notion of evolutionary pressure is a totally unproven hypothesis however convenient and desirable an idea it might be for evolutionary theory?"

"No, I would have to say" the professor admitted, "that no injustice would be done to the available evidence."

"Consequently," summarized the defense counsel, "currently, there really is no plausible, generally-accepted explanation of how or why DNA or RNA systems arose that showed a preference for left-handed amino acid isomers as well as right-handed carbohydrate isomers. Would you agree with this statement, Dr. Yardley?"

"Yes, at the present time, what you have said is the case," agreed the professor.

"Mr. Tappin, I'm going to exercise some discretion and intervene at this juncture," Judge Arnsberger indicated, "to propose that court be adjourned for lunch. Court will reconvene again at 2:00 p. m. this afternoon."

Monkeying Around with The Containment Blues

As Mr. Tappin rose from behind the defense table he took the material being handed to him by a member of his team. He started to walk toward Dr. Yardley, stopped and retraced his steps.

He leaned over and whispered something in the ear of his colleague. When he received an affirmative response, he straightened up.

On his way back to the area near the witness stand, he was busy inspecting the new batch of material. He continued to do so for a further ten seconds, or so, after stopping in front of the witness stand.

Finally, he said: "In your direct examination testimony you referred to an experiment by Fox in which urea [$CO(NH_2)_2$] and malic acid ($C_4H_6O_5$) were heated to 150 degrees Celsius under conditions free from water ... that is, which were anhydrous in nature. You indicated this experiment resulted in the synthesis of aspartic acid.

"In a further experiment, also performed by Fox, you talked about a recipe for generating polymers or bonded chains of amino acid. In this recipe, if one cooked the amino acid glutamic acid in an oil bath for one hour at 170 degrees Celsius, and, then, blended in a variety of other amino acids and cooked the whole mixture for a further three hours at the same 170 degrees Celsius, then one could produce a chain of amino acids consisting of up to a hundred units.

"In variations on this experiment, phosphoric acid was added, and the variables of time and temperature were played around with during different runs of the same experiment. This resulted in an increase in the amounts of neutral and basic amino acids that could be incorporated into the polymer chain of amino acids.

"You also described another experiment in which sunlight was passed through a solution of paraformaldehyde ($CH_2O)_3$, ammonia and ferric chloride. After a certain amount of time, this arrangement brought about the synthesis of the amino acids serine and asparagine.

"During direct examination testimony, you talked, as well, about an experiment by Oró in which hydrogen cyanide, ammonia and water were combined to produce, over a period of time, a number of different amino acids. In addition, a certain amount of the purine, nucleic base, adenine, showed up as a product in this experiment.

"You also discussed how when the foregoing set-up was altered somewhat, other kinds of molecules could be synthesized. For instance, if one combined cyanogen (C_2N_2) and cyanoacetylene (HC_3N) with hydrogen cyanide (HCN), then one could obtain other nucleic bases such as uracil, cytosine, guanine and thymine.

"Finally, in another experiment performed by Oró, you outlined, first, how he took some fatty acids, one of the fundamental building blocks of many important lipids, and, then, how he dried these fatty acids in the presence of phosphate and glycerol. In this manner, simple phospholipids, that are fundamental components of membranes in living organisms, were synthesized.

"I must admit," Mr. Tappin indicated, "on the one hand, I find all of this experimental ingenuity quite impressive. On the other hand, I also find such ingenuity potentially troublesome.

"More specifically, Dr. Yardley, different ingredients are taken from here and there and mixed together in certain ways, for particular lengths of time, under specified conditions of temperature, acidity, and so on. In other words, Professor, the requirements for these experiments are all different from one another, involving and depending on different conditions, reactants and treatment.

"Presumably, these experiments are intended to simulate prebiotic conditions and demonstrate how purely natural processes could lead to the synthesis of organic compounds that have potentially important implications for origin-of-life issues. However, just as was true in Miller's original origin-of-life, I'm having trouble understanding how these experiments simulate actual prebiotic conditions and processes.

"For example, Dr. Yardley, do we have any way of telling how prevalent such materials as urea, malic acid, paraformaldehyde, ferric chloride, cyanoacetylene, cyanogen, fatty acids, phosphate, and glycerol would have been in the Archean era?"

"We believe," answered the professor, "that most of the compounds you listed would have been available, some more so than others, during the Archean era. Most of these compounds are extremely simple in structural formula, and we believe they would have been formed relatively easily through natural chemical processes going on during that period of time."

"Dr. Yardley, correct me if I am wrong, but fatty acids are hardly simple hydrocarbons." Referring to the sheets in his hand, he added: "Let's see ... palmitic acid, which is one of the most abundant saturated fatty acids, has a formula of $CH_3(CH_2)_{14}COOH$. Oleic acid, which is one of the most common unsaturated fatty acids, has a formula of $CH_3(CH_2)_7CH:CH(CH_2)_7-COOH$."

"Wouldn't you agree, Professor, that oleic acid and palmitic acid have considerably more complexity than hydrogen cyanide (HCN), ammonia (NH_3) and methane (CH_4)?"

"Yes," Dr. Yardley acknowledged.

"I believe," suggested the defense counsel, "that in your direct examination testimony you said the Fischer-Tropsch reaction was involved in bringing about some of the steps necessary for the formation of fatty acids. Is my recall on this matter accurate, Dr. Yardley?"

"Yes, it is," stated the professor.

"Would you please review once more for the members of the jury, Professor, the general nature of the Fischer-Tropsch process," requested Mr. Tappin.

"One takes a gaseous form of carbon, like carbon monoxide (CO)," the professor explained, "together with water vapor, and, then one passes these over a hot iron-powder catalyst, at temperatures between 180 and 300 degrees Celsius and under anywhere from one to fifty atmospheres of pressure."

"Will one have fatty acids at the end of this process?" asked the lawyer.

"No," replied the professor. "After the foregoing procedure has been run, one must find a way to oxidize the hydrocarbon chains that have been generated by means of the Fischer-Tropsch mechanism."

"In your opinion, Dr. Yardley," asked the defense counsel, "how likely would a naturally occurring counterpart to the Fischer-Tropsch reaction be?"

"The fairest thing I can say" the professor suggested, "is that a naturally occurring counterpart to the Fischer-Tropsch reaction is extremely unlikely but not entirely inconceivable. When one adds to this the requirement of a further oxidation step, one is really pushing the envelope of credibility to the outer limits."

"In the Oró experiment mentioned earlier," indicated the lawyer, "from which phospholipids were synthesized -- two further ingredients were needed in addition to fatty acids ... namely, glycerol and phosphate. How available were these molecules likely to have been in prebiotic times?"

"This is hard to say. The structural formula for glycerol is $C_3H_8O_3$ and is normally formed from the decomposition of natural fats by means of an alkali compound or superheated steam.

"There might have been some series of natural chemical reactions during prebiotic times that was capable of synthesizing glycerol. The structural character of this compound is not so complex that the act of assuming the existence of such a hydrocarbon during the Archean era strains credibility.

"A phosphate, on the other hand, is produced by combining an alcohol group with any one of three phosphoric acids. For instance, orthophosphoric acid, which is quite stable, has the formula H_3PO_4.

"Phosphorus, one of the main ingredients of phosphates and phosphoric acids, is a fairly rare non-metallic element. Even at the best of times there are only trace amounts of phosphorus to be found in seawater, and the presence of phosphorus in the Earth's crust is quite limited relative to elements such as magnesium, iron, calcium, potassium, sodium and silicon.

"Phosphates are very rare in nature, although human beings are quite adept at dumping huge quantities of these compounds into the environment. However, as far as prebiotic times are concerned, there would be no obvious, plentiful source of phosphates, and, therefore, phosphates would not have been readily available to support, in a rigorous fashion, any reaction requiring them during the Archean era.

"This does not mean there were no phosphates in prebiotic times. It merely means their relative scarcity would have placed constraints on where, when, and how frequently phosphate-dependent reactions could have proceeded."

"Dr. Yardley, could one fairly say," inquired the lawyer, "that the plausible likelihood of not only producing, but, as well, bringing together, fatty acids, glycerol and phosphates in order to synthesize phospholipid compounds under prebiotic, Archean era conditions is seriously in question?"

"Yes," the professor agreed, "I think one would not be unfair if one were to characterize the situation in this fashion. This doesn't necessarily mean the whole thing is completely impossible, but at this point in time, in the light of what is known, many researchers can't imagine any series of plausible steps during prebiotic times that, one, would have led to the formation of the individual reactants involved in phospholipid synthesis, or, two, would have resulted in these ingredients coming together to make such a reaction possible."

"Therefore," reasoned the defense counsel, "to call Oró's phospholipid synthesis experiment a simulation that accurately reflects what went on under the Archean era's prebiotic conditions is really, potentially, quite misleading. Would you agree with this, Dr. Yardley?"

"Let's just say" the professor offered, "the indicated potential to be misleading is present, and one cannot treat the natural, prebiotic synthesis of glycerol, phosphates, fatty acids or phospholipids as foregone conclusions. At best, the issue lends itself to being highly contentious and argumentative."

"Dr. Yardley, let's return to the Fox polymerization experiment for a moment," Mr. Tappin suggested. "A recipe was used in that experiment that called for a variety of amino acids to be thrown into a mixing bowl of sorts. Subsequently, these ingredients were heated for some 3-4 hours in an oil bath at 170 degrees Celsius.

"In your direct examination testimony, Professor, you indicated many researchers believe the exposed surface of a sandy beach, or a mineral bed, or a strip of solidified lava, where temperatures might have reached up to 100 degrees Celsius, might have served as a crucible for certain condensation reactions during the Archean era. In another portion of your testimony, you spoke about hydrothermal vents in which the temperatures were in the vicinity of 350 degrees Celsius, but these took place under water, not in oil.

"You didn't specifically speak about the conditions around volcanoes in your testimony, Professor. Yet, since neither of the previously-mentioned possibilities really matches the required conditions of the Fox experiment, can one assume that, perhaps, the area in and around certain volcanoes is the only other candidate that, conceivably, might fit into the kind of scenario that Fox's proteinoid experiment is purporting to simulate?"

"Volcanic areas," the professor said, "seem to be the only possibility that comes readily to mind."

"Would you agree, Dr. Yardley," inquired the lawyer, "that finding a place in volcanic areas that provided an oil bath of precisely 170 degrees Celsius for just 3-4 hours would be ... let's be kind here ... a tricky project?"

"Yes," responded the professor, "I guess one might not find many places capable of meeting these precise conditions, but this is not the same thing as saying that these sorts of conditions couldn't or didn't, exist."

"Professor Yardley, in your testimony concerning the Fox experiment, you mentioned, I believe," recalled the defense counsel, "that not all of the bonds that linked together the amino acid monomers or units were peptide in character ... that only some of these bonds were peptide in character. Is this correct?"

"Yes," the professor replied.

"In living organisms on Earth, peptide bonds," the lawyer stipulated, "occur between the amino and carboxyl groups of neighboring amino acids, binding them together to form proteins. Isn't this so, Dr. Yardley?"

"That's right," the professor confirmed.

"Therefore," concluded the defense counsel, "the amino acid polymers or chains in Fox's experiment are not really proteins because they are not what we find in living organisms. Presumably, for precisely this reason, the polymers in Fox's experiment are called proteinoids and not proteins. Is this a fair way of putting things, Dr. Yardley?"

"I guess so," admitted the professor.

"Did any of these proteinoids exhibit substantial enzymatic characteristics?" inquired Mr. Tappin.

"Not really," the professor stated. "On the other hand, there might not be anything that prevents proteinoids from playing the other major role of proteins involving the morphology ... that is, the form and structure ... of organisms.

"Conceivably, a variety of ribozymes ... in other words, polymers of RNA with enzymatic properties ... might have served as the early enzymes of the protocell. Proteinoids could have filled the function of helping to give form to these protocells or to various organelles such as ribosomes or mitochondria, within the protocell."

"Is it not the case, Dr. Yardley," queried the lawyer, "that the bonds, whether peptide or otherwise, formed during condensation reactions in which water is removed from neighboring monomeric amino acids and, therefore, are called anhydride bonds ... isn't it the case these anhydride bonds are quite labile and, relatively speaking, easily broken."

"Yes, under certain conditions, this is true," the professor acknowledged.

"Would you agree, Dr. Yardley," asked Mr. Tappin, "that volcanic areas in which temperatures are 170 degrees Celsius, or higher, for prolonged periods of time, might be considered to have met the requirements alluded to by you through your use of the qualification: 'under certain condition', with respect to the labile nature of peptide bonds among amino acids?"

"Yes," admitted the professor.

"Are we not encountering here," wondered the lawyer, "yet another instance in which, under certain conditions, energy might be coupled to chemical reactants for short periods and in specific ways, to forge more complex arrangements of hydrocarbons, but when, under other circumstances, these same forms of energy can quickly turn the tables on the products of such reactions and, as a result, undo what these energy forms previously had helped to bring about?"

"Yes, this is a possibility," the professor agreed.

"In describing the Fox proteinoid polymerization experiment, Dr. Yardley, you said that, by playing around with the time and temperature variables, Fox was able to incorporate more neutral and basic amino acids into the proteinoid polymers synthesized through condensation reactions. Is this right?" Mr. Tappin inquired.

"Yes," affirmed the professor.

"In effect, Dr. Yardley, doesn't this mean," pressed the lawyer, "that if we are to consider the Fox experiment to be a simulation of Archean era conditions, then, not only must we assume there were specialized pockets in which amino acids could gather together in an oil bath for 3-4 hours at precisely 170 degrees Celsius, but there were also other pockets in these volcanic areas in which amino acids could be bathed in oil for slightly less, or slightly more, than 3-4 hours, at temperatures that were somewhat higher, or somewhat lower, than 170 degrees Celsius so that proteinoids with greater numbers of neutral and basic amino acids could be incorporated into these polymer chains?"

"Yes," stated the professor. "We believe the entire Archean era world was a prebiotic version of a modern laboratory in which there were many different kinds of evolutionary niche being explored. In these various pockets, millions, if not billions, of different sorts of experiment were being run across the several hundred million years required for protocells or primitive organisms to emerge.

"At this time, I should add," Dr. Yardley indicated, "there have been experiments in which polypeptide polymers have been observed to form in the absence of water when mixtures of amino acids were incubated at a temperature of 65 degrees Celsius for a period of 40 days. So, one doesn't have to be tied to the 170-degree Celsius figure of the early proteinoid experiments."

"Wouldn't you agree, Professor Yardley," suggested the lawyer, "that finding a little corner of the Archean era world that will allow one to incubate a mixture of amino acids at 65 degrees Celsius for precisely forty days, twenty-four hours a day, no more or no less, is really only a variation on the problem that is being discussed?"

"I suppose so," the professor responded, "but this latter experiment does introduce a broader spectrum of possibilities into the picture."

"Let us assume, for the moment," Mr. Tappin proposed, "that, as a result of some of the points brought out previously under cross-examination, the amino acids used in the simulation experiments of Fox, or this more recent 65 degree/40-day experiment, were not forthcoming from Strecker synthesis in the Archean era ocean. Given this assumption, how would these amino acids find their way into the mixing bowl pockets or crucibles of the different volcanic areas?"

"As a number of experiments have indicated," the professor stated, "there are a variety of alternative pathways to amino acid formation other than Strecker synthesis."

"Would," inquired the lawyer, "urea [$CO(NH_2)_2$], malic acid ($C_4H_6O_5$) and paraformaldehyde [$(CH_2O)_3$] ... which are just three of the reactants used in laboratory experiments in order to help synthesize a few, specific amino acids ... would these compounds have been readily available in the Archean era?"

"How readily the various compounds cited by you would have been available might be an issue of some debate," the professor offered, "but we believe there was a reasonably good chance such compounds would have been synthesized under various conditions during the Archean era."

"Would this last answer remain the same, Dr. Yardley," queried the defense counsel, "if one were to raise the same kind of question in conjunction with cyanogen (C_2N_2) and cyanoacetylene (HC_3N) that, together with hydrogen cyanide (HCN) have been used in laboratory experiments to synthesize nucleic bases such as uracil, cytosine, guanine and thymine?"

"Yes, my last answer would remain substantially the same," the professor stated.

"Dr. Yardley, do any of the alternative pathways to which we have alluded produce all of the amino acids?" Mr. Tappin asked. "In other words, in accordance with what has been established previously through testimony and cross-examination, aren't these pathways frequently quite specific in terms of the reactants, temperatures, and conditions that are necessary to generate certain kinds of amino acid?"

"This is often the case, yes," the professor confirmed, "but not always. Some methods have produced a number of different amino acids by varying the experimental conditions slightly, although, as you have indicated, no one method has generated all of the amino acids."

"If no one method has generated all of the amino acids," hypothesized the lawyer, "could one reasonably argue there might have been some physical distance that might have separated these pathways from one another since these alternative pathways often presuppose different precursor reactants, different temperatures, and so on?"

"I guess one could argue in this fashion," the professor acknowledged, "but I don't think one can assume great distances were necessarily involved. Many of these reactions could have happened in, and around, the same volcanic areas."

"Alternatively, Professor," the defense counsel pointed out, "one cannot necessarily assume relatively great distances were not involved either, can one?"

"No, one can't," Dr. Yardley conceded.

"If," Mr. Tappin postulated, "one assumes the Strecker synthesis process, followed by tidal movement to intertidal zones, was not the primary means of delivering amino acids to places where condensation reactions could take place, is there a secondary or backup account of how amino acids generated from different chemical pathways and under different conditions would have come together in Fox's prebiotic mixing bowl?"

"I suppose," the professor replied, "one would have to speak in terms of chance, random processes in order to account for how these kinds of events might be possible."

"Is this an explanation, Dr. Yardley, or an assumption?" asked Mr. Tappin.

"In other words, if one has no reliable baseline from which to construct distribution models that permit one to demonstrate how a series of unrelated and complex events might reasonably be anticipated to come together, what exactly is being explained? Isn't one merely assuming something has happened in a particular way and labeling that assumption with the name of 'chance events'?"

"Not entirely," the professor asserted. "If one were to take a large enough group of monkeys and put them together with a sufficiently large set of typewriters, then, mathematically, one could predict, with a fair amount of confidence, that, sooner or later, one of the monkeys would type a perfect copy of, say, Hamlet."

"What about," the lawyer wondered, " *The Glass Bead Game* by Hesse or, since we seem to be dealing with science fiction here, something by Isaac Asimov?"

"Objection Your Honor," Mr. Mayfield stated. "Learned counsel is being rather frivolous in his questioning at this point."

"Your Honor," Mr. Tappin countered, "since I have encountered the witness' argument before, under other circumstances, and since the example of Hamlet was often the work cited in this kind of argument, I was curious as to whether these monkeys were stuck in some sort of creative rut and were unable to write anything else."

"As was true in the case of the proverbial cat with the same propensity," Judge Arnsberger replied, "this sort of curiosity is not likely to have a long life time in my courtroom. You've made your point, Mr. Tappin, let's move on. The prosecution's objection is overruled."

"Your Honor," asked Dr. Yardley, "may I be permitted to answer the question?"

"Certainly," the judge responded, "but you are under no obligation to do so."

"I understand, Your Honor," acknowledged the professor, "but, nevertheless, I would like to address the question."

Turning back toward the defense counsel, the professor said: "In theory, there is no limit on the nature of the books that could be produced by these monkeys. So, Hesse's work or the *Foundation* series by Asimov, both would be possibilities, or, if you like, you can even throw in some Raymond Chandler."

"Dr. Yardley," inquired Mr. Tappin, "wouldn't one be able to predict, with considerably more confidence, and based on empirical evidence rather than on a mathematics rooted in contentious and unprovable assumptions, that, sooner or later, all of the typewriters would be destroyed, all the paper would have been used up, and the monkeys would have been dead long before so much as the thought, let alone the typed reality, of even a coherent paragraph of any kind would have occurred to these monkeys, whether considered collectively or individually?"

"Objection, Your Honor," Mr. Mayfield interjected.

Before Judge Arnsberger could speak, Mr. Tappin announced: "I'll withdraw the question, Your Honor."

"Let's assume," postulated the defense attorney, "the mathematical theory to which you are alluding is true. How large would the set of typewriters and group of monkeys have to be in order for a copy of, for example, Hamlet, to get written by one of the monkeys, and how long would all of this take?"

"We are dealing here with the mathematics of the infinite," stated the professor. "If one had an infinite number of typewriters, monkeys and paper, then, at some point, Hamlet would emerge.

"The interesting possibility in all of this is that, given such starting assumptions, Hamlet might very well get written within a finite length of time since there is no way to pin down where in the infinite series of events the desired copy of Hamlet would be forthcoming. The book might appear after 10,000 years or 10,000,000 million years or 100,000,000 million years, and even though these numbers are very large, they are finite, and, more importantly, they are reminiscent of the sort of time considerations involved in origin-of-life issues."

"This mathematical theory, Professor, seems to be assuming," Mr. Tappin suggested, "that in any given single striking action, all keys of the typewriter have an equal opportunity of being struck by any given monkey, with no single striking trial having any influence on the striking actions that precede or follow it. In other words, each striking action of the moment is entirely independent from all other striking actions, whether performed by the same monkey or by other monkeys. Would you agree with this Dr. Yardley?"

"Yes, I suppose so," the professor agreed.

"Your mathematical theory appears to be assuming, as well," the defense counsel continued, "that every possible sequence of key-striking events, eventually, will be represented by the activities of the monkeys. Furthermore, since the sequence of key-striking events that makes up or constitutes the work of Hamlet would be one such set of sequential key-striking events, then, one has opened the door for the possibility that at least one of the sets of independent key-striking events will give expression to a sequence that matches Hamlet word for word.

"Is the foregoing a fair way of describing the situation?" the lawyer asked.

"I believe" replied the professor, "the reasoning of the theory runs, more or less, along the lines you have indicated."

"Has anyone tested this mathematical theory empirically?" inquired the defense counsel.

"I'm not quite sure what you mean," the professor said.

"Has anyone, for instance," Mr. Tappin specified, "attempted to determine whether or not the assumption of independence with respect to key-striking action is warranted in the context of the activities of real rather than theoretical monkeys? Or, has anyone tried to discover whether all sets of sequential key-striking activity are equally represented or whether some sets are over-represented or underrepresented?"

"No, I don't think anyone has tried any of what you are suggesting," the professor responded.

"Has anyone attempted to discover," queried Mr. Tappin, "whether monkeys would continue to type from hour to hour, day to day, week to week, and month to month as a demonstration of their capacity, in principle, to be able to produce any kind of effort that would be comparable in length to the work of *Hamlet?*"

"Not really," answered the professor.

"Dr. Yardley, were there an infinite number of molecules on the surface, or in the atmosphere, of the Archean era Earth?" asked the lawyer. "Or, were there an infinite number of chemical reactions that went on during the Archean era? Or, was there an infinite amount of energy available to run those reactions?"

"No, of course not," the professor said.

"Then," Mr. Tappin proposed, "what might, or might not, happen in a universe of infinite monkeys, typewriters and paper, really doesn't constitute an appropriate way of modeling objects, processes and events that are finite in nature, does it?"

"Perhaps not," admitted the professor, "but the basic principle is, nonetheless, suggestive. Given large numbers of even finite chemical events, then, certain kinds of events might become more likely over the long run, although these same events might appear to be very unlikely in the short run."

"Wouldn't the projected likelihood of such events depend on the nature of those events?" the defense lawyer inquired. "Wouldn't one have to be able to provide some good reason why, in the long run, one might reasonably expect events with a specific character to occur that one would not anticipate would take place in the short run?

"More specifically, Professor, do we really have any reasons aside from, or independent of, the vague notion of chance events, which would permit us to suppose that in the long run we reasonably can expect a bunch of amino acids that are generated through different pathways and under different circumstances to all end up in the same place at the same time? Moreover, if we don't have anything independent of the notion of chance, random events with which to work, then, aren't we back where we started ... namely, isn't this a matter of assumption rather than a matter of scientific proof or demonstration?"

"Your Honor," stated Mr. Mayfield, "I must object. This question already has been asked of, and answered by, the witness. We are going over the same ground."

"Overruled," Judge Arnsberger proclaimed. "I'm going to allow the question."

Dr. Yardley was silent for about ten seconds or so. When he spoke, he said: "Stochastic models provide a way of setting parameters without presupposing any particular kind of metaphysics or ontology. These models offer an opportunity to explore and analyze what does happen against frameworks of expectation and anticipation based on the general properties and characteristics of natural phenomena.

"To say that some given event ... such as the coming together, at some point in time and space, of a variety of amino acids generated through separate pathways and conditions ... has a finite, although small, possibility of occurring is doing nothing more than to recognize that real events often are capable of reflecting different aspects of our stochastic models. The perfect bridge hand, or throwing 'x' number of consecutive passes at the gaming tables, or winning a lottery against huge odds, and so on, constitute, as far as our stochastic models are concerned, very rare events, but they do happen.

"In fact, the more runs of any given activity that take place, the greater, in general, will be the likelihood of seeing theoretical possibilities being realized or manifested in actual circumstances that one would not expect, on the basis of one's stochastic model, to occur with any degree of frequency. Although the chemical events taking place during prebiotic times might not have been infinite in number, nevertheless, the number of such reactions over the course of four to eight hundred million years is incredibly high.

"Given such large numbers, one might expect, at some point, that certain kinds of improbable events have a chance of taking place. I don't consider such an improbable event an assumption, however unlikely it might be, since its possible occurrence is rooted in a complex stochastic modeling process that acknowledges these kinds of event to be conceivable and capable of taking place in finite, real time."

"When you say, Dr. Yardley, that something is 'capable of taking place in finite, real time', are you saying," Mr. Tappin asked, "that this something must take place or necessarily will take place, or, are you merely saying the event in question could take place under the right circumstances?"

"I'm saying," the professor indicated, "that such an event could take place under the right circumstances and that such circumstances can be assigned some small, but finite, probability of actually occurring."

"What is the nature of this process of assigning some small but finite probability?" the lawyer asked.

"The nature of the assignment process would be shaped by the character of one's stochastic model," replied the professor. "Different models might assign different kinds of probability to this kind of situation."

"Are any of these assignment procedures based on empirical data?" inquired Mr. Tappin.

"Yes, they could be," the professor stated. "It depends on what one is talking about."

"How about," proposed the lawyer, "the coming-together of twenty left-handed-amino-acid-isomers of the sort that are observed to occur in Earth organisms?"

"Well," began the professor, "one would have to figure out how many different kinds of amino acids could have been synthesized under prebiotic conditions. One, then, might, or might not, multiply that number by two, depending on whether one believed ultraviolet light had been polarized slightly in its passage through the Earth's atmosphere and, as a result, had a tendency to decompose right-handed amino acid isomers.

"One also would have to try to work out frequency distribution tables for the different kinds of amino acids, including the 20 in which you are interested. These frequency distribution tables would depend on such things as production efficiency yields and energy efficiency yields for the various stages of amino acid formation that we discussed earlier in the context of the Strecker synthesis process.

"In addition, these frequency distribution tables would have to reflect, in some way, how many amino acids came from extraterrestrial sources. On the other hand, one would have to factor in losses due to pyrolysis, hydrolysis, photolysis, absorption by various clay materials, and so on.

"When one took all of these factors into consideration, one would be in a position to calculate theoretical values about what proportion of the total set of amino acids in existence at any given time were represented by the 20 left-handed amino acids you mentioned. This would provide some sort of stochastic baseline to apply to the real world and from which one's expectations concerning these possibilities would arise.

"Professor Yardley, has anyone worked all this out?" Mr. Tappin asked.

"Models have been developed that take various combinations of these factors into consideration," Dr. Yardley answered. "However, to the best of my knowledge, no one has taken all of these factors into consideration. At this time, we simply don't have the software, models, and computers capable of handling the complex dynamics that result from the interaction of all these variables."

"Would one, therefore, Professor, be incorrect in saying there is no complete model of what went on during the Archean era as far as amino acid formation is concerned?" inquired the lawyer?"

"No, this would not be incorrect," acknowledged the professor. "On the other hand, the very essence of science is a constant process of improving, revising, updating, modifying, and, sometimes, rejecting the models that are being constructed.

"Science doesn't purport to have the final answers," added the professor. "It is a work-in-progress, and, as such, it attempts to do the best it can with the materials that are available to it.

"As new material, techniques, ideas, and methods have become available the evolutionary model has been able to improve upon its past performance. The revisions and modifications that have come through this process of gradual, conceptual evolution have created a more rigorous model, but we continue to seek to improve it."

"Given what you have just said," hypothesized the defense counsel, "would one be fair, Dr. Yardley, if one were to say the following? If one does not wish to call the assignment of a probability concerning the likelihood of a bunch of amino acids coming together in the general vicinity of some volcano an 'assumption', then, could one fairly say the stochastic model responsible for assigning probabilities in this case stands in need of considerable revision?"

"I don't have a problem with this way of stating things," the professor indicated.

"Dr. Yardley, in all of our discussions up to this point, concerning the different kinds of experiments that have been conducted in relation to origin-of life issues, is it not the case that the various experiments were run with purified compounds under conditions in which there was no chemical competition going on among different kinds of compounds to determine which compounds would form covalent bonds with which compounds?" Mr. Tappin inquired.

"I would say so, yes," the professor replied.

"Would you agree, then, Dr. Yardley," asked the lawyer, "that one might have difficulty understanding how simple condensation cycles of heating and drying might bring about a very selective synthesis of pure polymers, such as proteins, DNA, and RNA -- with the right kinds of bonds, optical activity, and monomer composition -- from amongst the highly complex mixture of hydrocarbons that might have been available as reactants in the Archean era world?"

"I would agree there is a challenge here for evolutionary theory," admitted the professor, "because there still are quite a few things we don't, yet, understand. I would not agree this challenge necessarily constitutes an insurmountable barrier to our being able to understand these issues eventually.

"Our knowledge base," pointed out the professor, "is developing exponentially. Furthermore, the interim periods required for our knowledge to double are becoming increasingly shorter.

"Phenomena that were inexplicable a few years ago are now being understood. To acknowledge the existence of a problem or challenge is to participate in the natural order of things in the world of science."

"Professor, consider the following hypothetical situation," requested Mr. Tappin. "Suppose there were a relatively dilute, Archean era, ocean solution of phosphates, carbohydrates, pyrimidines, purines, fatty acids, amino acids, and various kinds of other simpler hydrocarbons.

"Let us further suppose, Dr. Yardley," added the lawyer, "that some of this seawater solution finds its way, via tides and the wind, to some intertidal zonal, or lava, surface. What is likely to happen once this dilute solution starts to get heated from the sun and/or volcanic-related activity?"

The professor considered the hypothetical situation briefly and began to speak. "Probably, as evaporation proceeded, then, at some point, sodium chloride crystals would form. Bivalent cations, or positively charged ions and radicals, would interact with organic anions, or negatively charged hydrocarbon groups. Finally, there would be a very large number, and variety, of covalent bonds that would join together different functional groups in virtually every conceivable combination."

| Evolution Unredacted |

"Would you expect," the counsel for the defense inquired, "that such a mixture of ions and covalent bonds would organize itself thermodynamically into a working protocell?"

"If you are asking me," posited the professor, "whether I would expect something interesting to happen in the single exposed lava surface or intertidal puddle that is being examined hypothetically, then I would have to say no, I would not expect such a mixture to organize itself into a working protocell. However, if you were asking me about my expectations in relation to billions of such exposed surfaces and/or intertidal puddles, then I would have to say, yes, I would begin to feel confident in my expectations that at least one of these prebiotic crucibles would be capable of thermodynamically and kinetically organizing itself into something very interesting as far as the origin-of-life issue is concerned."

"Do we have anything," the lawyer queried, "beside your rising level of felt confidence in such expectations that is likely to persuade us there is something inevitable or necessary about the possibility that, at some time and at some place, there must be a protocell that must emerge from the prebiotic mists? After all, Professor, if you are relying on billions and billions of exposed lava surfaces and intertidal puddles to give rise to at least one interesting protocell or near-protocell, then, the prima facie odds against this sort of event happening are billions and billions to one, wouldn't you say?"

"As I indicated earlier," the professor replied, "the more opportunities there are for experimentation with different combinations of possibility, then, the greater is the probability that one of these sets of combinations will possess and exhibit the sort of characteristics and properties in which one is interested as far as origin-of-life issues are concerned."

"Dr. Yardley, you seem to be assuming," the lawyer suggested, "that all of these billions and billions of prebiotic crucibles will necessarily be exploring all conceivable possibilities. However, what guarantee do we have that these mini-laboratories, even if they are in the hundreds of billions and trillions, will be sufficient to explore all the possible combinations available to the molecules in the dilute solutions that have washed up on various exposed surfaces or into some intertidal puddle?"

"Naturally," responded the professor, "there can be no such guarantee."

"Moreover," Mr. Tappin continued without pausing, "what guarantee do we have that even if, on the basis of thermodynamic theory, a given combination is considered possible that, therefore, from a kinetic perspective, every such thermodynamically conceivable combination will actually occur."

"Again," said the professor, "there can be no guarantee in such matters."

"Or," the lawyer added, "how do we know there won't be a tendency in such mini-laboratories, due to various thermodynamic or kinetic considerations, to repeat, again and again, some finite, but large, set of prebiotic experiments at the expense of other possibilities, and, in the process, consume a great deal of the resources of materials, space, energy and time that are available?"

"All I can say," remarked the professor "is that, in general terms, you have raised a number of valid issues that need to be addressed. However, the fact these problems have been raised doesn't preclude the possibility of discovering either answers to your challenges or of finding ways that open up the possibility of side-stepping or circumventing these problems in some way."

"At the present time, Dr. Yardley, does evolutionary biology have any remotely satisfying answer for the problems being raised here -- yes or no?" specified the lawyer.

"I would have to say no," answered the professor.

Returning to the defense table, Mr. Tappin went through the, by now, well-established ritual of exchanging new material for used material with his colleague. As the lawyer turned toward the witness, he began speaking.

"Professor Yardley, during an earlier part of cross-examination, we talked about the difficulty of plausibly accounting for the generation, and bringing together, of compounds such as fatty acids, phosphates and glycerol in order to try to synthesize phospholipids, one of the primary components of many kinds of cell membrane. Before proceeding to talk about cell membranes in a little more detail, there is one further point that I would like to address.

"I believe phosphatidic acids are the simplest class of phospholipids," the lawyer said. "Is this correct?"

"Yes," replied the professor.

"Moreover," Mr. Tappin added, is it also the case that derivatives of phosphatidic acid, such as lecithin, tend to exist in cells primarily in an optical isomeric form that is in a left-handed rather than in a right- handed isomeric configuration?"

"That's right," the professor confirmed.

"Consequently, Professor," Mr. Tappin concluded, once again, evolutionary theory is confronted with the problem of having to come up with an explanation for how such a preference arose with respect to optical isomers, just as in the case of proteins, as well as of ribonucleic acids. Would you agree with this assessment of the situation?"

"Yes, I would," acknowledged the professor.

"If, Dr. Yardley, as presently seems to be the case based on present knowledge, there is no readily apparent, natural pathway by which to generate phospholipids, how do evolutionary biologists propose to account for the development of cell membranes?" inquired the lawyer.

"There are a number of different possibilities," the professor stated. "In my earlier testimony, I touched on a number of these, including carbonaceous chondrites, proteinoid micro spheres and transitional liposome-like structures."

"Would you expand a little, Dr. Yardley, on the possible role of carbonaceous chondrites with respect to cell membrane formation?" the lawyer requested.

"There are several ways to look at the findings vis-à-vis carbonaceous chondrites," the professor began. "One of these ways involves the discovery of amphiphilic compounds, and the other possibility deals with the hydrocarbons that are found in some of these meteorites.

"Amphiphilic compounds," explained the professor, "have both: hydrophilic, or water-loving, as well as hydrophobic, or water-hating, components. These compounds have been observed to spontaneously form membraneous-like boundary structures when placed in an aqueous environment.

"When placed in water, the hydrophobic parts of these compounds tend to curl up in order to minimize contact with water. In the process of curling up, a vesicle or protected, interior space is created within which various kinds of chemical reaction might take place under the right circumstance."

"Dr. Yardley, before you continue on," Mr. Tappin interjected, "I would be interested to know if tests have been conducted to determine if these amphiphilic compounds exhibited any phospholipid-like properties?"

"Samples of these compounds were studied by means of an electron microscope," responded the professor. "One of the purposes of this analysis was to determine if a membranous structure was present in these compounds.

"These studies did detect the presence of a membranous structure approximately 10 nanometers, or 10 billionths of a meter, in thickness. This is consistent with the upper boundary size of the cell membranes of many organisms.

"In addition to the electron microscope studies, tests were performed in order to examine the ability of these membranous structures to encapsulate polar solutes, or water-soluble molecules, in a manner that was the same as, or similar to, cellular membranes in living organisms. A dye was used in this study and the researchers found that the amphiphilic material from carbonaceous chondrites had the ability to encapsulate polar solutes with approximately one-tenth of one percent of the encapsulation efficiency of the phospholipids found in living organisms."

"In other words, Dr. Yardley, although these extraterrestrial compounds could form membrane-like structures with about the same thickness as the cell membranes of living organisms, they were almost nothing like phospholipids in this, presumably, important area of being able to encapsulate polar solutes. Is this correct?" the defense counsel asked.

"Essentially, yes," the professor responded.

"You also mentioned, Professor, the hydrocarbon-related possibility associated with the carbonaceous chondrites," the lawyer said. "What exactly does this involve?"

"Around 1970," the professor pointed out, "several researchers studied seven carbonaceous chondrite meteorites. They discovered chains of hydrocarbons consisting of between 10 and 23 carbon atoms - a finding that was consistent with what also had been observed in the Murchison meteorite.

"This is comparable, in some respects, to the 12 to 20 carbon atoms contained in fatty acids, one of the main components of the lipids found in the phospholipids that make up most cell membranes. In the absence of any plausible natural prebiotic method of synthesizing fatty acids, such chains might have served as a source for the type of hydrocarbons that make up fatty acids in lipids and, therefore, cell membranes."

"Dr. Yardley, isn't it the case," asked the defense counsel, "that fatty acids contain chains of hydrocarbons consisting of even numbers of carbon atoms?"

"That's right," the professor acknowledged."

"Therefore," said the lawyer, "not only are some of the hydrocarbon chains, ranging in length from 10 to 23 carbon atoms, which are found in the meteorites, both too short or too long, relative to those hydrocarbon chains that range in length from 12 to 20 carbon atoms that are found in fatty acids, but if the meteorite hydrocarbon chains contain odd numbers of carbon atoms, then, this would be another dissimilarity between the meteorite hydrocarbons and fatty acid hydrocarbons. Is this correct, Dr. Yardley?"

"Yes," the professor replied.

"In effect, if one tried to view these differences in the best possible light," stated the lawyer, "one would have to assume that, somehow, carbon atoms either would have to be added to, or removed from, many of the hydrocarbon chains found in the meteorites. Would you agree with this, Dr. Yardley?"

"This seems reasonable," indicated the professor.

"Furthermore," Mr. Tappin continued, "isn't it the case that the hydrocarbon chains found in the meteorites would have to be oxidized before those hydrocarbon chains, with the right lengths of even numbered carbon atoms, could be considered to be fatty acids?"

"Most probably," the professor answered.

"In addition," Mr. Tappin pressed, "even if one were to concede that fatty acids might arise in the prebiotic Archean era world in this extraterrestrial fashion, one still would have to find a way to bring these fatty acids together with phosphates and glycerol, under the right conditions, in order to synthesize phospholipids. And, given that phosphates, in particular, are likely to be extremely rare compounds in the Archean era, then, Dr. Yardley, wouldn't one have to consider this whole sequence of events to be very, very improbable?" the lawyer asked.

"I imagine this would be the case," affirmed the professor.

"Finally, in the light of previously established testimony," the lawyer stipulated, "one cannot assume meteorites would represent a very substantial source of these kinds of hydrocarbon chains, nor can one assume these hydrocarbon chains necessarily would survive post- impact, prolonged exposure to ultraviolet photolysis or, perhaps, even heat, in the form of, possibly, relatively high surface temperatures or volcanic activity. Isn't this so, Professor?"

"Yes," Dr. Yardley agreed, "one cannot assume these sorts of thing to be automatic or given."

"As far as the possible role of proteinoids is concerned in relation to membrane functioning," the defense counsel queried, "is there any evidence, Dr. Yardley, that proteinoids have the necessary properties to form active transport systems, or establish ion pump mechanisms, or to provide transmembrane channel ways, as proteins do in the membrane complexes of living organisms?"

"At the present time, I believe there is little, if any, evidence to suggest proteinoids have the kinds of capability to which you are referring," replied the professor. "Nevertheless, the absence of evidence in the few laboratory experiments that have been performed to date does not preclude the possibility that during the Archean era, proteinoids with some of these sorts of functional capacity might have been synthesized naturally."

"Is there any evidence, Dr. Yardley, that the proteinoids have the necessary sort of sequential arrangements of hydrophobic and hydrophilic amino acids that, upon folding into their tertiary or folded structure by means of thermodynamic forces, will enable their folded hydrophobic portions to be located in the interior portions of the phospholipid bilayer and, consequently, match up with the hydrophobic hydrocarbons of the lipid molecules, as is the case in the transmembrane proteins of living organisms?"

"At this time, I know of no such evidence," Dr. Yardley admitted.

"Previously, Professor," pointed out Mr. Tappin, "you talked about liposomes. You described them as small vesicles made up of lipid bilayers that might have served as a transitional membrane-like structure.

"To talk about liposomes, of course, is assuming that the issue of lipid formation had been resolved in the Archean era. Would you agree with this, Dr. Yardley?"

"Yes," the professor said.

"While elaborating on the structural character of liposomes," said the lawyer, "you spoke about properties such as the ability to reverse breakage of the bilayer by spontaneously resealing any gaps that occur as a result of, say, mechanical agitation or shaking. In addition, Dr. Yardley, you mentioned liposome properties such as being able to trap solutes that might happen to be nearby when these vesicles are dried, as well as liposome qualities of growth, division and multiplication - all of which are reminiscent of what goes on in living organisms.

"Growth, division and multiplication, Professor, all suggest having access to a supply, regular or irregular, of lipid molecules. Consequently, wouldn't you agree these properties of growth, and so on, all presuppose that additional lipid molecules will be available that, in turn, means that, once again, the question of lipid availability in the Archean era would have to be addressed?"

"Yes," iterated the professor.

"Do liposomes control their own growth, division and multiplication, Dr. Yardley, or is this alleged growth, division and multiplication something that sometimes occurs to liposomes as a result of external forces impinging on the liposomes, or as a result of, say, osmotic lysis ... that is, the rupture of the liposome due to an inward diffusion of salt and water in the process of establishing an equilibrium between internal and external environments of a liposome?

"If," postulated the professor, "you are asking me whether the liposome can be said to be alive in some sense, then, clearly, the liposome cannot be described as being alive, nor does it control its growth, division and multiplication in the same sense that a biological organism actively controls these processes. On the other hand, the capacities of a membrane structure to reverse breakage, expand in size, and be able to participate in processes of division and multiplication, are fundamental stepping stones on the road toward becoming part of the life phenomenon."

"If my understanding on the matter is correct, Dr. Yardley, living cells are, within certain limits, able to maintain an internal electrical or ionic potential that is different from the surrounding environment. In fact, some people have suggested that the ability of a bounded, or membrane-enclosed, system to maintain this kind of differentiated energetic relationship with the environment is one of the most recognizable attributes of a living organism. Would you agree with this way of characterizing the situation?"

"Yes," the professor confirmed.

"Are liposomes capable of maintaining this kind of differentiated energetic relationship with the environment?" asked the lawyer.

"No," Dr. Yardley stated. "As I indicated before, liposomes are not living organisms."

"What happens, Professor, if some sort of potential difference arises between the internal and external regions of a liposome?"

"Lipid structures," Dr. Yardley stated, "tend to show considerable permeability to water, as well as a small amount of permeability to positively and negatively charged ions of low molecular weight, although these ions diffuse across the membrane at a rate that is about one billion times slower than is the case for water molecules. Therefore, whenever there is disequilibrium between the inner and outer environments of the liposome, osmotic diffusion occurs, and this tends to eliminate the disequilibrium.

"If these potential differences are slight, then, equilibrium might be re-established with no appreciable effect on the bilayer structure of the liposome's membrane. If the potential differences are great, say, in favor of the external environment relative to the internal environment of the liposome, then, the liposome will swell with the osmotic diffusion of ions and water into the vesicle's interior and, eventually, might undergo lysis or rupture."

"Dr. Yardley, what would a liposome-like structure need in order to get around this osmotic problem?" Mr. Tappin inquired.

"One would need," replied the professor, "either some kind of rigid wall capable of resisting the stresses of lysis, or one would need a system capable, as required by circumstances, of pumping ions in an out of the interior of the structure, or one would need some combination of rigid walls and an ion pump."

"When you say 'rigid wall', this, presumably, refers to things like cellulose in plants," queried the lawyer.

"Yes," the professor answered. "However, fungi, bacteria, and algae have evolved a variety of rigid structures besides cellulose to handle the problem of osmotic lysis.

"Some of these alternative strategies involve combinations of polysaccharide molecules that are different from cellulose. Other strategies for creating rigidity in membrane walls also have arisen, involving, for example, silica, lime and chitin ... an amorphous polysaccharide that is intermediate between proteins and carbohydrates ... in conjunction with, say, various carbohydrate matrices."

"Would one be fair, Dr. Yardley, if one were to say that phospholipids require the presence of particular kinds of protein in order to have ion pumping capabilities so that even if one were to assume phospholipids were laying around, so to speak, in the Archean era, nonetheless, the mere presence of phospholipids, in and of themselves, would not solve the osmosis problem?"

"Yes, that's right," indicated the professor.

"Therefore," the lawyer said, "attaching just any old kind of proteinoids, or even proteins for that matter, to phospholipids will not necessarily establish an ion-pumping capability, unless these proteinoids or proteins have the right kind of sequential, structural, and tertiary folding properties that are suited to transporting particular kinds of ions into and out of the membrane-enclosed structure. Is this right, Dr. Yardley?"

"I would say so," the professor replied.

"Presumably," Mr. Tappin hypothesized, "various kinds of proteinoids or proteins would be necessary to handle the transport or pumping of different kinds of ions such as sodium, magnesium, potassium, calcium, and so on. Would you agree with this, Professor?"

"Yes," Dr. Yardley said.

"This capacity of a membrane system to actively participate in accepting some things while excluding others is referred to as 'selective permeability', isn't it?" the lawyer asked.

"That's correct," acknowledged the professor.

"Besides ions, Dr. Yardley, what other kinds of capability," the defense counsel inquired, "would need to be actively included or excluded if a membrane-enclosed structure were to possess the full range of functional characteristics exhibited by the membrane systems of living organisms?"

"Organisms would need some means of actively transporting nutrients into the interior of the cell," the professor stated. "Simultaneously, organisms would need a means of not only getting rid of toxic materials that might be accumulating as a result of the catabolic and anabolic ... that is, respectively, the tearing down and synthesizing ... processes going on in the cell in relation to such nutrients, but there would have to be some way for this active transport system to be able to selectively differentiate toxic materials from metabolites being used in the cell."

"Presumably," the lawyer reasoned, "different kinds of transport mechanisms across the membrane and/or channel ways through the membrane would be needed in order to bring different kinds of nutrient into the cell, as well as carry various sorts of toxic material out of the cell. Would you agree with this, Dr. Yardley?"

"Yes, I would," affirmed the professor.

Mr. Tappin asked: "Why couldn't nutrients and toxic substances just enter and leave the cell by means of osmotic diffusion, in the same way water and low molecular weight ions do in liposomes?"

Dr. Yardley explained: "The phospholipid molecules that form the bilayer structure characteristic of membranes, constitute a hydrophobic permeability barrier to all hydrophilic, or water loving, materials, as well as to high molecular weight ions. Passive diffusion, or osmosis, will not carry those kinds of compounds across the permeability barrier formed by the phospholipid bilayer, and, therefore, active forms of transport must be used, or channel ways must be provided that will allow unimpeded passage through the hydrophobic interior of the bilayer membrane structure."

"What would happen," Mr. Tappin hypothesized, "if the nutrients transported across the membrane were not coordinated with the organism's ability to catabolically tear down, and then anabolically build up necessary molecules using these kinds of nutrient?"

"The organism would starve to death," responded the professor.

"In other words," continued the lawyer, "being able to actively transport nutrients across the membrane's permeability barrier is not enough. These nutrients must be of the right kind, and, therefore, would one be right in supposing, Dr. Yardley, that this particular transport mechanism must be able to preferentially select those nutrients that will be of use to the organism?"

"Yes, I suppose this would be the case," said the professor.

"Isn't it true," queried the lawyer, " that modern bacterial organisms tend to divide about every twenty minutes or so, and, consequently, they need to transport enough phosphates, of one sort or another, across their membranes, in the interval between divisions, to be able to double the supply of these molecules that are crucial to the process of synthesizing the increased amount of ribonucleic acids required for cell division?"

"Yes, that is right," the professor indicated.

"Moreover, isn't it the case, Dr. Yardley," asked the defense counsel, "that because phosphates tend to be ionized, a specialized carrier enzyme is necessary for the capturing and transporting of phosphates across the permeability barrier formed by the cell membrane of these bacteria?"

"Yes," agreed the professor.

"Consequently," the lawyer concluded, "to look after processes of selective permeability ... such as ion-pumping, nutrient or toxic transport, along with phosphate acquisition and carrier requirements ... one needs a variety of proteinoids or proteins with specialized amino acid sequences to give one the structural characteristics, hydrophobic or hydrophilic properties, and tertiary folding patterns that meet such a diverse array of cellular needs. Therefore, not just any kind of proteinoid or protein structure will serve such purposes, is that right Dr. Yardley?"

"As far as we know, this is the way things work," the professor confirmed.

"Do membranes provide functions other than the ones already mentioned, Dr. Yardley – other than, that is, ion-pumping and active-transporting, mechanisms of one kind or another?" the lawyer inquired.

"The ability to maintain a differentiated energetic potential between the interior and exterior environments of the cell," pointed out the professor, "establishes an ion gradient. This gradient represents a mother lode of energy that can be mined in various ways to serve a number of cell functions, including coupled transport of nutrients that already has been touched on to some extent and the production of compounds like adenosine triphosphate (ATP) which becomes a mobile means of supplying energy to chemical processes going on throughout the cell.

"For many years," the professor added, "scientists have known that if one heats and then dries a phosphate solution, an anhydride bond forms between pairs of phosphate molecules. This anhydride bond is able to store the energy that is released by the heating and drying process.

"The pair of phosphate molecules that are bonded by the anhydride bond are known as pyrophosphate molecules. Adenosine triphosphate, along with a number of other kinds of phosphate compounds such as creatine phosphate and phosphoenolpyruvate, contain pyrophosphate bonds that are capable of storing energy.

"Essentially, in the case of the potential electrical difference that has been established across the membrane's permeability barrier, the ion gradient becomes the source for generating the energy that is stored in the pyrophosphate bonds of ATP rather than through the energy that is released by the aforementioned laboratory method of heating and drying of a phosphate solution."

"So," the defense counsel proposed, "in order to have a protocell begin to self-assemble, not only do we need to come up with a solution of phosphates in the Archean era, we also need to find a way to generate, at a minimum, the anhydride bonds of pyrophosphates so that we have a means of storing energy generated by the ion gradient associated with the cell membrane ... providing, of course, we can manage to find a way to get these pyrophosphate bonds into the interior of the bounded environment formed by a complex of phospholipids and proteinoids. Does the foregoing scenario cover, in broad terms, this aspect of the evolutionary perspective, Dr. Yardley?"

"In broad terms, yes," replied the professor.

"Stripped down to its bare essentials, Dr. Yardley, would one be right to say," Mr. Tappin asked, "that the mining of the energy contained in the ion gradient being maintained by the potential electrical difference between the interior and the exterior of the cell ... would one be correct if one were to describe this mining process as the rolling, so to speak, of electrons and/or protons down the gradient in order to gain the energy generated by the downhill movement of these charged particles along the ion or proton gradient?"

"This is, more or less, accurate," acknowledged the professor "although, as you indicated, your description is obviously an extremely simplified version of what actually occurs in the energy producing reactions that take place along the ion gradient established by the potential electrical difference across the cell membrane."

"Dr. Yardley, in living organisms, isn't this process of electron or proton translocation along the electrical gradient that extends across the membrane, handled by specific enzymes or proteins?" inquired the lawyer.

"That's correct," the professor said.

"Therefore, in addition to the specialized proteinoids or proteins needed for the pumping of ions, as well as the transport of compounds such as phosphates, nutrients and toxic materials, one also needs specialized proteinoids or proteins capable of translocating electrons or protons across the membrane's ion gradient in order to be able to transfer the energy potential of that gradient to pyrophosphate bonds in compounds such as adenosine triphosphate. Is this the case, Dr. Yardley?"

"Yes, it is," affirmed the professor.

"Given," postulated the lawyer, "a phospholipid bilayer that is impenetrable to all ionic molecules except ones of very low molecular weight, and given that many proteins contain not only ionic side chains but hydrophilic components, how does evolutionary theory account for the process that would allow proteins to become embedded in a permeability barrier that, due to its hydrophobic character, one might assume would be resistant to such a process?"

"We believe," Dr. Yardley stated, "there is some sort of thermodynamic driving force that would allow the proteins and the phospholipids to overcome the repulsive forces acting between the two kinds of molecule. This chemical antagonism is inherently unstable.

"Conceivably, this condition of disequilibrium could be resolved if there were some, as yet undiscovered, thermodynamic process that allowed the energy of the system to be re-distributed in a more stable arrangement. Presumably, the embedding action might take place during this process involving the thermodynamically driven ... and, therefore, spontaneous ... redistribution of the energy toward a more stable ground state."

"You did say, Professor, this thermodynamic mechanism for the insertion of proteins into phospholipid bilayers was both theoretical and, as of yet, undiscovered, is this right?" queried the lawyer.

"Yes, I did," Dr. Yardley admitted. "However, the fact proteins are found embedded in phospholipid bilayers in living organisms, despite the inherent chemical antagonisms that are involved and the fact we have not seen any evidence of a kinetic or non-thermodynamic mechanism to account for this state of affairs, then, the thermodynamic hypothesis outlined above, although theoretical and unproven, is not as speculative and arbitrary as you might think."

"Has anyone," Mr. Tappin asked, "come up with a non-protein related way of mining the energy of the ion gradient that exists in conjunction with the cell membrane?"

"Over the years, a lot of different theories have been proposed in this regard," the professor remarked. "These usually concern variation on themes involving some kind of electron tunneling, ion migration, or proton transfer.

"So far, however, there doesn't appear to be a plausible way of making these mechanisms capable of working in any consistent, reliable fashion, or capable of generating the levels of energy that would be required to maintain membrane functioning, not to mention many other cellular processes. In addition, even if one could come up with a viable, non-protein-related mechanism for mining energy from the membrane's ion gradient, there is no way of either storing the energy once it reaches the interior of the cell, nor is there any way of transferring the charge in order to chemically activate other molecules involved in cell processes, since, as far as is known, both the storage of charge as well as the charge-transfer processes are effected by proteins, although the energy storage compound, itself, is often some kind of a nucleotide rather than a protein."

"Dr. Yardley, would you agree," inquired Mr. Tappin, "that even if one could come up with a plausible prebiotic theory for, one, the migration of charge across the permeability barrier of the membrane, two, the storage of charge, and, three, the transfer of charge, all of which we will assume are capable of operating quite independently of proteins, wouldn't one still be faced with the problem of having to explain how the non-protein system evolved to produce the protein-based system that now helps govern charge-migration, charge-storage and charge-transfer in the biological organisms with which we are presently familiar?"

"Yes," acknowledged the professor. "I don't see how one could avoid having to address this problem under such circumstances.

"In fact, in my opinion, this is precisely the sort of difficulty that emerges in relation to theories of the origin-of-life that focus on the possible role of clay minerals. The proponents of these theories talk about the capacity of clay surfaces to carry out some of the functions important to life ... such as exhibiting a few catalytic properties that can help bring about certain stages in the polymerization of some of the nucleotides in nucleic acids, as well as some peptide chaining; or, providing a surface on which concentration reactions can take place; or, offering a means to compartmentalize and organize different metabolic pathway; as well as having the potential to store, and replicate, certain kinds of information on crystalline patterns, somewhat reminiscent of genetic system. However, in point of fact, even if one were to ignore all the problems and rather severe limitations that surround such capabilities in mineral clays, like kaolin and montmorillonite, nonetheless, these theorists have no way of explaining how life, as we understand it, came into being.

"In effect, they avoid the real problems surrounding origin-of-life issues by trying to define life in another, very limited and superficial way. As a result, they tend to multiply the theoretical problems because not only must they account for the rise of such clay mineral photocells, these theorists also must come up with a plausible theory of transition that accounts for the genetic takeover of these clay mineral systems by protocells that are not based on clay minerals ... unless, of course, such clay mineral protocells are not part of our evolutionary lineage, in which case, whether the theory is right or wrong, it really has nothing to do with life as we understand it.

"Above and beyond the foregoing, there is a further problem concerning the viability of a clay mineral hypothesis for the origin of life. Many clays -- including kaolin -- tend to be extremely rare in pre-Cambrian sediments.

"This fact does not constitute a fatal blow to these kinds of hypothesis. On the other hand, such a fact does tend to lessen the chances of such a hypothesis being correct.

"Quite frequently, one will find various kinds of inorganic conjectures thrown into the picture in an attempt to augment or complement the clay mineral origin-of-life hypothesis. For instance, relatively recently there was a conjecture by a European theorist that is based on the manner in which iron sulfides, like pyrite, contain free energy when the iron becomes reduced to a ferrous state.

"Using such an observation as a launching pad, this theorist postulated that, possibly, if one could find a way of coupling this free energy to possible reactants in a protocell-like environment, then, an important component in the formation of one or more primitive metabolic pathways would have been established. When one added that this kind of energy source might tend to be found in close contact with, say, clay mineral surfaces that, among other things, were capable of bringing about concentration reactions, such a conjecture became quite attractive to some people.

"However," Dr. Yardley concluded, "no plausible, dependable means has been found for accounting how the charge-transfer, or coupling, process will take place in conjunction with potential chemical reactants in a protocell-like environment. Therefore, the iron sulfides conjecture remains nothing but an unrealized conjecture.

"Similarly, some people have proposed that when the various components of nucleotides ... ribose, phosphate, and a nucleic base of one kind or another ... are adsorbed onto the surface of some clay mineral, then, perhaps, the specific character of the mineral might have brought these components together in particular orientations. Unfortunately, for this kind of proposal, none of the minerals that have been tested to date have exhibited the requisite specificity to be able to generate nucleotides with the sort of structural character that are observed in living organisms."

"In conjunction with the previous discussion of membrane activity and functions," Mr. Tappin specified, "isn't it the case that various classes of pigments might be involved with the processes of photosynthesis that take place in, and about, the thylakoid membranes in photosynthetic bacteria and blue-green algae, as well as the chloroplasts of plants?"

"That's right," answered Dr. Yardley.

"What role does porphyrin play in all of this?" the defense lawyer asked.

"Porphyrins," explained the professor, "are one of a group of pigments that are widely distributed among different kinds of organisms. They are derived from a porphin molecule that is a ring structure made up of four pyrrole nuclei (C_4H_4NH) linked together by carbon atoms.

"The nitrogen atom in porphins often tends to form very strong and stable bonds with metallic ions such as magnesium or iron. This kind of bonded group is referred to as a chelate.

"Chlorophyll, which is present in all photosynthetic organisms, consists of a porphin group with a magnesium ion at its center. In addition, different kinds of chlorophyll have various kinds of side chains attached to them.

"Generally speaking, pigments are divided into two broad classes known as accessory and principle pigments. Accessory pigments tend to gather light energy and pass it onto the principle pigment that, for the most part, is either chlorophyll 'A' or one of the forms of chlorophyll occurring in certain bacteria.

"There are, however, other classes of non-chlorophyll pigments such as carotenoid and phycobilin. These other classes of pigments tend to have accessory, rather than principle, roles in photosynthetic systems."

"Professor Yardley, to the best of your knowledge," inquired the lawyer, "is there any plausible prebiotic pathway of synthesis that might give rise to the Porphyrins that are at the heart of the chlorophyll contained in all photosynthetic organisms?"

"None is known at the present time," replied the professor. "Nonetheless, as I indicated in previous testimony, on occasion, pigment-like molecules have been found in the organic residue of some carbonaceous chondrites."

"Even if," Mr. Tappin postulated, "we were to assume these pigment-like molecules had a full capacity to accept and transfer light energy, and even if we were to assume these extraterrestrial pigments were in plentiful supply and did not get degraded through photolysis and so on, and even if one were to assume that, somehow, these pigment-like molecules were to find their way into a protocell system, wouldn't one still be faced with the problems of explaining how porphin-containing chlorophyll came into existence and how these pigment-like molecules became coordinated with chlorophyll molecules in various kinds of photosynthetic systems?"

"Yes," the professor conceded, "one still would be left with having to account for such things."

"Furthermore, Dr. Yardley, in the photosynthetic systems with which we currently are familiar, doesn't the transfer of energy charge from accessory to principle pigments take place by means of an electron transport system made up of a series of protein enzymes, and, therefore, even if one were to accept the idea of an extraterrestrial pigment-like molecule playing a role in the formation of early photocells, wouldn't one still need to account for the rise of the requisite support system of enzymes that had the ability to serve as a specific transport mechanism in relation to the movement of electrons to their final acceptor destination in the protocell?"

"Yes," the professor acknowledged, "these sorts of phenomena would remain as problems to be explained ... but even in the assumptions that you have cited there are also chemosynthetic autotrophic organisms that derive their carbon and energy in a quite different manner from photosynthetic autotrophic organisms. Conceivably, these chemosynthetic autotrophs, and not photosynthetic autotrophs, were the first photocells to exhibit the properties of life."

"If I understand what you are saying, Dr. Yardley, wouldn't evolutionary biology now have two problems to solve rather than one?" suggested the defense counselor. "The origin of two different kinds of autotrophs would have to be accounted for ... one which is chemosynthetic in nature and one which is photosynthetic in nature. Is this the case?"

"It is," stated the professor," unless one of the two systems was the prototype from which the other eventually was derived through an evolutionary process."

"If this were the case, wouldn't one still be faced with two problems?" Mr. Tappin challenged. The first problem would be to provide a plausible explanation for either photosynthetic or chemosynthetic autotrophs, depending on which one an individual considered to have arisen initially. The second problem would be to provide a plausible explanation for the sort of transitional steps that would have permitted a very different kind of autotrophic system to be derived from the first autotrophic system. Isn't this the situation, Professor, with which evolutionary biology would be, and is, faced?"

"Yes, I suppose it would be, and I suppose it is," Dr. Yardley responded.

"Mr. Tappin," stated Judge Arnsberger, "once more, I must interrupt your cross-examination. The dinner hour is at hand, and I feel we all could use a break from these deliberations.

"Please remember, all of my previous instructions to the jury remain in effect. These court proceedings will be adjourned until 7:30 p.m. this evening."

The Science of Presumption Can Be a Beautiful Thing

"Dr. Yardley," stated Mr. Tappin, "you have testified that ribose is a 5-carbon monosaccharide or pentose sugar monomer. In addition, you said this sugar, along with phosphates and nucleic bases, are fundamental building blocks of nucleic acids, and nucleic acids are the carriers of genetic information.

"How do evolutionary theorists account for the synthesis of ribose sugars in the prebiotic Archean era?" asked the defense counsel."

"Many researchers feel," the professor replied, "that a process known as the formose reaction might have been the most plausible means for synthesizing a variety of sugars including ribose. Essentially, this involves a base-catalyzed condensation reaction of formaldehyde."

"Leaving aside for the moment," said the lawyer, "the previously established point concerning the possible, relative unavailability of formaldehyde in a prebiotic environment due to, among other things, ultraviolet photolysis, would you describe in a little more detail the nature of the formose reaction."

"If," began the professor, "one takes a strong alkali agent such as thallium hydroxide or lead hydroxide and treats formaldehyde with one or the other of these agents, one can generate a variety of sugars. On the other hand, one also can use agents like alumina ... that is, aluminum oxide (Al_2O_2), as well as calcium carbonate or barium hydroxide.

"Following an induction period ... which might last for many hours and in which products such as glycolaldehyde, glyceraldehyde and dihydroxyacetone are formed ... a variety of sugars are synthesized. These include tetroses, pentoses and hexoses, or, respectively, 4 -, 5-, and 6-carbon sugars.

"The formose reaction is autocatalytic in nature which means that once the induction period is over, the reaction proceeds to completion rather quickly. In addition, if the reaction is stopped at the appropriate stage, yields of up to 50% of some of the higher sugars are possible."

"Dr. Yardley, since, presumably, there was no one around in prebiotic times to stop the formose reaction at the appropriate stage, can one reasonably assume that the yields would have been considerably less than the 50 percent figure you have cited?" Mr. Tappin inquired.

"Yes, I guess so," indicated the professor. "On the other hand, there could have been forces active in the prebiotic environment that might have disrupted the reaction before it went to completion."

"I won't pursue this Archean era version of a mugging by unknown assailants," the defense counsel remarked, "but I would like to pursue the issue of the alkali agents that might be used in the formose reaction. How common would, respectively, thallium, lead, and barium hydroxide have been during the Archean era?"

"This is relatively difficult to say," the professor responded. "Perhaps the most accurate thing I can say is these hydroxides probably would have been far less plentiful than either aluminum oxide, which is very common in the silicates that make up a large portion of the Earth's crust, or calcium carbonate - that is, limestone, which also would have been quite plentiful in the prebiotic period."

"Is there," Mr. Tappin asked, "only one kind of pentose sugar -- such as ribose -- which is synthesized during the formose reaction?"

"No," replied the professor. There are a number of pentoses that are formed during this reaction, and each of these pentose sugars are produced in varying amounts.

"For example, in addition to ribose, one also will find xylose, lyxose, and arabinose. These other pentoses involve various kinds of inversion of one or more of the hydroxyl groups of ribose."

"What proportion of all the different kinds of tetrose, pentose, and hexose sugars formed during the formose reaction," queried the defense counsel "are the ribose variety of sugar?"

"Ribose forms a very small portion of the overall yield of sugars," the professor stated.

"Do the other pentose sugars beside ribose get synthesized in amounts that are comparable to, if not more than, the ribose yields?" inquired the lawyer.

"Yes, they do," answered the professor.

"What sorts of concentration levels of formaldehyde are minimally necessary for the formose reaction to proceed?" Mr. Tappin wondered.

"As far as we know," the professor stipulated, "the formose reaction does not seem to proceed if the solute level of formaldehyde falls much below one-hundredth of a mole per liter of solution."

"Given," postulated the lawyer, "what has been said before about the possible scarcity of formaldehyde in the Archean era ... and, perhaps, even in the best of circumstances ... aren't expectations for the existence of such high solute concentrations of formaldehyde during prebiotic times rather inflated and optimistic?"

"Yes, realization of these levels of formaldehyde concentration during the Archean era could be a significant obstacle to the formation of ribose," confirmed the professor.

"Dr. Yardley, how stable are sugars in aqueous solution?"

"Not very," the professor replied, "especially if the pH value is above 7. Under these circumstances, sugars tend to be degraded over a period of time that is not much longer than what is required to synthesize such molecules."

"Previously, Professor, you stated that evolutionary researchers usually consider the pH of the Archean era ocean to have been 8 -- plus or minus one. Consequently, would you agree, Dr. Yardley, the pH of the Archean era ocean had a very good chance of exceeding a pH of 7 and, therefore, readily could have led to the destruction of whatever small amounts of ribose were synthesized almost as quickly as these molecules were formed."

"Yes, there could have been a very good chance this happened if the pH of the Archean era ocean was much above 7," affirmed the professor."

"Other than the issue of isomers with different-handed optical activity, does ribose come in more than one form?" the defense counsel inquired.

"Yes, it does," the professor replied. "There are three forms in all. "In addition to a form known as ribopyranose," he explained, "there are two ringed forms of ribose. These are referred to as alpha and beta-ribofuranose."

"Do all three of these forms of ribose appear in the nucleic acids that occur in living organisms?" asked Mr. Tappin.

"No," stated the professor. "The only form of ribose that occurs in living organisms is beta-ribofuranose."

"Nucleosides," stated the lawyer, "are one step removed from a full-fledged nucleic acid due to the absence of a phosphate group, and nucleosides consist of bonding together one of the five nucleic bases with a beta-ribofuranose. Have I got this right?"

"Yes," the professor indicated.

"Could other sugars, such as some of the non-ribose pentoses, bond with the five nucleic bases?" inquired the defense counsel.

"Yes," Dr. Yardley confirmed.

"Presumably," surmised the lawyer, "all three forms of ribose also could form bonds with the nucleic bases. Is this correct?"

"Yes, that is right," said the professor.

"Consequently," Mr. Tappin concluded, "any one of a number of pentose sugars, or different forms of ribose, or optical isomers could bond with the nucleic bases and form one species, or another, of a nucleoside. Yet, only one of the nucleosides, amongst this mixture of possible nucleosides, has any functional value in living organisms. Would you agree this is the case, Professor?"

"I would," Dr. Yardley acknowledged.

"How," the lawyer queried, "did the one nucleoside that would have functional value once living organisms arose come to be selected from the multiplicity of very similar choices available in the Archean era environment?"

"We are not sure," Dr. Yardley admitted. "Obviously, whatever the mechanism of selection, the beta-ribofuranose nucleoside had selective value."

"What exactly do you mean, Professor, by the notion of selective value?" asked the defense counsel.

"The beta-ribofuranose nucleoside worked," the professor responded. "It fit in with the rest of the protocell system and, presumably, played a fundamental role in forming a self-sustaining, and self-perpetuating, system."

"Wouldn't you say this is a matter of twenty-twenty hindsight?" challenged Mr. Tappin. "Before one reached the stage of establishing even a primitive protocell, one would have to assume the beta-ribofuranose nucleoside is being selected.

"One cannot use the functioning of a system," argued the lawyer, "which has not yet been established as the reason for why such a molecule is being selected. So, why is this particular molecule, among all the other possibilities, being selected for, prior to the existence of a working protocell?"

"One can only assume," the professor stated, "that this particular nucleoside must have satisfied certain thermodynamic and kinetic contingencies which existed during the Archean era."

"Are the identities of these contingencies to be kept anonymous at this time, Professor?"

"I'm afraid so," acknowledged the professor. "I should point out, however, that Albert Eschenmoser, of the Swiss Federal Institute of Technology, has made several contributions relatively recently that bear on some of the issues we have been discussing."

"Yes, please go on," the lawyer requested.

"First of all," Dr. Yardley stated, "Eschenmoser constructed a molecule, known as pyranosyl RNA. This compound contains a modified form of naturally occurring ribose.

"The ribose that occurs in normal RNA contains a five-member ring, consisting of 4 carbon atoms and one oxygen atom. The ribose molecule that forms part of Eschenmoser's pyranosyl RNA compound has been constructed to allow an extra carbon atom in the ring.

"Like normal RNA, complementary strands of pyranosyl RNA are capable of joining together by means of Watson-Crick hydrogen bonding. Furthermore, the use of pyranosyl RNA, with its modified form of ribose, prevents fewer unwanted variations of nucleoside structure from among the multiplicity of available possibilities than does normal RNA.

"In addition, double-strands of pyranosyl RNA do not twist around one another, as is the case with the normal forms of double-stranded RNA. This quality could be extremely important if enzymes were not available, unlike the situation currently, to unwind these strands so that replication could take place."

"Dr. Yardley, as far as you know, does pyranosyl-RNA exist outside the laboratory?" the defense counsel asked.

"No," the professor admitted.

"Would I be fair in saying, Professor," Mr. Tappin queried, "that although one might agree the pyranosyl RNA molecule that has been created in the laboratory is very interesting and suggestive of possibilities, nevertheless, this molecule really is of little practical import to origin-of-life issues if it, or something similar to it, did not exist in the Archean era?"

"Yes, this would be a fair way of saying things," agreed the professor.

"Moreover," added Mr. Tappin, "even if one were to suppose such a molecule as pyranosyl RNA existed in prebiotic times, one would have to explain why, and how, a molecule ... namely, normal RNA ... which, from a number of different perspectives, did not have anywhere near the selective value of pyranosyl RNA, would have come to replace the latter molecule. Would you say these are fair issues to ask?"

"I would assume so," the professor offered.

"Can either of these problems be resolved at the present time," inquired Mr. Tappin.

"Not satisfactorily," responded the professor.

"You stated earlier Dr. Yardley that this fellow Eschenmoser had made several contributions that bear on the issue being discussed. What is the other one?"

"Around 1994," said the professor, "Eschenmoser discovered a way of limiting the kinds of sugars that are synthesized during the formose reaction. Without getting into the technical details of the experiment, essentially, he replaced one of the normal intermediates of the formose reaction with a similar phosphorylated molecule, and, then, he permitted the subsequent steps of the reaction to proceed as normal."

"Excuse me," Dr. Yardley, "am I right in believing that a phosphorylated molecule is a compound to which a phosphate group has been added and which, under certain circumstances, might be capable of storing energy if particular kinds of pyrophosphate bonds are present?"

"Essentially, yes," the professor said.

"Under certain conditions, when this kind of substitution was made, the primary end product of the formose reaction was a phosphorylated derivative of ribose. This substitution process, therefore, represents a possible way of getting around the selectivity problem that arises as a result of the multiplicity of competing sugar forms that exists when one permits the formose reaction to proceed as usual."

Checking the papers in his hand before speaking, Mr. Tappin said: "In the experiment just described, Professor, wouldn't the phosphate group on the synthesized ribose derivative have to be rearranged upon completion of Eschenmoser's altered pathway for the formose reaction in order to be the same as the phosphorylated ribose that is found in normal nucleotides?"

"Yes, that's true," the professor acknowledged.

"In addition," the defense counsel observed, "doesn't the Eschenmoser experiment leave one with a slight problem of needing to explain how one is going to bring about this substitution process under prebiotic conditions when, presumably, there is no Archean era counterpart to Albert Eschenmoser, or his lab assistants, who would be available to make the substitution? Moreover, doesn't all of this assume that the closely related phosphorylated molecule that is to be substituted for the normal intermediate of the ribose-forming reaction is going to be available to be inserted into the formose reaction at just the right moment?"

"I guess so," replied the professor.

"Would you agree, Dr. Yardley," queried the lawyer, "that although there has been some success in synthesizing adenosine and guanosine nucleosides when purified mixtures of ribose and purine bases have been heated in the presence of certain inorganic salts, these same successes are not observed with pyrimidine nucleosides, such as uracil and cytosine, under any conditions that could be considered to be plausible in the Archean era?"

"Yes, that is correct," the professor confirmed.

"Apparently, then," summarized Mr. Tappin, "at the present time there is no known, plausible pathways under prebiotic conditions for synthesizing more than half of the five nucleosides that are fundamental to the storage of genetic information in both DNA and RNA. Is this more or less the state of things in evolutionary theory, Professor?"

"More or less," Dr. Yardley stated.

"To further confuse matters," added the lawyer, "even in the case of the synthesis of the nucleic purine bases, adenine and guanine, one is likely to find other kinds of bases such as hypoxanthine, diaminopurine and a variety of related molecules accompanying the synthesis of the specific purine bases that are important to the nucleic acids which occur in living organisms. So, wouldn't you agree, Dr. Yardley, that, here too, the Archean era, through natural chemical processes, is likely to have generated a variety of cross-linked polymers that somehow would have to be selected against in order to work toward the kind of life form which resembles that with which we are familiar today?"

"Yes, I would agree with this," Dr. Yardley said.

"Would you agree Professor," asked Mr. Tappin, "that all of the problems that have been discussed in relation to the formation of nucleosides would carry over into the formation of nucleotides during which a phosphate component is added to the nucleoside combination of ribose and one of the five nucleic bases? In other words, wouldn't there be a substantial array of abnormal nucleotides consisting of various pentoses other than ribose, as well as forms of ribose other than the right-handed optical isomer of beta-ribofuranose, and, if this is the case, wouldn't these interfere with both catalytic processes as well as RNA replication?"

"Yes, one would have to assume this very well could have been the case," affirmed the professor.

"Dr. Yardley, beside the abnormal nucleotides that would form as a result of the presence of different pentoses, ribose forms and optical isomers, wouldn't there also be an assortment of abnormal phosphate bonds that could arise? In other words, isn't it true that beyond the normal, 5-prime- phosphate bond that occurs during one of the stages leading to the formation of the sorts of nucleic acid found in living organisms, one also might obtain problematic bonding arrangements such as: 2-prime-phosphate bonds; or, 3-prime-phosphate bonds; or, 2-prime-3 prime-cyclic phosphates; or, 2-prime-5 prime-biphosphate; or, 3-prime-5-prime-biphosphates?"

"This is true," affirmed the professor.

"Would you also agree, Dr. Yardley," added Mr. Tappin, "that, in the light of current knowledge, the Archean era is much more likely to have consisted of such a mixture of phosphate bonds, pentoses, different forms of ribose, as well as a racemic aggregation of optical isomers, rather than having consisted of the purified solutions with which laboratory experiments are run?"

"Yes," said the professor.

"In addition, Dr. Yardley, would you agree that despite all the problems that exist in relation to the formation of ribonucleic acids, nevertheless, RNA is more easily synthesized than is deoxyribonucleic acid? In fact, can we not say that one of the considerations which led to the rise of the RNA-world hypothesis was rooted in the way RNA is much more easily synthesized than is DNA?"

"The answer to both of your questions is 'yes'," responded the professor.

"In your opinion, Dr. Yardley," queried the defense counsel, "even if much of the RNA-world hypothesis turned out to be true, wouldn't evolutionary theorists still be faced with the problem of proposing a plausible prebiotic mechanism for the synthesis of DNA?"

"I believe this would be the case, yes," the professor admitted.

"On the other hand," Mr. Tappin indicated, "although RNA is more easily synthesized than DNA, DNA is much less susceptible to hydrolysis, or breakdown in an aqueous environment, than is the case with RNA. If my information is correct, isn't it true, Professor, that at room temperature RNA breaks down at a rate that is roughly 100 times faster than does DNA, and, within certain limits, this differential rate of breakdown climbs somewhat with increases in temperature above room temperature?"

"This is basically right," stated the professor, "except that depending on the temperatures you are talking about, both DNA and RNA tend to decompose more readily at elevated temperatures."

"Dr. Yardley, assuming my understanding of things is right, if one starts with a single polymer or chain of RNA in solution, a complementary strand easily can be generated by adding free, unpolymerized nucleotides to the solution, since, subsequently, these free nucleotides will line up opposite their pairing partner on the original RNA strand ... that is, uracil with adenine and cytosine with guanine. Moreover, the original strand and its complement will form, in the absence of enzymes, a double helical structure by means of the spontaneous hydrogen bonding of these Watson-Crick pairings. Is all of this correct?"

"Yes," the professor replied.

"Yet," the defense counsel stipulated, "the foregoing scenario assumes, does it not, Professor, that all of the free nucleotides that are being added to form the complementary strand must exhibit the same optical properties or handedness as the original strand of RNA?"

"That's correct," Dr. Yardley affirmed.

"In other words," indicated the lawyer, "if one places both left- handed and right-handed optical isomers of various free nucleotides into the solution, then, the presence of both left- and right-handed isomeric forms of the nucleotides will inhibit the formation of a complementary strand capable of bonding with the original strand through Watson-Crick pairings. Isn't this so, Dr. Yardley?"

"Yes, it is," acknowledged the professor.

"Furthermore," Mr. Tappin continued, "according to the information that is available to me, despite years of experimental efforts by hundreds, if not thousands, of scientists and researchers, no one has been able to find a way to replicate or copy a complementary strand of nucleic acids without the assistance of enzymes. Consequently, would you agree Dr. Yardley, that although scientists can generate, in the absence of proteins, a complementary strand for an original strand of RNA, these same scientists cannot copy the complementary strand without the right kinds of enzyme being present?"

"Although, in general, much of what you have said is true," the professor indicated, "I wouldn't agree with your statement without adding at least one qualifying remark. More specifically, two researchers, by the name of McHale and Usher, have demonstrated that when strands of RNA oligonucleotides, consisting of 10 polymerized units or less, are dried and heated in temperatures that approximate sunlight, these RNA oligonucleotides will line up along a complementary template and form polymers or bonded chains similar to the process of replication that occurs in living cells."

"Correct me if I'm wrong, Dr. Yardley, but I believe," suggested the lawyer, "there are a number of differences between the experiment you are describing and the conditions one is likely to be working with in an Archean era environment. First of all, wouldn't you agree, Professor, the experiment to which you are alluding is presupposing what has not, yet, been able to be satisfactorily demonstrated by evolutionary science -- namely, that normal RNA nucleotides would have been synthesized and selected out in pure, concentrated forms from amongst the motley array of possibilities involving: pentose sugars, different forms of ribose, optical isomers, alternative phosphate bonding possibilities, lack of pyrimidine bases, as well as a variety of odd purine bases in addition to adenine and guanine?"

"Yes, that is correct," the professor responded.

"Isn't it also the case, Dr. Yardley," queried the defense counsel, "that in living cells there is an unwinding protein that is able to help separate the individual strands of the double-helix form of nucleic acids that is being held together by Watson-Crick hydrogen bond pairings. In fact, in your previous discussion of Eschenmoser's laboratory creation, pyranosyl RNA, wasn't one of the attractive features of this molecule the fact it offered a possible way around needing a protein to unwind the double-helix structure of nucleic acids?"

"That's right," said the professor.

"Consequently, isn't the McHale-Usher experiment presupposing," Mr. Tappin asserted, "that there was a means, under Archean era conditions, to unwind the strands that spontaneously tend to form double-helix structures through Watson-Crick pairings in order for there to be a complementary, single-stranded template with which to work?"

"This would seem to be the case," the professor agreed.

"To the best of your knowledge, Dr. Yardley, has any ribozyme ... that is, an RNA polymer with catalytic activity -- been discovered that has the required unwinding capacity that appears to be presupposed by the McHale-Usher experiment?"

"Not as far as I know," answered the professor.

"Furthermore," the lawyer added, "given that the experiment was successful with short polymers of 10 units or less, one is left wondering why the same kind of experiment has not been successful in the replication of much, much longer polymers of nucleic acid as would be required in fully functioning, living cells. In fact, Professor, isn't it the case that part of the lack of experimental success with respect to being able to polymerize long sequences of RNA molecules is due to the instability of the RNA molecule? In other words, isn't it true that the rate of RNA polymerization must take place fast enough to compete with the rate of random, hydrolytic decomposition of the same RNA molecules, and this is difficult to achieve in the absence of protein enzymes that have the capacity to increase reaction rates by magnitudes of between one million and one billion times?"

"Yes, I guess so," responded the professor, "but, if nothing else, I believe the McHale-Usher experiment is very suggestive and carries a lot of implications for the origin-of-life issue."

"Finally, Dr. Yardley, wouldn't you agree," Mr. Tappin inquired, "that the experiment in question is assuming the following. Even if one, or more, normal RNA oligonucleotides somehow found their way into existence under Archean era conditions, nevertheless, the researchers do not seem to be allowing for the possibility of the degradation or decomposition of these molecules through hydrolysis, ultraviolet photolysis or pyrolysis?"

"Quite frankly," replied the professor, "I'm not sure I would agree the researchers should have to take any of these factors into consideration. The experiment was intended to show a possibility rather than be a definitive way of resolving all conceivable problems facing evolutionary theory."

"Fair enough," responded the defense counsel, "but would you agree, in turn, that even if McHale and Usher do not have to take any of these various, nevertheless, if evolutionary theory is to provide a plausible account for the origin-of-life through natural processes, then, this theory must be able to resolve the problems that are being raised in relation to the McHale-Usher experiment. After all, just as there are positive implications that follow from the McHale-Usher experiment, are there not also a number of negative or problematic implications that are inherent in that same experiment?"

"I guess I can live with this way of stating things," offered the professor.

"During direct examination testimony, Dr. Yardley, you spoke about a number of different ribozymes or sequences of RNA with catalytic properties. If I remember correctly, these properties involved such activities as the cutting and splicing of specific RNA sequences, as well as assuming some limited characteristics of a polymerase by helping to bring about the formation of the bonds that link together certain kinds of polymer chains. Is this right?" the lawyer asked.

"Yes," affirmed the professor.

Mr. Tappin briefly looked through the material he had been holding in his hands while conducting the cross-examination. After five or ten seconds of searching, he pulled out a sheet of paper and placed it on top of the material in his hands.

Eventually, he said: "Dr. Yardley, in doing research concerning some of the experiments dealing with ribozymes, I came across something about which I'm curious. Perhaps, you can help me out.

"At one stage during the particular study that I have in mind," explained the lawyer, "the researchers were interested in determining whether the catalytic specificity exhibited by a naturally occurring ribozyme could be overcome or altered. More precisely, these researchers wanted to see if the ribozyme could be induced to interact equally effectively with a variety of base sequence combinations rather than just the limited nucleic sequences for which the ribozyme, under normal circumstances, seemed to show an inherent, interactive preference.

"In order to overcome the inherent sequence specificity of the ribozyme, the researchers began exploring the possible effects that a variety of polyamines might have on the ribozyme. Although, undoubtedly, Professor, you know what a polyamine is, for the benefit of the jurors, a polyamine, as the name suggests, is a compound that contains two or more amino groups.

"Now," the defense counsel continued, "the simplest of polyamines, such as putrescine $[NH_2(CH_2)_4NH_2]$ and spermine $[NH_2(CH_2)_3NH(CH_2)_4-NH(CH_2)_3NH_2]$ are far more complex than compounds such as hydrogen cyanide (HCN), methane (CH_4), formaldehyde (CH_2O), or ammonia (NH_3). Yet, there is considerable discussion concerning the extent of the availability of even these latter, simple hydrocarbons during Archean era times.

"There were ten polyamines that were tested during the experiment. Only one of these polyamines, spermadine, which is of moderate complexity relative to other polyamines, was found to be capable of inducing the ribozyme to overcome its inherent base sequence specificity.

"Once again, Professor, as was true in relation to the original origin-of-life experiment of Miller, or any of Fox's proteinoid experiments, or Eschenmoser's pyranosyl RNA molecule, and numerous other experiments that supposedly simulate the conditions of the prebiotic Archean era, I question the value of such experiments as far as their implications for origin-of-life issues are concerned. How much spermadine, Dr. Yardley, was there in the Archean era world?"

"The short answer to your question," replied the professor "is that I don't know. Although polyamines might be more complex than the simpler compounds from which various origin-of-life scenarios usually begin, the quality of complexity does not, in and of itself, automatically preclude the possibility that polyamines could not have been synthesized under prebiotic conditions.

"As I indicated previously," pointed out the professor, "just because an experiment is performed that does not necessarily faithfully simulate certain aspects of the conditions of the Archean era, this does not mean such an experiment cannot have implications for what might have gone on during prebiotic times. For example, even if one were to assume that spermadine didn't exist during the Archean era, the fact that, under certain conditions, ribozymes can be induced to broaden their catalytic activity, raises the possibility there might have been other agents that did exist during the Archean era and that might have had an effect on ribozymes similar to the action of spermadine.

"If we didn't know about what spermadine helps make possible, we might not have a reason to go looking any further to determine whether there might have been a more plausible prebiotic method for bringing about the same kind of result that spermadine does. In all likelihood, the experiment to which you refer was not, in any technical sense, intended to serve as a simulation experiment, but, nevertheless, this experiment provides evidence that helps shape theory and future experiments as well as strengthens the overall evolutionary model."

"Would you say, Dr. Yardley that the spermadine experiment constitutes evidence in support of evolutionary theory?" Mr. Tappin inquired.

"If you are asking me," the professor replied, "whether this experiment constitutes a sort of 'smoking gun' that brings us to the brink of completing an unbroken chain of evidence that overwhelmingly and undeniably demonstrates the truth of an evolutionary explanation for the origin-of-life, then, my answer is the spermadine experiment does not provide the kind of evidence in support of evolutionary theory that you are seeking. If, on the other hand, you are asking me whether the spermadine experiment provides information that helps to shape, color, modulate, and orient evolutionary theory, then my answer is that this experiment does constitute evidence in support of evolutionary theory."

"Actually, Dr. Yardley," Mr. Tappin responded, "I'm asking neither kind of question. The question that I'm posing is more like the following: given that legitimate questions can be raised about the availability of polyamines such as spermadine in the Archean era, does the fact a ribozyme can be experimentally induced to overcome its inherent sequence specificity under artificial, and prebiotically unrealistic, conditions, really bring us any closer to answering the question of how life came into being, especially in view of the very strong possibility that ribozymes might not have been capable of being synthesized in the prebiotic world?

"In other words, Professor, many evolutionary researchers seem to be saying: if such and such a set of conditions holds, then, such and such a outcome is possible, and if we assume that these condition s did hold during the Archean era, then, this constitutes evidence in support of evolutionary theory. Yet, the question that really needs to be asked and answered is this: do we have any plausible means of demonstrating the likelihood that such a set of conditions existed and that such an outcome did, in fact, take place during the Archean era?"

"All of evolutionary theory," Dr. Yardley asserted, "is about establishing and demonstrating how some conditions, events, processes and outcomes might have been more likely than other conditions, events, processes and outcomes."

"That might well be true, Professor, but there seems to be a heavy fog warning that is being posted with respect to conceptual travel in the areas of 'demonstration' and 'likelihood'," the defense counsel replied. "For instance, you previously said the spermadine experiment can be considered to constitute evidence in support of evolutionary theory because, irrespective of whether it is right or wrong, the findings of the experiment can be used to help shape and modulate that theory, and, yet, at the same time, the spermadine experiment might have nothing to do with the Archean era, and, therefore, by implication, the spermadine experiment might have nothing to do with one of the most important questions facing evolutionary theory ... namely, how did life come into being.

"In effect, I'm having a little trouble, Dr. Yardley, understanding how you propose to reconcile these seemingly antagonistic elements. If, and the viability of this 'if' needs to be examined ... if one can raise questions which cast serious doubt on the degree of relevance of the spermadine experiment with respect to helping us resolve the origin-of-life issue, then, how does it serve as evidence for evolutionary theory?"

"Science," suggested the professor, "is about empirically and conceptually exploring possibilities concerning the physical/material world in a methodical, rigorous fashion. Within certain limits, whatever an experiment permits us to eliminate in the way of possibility, we eliminate. Similarly, within certain limits, whatever an experiment permits us to retain in the way of possibility, we retain.

"Over time, the relationship between what has been eliminated and what is retained takes on a structural form. We describe this relationship through the concrete vocabulary of hypothesis, conjecture, experiment, methodology, data, evidence, analysis, principles, laws, theory, and model.

"Unfortunately, at any given time, there is often a certain amount of ambiguity that surrounds the issue of what justifiably can be eliminated or retained as a function of the empirical data and experimental results that might be in our possession. The spermadine experiment gives expression to a certain amount of this sort of ambiguity.

"On the one hand, as you rightly point out" affirmed the professor, "we don't know whether spermadine, or ribozymes for that matter, existed during the Archean era, although there is evidence that can be offered both for, and against, such possibilities. Even if we eliminate the ontological possibilities of spermadine and ribozymes from the picture, we still can retain the idea that something like them might have existed and that, if they did, would help resolve certain kinds of problem, so, we proceed to try to determine whether we should eliminate or retain such conceptual possibilities on the basis of forthcoming empirical data and conceptual reflection.

"On the other hand, if spermadine and ribozymes did exist during the Archean era -- a possibility concerning which, once again, evidence can be offered both for and against ... then, the spermadine experiment is revealing a very interesting possibility that ought to be retained and explored further. Now, although the available evidence does suggest there are a variety of factors that help mitigate against continuing to retain either spermadine or ribozymes as viable, plausible pieces of the origin-of-life puzzle, nonetheless, we have not yet reached a point where these possibilities can justifiably be eliminated from the picture.

"Quite frequently, there is a constant dialectic and tug-of-war going on between how we feel about what, both conceptually and empirically, should be eliminated and what should be retained at any given time. Consequently, despite the fact something might have a theoretical status, vis-à-vis elimination and retention, which is ambiguous, nonetheless, such an ambiguous element still can come to have a shaping influence on one's theories, models, conjectures and hypotheses, even while there are other factors that serve as contraindications to this shaping influence."

"What happens," hypothesized the lawyer, "if your feelings about the proper relationship between what is to be eliminated and what is to be retained are at odds with my feelings about the proper relationship between what is to be eliminated and what is to be eliminated?"

"Then," the professor said with a shrug of his shoulders, "we have a difference of opinion."

"Is there any way to resolve such a difference of opinion," the defense counsel asked.

"Yes and no," answered the professor. "One can try to do more science until the balance of evidence seems to point more in the direction of one kind of relationship of elimination/retention rather than some other such relationship. However, this often is easier said than done, and, moreover, there frequently are other ideas about the proper relationship between what should be eliminated and retained that arise in the meantime and complicate any straightforward resolution of the original difference of opinion.

"Progress does occur in the sense that despite a variety of differences of opinion about numerous issues concerning what should be retained and what should be eliminated, a broad consensus develops about some of the things, both empirical and conceptual, that should be eliminated and some of the things that should be retained. Even here, however, one finds some people who are resistant to either eliminating possibilities or retaining possibilities despite the presence of a general consensus among many researchers on such matters."

"Does the existence of a consensus," queried the lawyer, "necessarily mean this decision on what, in broad terms, should be eliminated or retained is, in some sense, a correct one?"

"Not at all," Dr. Yardley stated. "Yet, one could say that where such consensus exists, there usually is considerable justification that can be offered ... through empirical observations, experimental results and conceptual analysis ... in support of such decisions, and, therefore, anyone who wishes to oppose these kinds of decision will be swimming against the tide of an informed consensus of opinion.

"Of course, historically, conceptual revolutions often have come in the form of one or more people who believed the wrong consensus decisions had been made about the possibilities that are being eliminated, retained or even entertained. Apparently, your client, Mr. Corrigan, is an individual who feels consensus opinion concerning evolutionary theory is wrong-headed, but whether his opposition will result in a revolution or merely fall by the wayside as a very minor historical oddity will be decided, to some extent, by what the present jury and other similar forums of public opinion decide."

"I've noticed," Mr. Tappin observed, "there doesn't seem to be a lot of talk about the notion of truth in your characterization of science. Given that many people normally link issues of scientific evidence and demonstration with the idea of having, to some extent, proven that something is true, I'm wondering if you might elaborate a little on this aspect of science."

"Naturally," Dr. Yardley replied, "researchers hope that, in some way, elements of reality are faithfully captured in what is retained by the scientific community. Similarly, researchers hope everything that we eliminate is being thrown out because it lacks this quality of faithfulness or reflectivity when compared with experience, experiment, analysis and so on.

"In fact, generally speaking, there are only two kinds of mistake that can be made in science. On the one hand, we can retain something that, in reality, turns out to be incorrect, erroneous, false, and, therefore, in some sense, distortive with respect to our experience concerning what is. On the other hand, we can eliminate something that, in reality, turns out to correct, accurate, true, and, therefore, is, in some sense, reflective of our experience of what is.

"The problem in all of this is that, quite frequently, there are distortive elements mixed in with the reflective features that are retained, just as there often are reflective elements mixed in with the distortive features that are eliminated. This adds to the ambiguity of the situation to which I alluded earlier, and this also helps to explain why researchers are not inclined to rush to judgment about what should be retained or eliminated, and also why some individuals are reluctant to eliminate certain possibilities despite a contrary judgment by the consensus of opinion of the scientific community.

"Oddly enough, at least from the perspective of some people, scientists are more inclined to want to talk about the beauty of a theory rather than its truth. Etched deep in the psyche of many a scientist is the belief that whatever truth or reality might ultimately turn out to be, it will be beautiful as well.

"Because the truth is not always easy to come by or discover, scientists often use the beauty of a theory as a possible index or sign of the presence of truth within the theory. Like so many bag-people, researchers furiously rifle through the garbage cans of empirical data in search of the nuggets of truth that are to be retained while we wait for the dump trucks of history to remove the remaining refuse, and, often times, the only thing that sustains our search is the beauty of the receptacles through which we are foraging and the belief that such beauty is, at least in part, derived from the sweet smell and colors of truth contained somewhere in the garbage cans through which we are searching."

"What is meant by the notion of the beauty of a theory," the defense counsel inquired.

"The beauty of a theory is not always easy to pin down. A lot of the time, researchers recognize such beauty when they encounter it, but they would be hard pressed, if asked, to delineate the nature of such beauty prior to, and sometimes even after, the actual encounter experience.

"There are, however, some classic indices usually associated with the beauty of a theory. For instance, a beautiful theory often tends to be able to lend a directed and consistent sense of meaning and organizational orientation to disparate sets of data, observations, ideas, experiments, and findings.

"Normally speaking, the data of life look like a scatter diagram with the temporal, spatial and qualitative co-ordinates of experience appearing as just so many unconnected and unrelated points. Then, someone comes along with a theory that shows a way of connecting many of the plotted points of experience in a very consistent, meaningful and organized manner, sort of like when one comes up with a regression line to give linear expression to the various tendencies contained within the scatter diagram at which one has been staring and trying to make sense out of its many data points.

"When one sees conceptual order emerge out of seeming chaos and disorder, the experience is a very aesthetic one. The beauty being given expression through this aesthetic dimension is very compelling and alluring.

"Another qualitative index of a beautiful theory revolves around the notion of simplicity. The capacity of a theory to take a few fundamental ideas and weave them together into complex patterns that can encompass an ever-expanding horizon of experiences, possibilities, and so on, has the aura of beauty about it.

"No matter how complicated things become, one always can return to the few simple ideas out of which the theoretical tapestry has been woven and, thereby, develop a deep aesthetic appreciation for how the whole pattern has arisen as a function of those underlying ideas. Under such circumstances, one's understanding might be fuzzy with respect to the details and minutiae of theoretical complexity, but grasping the simple elements and forces that bind, and animate, the complexity, allows one to be able to orient oneself in the midst of uncertainty.

"This dimension of simplicity has a quality of beauty about it. When researchers encounter this property, we tend to be very attracted by it.

"A third index of a theory's beauty revolves around the heuristic value and power of such a theory. This quality is intimately connected to the two previous facets of theoretical beauty, namely its dimensions of simplicity and organizational capacity.

"When one combines organizational strength with simplicity, this tends to lead to a conceptual dialectic and dynamic that becomes very fruitful with respect to the possibilities, ideas, experiments, hypotheses and explorations that are set in motion by this kind of dialectic and dynamic. The more fruitful a theory is in these respects, the more powerful, stimulating, productive, and valuable the theory becomes.

"This heuristic component of a theory -- that is, its conceptual and experimental fruitfulness, and, therefore, its power ... is, obviously, very desirable. When researchers encounter it, we tend to find it to be a thing of beauty.

"A fourth index of beauty in scientific thinking revolves around the notion of symmetry. This property deals with the capacity of a theory to allow different parameters and variables within that system to undergo operational transformations without the essential aspects of the theory being altered, so that observers in various frameworks will agree these essential features remain the same across the transformations, and, therefore, those features are considered to have been conserved.

"Finally," the professor concluded, "there is an aura of integrity and nobility about a theory that possesses beauty. A beautiful theory tends to stand against the onslaught of confusion, error, darkness, ignorance, and corruption that surround us ... repelling, in an eloquent and elegant fashion, the potential forces of conceptual and social dissolution.

"All in all, the aesthetics of a beautiful theory allow researchers to develop a feeling for some of the realities with which they are attempting to deal. By following this aesthetic pull, researchers are quite frequently led to closer approximations of, or better reflections of, the truths that often are aligned closely to the presence of beauty in a theory.

"I suppose, in many ways, researchers believe it is not possible for a theory to exhibit the various dimensions of beauty, such as organizational meaning, simplicity, heuristic value, symmetry and integrity, without the truth being involved in some fashion. Consequently, seen from this perspective, science really becomes a rigorous, methodical exploration for the elements of truth or reality that researchers believe are being reflected in, and, consequently, that are responsible for, a given theory's beauty."

"Dr. Yardley, couldn't one argue," Mr. Tappin postulated, "that throughout history, including the history of science, there have been a succession of aesthetic theories of truth, if you will, which have been quite captivating and alluring during their time, but, with the passage of time, the beauty of these theories has faded?"

"Yes, this frequently has been the case," acknowledged the professor.

"Moreover," the defense counsel continued, "don't we all, whether or not we are scientists, constantly have to grapple with the possibility that what we find beautiful might, in reality, be a counterfeit, or an illusion, or purely a subjective projection being imposed onto the character of experience or reality?"

"Yes," the professor said.

"Furthermore, Dr. Yardley, would you agree," the lawyer asked, "that, perhaps, on occasion, the reason why we find a theory beautiful is because it serves our personal interests, needs and aspirations, rather than because the theory's beauty is an index for, or sign of, the presence of truth."

"Again, I would agree, in principle, with what you are saying," affirmed the professor.

"In addition," Mr. Tappin pressed, "isn't it possible that what we take to be the reflective beauty of truth and reality is but the reflection of a scientific, political, religious, cultural and/or philosophical conception of beauty and truth into which we have been initiated or indoctrinated by the formal and informal aspects of the educational processes to which we have been exposed during our lives?"

"Of course, this is a possibility," remarked the professor.

"Lastly, Dr. Yardley, don't myths have many of the same kinds of properties that you have outlined with respect to the idea of beauty?

In other words, don't myths have the capacity to offer organized systems of: directed meaning, simplicity, heuristic value, symmetry, and a certain kind of integrity and nobility of purpose?"

"Yes, I suppose so," the professor responded, "but I believe the qualities of beauty in science are a lot more sophisticated, methodologically sound, and analytically rigorous than anything that might be generated through myths."

"Maybe you feel this way, Dr. Yardley, because you are firmly caught up in the myths of science. Isn't this possible?"

"Perhaps," stated the professor.

Reviewing the material in his hands, Mr. Tappin asserted: "In earlier testimony, we have established that, so far as is known, there is no ribozyme capable of unwinding double helical structures that have assumed a stable state through Watson-Crick pairing. In similar fashion, Professor, is there any naturally occurring ribozyme that has proven to be capable of serving as the RNA-world's counterpart to the exonuclease proteins that are able to eliminate errors during the replication of nucleic acid polymers?"

"Not so far," Dr. Yardley indicated.

"What happens if there is no means of maintaining replicational fidelity from one generation to the next?" Mr. Tappin asked.

"Within limits," Dr. Yardley pointed out, "a system can tolerate a certain amount of replicational infidelity. A lot depends on where such errors occur since some pathways and functions are a lot more crucial than are others.

"In addition, under some circumstances, errors in replication actually serve a positive function. Such errors become the mutations through which new evolutionary possibilities might be introduced into the system.

"However, when the replicational fidelity of a genetic system falls below a certain level, then, vital information is lost, not only with respect to the individual, but also in relation to the species population as well. Generally speaking, any kind of replicational process that falls much below, say, a 96-99 percent fidelity rate per nucleic acid residue is very likely, sooner or later, to run into problems that will challenge the continued existence of the kinds of pathways, reactions, structures, activities and functions that are being underwritten by such a replicational process."

"If the RNA-world hypothesis is to be taken seriously," postulated the defense counsel, "wouldn't it have to be able to propose some plausible way to ensure that the fidelity of replication from one RNA generation to the next could be maintained? In fact, wouldn't such a capacity be of the utmost importance given the vast range of abnormal nucleotides and nucleosides that are likely to be roaming about in an Archean era environment?"

"Yes," agreed the professor, "an exonuclease-like capability would be very important to an RNA-world, just as such a capacity is crucial to the DNA-world in which we live."

"I'm sorry, Professor, could you briefly explain what an exonuclease is," Mr. Tappin requested.

"Perhaps, the easiest way to describe the function of this kind of molecule" responded the professor, "is to say they are able to identify and eliminate the vast majority of errors that might arise during, say, the process of replication."

"Thank you," the lawyer acknowledged, and, then, he proceeded to ask: "Can one assume, Dr. Yardley, that a plausible RNA-world hypothesis would require substantially fewer kinds of functions ... such as, but not limited to, the just mentioned exonuclease ... than the DNA-world requires in the way of structural and enzymatic proteins?"

"No, I wouldn't think so," the professor replied.

"Yet," challenged the defense counsel, "only a very few, limited ribozymes have been discovered so far. How do these few discoveries lend much plausibility to a RNA-world hypothesis?"

"First of all," Dr. Yardley responded, "these discoveries are important because of their implications. The fact there might be few ribozymes in existence today does not preclude these molecules from having been a dominant force at some early stage of evolutionary history.

"Secondly, and related to the first point, the ribozymes we have been finding might merely be the left-over remnants of the order of things that once was, just as our appendix might be an evolutionary remnant of an organ or process that once had a function at some point in our evolutionary past. These sorts of evolutionary relic are found throughout the animal and plant worlds.

"Thirdly, the discovery of ribozymes opened up a lot of conceptual possibilities that helped set the stage for a variety of exploratory probes, both experimental and theoretical in character. A lot of important work has come out of the RNA-world hypothesis that has helped to expand the horizons of the evolutionary model in a number of ways.

"Admittedly, there are quite a few outstanding problems facing the RNA-world hypothesis. However, even if this hypothesis is eventually rejected or abandoned, science and evolutionary theory will have benefited by going through the rigorous processes of questioning, experimenting, analyzing, and reflecting that have been necessary in order to properly consider the possible tenability or value of such a hypothesis."

"Gentlemen," interjected Judge Arnsberger, "I feel the time has come to put the discussion to bed for the night. We'll pick things up again tomorrow morning at 10:00 a.m.

"I trust the jurors will continue to behave themselves with respect to the restrictions that have been placed on their discussing the case with anyone. Court is adjourned."

Transposable Conceptual Elements

Mr. Tappin studied the papers in his hands for about five or ten seconds. When he had finished, he asked: "In the Cech and Zang study involving a particular kind of ribosomal activity, one comes across references to something known as 'L-19 IVS RNA'. What is this?"

"This is the working name," Dr. Yardley explained, "for a large molecule of ribosomal RNA. The L-19 portion of the designation refers to the 19 nucleotides that have been removed from an original sequence of 395 nucleotides by the catalytic self-splicing action of this molecule.

"Because the original sequence catalytically operates on, or intervenes with respect to, itself, it is referred to as an intervening sequence. This is the IVS component of the working name."

"What function is served when the 395-nucleotide polymer cuts off 19 nucleotides from itself?" the lawyer inquired.

"Apparently," replied the professor, "this provides a more accessible binding site on the L-19 IVS RNA molecule to which several other oligonucleotides, or short sequences of nucleic acid, can be brought together to form a bond through what is known as a transesterification reaction. In effect, the L-19 IVS RNA enhances the rate of hydrolysis that is characteristic of this sort of reaction by a factor of 10^{10} ... or 10 billion times.

"This kind of transesterification reaction has never been observed to occur between two free oligonucleotides. Consequently, the presence of a protein enzyme or, as in the present case, an RNA ribozyme is of paramount importance if such reactions are going to occur."

"How large," asked the lawyer "is the binding site that is made available by the cleaving of the 19 nucleotides from the original 395 nucleotide IVS RNA molecule?"

"We believe it to be about 7 nucleotides, or so, in length," the professor answered.

"If the binding site is only 7 nucleotides in length," the defense counsel reasoned, "why is there a need for the other 388 nucleotides? Why doesn't the original IVS RNA molecule simply cleave off all but the 7 nucleotides that constitute the binding site?"

"First of all," pointed out the professor, "the 395-nucleotide sequence supervises the initial, precise process that eliminates the 19 nucleotides that render the binding site more accessible to the nucleotides that are to be chemically bonded together. Secondly, the remaining L-19 IVS nucleotide sequence also supervises, so to speak, the bringing together of nucleotides and, in doing so, is required to recognize three or more nucleotides in order to establish a reaction site.

"Consequently, the L-19 IVS RNA molecule has more base-sequence specificity for single-stranded RNA than many, if not most, protein enzymes that are involved in similar kinds of reactions under other cellular circumstances. In fact, this specificity might even rival the specificity of various DNA restriction endonuclease protein enzymes that key in on, and cleave, very specific bonds such as those occurring during the unwinding process of the double-helix structure that is preparing for replication.

"Various kinds of base-deletions studies have been done in relation to IVS RNA to determine just how much of the original 395 nucleotides are necessary for efficient cleavage-ligation activity. On the basis of these kinds of study, at least 300 nucleotides appear to be minimally required in order for efficient catalytic activity to be manifested."

"Does this mean" the defense counsel queried, "that all ribozymes would have to be this large in order to be effective catalysts?"

"At this point," the professor indicated, "we are not quite sure.

There are molecules known as group-I introns whose core structure consists of about 100 nucleotides and that exhibit considerable catalytic activity.

"As a result, seemingly, not every ribozyme necessarily has to be as big as, say, the 300 nucleotides that appear to be minimally necessary for effective IVS RNA functioning. There might be a range of possible ribozyme sizes depending on function and so on, but, at the present time, we do not know what the upper and lower limits of this range might be."

"Given the catalytic specificity of these ribozymes," postulated Mr. Tappin, "even if we were to select, say, a group-I intron consisting of 100 nucleotides, wouldn't the odds of generating this kind of specific sequence on a random basis be, at a minimum, 4^{100}, since there are four nucleic bases that could occupy any one of the 100 nucleotide positions in the entire sequence?"

"Yes, this is correct," the professor confirmed.

"Similarly, for the, let us say, 300 nucleotide IVS RNA molecule," the defense counsel added, "the odds of generating such a specific sequence on a purely random basis would be 4^{300}. Is this right, Dr. Yardley?"

Evolution Unredacted

"Yes," said the professor.

"Previously, Dr. Yardley, you have suggested the entire Archean era was filled with mini-prebiotic laboratories. Let us suppose we were to give those laboratories about 400 million years to come up with the correct sequence for a ribozyme consisting of 100 nucleotides -- the 400 million years being near the figure you cited in direct examination testimony for the length of time during which life is likely to have originated on Earth.

"Let us further suppose all activity in these mini-prebiotic laboratories stopped except work that was directed toward coming up with the right sequence for one specific ribozyme catalyst consisting of 100 nucleotides. How many experiments, Dr. Yardley, would have to be performed per day, over the course of the allotted 400 million years, in order to exhaust the 4^{100} combinations of nucleotide sequences that are possible?"

The professor was silent for about 15 seconds and, then, said: "Probably, in the vicinity of 3×10^{88} experiments per day."

"I've read somewhere, Professor," stated the lawyer, "I forget where, that the surface of the Earth covers about 196,938,800 million square miles. Assuming this figure to be correct and if we were to assume that every square mile of the Earth were to be dedicated to trying out experimental combinations of 100 nucleotides to come up with the specific sequence of our Group-I ribozyme, how many experiments would have to be performed per square mile in order to exhaust the possible combinations?"

"About 2×10^{92} experiments per square mile," replied the professor.

"Of course," Mr. Tappin indicated," we have been assuming in all of the foregoing that we are dealing with the same kind of nucleotides that occur in living organisms. If we add in the assortment of different pentose sugars, ribose forms, optical isomers, odd nucleic bases, and phosphate bonds that are likely to have been hanging around during Archean era times, then, Professor, won't we have to significantly revise all of the foregoing figures in an upward direction in order to factor in the increased possibilities for combining 100 nucleotides in a specific sequence?"

"Yes, we would," Dr. Yardley responded.

Mr. Tappin held up the papers in his hand. "Professor Yardley, according to the information available to me, an Escherichia coli bacterium contains 4 million base pairs of nucleic acid. Let us assume, arbitrarily, that the first self-sustaining life form had only one-quarter as many base pairs ... that is, 1 million base pairs. Would this be a fair assumption?"

"Nobody really knows," stipulated the professor. "No one knows how few ribozymes or enzymes one needs in order to have a self- sustaining, self-replicating organism or protocell.

"Obviously, one needs more genetic information than is carried by a virus since such entities presuppose the existence of a host's replicating capabilities in order to produce new generations of the virus. However, precisely how much more would be minimally necessary is, at the present time, an open theoretical question."

"Let's assume," the defense counsel proposed, "that the average ribozyme is 100 nucleotides in length. In an RNA-world scenario, how many ribozymes do you feel, Dr. Yardley, would be reasonably necessary to look after the catabolic and anabolic pathways of a minimally functioning protocell capable, I would presume, of, to varying degrees: self-replication, division, growth, membrane transport, ion pumping, energy storage, charge transfer, ribosomal activity and the like?"

"I only would be blindly guessing," the professor stated. "Maybe, somewhere between: 100 and 200 ribozymal genes."

"All right," Mr. Tappin suggested, "let's take the lower boundary figure of 100 ribozymal genes. This means, in effect, the mini-prebiotic laboratories would have to find a collective way of bringing together in one place and at one time, a specific sequence of 10,000 nucleotides, 100 times, which is 1 million base pairs -- the number of base pairs we are assuming to have been in our pre- E. Coli life form -- divided by our arbitrary and average figure of 100 ribozyme genes.

"If we were to assume there were a naturally occurring pathway for synthesizing ribonucleic acids, and if we were to assume there were a plausible means of polymerizing these nucleotides under prebiotic conditions, and if we were to assume there were no cross-bonding of pentoses, odd nucleic bases, phosphates, optical isomers, or different forms of ribose, then there are $4^{10,000}$ possible combinations for a series of sequences adding up to 10,000 polymerized nucleotides. Now, none of the foregoing takes into consideration the fact that even given such a specific sequence of nucleotides, the order in which the ribozymal genes are activated and deactivated, as well as when, or for how long, this process of turning the genes on and off takes place, all of this has to be factored into calculating a baseline probability figure.

"Consequently, the 4^{10000} figure is, very much, a lower boundary figure for calculating the odds of generating such an arrangement of nucleotides if one were to assume chance factors were to be the only determinate in inventing such an effectively, functioning system capable of self-replication. Would you agree with this, Dr. Yardley?"

"These are your figures, Mr. Tappin," indicated the professor, "but, for the sake of argument, I'm willing to live with them."

"Given the foregoing, Dr. Yardley, would you be surprised," asked the lawyer, "if I were to tell you there would have not been enough time, space, energy or organic materials on Earth for the mini-prebiotic laboratories to experimentally search through even an extremely minuscule fraction of the total possible combinations that rise from a protocell organism with a genetic repository of 10,000 nucleotides during the 400,000,000 year, or so, period in which life is thought to have originated according to evolutionary theory?"

"Look," the professor asserted, "sometimes one can get figures and numbers to dance almost any tune one likes. There are many possibilities that are not being taken into consideration by your calculations."

"Such as ..."the defense counsel queried?

"There might have been," the professor proposed, "selective forces and conditions operative in the Archean era that might have placed severe constraints on many of the combinations and, as a result, preempted the need for an extended search. For instance, if some given prebiotic experiment produced results that were compatible with the existing thermodynamic and kinetic conditions of a particular evolutionary niche, then, such a result would tend to be selected over other prebiotic experimental results that either were not compatible with existing conditions, or were not compatible to the same extent and, therefore, were at a selective disadvantage as far as thermodynamic and kinetic forces were concerned.

"If one extrapolates this process across the course of hundreds of millions of years, then, there is a finite, but extremely large, set of intermediate steps, all of which could have been selected by available thermodynamic and kinetic conditions. Looking backward, after billions of these steps have occurred, one might have difficulty in understanding how one has got to where one is.

"One also might become overwhelmed when one considers the vast numbers of prebiotic reactions that were experimented with but that were not compatible with the shifting fortunes of thermodynamically and kinetically favorable conditions. Finally, one might be totally amazed when one performs the calculations and discovers what the odds were against this happening on a purely theoretical basis, but the theory on which such calculations is based has not, and, probably cannot, take into account the way a series of thermodynamic and kinetic conditions have selected for a succession of results that has permitted the improbable to be overcome.

"In addition, you are assuming the entire set of combinatorial possibilities would have to be searched before the correct sequence was found. Conceivably, a functional solution could have been discovered at any juncture of the search, and there is no way of predicting when this juncture will be reached.

"Odds become meaningless to the person who is struck by lightning or who wins a lottery. Similarly, no matter how improbable the theoretical odds are concerning a sequence of 10,000 nucleotides, if the prebiotic version of a jackpot occurs, all we can do is to say that on the basis of theoretical calculations we were not expecting such an ontological event to occur and that one would fully anticipate such a rare event to be extremely unlikely to happen again ... although who knows, lightning sometimes does strike twice."

"Am I to understand, Dr. Yardley," Mr. Tappin wondered, "that if I were to remove all but one bullet from a revolver, spin the chamber, hand it to you, and tell you to point the muzzle toward your brain and pull the trigger five times in succession, you would do so because we can assume the bullet has, relative to the formation of a 10,000 sequence nucleotide, an inordinately good chance of showing up in the last chamber?"

"No, this is not what I'm saying," remarked the professor. "Your counter-example is not the same thing."

"Why isn't it?" inquired the lawyer. "Is it because, my counter-example, unlike your various assumptions about possibilities during the Archean era, is capable of demonstration?

| Evolution Unredacted |

"What evidential, demonstrable, rigorous reasons do I have," Mr. Tappin continued, "for holding onto, or retaining, the idea of a natural account for the evolution of life from prebiotic beginnings, when the best you seem to be able to give me, at this point, is that the first protocell might have popped into existence despite the calculated odds against such an event happening?

"More specifically, in your scenario about the shifting tides of thermodynamic and kinetic fortune, Dr. Yardley, which, supposedly, have been selecting out, on a consistent basis, certain prebiotic reactions in preference to other possible reactions, you are, in effect, assuming your conclusions. You have assumed that, once upon a time, there was a sequence of thermodynamic and kinetic conditions that had precisely the properties that were necessary to generate and select an extremely long but finite series of reactions that culminated in the life forms we see before us today and that did so through entirely natural means.

"Yet, whenever one begins to examine some of these alleged thermodynamic and kinetic conditions of natural selection with which evolutionary theory is littered, they fall apart before one's eyes. They don't stand up to any kind of careful, reflective consideration.

"The concrete examples that are being offered to the general public by your theory or model, Professor, and some of which we have been exploring during this cross-examination, are intended to serve as a sampling of the kind of thermodynamic and kinetic processes that form the underpinning of evolution's alleged reality and truth. Yet, if these concrete samples don't stand up to examination, then, why should we extend a line of free intellectual credit to evolutionary biologists that permits them to take advantage of the trust that has been invested, at an ever-accelerating rate, in the evolutionary project for the last 140- plus years?"

"As a great evolutionary scientist once said," replied the professor, " 'nothing in biology makes sense except in the light of evolutionary theory'. The theory brings together an incredible wealth of data that cuts across many disciplines such as cosmology, meteorology, geology, hydrology, paleontology, molecular biology, organic chemistry, biochemistry, microbiology, thermodynamics, population genetics, ecology, anthropology, sociobiology and so on.

"Evolutionary theory has great beauty in its dimensions of simplicity, heuristic value, symmetry, integrity, and organizing power. It renders meaningful what would otherwise be inexplicable and disparate pieces of data.

"There is no other scientific account concerning the origins and development of life forms that can compete with, or is a competitor of, modern evolutionary thought. The consensus of the best minds of our time is that irrespective of whatever relatively minor squabbles separate one theoretician from another, or one researcher from another, in broad outline and in its general principles, evolutionary theory has been established beyond all reasonable doubt.

"Focusing on what lends itself to disputation, rather than concentrating on strengths, and engaging in endless rounds of philosophical nitpicking, rather than getting busy with filling in the blanks, are easy, Mr. Tappin. The reality of the matter is, however, that if one rejects evolutionary theory, what is the alternative?"

"The alternative, Professor, is honesty. When you don't know something, admit it, instead of trying to cover up ignorance with a theory that seems to make a lot of sense when viewed from afar, but, when examined from a closer vantage point, one becomes aware of the fact that much of the beauty of this theory is only skin deep.

"I have no doubt there are many, many truths to be found in evolutionary theory, but the problem is, evolution just doesn't seem to be one of them. I see no reason why evolutionary theory should be granted a license to get away with sloppy thinking and presumption when science has never been willing to extend the same latitude to philosophy, religion or mythology.

"Dressing something up in technical language and surrounding it with the pomp of a false rigor, cannot conceal the naked truth. On all too many occasions the evolutionary emperor has little more to wear than the rather threadbare, and all too revealing, cloth of mental presumptions.

"Furthermore, demanding that a critic provide an alternative to evolutionary theory is a little like a prosecutor expecting a defense lawyer in a murder trial to come up with the killer's identity in addition to proving one's client to be innocent. The reality of the matter is, coming up with an alternative to evolutionary theory is not my responsibility since I did not profess to have a solution in the first place."

"Your Honor," intervened Mr. Mayfield, "is there a question in all of this? My esteemed colleague is badgering the witness."

"Yes, Mr. Tappin," noted the judge, "I think this has gone on long enough."

"Very well, Your Honor," the defense counsel acknowledged.

"Before concluding the cross-examination, Dr. Yardley, there are a few more questions that I would like to ask. Please be patient with me for a little longer.

"Let us suppose Professor that the RNA-world hypothesis is true in the sense that whatever ribozymes were necessary to underwrite a fully functioning and self-replicating organism or protocell had, somehow, come into being. How do we explain the transition to a DNA - world in which amino acids are being encoded for rather than nucleotide sequences?

"In other words, although we might assume the kinds of enzymatic functions that ribozymes and proteins perform are similar in character, in a RNA-world the nucleotides in the genome stand for themselves, they don't stand as a code for something beyond themselves as is the case in the DNA-world. In effect, this means a whole new set of nucleotide sequences must be generated when we change over from the RNA-world to the DNA-world, because, in the forms of life with which we are familiar, DNA codes for amino acid sequences not ribozymal nucleotide sequences, and the nucleotide sequences that confer catalytic activity on ribozymes will not necessarily confer catalytic activity on amino acid sequences.

"For instance, if we take the example of the previously discussed IVS RNA ribozymal molecule, then, its 395 nucleotides would have to be divided by 3, in accordance with the requirements of amino acid, formation rules in the genetic code, and this would give a sequence of about 121 amino acids. Not only is this sequence of 121 amino acids unlikely to have the same enzymatic properties as the 395 ribozymic sequence of nucleotides, but there is no guarantee that the 121 amino acid sequence being coded for by the original ribozymal 395 nucleotides, subsequent to the transition to a DNA-world, would have any enzymatic or structural function whatsoever.

"Furthermore, and in an attempt to ensure that what I'm getting at is, hopefully, entirely clear, we will assume there is no trouble in the first stage of transition from the RNA-world to the DNA-world. We are assuming that everything that previously had been stored in RNA, is now being stored by

DNA, so that, during this first stage of transition, DNA can now bring about all of the synthesis of ribozymes that had been handled by RNA-dominated activity in the RNA-world.

"As I see it, the problems that the transition from: an RNA-world, to: a DNA-world tend to pose for evolutionary theory would begin to arise during subsequent stages of the transition process. Even if one assumes continued full ribozymal activity after the first stage of this transition has been completed, how does DNA come to begin coding for amino acid sequences rather than nucleotide sequences, and how does the organism continue to function when the DNA sequences that previously had been coding for ribozymes during the first stage of transition, no longer are doing this?

"Are we to assume Dr. Yardley that yet another incredibly serendipitous event in evolutionary history occurs just in the nick of time? Are we to assume, in other words, that just as each ribozymic nucleotide sequence is lost from the DNA's genetic repository, then, simultaneously, and, yet, quite independently, an encoded nucleotide sequence for a protein comes into being, with precisely the same kind of enzymatic function as the ribozymatic nucleotide that is being lost?"

"Actually, you seem to be assuming," the professor replied, "that the RNA-world hypothesis is the only theoretical game in town. This issue of transition to which you are referring would be a problem only if a DNA-world did in fact arise out of a RNA-world.

"People need to understand that researchers often adopt a given hypothesis on a trial basis and proceed to give it a work out in order to see how it responds under various theoretical and experimental conditions. During this testing period, one tends to finds things about the theory that are appealing as well as features that one dislikes.

"Almost any hypothesis involves tradeoffs between advantages and disadvantages. Researchers might retain a hypothesis because the problems it solves are considered to be more crucial than the problems the hypothesis creates.

"A scientist might develop a working relationship with a hypothesis not because the individual believes the hypothesis is, in some ultimate sense, true, but because the ideas contained in the hypothesis have heuristic qualities that help suggest theoretical possibilities and experiments or help organize and direct thinking in some fruitful manner. A researcher might stay with this kind of hypothesis until something more useful or less problematic or more elegant comes along.

"Although the RNA-world hypothesis solves a number of problems if one adopts it, there are a number of problems that it generates as well. The transition issue to which you alluded earlier is just one of these difficulties.

"In effect, the RNA-world hypothesis requires the genetic wheel, so to speak, to be invented twice. On the first time through, ribozyme - 'nucleotides' are switched over to DNA-'ribozymes' or, one might say, 'dibozymes', while during the second revolution, nucleotides must code for amino acids.

"There are a number of evolutionary researchers and theorists who feel the double-invention aspect of this hypothesis lacks elegance and simplicity. Such people believe that whatever problems might surround the issue of DNA synthesis in Archean era times ... which, remember, was one of the considerations that helped launch the RNA-world hypothesis in the first place ... nonetheless, such unanswered questions, ultimately, might prove to be more conducive to resolution than are some of the difficulties with which we are left in the wake of the RNA-world hypothesis.

"A further possibility is that some sort of hybrid system arose, combining certain elements of both RNA-world and DNA-world scenarios. Conceivably, for example, some of the huge quantities of so-called surplus or junk genetic material that have been discovered in a variety of species, including human beings, and that appears to have no specific function, might have served, at one time, as a kind of laboratory in which various coding schemes were experimented with until something that worked arose, and, gradually, this was introduced into the operations of the cell."

"Is there any evidence," asked the lawyer, "which lends support to the idea that this surplus or junk genetic material might have played a role in helping the first protocell come into existence?"

"Not that I'm aware of," the professor stated. "However, the night is young, so to speak, in the world of evolutionary biology.

"Quantum theory and relativistic physics didn't come into the scientific picture until more than 200 years had elapsed since the Newtonian revolution helped set the stage for much of modern science. Given that only 140-plus years have passed since Darwin helped set the stage for modern biology, I believe many of the questions that you are asking, Mr. Tappin, stand a very good chance of being answered during the next 60 years."

"What you say Dr. Yardley might turn out to be the case," remarked the defense counsel. "Yet, the point that needs to be emphasized is that, at the present moment, evolutionary biology does not have answers to some fundamental questions that affect the plausibility of the origin-of-life problem.

"For instance," posited the lawyer, "if one rejects the RNA-world hypothesis and maintains the first protocell was a DNA based organism, then, presumably, one will have to come up with a plausible account of how the DNA coding system arose. Are we to assume, once again, that randomness has worked its magic and, one fine day, everything suddenly fell into place?

"Moreover, even if we were to allow the randomness assumption to stand, what about the problem of having to explain how a protocell was able to survive sufficiently long for all of this to come together? In fact, this leads to a key issue ... which came first, a working protocell or a working set of genetic instructions?

"Seemingly, Dr. Yardley, no matter which way one goes with these questions, evolutionary biology faces major problems. Wouldn't you agree?"

"One possibility," the professor suggested, "that you might be overlooking is that the problem is not an either-or issue. The idea of co-evolution offers a third alternative.

"Perhaps, the first working protocell joined forces with a developing set of nucleotide sequences, whether RNA or DNA or both, and the two assisted one another in various ways. Perhaps, in the beginning, this mutual assistance only might have been in some minimalist fashion, but, over time, this working relationship might have become refined and more complex."

"Let me see if I understand this, Professor Yardley," replied the defense counsel. "Are you suggesting that, first, a minimally working protocell, somehow, arose spontaneously through a self-assembly process, and, quite apart from this, a mass of nucleic acids, with some kind of minimal or primitive genetic abilities, arose in one of the mini - prebiotic laboratories, and, then somehow, the protocell and the primitive genetic system came together to form a system that became integrated over time such that the genetic instructions that arose in the DNA/RNA system reflected all of the characteristics of the original protocell? Is this what you mean by the idea of co-evolution?"

"Well, I believe," the professor stated, "the idea has suffered somewhat in your translation of it. Nevertheless, in very crude general terms, you have managed to capture some of the spirit of the co-evolution hypothesis?"

"Doesn't this," queried Mr. Tappin, "raise a variation on the same kind of problem that confronts the RNA-world hypothesis? Isn't one asking for the wheel of life to be invented twice?

"More specifically, on the one hand, life is said to be arising in relation to the self-organizing protocell that we are assuming is spontaneously gathering together and assembling all the requisite parts of a cell that, supposedly, are being synthesized in the Archean era environment. On the other hand, life also is arising in the form of a set of genetic blueprints that is capable, among other things, of self- replication.

"In addition, apparently, we are being asked to suppose that the protocell and genetic system: meet; fall in love; join forces; and, somehow, gradually work out their differences over the course of their lifetimes, so that, in the end, their beings have become so inextricably intertwined, not only can't we tell where the protocell begins and the genetic system ends, but the genetic blueprint, somehow, has come to be able to carry an image of the structural architecture of the protocell, much like a lover carries a photo of the beloved. Is this about it, Dr. Yardley?"

"The imagery is somewhat overwrought but serviceable, I suppose, in a very broad sense," replied Dr. Yardley. "In fact, something very similar to the foregoing has been proposed in another context within evolutionary theory.

"In trying to account for how eukaryotic life forms arose from prokaryotic organisms, Lynn Margulis developed what has come to be known as the symbiotic theory of evolution. In this theory, a variety of prokaryotic life forms join together in a symbiotic relationship that, eventually, over time, and through a complicated sequence of increasingly integrated steps of co-evolution, became a new life form -- that is, the eukaryote, whose different internal organelles, such as the mitochondria, might be remnants of what remains of a symbiotic evolutionary history.

"In effect, the coming together of protocells and some sort of primitive system of self-replicating nucleic acids might just be an earlier, cruder version of what could have happened later on with symbiotic co - evolution when the next giant step of evolutionary transformation occurred and the jump from prokaryotic to eukaryotic life was accomplished. One often sees this kind of repeated use of a creative evolutionary strategy take place under different circumstances and at different junctures of evolutionary history."

"I hate to be impolitic about this, Professor," apologized Mr. Tappin, "but I suppose no one has managed to come up with a plausible step-by-step account of how all this is supposed to have happened. Are we dealing here, once again, with that elusive, shadowy and mysterious agent of evolutionary transformation, Mr. Lucky?"

Smiling, Dr. Yardley said: "He's really not such a bad fellow when you get to know him. He's full of strange, wonderful and unexpected things, although his quality of unpredictability can be quite frustrating to deal with for those who are impatient and demand closure on issues right away."

"I have one final question to ask," asserted the lawyer. "As I understand things, there are some 100,000 genes that are encoded in human beings. These genes consist of tens, if not hundreds, of millions of nucleotides."

"Let us assume, Dr. Yardley, as evolutionary theory must, that there is an unbroken chain of genetic lineage reaching back to the original protocell with which this all started. Given this, is there any mechanism in evolutionary biology ... other than the idea of chance, random events -- that can explain how these 100,000 genes, all of which code for different kinds of enzymatic and structural proteins, came into existence? This question is especially important in view of the fact that natural selection can only operate after a gene has arisen, and, therefore, cannot be cited as a cause for the origin of such genes, unless one wishes to argue that quite independently of the function that such a completed gene serves, each and every step of molecular change leading to this gene also was specifically selected by the environment for reasons that we currently can't fathom."

"Relatively recently," Dr. Yardley responded, "the idea of jumping genes or transposable genetic elements ... transposons, for short -- have caused quite a lot of stir in some parts of the evolutionary community. There is a growing body of evidence suggesting these transposable genetic elements might not only move around from one chromosome to another within an individual or a species, but transposons might even be capable of jumping from one species to another.

"Transposable genetic elements seem capable not only of altering the way genes are given expression, but they appear to be capable of becoming inserted into, and integrated with, different genetic systems. If this is the case, then, transposons might constitute a significant medium for potential evolutionary change.

"Although there is still considerable discussion concerning the possible origins of transposons, one hypothesis suggests these transposable genetic elements might be the remnants of viruses that, at one time or another, had integrated some, or all, of their genes into the genome of their hosts. One reason for supposing the virus-transposon theory of origin might have some merit concerns a commonality that seems to be shared by both some viruses and some transposons.

"There are certain viruses possessing a gene for an enzyme known as reverse transcriptase. Essentially, this enzyme permits such viruses to transcribe RNA into DNA.

"Transposons also appear to employ a similar kind of reverse transcriptase mechanism. Sometimes these transposable genetic elements are capable of generating their own enzymes of this sort, and sometimes these transposons will borrow such enzymes from elsewhere.

"There is some evidence indicating transposons often seem to bring about macro mutations. By this, I mean that when jumping genes become inserted into, and integrated with, other genes, then, one tends to observe substantial alterations in the way phenotype, or the total package of physical characteristics of an organism, might manifest itself.

"There even is some evidence being hotly debated which raises the possibility that, under some conditions of environmental stress, certain species of bacteria might enter into a sort of hyper mutable state. In this state, the claim is being made that a variety of mutated offspring are generated in the apparent attempt to overcome, for instance, the species' inability to digest the only available food source in a given environment.

"According to certain researchers, if one, or more, of the mutated offspring happens to come up with the right solution, the colony survives. On the other hand, if there is no such solution forthcoming, then, assuming that the environmental circumstances do not change, the colony dies.

"I don't know, Mr. Tappin, if you would consider this notion of transposable genetic elements to be a non-random element. Nevertheless, there are some intriguing possibilities that arise from this in the context of evolutionary biology."

"Do these transposons," inquired the lawyer, "merely affect how, or if, certain existing genes are given expression, or do the transposons generate new genes in the sense of introducing a totally new enzyme or protein into the phenotypic milieu?"

"The jury is still out on that one," answered the professor. "Even in those bacteria studies which suggest that a new capacity to digest an, heretofore, indigestible nutrient has arisen, no one is entirely sure about what is going on, and there are quite a few scientists who have criticized such studies for insufficient controls as well as for faulty statistical methodology.

"Because we don't know what, if anything, is taking place, we cannot develop any theory about what the possible parameters are that might govern, or regulate, or limit the kinds of evolutionary changes that might be capable of taking place. We might be dealing with a significant force, or a very minor force, for evolutionary change. We just don't know enough at this point."

"Let us suppose" postulated Mr. Tappin "for the sake of argument, that transposable genetic elements have the capacity, on occasion, to code for the introduction of new structural proteins or enzymes. Do genes act in isolation, or do genes tend to work in concert with one another ... not only in the case of gene regulation and expression but also in terms of establishing catabolic and anabolic pathways that involve the action of a number of different enzymes in order to achieve some biologically useful result?"

"Genes tend to presuppose other genes," the professor responded, "since a single protein or enzyme by itself will have, for the most part, a limited capability to bring about any sort of useful biological result. Viruses are a very good example of this since even though they have a few genes, they do not have enough genes to establish, without the help of a host, a means of replication or reproduction. Among other things, they lack: energy storage as well as charge transfer capabilities; ribosomes; enzymes required for the synthesis of various products fundamental to the maintenance of life, and transfer-RNA."

"If," hypothesized the defense counsel, "the bacteria in the experiments to which you referred were able to enter into a hyper mutable state, in your opinion, Dr. Yardley, do you think a whole bunch of new gene are being generated, or do you feel just certain nucleotide sequences were changed in an existing gene that, in one of the mutants, created a gene capable of coding for an enzyme that fit into an existing metabolic pathway?"

"I think the more likely possibility," the professor stated, "is to suppose there was some kind of mechanism for lifting normal regulatory controls on the replication of a limited number of genes, and, possibly, no more than one particular gene. Changing things holus - bolus would not be a good strategy in an organism that already enjoyed evolutionary success. Furthermore, the more changes that are made, the less likely will these changes be capable of being harmonized or integrated with either one another or with the rest of the existing biological system."

"Therefore, can one assume," inquired the lawyer, "that in those cases where an immediate phenotypic change is brought about by the activity of transposable genetic elements, this is because such elements either are affecting the way an existing gene system is being regulated or given expression or because the jumping gene, if it is a new gene in its own right, codes for a protein that has compatibility with an existing pathway?"

"I would say this is a fairly good assumption," the professor affirmed.

"Would one be unreasonable to assume," Mr. Tappin continued, "that in cases where transposable genetic elements don't affect existing gene regulation or expression and do not fit into any of the existing metabolic pathways, then, there is a good chance the new gene would lack the necessary supporting elements of regulation and expression by other genes to be given phenotypic form or manifestation."

"No, I don't believe this would be an unreasonable assumption," the professor indicated.

"Would you agree," asked the defense counsel, "that in order to become phenotypically manifested, the new gene would have to wait for the necessary gene support to arise through the arrival of, say, other kinds of transposable genetic elements? Moreover, would you agree these new arrivals would have to possess the sorts of nucleotide sequence that could code for the regulated and integrated expression of a series of functionally related genes capable of bringing about some sort of coherent phenotypic expression through protein activity?"

"The answer to both of your questions," said the professor "is yes."

"How many genes," wondered the lawyer, "are necessary for an average, integrated unit to be able to be given phenotypic expression."

"This is very hard to say," offered the professor. "The genetic and phenotypic feedback systems affecting gene regulation and expression can be quite complex."

"However, the basic operon model usually consists of, at least, 4 genes. These are known as: an operator gene; a regulatory or repressor gene; an inducer or promoter gene; and, a structural gene that codes for one or more proteins involved in other kinds of biological functioning such as helping to establish some sort of catabolic or anabolic pathway.

"These genes are involved in a set of feed-back relationships. Under certain conditions, a regulatory or repressor gene gets expressed and prevents the operator gene from setting in motion the steps required for the active expression of the different proteins coded for by the structural gene. Under other conditions, an inducer or promoter gene gets activated and produces a protein that can trigger the operator gene to begin operations. In effect, different genes in the operon get turned on and off at various times depending on circumstances."

"On average, Professor, how many nucleotides would be required to encode the information for these four genes?" the lawyer asked.

"This, again, is difficult to say," the professor replied. "Proteins range in size from relatively small ones like insulin that consist of a couple of chains having 21 and 30 amino acid residues, respectively, to monster proteins consisting of multiple chains of amino acid sequences that run into the hundreds of residues per chain.

"Since each amino acid is encoded by a sequence of three nucleotides, even insulin would consist of, at least, 153 nucleotides. In addition, one has to take into consideration there usually are more nucleotides in a gene than what is required to code for just the amino acid sequence.

"For instance, before a particular, transcribed protein code or message leaves a given nucleotide sequence in the form of messenger - RNA, there often are one or more introns, or sequences of nucleotide bases, which get transcribed but are excised or eliminated before the m-RNA, or messenger-RNA, leaves to be translated into proteins at various ribosomal sites."

"Continuing on with the arbitrary nature of this exercise," stipulated the lawyer, "let's set our hypothetical average for a protein, whether structural or enzymatic, at 50 amino acid residues. Would you agree this is likely to be on the low side of the reality of things?"

"Yes," the professor confirmed.

"Let's, also arbitrarily, set the entire number of nucleotide sequences for the four-gene unit at 1000," Mr. Tappin stated. "This figure will take into consideration an assumption that allows for an operator gene, a regulatory/repressor gene, an inducer gene, as well as three proteins within the structural gene component of our hypothetical operon. This figure will permit us to throw in some extra nucleotides -- say, around 100 ... to take care of minor administrative and organizational duties that probably are part of the operon's effective functioning, somewhat like the excised introns to which you referred earlier.

"Now, Dr. Yardley, if we hook up these, admittedly, arbitrary figures and apply them to the 20-25,000 genes that make up the human genome, and that do not take into account the additional 97 percent of surplus/junk DNA of unknown function, then, we are confronted by the following. From the time the last common ancestor arose on: 20 to 25,000 different occasions, an operon of some 1000 nucleotides arose, in order to culminate in a human being.

"These operons would have been involved in establishing a wide variety of new, not previously evolved functions. Thus, as one goes from, say, the time of no immune system to an increasingly complex immune system, a number of new operons would have had to be generated in order to look after the catabolic and anabolic cellular activities that would have to underwrite these biological capabilities.

"Similarly, the emergence of such things as different kinds of hormonal functioning, or embryological activity or enhanced neurophysiological capabilities, would all require the evolutionary biologist to be able to account for the appearance of the new sets of operons that would help manage the regulation and expression of these systems. Indeed, every new organ, organelle or metabolic system presupposes that at some point in evolutionary history, one or more operons somehow came into existence in order to underwrite processes that had not been in existence heretofore.

"Notwithstanding my arbitrary way of alluding to all of this, in order to account for the complexity of a human being, then, on 25,000 different occasions, over a period of some 3.5 to 3.85 billion years, a specific sequence of 1000 nucleotides was, somehow, selected from a total of 4^{1000} possible nucleotide combinations. On average, I believe this means that once every, very roughly, 100,000 years, this search for a functional sequence of 1000 nucleotides with phenotypic survival or selective value must be solved in the midst of 4^{1000} possibilities.

"One could work out, I'm sure, how many mutational experiments would be necessary, on a moment to moment basis, over the course of 100,000 years to explore even a small fraction of the total combinatory possibilities that are available with respect to one operon, consisting of 4 genes and 1000 nucleotides. I will, as they say, leave it as a homework exercise and, I believe anyone who cares to perform the calculations will conclude this seems to be carrying the idea of hyper mutability beyond the realms of believability.

"My question, Dr. Yardley, is this. How exactly does the idea of transposable genetic elements reduce the element of randomness that seems to saturate the foregoing figures? That is, how does the notion of the transposon, as an alleged agent of evolutionary change, but whose origins are lost in the swirling mists of chance events, permit one and all to see through the mysterious shadows cast by the impenetrable nature of randomness and understand, as you previously suggested is the case, that beyond any reasonable doubt, evolutionary theory is, indeed, true?"

While the professor was reflecting on the question, Mr. Tappin held up his hand in a sort of halting motion. The lawyer said: "To show you what a sporting fellow I am, Dr. Yardley, and to indicate what I think about the idea of chance, I'm going to give you a chance and withdraw the question.

"Your Honor, my cross-examination of this witness has concluded.

"Mr. Mayfield," indicated Judge Arnsberger, "you may call your next witness.

"Your Honor," replied the prosecuting attorney, "we have no further witnesses scheduled to appear. In the matter of the People versus Wayne Robert Corrigan, the prosecution rests."

"Are you prepared to proceed at this time with your first witness for the defense, Mr. Tappin?" inquired the judge.

"Your Honor," he responded, "the defense is prepared to rest its case. Furthermore, if it pleases the court, we would like to move that the charges against our client, Mr. Corrigan, be dropped for lack of sufficient evidence."

"The motion doesn't please the court," Judge Arnsberger asserted. "Whenever possible, Mr. Tappin, I prefer to leave judgments concerning matters before the court in the hands of the people. I believe this is why we have a jury system. They are the fact finders and ones who weigh the evidence, not the judge. Motion denied."

"Are counsels for the prosecution and defense ready for summation?" the judge asked.

"The people are ready, Your Honor," Mr. Mayfield stated.

"The defense, also, is prepared, at this time, to proceed with closing remarks, Your Honor," said Mr. Tappin.

"Mr. Mayfield," Judge Arnsberger announced, "the floor is yours."

Closing Arguments

The prosecutor rose from behind his table and approached the jurors with a smile on his face. When he was a few feet away from the end of the jury area nearest his table, he stopped.

He surveyed the jurors for a few seconds, and, then, he walked to a mid-point several feet removed from the jurors. Slowly at first, but quickly picking up a little speed, his words began to flow.

"Ladies and gentlemen of the jury, you have been very patient and attentive during the last several days of testimony and cross - examination. However, in many ways, your most important task lies ahead of you.

"Now, you not only have the responsibility of making sense of a great deal of information and technical argument, but you also have a duty to come to a judgment about the guilt or innocence of a fellow human being. Such activities can neither be taken lightly by, nor can they rest lightly with, any of you.

"I, as the prosecuting attorney, also have duties and responsibilities. Both during the trial, as well as currently, during these closing remarks, I have had the job of putting forth a case, within the limits imposed upon all participants at the outset of this trial, that would provide you, the members of the jury, with sufficient reason to come to the only conclusion that I believe reflects the totality of evidence ... namely, that Wayne Robert Corrigan is guilty of teaching material that contravenes well-established principles of evolutionary theory, as well as, of scientific methodology.

"During the evidential portions of the trial, I have attempted to fulfill my task in several ways. First, you have been supplied with materials that constitute the written part of Mr. Corrigan's curriculum, and I believe these materials speak for themselves.

"Secondly, I sought, and secured, the co-operation of one of the world's leading evolutionary biologists, Dr. Alan Ross Yardley. For several days, we all have been enjoying the benefits of being able to listen to an eminently qualified expert talk eloquently, precisely and movingly about his discipline.

"Nonetheless, without, in any way, wishing to diminish the quality or value of Dr. Yardley's participation in these proceedings, there are two points, ladies and gentlemen of the jury, that I would like to bring to your attention. Moreover, these are points with which, I am completely certain, Dr. Yardley would concur were he to be asked his opinion on the matter.

"First, Dr. Yardley is but one individual, among thousands of very qualified and gifted professional scientists, who has been called to give testimony in this case. With all due respect to Dr. Yardley's eminence as a scholar, many, many people could have been asked to give testimony, and each of them would have been able to provide the same kind of standard of excellence and expertise as has Dr. Yardley.

"They could do this because they are part of a community of scientists and researchers who have dedicated their lives and talents to the pursuit of what can be known by human beings on the basis of a disciplined, rigorous and methodical application of reason to human experience in the context of a physical and material world. Any of these researchers and scientists could have substituted for Dr. Yardley because they all are contributors to, as well as inheritors of, the treasury of accumulated knowledge and wisdom that has been struggled for, through tireless efforts, in the uncharted and, at times, dangerous territories at the frontiers of human understanding.

"One of the reasons these sorts of struggle can be dangerous is because when knowledge and wisdom come, lives that are ruled by ignorance, superstition, and habit are threatened. Under such circumstances, historically, the tendency of vested interests that feel threatened is to be reactionary and strike out in harmful ways at those who would have the temerity to throw back the curtains of conceptual darkness that are preventing light from coming into the life of the mind.

"We owe people like Dr. Yardley a debt of gratitude for the way they have stood their intellectual and moral ground, for more than one hundred and forty years, against people, like Mr. Corrigan, who seek to hold on to the familiar and convenient at the expense of the truth. Courageous individuals from: Charles Darwin, to: Allan Yardley have risked much in order to help humanity to transcend its tendency to become locked into non-productive patterns of intellectual inertia and lethargy."

Mr. Mayfield stepped back a few paces from the place where he had been standing. He began to walk slowly, back and forth, in front of the jury area, using his arms to help give animated expression and emphasis to his words.

"The other point, ladies and gentlemen of the jury, to which I wish to draw your attention, again with no wish to cast aspersions upon the quality of Dr. Yardley's wonderful testimony about, and defense of, evolutionary theory, concerns the following. I don't believe I can adequately stress the importance of understanding that the evidence which was forthcoming from Dr. Yardley during direct examination and cross-examination is but a tiny subset of the amount of information, knowledge, data, experiments, analysis and reflection that bears upon the issue of evolution.

"When one brings together, in dynamic juxtaposition, firstly, the dedicated expertise of the community of scientists and researchers who were the focus of my first point, as well as, secondly, the wealth of understanding concerning evolution that has matured over the last century and a half, that was the focus of my second point, then, one cannot help being deeply affected by the strength, depth, richness and sophistication of evolutionary thought. Those individuals, who are among the best, the brightest, the most skeptical, the most rigorously analytical and demanding minds in the history of humankind, have established an inter-subjective consensus concerning the truths at the heart of evolutionary theory.

"I suppose one can forgive the fact there are people who, perhaps as a result of an inadequate or poor quality of education, ask the question: if evolutionary theory is so true, why is it still only a theory? Why don't we raise the epistemological status of evolutionary thought?

"Ironically, the reason for retaining the moniker of 'theory' actually has more to do with the integrity of the scientific process than it does with any presumed, tacit admission there is something inherently wrong with this discipline. Although the observational data, facts, experimental results, principles and laws that form the substantive foundations of evolution have been established scientifically and are agreed upon by the community of evolutionary researchers, both past and present, nevertheless, and evolutionary biologists are the first to admit this, there still is much work that needs to be done in order to discover the many things that continue to elude our understanding at the present time.

"None of these unknowns is expected to undermine anything that has been established, and agreed upon, to date. If anything, when these unknowns are discovered and added to the treasure house of our knowledge concerning the process of evolution, they merely will deepen our appreciation and understanding of the complexity and intricacy of nature as it manifests itself through, among other things, evolutionary phenomena.

"Scientists know that as our understanding changes, modifications have to be made in the conceptual framework, model or theory that is being held up as a mirror, of sorts, to natural events. Among scientists there is an awareness of the difference between our understanding of something and the reality, whatever this term might ultimately mean, of that to which our theories are attempting to make identifying reference through descriptions, explanations and so on.

"As Dr. Yardley indicated at one point during cross-examination, science is a work in progress. One can acknowledge this unavoidable truth and, nonetheless, maintain that the conceptual changes that are inevitable are played against a background of fundamental truths whose essence does not change even if the vocabulary through which they are given expression might change with time.

"Quantum and relativistic theory did not alter the truths that had been established previously by science. Rather, these revolutions changed the way we understood, and made use of, what already had been established and known, and, in addition, these ways of thinking helped bring about tremendous contributions to, and the growth of, the repository of human knowledge.

"What disappeared from the intellectual scene in the wake of these revolutions as they passed through the physics and scientific communities, were the ideas, hypotheses, conjectures, theories and models that were rooted in the untenable ways of organizing and interpreting the knowns of science. These revolutions showed more viable, more heuristically valuable, more elegantly fundamental, and more beautiful ways of organizing the knowns of science.

"Evolutionary thought is in the process of effecting the same kind of changes in a variety of biological and associated disciplines. Yet, there are many social, religious, political and philosophical forces that are attempting to resist, and interfere with, efforts to proceed with the exploration, and expanding, of the horizons of human understanding.

"We need to be very clear in our focus on these matters. We need to understand that if people, like the defendant, are permitted to teach anything they like ... no matter how much it might fly in the face of well-established scientific facts, principles and knowledge ... then, we are doing a great disservice to our children and future generations of children.

"To permit the Wrong-way-Corrigans of the world to ply their trade in our classrooms will lead to the development of only confusion, ignorance, and scientific illiteracy among students. To allow individuals such as Mr. Corrigan to indoctrinate children with a dogmatism that can only corrupt and diminish human potential, is to abandon the fiduciary responsibility to humanity that each and every one of us has by virtue of coming into this world as human beings.

"People such as Wayne Corrigan wish to interpose themselves between their students and the community of scientists and say: I know better than these experts and professionals who have dedicated their lives to mastering their disciplines. People like Wayne Corrigan have dropped the gauntlet before society and belligerently pronounced: I refuse to pass on the legacy of understanding and knowledge that has been bequeathed to students by the researchers of the scientific community.

"The Wayne Corrigans who live among us have a tendency to envision themselves as courageous individuals who are fighting the lonely battle against the forces of repression that, in this case, are allegedly being perpetrated by science, in general, and evolutionary theory, in particular. In reality, all too frequently, these individuals are merely caught up in their own megalomania and wish to entangle everyone else, and especially vulnerable students, in their delusions as well.

"Individuals like Wayne Corrigan have no viable alternatives to offer to evolutionary theory. Instead, they prefer for all of humanity to sit idly about, twiddling its collective thumbs, and saying: but you evolutionary scientists haven't proved this relatively minor point or you haven't demonstrated that minor point.

"They ignore the scope, power, value, beauty, elegance, richness, and productive capacity of evolutionary thought, and, consequently, these individuals wish to jettison these important aspects of our cultural manifest, as so much jetsam, in order to save the conceptual ship that they believe is in imminent danger of floundering amidst the rocks of moral turpitude that they associate with scientific activity.

"They are like Don Quixote's evil twins who are flailing away at imaginary windmills but who do so for something other than noble ... albeit, rather excessively and romantically misguided ... idealistic purposes. Instead, they won't be happy until everyone thinks in the same profoundly limited and superficial fashion as they do.

"Ladies and gentlemen of the jury, the choices before us are fundamental in character. We can proceed into the unknown with people of rigorous and methodical dedication like Dr. Yardley who might not have all the answers but who are committed to finding them, or we can proceed into the future by returning to a regressive and dogmatic past like people such as Mr. Corrigan who believe they have all the answers, and, therefore, there is nothing left to discover.

"I have confidence in your ability to make the correct and courageous choice and, consequently, I believe you will endorse the People's belief that Mr. Corrigan is, indeed, guilty of teaching material that conflicts with established principles of both science and evolutionary theory. I beseech you to find Mr. Corrigan guilty and establish a precedent for which history and our children will be eternally in your debt.

"Ladies and gentlemen of the jury, I wish to thank you, again, for your time and consideration. I know you will faithfully fulfill your duties and responsibilities with respect to the matter that is before this court."

Mr. Mayfield nodded his acknowledgement of thanks and returned to his seat. As he sat down, he poured himself a glass of water and began to drink from it.

"All right, Mr. Tappin," Judge Arnsberger said, "you may offer your summation."

The defense counsel rose from his chair and began speaking almost as soon as he was standing. He continued to speak as he gradually made his way to the general area of the jury.

"Mr. Mayfield would have us all believe the issue that is to be decided by you, the members of the jury, is whether or not Mr. Corrigan has taught students in a manner that is in conflict with the principles of evolutionary theory and scientific methodology. One of the problems with this perspective of the prosecuting attorney is that no one, least of all him, has been able to demonstrate just which specific principles of either evolutionary theory or scientific methodology are, allegedly, being contravened by Mr. Corrigan.

"When you look through the curriculum materials that have been introduced as People's Exhibit 'A', you will find that Mr. Corrigan is advocating nothing except the following. An individual should not accept conclusions, scientific or otherwise, until there is a demonstrable chain of evidence that is capable of lending plausible support to the claimed link between the premises of an argument and the conclusions that are said to follow from those premises.

"In addition, you will find in those curriculum materials that a wide number of methods have been developed and elaborated that are designed to help students engage evidential claims from a variety of analytical, reflective, contemplative, experiential and interpretive vantage points. Those curriculum materials, in fact, constitute, and I'm sure you will agree, once you have had an opportunity to examine those materials, a rather intense investigation into the varieties, possibilities, and problems of methodology.

"Difficulties arise, however, at least as far as Mr. Mayfield is concerned, because Mr. Corrigan has the audacity to suggest scientific methodology is not the be all, and end all, of epistemology. Mr. Corrigan, in other words, is questioning the legitimacy of the tendency of many scientists to arrogate to themselves the role of being final arbiters in all matters involving analysis of, critical reflections on, and interpretations about the meaning, value, significance, tenability, truth and rigor of scientific statements.

"For someone to say that such-and-such is what scientists do or that so-and-so is what scientists have agreed upon, is one thing. To make the claim that because this is what scientists do and this is what scientists agree upon, then, therefore, ... especially someone who has not gone through the validation and accreditation process of professional science ... who is critical of what scientists do or say must be dismissed as a fanatic, is an entirely different matter.

"Science is but one approach to dealing with, and understanding, various facets of the phenomenon of lived experience, and, quite frankly, it is an extremely limited way of trying to understand the breadth and depth of what is entailed by being human. Science is but one kind of activity among many possibilities such as law, art, music, literature, philosophy, religion and mysticism that are all capable of deepening human awareness of the many, many factors that can affect how we perceive, interpret, value and act upon experience, including scientific experience.

"Science, in fact, knows little or nothing about a variety of tools that it presupposes in all of its endeavors. More specifically, science knows virtually nothing at all about the processes of consciousness, creativity, thought, insight, interpretation, or understanding that frame, color, orient and shape every cubic nanometer of scientific activity.

"Moreover, individual scientists are as vulnerable to bias, prejudice, error, distortion, and dogmatism as any other group of people. In addition, collectively, scientists have demonstrated throughout their illustrious history that just because the generality of scientists agree upon something is no guarantee of the truth of whatever it might be on which agreement has been reached.

"Many of the most vociferous opponents of Galileo, Copernicus, Kepler, Newton, Darwin, Planck, Einstein and so on, were eminent and respected scientists of their days. Scientific revolutions are called this because of the vast upheaval that they introduced into the thinking, methods, ideas, practices and understanding of, among others, scientists ... many of whom were extremely resistant to what was being proposed by a given revolution in science.

"The activities of scientists that help shape the nature of science have their share of politics, pettiness, lack of vision, inertia and blindness. Furthermore, scientists and science are not autonomous entities that are independent of the cultural, social, political, economic and religious milieu in which they operate.

"Scientists, each according to her or his ability, might be dedicated to truth. Yet, many of them also tend to become entangled in a variety of: associations, networks, vested interests, processes of marginalization, and value judgment s that frequently have fundamental effects on who and what gets funded, published, hired, and taught.

"Mr. Corrigan believes in teaching his students to be skeptical of, but open to, a variety of possibilities. He encourages his students to be: analytical, reflective, contemplative, critical, fair, honest, creative, eclectic, practical, idealistic, thorough, experimental, as well as dispassionate but committed.

"He wants his students to become aware of their own assumptions, prejudices, and biases. He tries to help his students come to a fundamental realization that the dynamics of perception and interpretation are shaped and colored by a lot of individual, professional, cultural, historical and philosophical factors.

"Mr. Corrigan is interested in trying to instill in his students a deep awe, respect and love for the pursuit of truth and understanding. He does whatever he can to inspire his students to work toward acquiring a sense of joy and excitement concerning the exploration of human existence.

"He teaches his students to take the issues of methodology seriously and not to leave the subject matter in the classroom. He wants his students to understand that a judicious methodology has implications for self, life, meaning, values, and community.

"If any of this is in conflict with the principles of science, then, perhaps, the time has come to get rid of those aspects of science that are in conflict with the kinds of thing that Mr. Corrigan is attempting to teach his students. If anything, he has run into difficulties because he has held up a mirror to the way evolutionary scientists go about plying their trade and questioned whether such practices constitute satisfactory epistemology, let alone sound science.

"What does it mean to say a given chain of evidence is a plausible one? The curriculum materials, that constitute People's Exhibit 'A', attempt to explain why Mr. Corrigan believes there are major, not minor, problems with the chain of so-called evidence that is cited by many scientists and biologists as justification for the conclusion that natural evolutionary processes adequately account for, among other things, the origin-of-life.

"During cross-examination a very extensive sampling of evolutionary thought has been investigated in some detail. We have taken a look at: cosmological theories concerning the origin of Earth; asteroid bombardments; interstellar dust clouds; interplanetary dust particles; carbonaceous chondrites; differentiation of the Earth's magnetic core; ocean formation; atmospheres of various kinds of reducing and non-reducing composition; ocean-vaporizing impacts; photic-zone vaporizations; interpretation of Carbon [12] and [13] isotopes in the Isua rock formation; the faint early sun paradox; run-away greenhouse effects; ocean pH values; ultraviolet radiation; shock-wave synthesis; processes of photolysis, hydrolysis, and pyrolysis; possible synthetic pathways for hydrogen cyanide, formaldehyde, amino acids, ribose sugars, nucleic bases, phosphates, fatty acids, and phospholipids; the nature of membrane functioning; porphyrin pigments; issues of chirality or handedness; cross-bonding potential in prebiotic condensation reactions; Strecker synthesis in the Archean era ocean; Fischer-Tropsch mechanisms; formose reactions; alleged simulation experiments; problems of polymerization involving proteins, DNA, and RNA; issues of replication; the RNA-world hypothesis; ribozymes; natural selection; evolutionary pressure; transposable genetic elements; the possible role of random events; origins of the genetic code; protocell formation; co-evolution, and the operon model.

"At each and every stage of our investigation there were major, unresolved questions concerning the tenability and plausibility of the evolutionary model. There is no consistent, rigorous chain of evidence that starts from first principles concerning known facts about the natural processes of cosmology, geology, hydrology, meteorology, thermodynamics, inorganic chemistry, or organic chemistry, and which permits one to see that, in principle, if not in broad detail, there is a plausible path that is capable of leading any reasonable individual to understand how life originated through purely natural processes under what we believe to have been Archean era conditions.

"This is not a matter of two people looking at the same glass of water, and one person seeing it as being half empty, while the other individual perceives it to be half full. This is a matter of too many assumptions, problems, questions, ambiguities, uncertainties, unresolved dilemmas, and unbridgeable, at least at this time, conceptual chasms.

"Now, an evolutionary researcher might look at the mounds of data, experimental results, technical models, or mathematical formulae and believe this is all great science. In reality, however, this kind of science not only fails to demonstrate the validity, or even tenability, of a plausible evolutionary account concerning the origin-of-life, but anyone, given what is currently known and understood, who should try to claim that the available evidence supports, beyond a reasonable doubt, a natural account of the origin-of-life, is engaging in both bad science as well as terrible epistemology.

"There have been a great many books, articles and so on that have been written by scientists and others who have been severely critical, and rightly so in my opinion, of the attempt by creationists to try to pass creationism off as a science. People degrade the magnificence of creation by reducing it to the very limited, narrow preoccupations of the world of physical, material science.

"On the other hand, almost no criticism has been directed toward scientists for attempting to pass off evolutionary theory as a disguised form of faith. Ultimately, however, evolutionary accounts of the origin-of-life require one to have faith in the great deities of chance and assumption.

"The deities of chance and assumption render all things possible. Whatever your theoretical problems might be, these deities can resolve them.

"There is no process, reaction, event, or possibility for which provisions cannot be forthcoming from the infinite powers of chance and assumption. Whatever theoretical rivers need to be forded, or whatever conceptual mountains must be scaled, or whatever evidential chasms need to be bridged, the deities of assumption of chance are present and waiting for the faithful to call out in supplication.

"The miraculous, the inexplicable, the amazing, and the incredible become the commonplace by the grace of the holy writ of the law of large numbers and the givens of assumption. Seek, and you shall find; ask, and it shall be given to you; knock, and all doors will become open to you.

"Little is required of you to adopt this faith. All you need to do is take advantage of the opportunities that chance provides and assume everything turns out okay.

"The faith is simplicity itself. For those who are prepared to submit, beauty and meaning shall flow into their lives like manna from heaven.

"The litanies are easy to learn. Just say: 'if', 'given', 'possibly', 'conceivably', 'assuming', 'probably', 'theoretically', and 'plausibly', and all manner of things will be added unto you.

"I would love to play golf with evolutionists. I can see myself standing at the 18th hole at Augusta and saying: 'If we assume that I hit this just right and, then, get a few, chance, lucky bounces in my favor, I believe I could hole out ... what do you think?' I'm sure they would treat it as a 'gimme'.

"We could play a round of golf and never leave the clubhouse. All we would have to do is assume our way through eighteen holes.

"We could shoot 18, or less, every time out. Who could argue with us since we would have consensual validation on our side?

"Of course, people might begin to suspect something was amiss with our consistently, incredibly low scores. They even might want to make a federal case out of it.

"If this were to happen, however, neither I nor my evolutionary golfing buddies would have anything to fear. We would just get Dr. Yardley to testify on our behalf as to how plausible our account was despite the numerous improbabilities, problems and questions surrounding our theory of what was happening out on the golf course.

"Mr. Corrigan has made the mistake of stepping on the toes of those who are deeply committed to their faith. These zealots take umbrage with anyone who would ridicule their faith as being merely a myth told to impressionable children in order to help the youngsters make sense of, and feel at home in, a bewildering, mysterious, and, sometimes, frightening universe.

"Their deities are jealous gods who do not tolerate worship at any other alter. The guardians of the faith and the keepers of the ark of the chance covenant with assumption are quite certain that all those who do not bow down to the idol of evolution will surely be condemned to an eternal doom in the outer darkness where there will be much wailing and gnashing of teeth.

"Ladies and gentlemen of the jury, you have an opportunity to stop this nonsense. Someday, somebody might come along and be able to demonstrate, in a plausible fashion, and beyond any reasonable doubt, that there is an unbroken chain of evidence that links first principles of science with a purely natural account of the origin-of-life."

"Today is not such a day, and no one, so far, has provided anything remotely approaching such a plausible chain of evidence. Consequently, we need to demand that the classrooms of our nation be made safe for the teaching of science uncontaminated by matters of faith.

"The teaching of biology is a wonderful thing. However, including evolutionary theory as part of the biology curriculum is a violation of the Constitution's separation clause between the state and matters of faith.

"I ask you to find Mr. Corrigan innocent of all charges. I ask you to make the sort of judgment on this issue before the court that you know in your heart is the right thing to do. Thank you!"

Section IV. Some Evolutionary Considerations Concerning The Establishment Clause of the First Amendment and Article IV, Section 4 of the Constitution

Analysis

In the preface to *But is it Science? : The Philosophical Question in the Creation/Evolution Controversy* edited by Robert T. Pennock and Michael Ruse, the two editors indicate that while the U.S. Constitution prohibits the teaching of religion – since doing so gives expression to a form of establishing a system of religious belief and, thereby, contravenes the 1st Amendment – nevertheless, that same fundamental document does not prohibit the teaching of science, even if the quality of the latter should be bad. Over a period of several decades, at least three cases wormed their way through various facets of the legal system and each of those cases led to judicial decisions that, apparently, verified the perspective that was being advanced by Pennock and Ruse.

Among the cases that seem to confirm the foregoing claim of Pennock and Ruse are: *McLean v. Arkansas*, 1982, as well as the 1987 *Edwards v. Aguillard* decision that took place in Louisiana and, eventually, went to the U.S. Supreme Court. In addition, the *Kitzmiller et al v. Dover Area School Board* judgment was rendered in Pennsylvania around 2005.

However, upon examination, the idea that science does not violate provisions of the U.S. Constitution seems fraught with difficulties. Indeed, the title of the book of readings edited by Pennock and Ruse might be focusing on the wrong philosophical question.

More specifically, instead of asking whether or not creationist science or the doctrine of intelligent design qualify as science – even bad science – perhaps the philosophical question that needs to be asked is: 'But is it true?' In this instance, the "it" that is being questioned with respect to some degree of truth could either be, on the one hand, creation science and the thesis of intelligent design, or, on the other hand, evolution ... or, perhaps, both sides of that controversy need to be engaged in a critically reflective manner.

Let us suppose that one accepts the collective conclusions of the aforementioned three legal proceedings. In other words, let us assume that creation science and the thesis of intelligent design do not qualify as science but give expression – each in its own way -- to the teaching of religion and, as well, that the theory of evolution does qualify as being scientific in nature. Does this end the matter?

Not necessarily! The theory of evolution might satisfy the conditions of being scientific, but if essential features of that theory cannot be shown to be true, then one might wonder why students should be required to learn its details.

Of course, an obvious response to the foregoing issue would be to point out that science is a methodological process that historically can be shown to have assisted human beings to establish better and better understandings concerning the nature of certain aspects of reality. Consequently, a student should be exposed to scientific methods, together with the results arising from those methods, so that an individual can gain facility and competence with respect to being able to critically engage both scientific methods and results, thereby, enhancing a person's chances of being able to deal with various facets of life in a constructive, rational, informed, and insightful fashion.

Nonetheless, even though there is plenty of historical evidence to indicate that a great many truths have been established through the process of science, there is also considerable historical evidence to demonstrate that an array of false ideas have populated the annals of science. Among the false theories that were accepted by a majority of the scientific community – sometimes for substantial periods of time – were: Ptolemaic astronomy; phlogiston theory; Caloric theory of chemistry; spontaneous generation; Lamarckian evolution; the blank slate (tabula rasa) model of mind; Phrenology; steady state theory of the universe (or, possibly, the Big Bang … depending on which cosmological version of the universe turns out to be correct); and various editions of string theory.

Moreover, even if we leave aside issues concerning the manner in which certain false theories have dominated the practice of science from time to time, and even though scientific methodology offers a means through which to constantly seek to improve one's understanding of some given phenomenon, the fact of the matter is that scientists tend to be wrong more often than they are right. Indeed, the history of science provides an account of how researchers – both individually and collectively – struggle to escape from a condition of ignorance concerning various physical phenomena and work their way through resolving an array of problems that – hopefully – eventually puts them in a position to fashion a tenable understanding concerning such phenomena that, in time, gets modified or overthrown to better reflect empirical observations, both old and new.

Over the years, human understanding concerning quantum physics, chemistry, gravitation, thermodynamics, materials science, biology, astrophysics, mathematics and a host of other disciplines have all gone through a series of changes – some small and some quite considerable. Our current grasp of the foregoing areas – and many others -- is built on a multiplicity of mistaken ideas that were reshaped or replaced by a series of insights and discoveries that appeared to bring us closer to certain truths than previous ways of understanding were able to do that were, in turn, replaced and reshaped by an array of subsequent insights, discoveries, and observations.

An essential part of science revolves about becoming involved in a rigorous process of discernment in which that which is true or truer must be differentiated from that which is false. This is accomplished through observation, measurement, experimentation, analysis, critical reflection and so on.

Given the foregoing considerations, one might ask: Is evolutionary theory an example of a science that leads to a true or a false understanding of reality? Although the vast majority of scientists in the world today accept one version, or another, of a neo-Darwinian evolutionary model, I believe that enough problematic features have been put forth in my book: *Evolution Unredacted* to, at the very least, call into question the tenability of many facets of evolutionary theory, and, as a result, lend some degree of legitimacy to the idea that a student might have a right to resist, and not be subjected to, the doctrinaire teachings of evolutionary theory.

Among other things, the theory of evolution cannot provide a step-by-step account concerning: The emergence of the first protocell; the origins of the genetic code; the transition from: Chemotrophs to cyanobacteria and/or Archaea organisms (many of the latter life forms are extremophiles) – or vice versa; the transition from: Anaerobic to aerobic organisms; the transition from: Prokaryotic to Eukaryotic life forms; the origins of metabolic systems specializing in, for example, respiration, endocrine activity, immune responses, nervous functioning, sexual reproduction, consciousness, memory, reason, intelligence, language, and creativity.

Does the theory of evolution offer accounts that purport to explain all of the above sorts of transitions? Yes, it does.

However, none of those accounts has been proven to be true. All of those accounts are missing key pieces of evidence that are capable of substantiating that those models, hypotheses, and ideas are unquestionably true.

On the one hand, evidence exists that supports the possibility that in certain cases, species might have been formed through a process of, say, isolating different portions of a population that, over time, leads to the appearance of new variations that are no longer able to produce viable offspring with members of the original population. Nonetheless, one cannot demonstrate with real scientific rigor that the sorts of processes be alluded to above are responsible for the origins of <u>all</u> species.

The theory of evolution encompasses a great many factual observations and discoveries. Yet, at the same time, it gives expression to a model in which speculation and assumption continue to play a major role, and, as a result, despite all of the propaganda being issued by various evolutionary scientists, many facets of the theory of evolution are a long way from having been verified and, quite frankly, might never be capable of being verified.

Moreover, even if one puts aside all of the scientific inadequacies of the theory of evolution, there are a variety of constitutional issues that need to be explored. In other words, although evolutionary theory might be classified as a science, nevertheless, there might be a partisan quality to its framework that could be at odds with the requirements of Article IV, Section 4 of the United States Constitution (more on this shortly). In addition, one could raise the possibility that there also is a religious dimension to the theory of evolution (more on this shortly) and, if so, then, science, or not, such a theory might well be in contravention of the establishment clause of the 1st Amendment.

Article IV, Section 4 of the U.S. Constitution indicates that the federal government "shall guarantee to every state a republican form of government, and shall protect each of them against invasion;" Republicanism is a moral philosophy of the Enlightenment that generated a great deal of interest within colonial America and helped shape the fabric of the Constitutional process.

In order to qualify as being republican in nature, judgments and actions had to exhibit a variety of qualities. More specifically, to be considered republican in nature, actions and judgments had to exhibit: Integrity, objectivity, independence, non-partisanship, equitability, fairness, disinterestedness, nobility, and be devoid of elements that served the individual interests of the person performing a given action or making a particular judgment rather than serving the collective interests of society.

The collective interests of society are summed up in the Preamble to the Constitution. Those collective interests include: Forming a more perfect union; establishing justice; insuring domestic tranquility; providing for the common defense, promoting the general welfare, and securing the blessings of liberty for ourselves and our posterity.

The theory of evolution fails to be objective, independent, and non-partisan in a variety of ways. More specifically, that theory is being advanced as a true account concerning the random, material origins of species despite the fact that: (1) no one has been able to prove that <u>all</u> species (as opposed to <u>some</u> species) are the result of neo-Darwinian dynamics; (2) no one has been able to demonstrate that reality is inherently random, and (3) no one has been able to prove that consciousness, reason, memory, logic, intelligence, understanding, language, creativity, talent (e.g., musical, artistic, mathematical, etc.), and spirituality are purely material phenomena.

Furthermore, the theory of evolution is replete with elements having to do with notions of randomness and the material basis of reality that might be serving the hermeneutical and political interests of those who are propagating the theory of evolution rather than the collective interests of society, and, therefore, are not necessarily promoting the general welfare of the country ... especially if the aforementioned elements involving randomness turn out to be wrong. While such ideational elements have not, yet, been proven to be incorrect, they also have not, yet, been demonstrated to be a correct description of reality, and, therefore, requiring students to learn the theory of evolution would appear to undermine principles of equitability and fairness that constitute integral dimensions of the principle of republicanism that has been guaranteed to each state of the union, and, therefore, under the provisions of the 9th and 10th Amendments, to all the people of those states.

As noted previously, Article IV, Section 4 of the Constitution not only guarantees a republican form of government to every state but, as well, promises to "… protect each of" the states from invasion. Presumably, the protections to which the Constitution might be alluding do not involve just physical threats but could also be extended to protections against certain kinds of philosophical, hermeneutical, and conceptual systems that seek to invade the minds and hearts of the people of the United States through institutions of learning and, thereby, acquire political and legal control of the citizenry and, in the process, undermine the guarantee of a republican form of government.

Notwithstanding the foregoing considerations, teaching the theory of evolution in public schools might also be in contravention of the establishment clause of the 1st Amendment. After all, some individuals have traced the etymological roots of the word religion back to a Latin word – re-li-gare -- that conveys a process of binding or tying.

Any conceptual system constitutes a way of binding or tying a person's understanding to one, or another, understanding of reality. Consequently, the theory of evolution is a conceptual system that tends to tie and bind a person's understanding to various kinds of assumptions, ideas, beliefs, and values in an organized fashion.

Other individuals feel that the notion of religion might also be etymologically linked to another Latin word: "re-li-gi-o-nem". This latter term gives expression to a sense of reverence toward whatever might be considered to be sacred in nature – E.g., the truth, or qualities of compassion, love, forgiveness, meaning, purpose, and so on.

The sacred need not be tied to the notion of Divinity. For instance, Buddhism is considered to be a religion, yet that spiritual tradition often is understood to be based on teachings that tend not to be God-centric in character but, instead, embrace an array of methods, principles, and values that are engaged in a reverential, and, therefore, sacred fashion.

Those who are proponents of evolutionary theory tend to defend their perspective as being inviolable, true, sacrosanct, as well as being worthy of commitment and deep respect. Moreover, such individuals tend to treat the principles, values, and ideas of evolution with attitudes and behaviors that appear to be indistinguishable from individuals who have reverence toward certain religious ideas, principles, or values and consider those themes to be sacred and inviolable.

Referring to the theory of evolution in terms of science does not extinguish the qualities of: Reverence, sacredness, commitment, binding, and tying that are present in the understanding of many of those who are advocates for that theory. Placing the theory of evolution under the rubric of science does not remove the properties of assumption, speculation, belief, interpretation, faith (sometimes referred to as a degree of confidence), and philosophy that tend to flow through that theory.

Given the foregoing considerations, then, surely, teaching the theory of evolution would seem to qualify as an attempt to establish a religious-like belief system. All of the elements of religion – namely, a sense of: Reverence, sacredness, faith, interpretation, inviolability, the sacrosanct, commitment, binding, universality, essentialness, and so on – are present in those who are proponents of, and advocates for, the theory of evolution.

There are several other possible etymological dimensions in the notion of religion that potentially tie that word to the theory of evolution. One of these dimensions is linked to Cicero's way of using the term 're-le-gere', while another etymological derivation of religion gives emphasis to an Old French sense in which the notion of religion refers to a process through which a community exhibits collective devotion to certain ideas.

Cicero's aforementioned manner of engaging the idea of "re-le-gere" involves a methodology through which an individual goes over a given text on a number of different occasions. Presumably, the process of reading and re-reading a given text is a way of exercising due diligence with respect to trying to determine, among other things, the truth concerning the meaning of that text.

Similarly, proponents of evolutionary theory also tend to go over, again and again, the observations, measurements, experiments, and so on associated with that theory in order to try to determine the meaning and truth that might be entailed by those activities. Whether the text being studied is a book or the language of nature seems irrelevant.

Furthermore, Cicero's manner of approaching the process of "re-le-gere" tends to imply that the process of critically reflecting on the meaning of a given text – whether written or having to do with the nature of reality -- is intended to serve as a way of providing one with an opportunity to work toward distinguishing between, on the one hand, the actual meaning of something and, on the other hand, meanings that might be arbitrarily imposed on a text by the individual engaging that material. If so, then, this also reflects the tendency of science to go over something again and again in order to try to discern the difference between, on the one hand, the actual truth of something and, on the other hand, false beliefs concerning the nature of some aspect of experience and, consequently, appears to bind the theory of evolution to religion in, yet, another way.

Moreover, just as religious communities tend to be devoted to the principles, values, and practices which bind the members of that community together in relation to what they believe constitutes the truth of Being, so too, the members of those communities that accept the theory of evolution reflect many of the qualities that characterize the Old French etymological derivation of the term religion. In other words, members of a community of believers involving evolutionary theory are tied together by a common sense of purpose, meaning, valuation, understanding, belief, and truth concerning the principles, ideas, values, and practices entailed by the theory of evolution in ways that parallel what goes on within so-called religious communities.

Therefore, one cannot automatically assume that just because the theory of evolution is referred to as being, or categorized as being, scientific, then, this kind of classification prevents that theory from also giving expression to a variety of religious-like qualities. To whatever extent the theory of evolution entails the foregoing sorts of religious elements, then, that theory also would appear to contravene the establishment clause of the 1st Amendment.

Thus, there seems to be a conflict between the theory of evolution and the U.S. Constitution not only in relation to the 1st Amendment, but, as well, in relation to Article IV, Section 4 of that document. As a result, the editors of: *But Is It Science? -- The Philosophical Question In the Creation/Evolution Controversy* – have put things in a misleading manner since the issue is not whether one can consider the theory of evolution to be scientific in nature – which, in certain ways, it might be – but, instead, the issue is whether, or not, a person recognizes the religious and non-republican elements that are present in the theory of evolution and, as a result, one is prepared to remain consistent by seeking to ensure that such a theory – along with other religious-like systems of thought – are prevented from being taught in public schools because that theory is in contravention of various provisions of the U.S. Constitution.

The previously mentioned *McLean v. Arkansas Board of Education* legal proceeding arose in conjunction with Act 590 that the governor of Arkansas had signed into law on March 19, 1981. The title of that act was: "Balanced Treatment for Creation Science and Evolution Science," and as the act's name suggests, the law required public schools in Arkansas to offer programs that provided balanced treatments of creation science and evolutionary science.

A number of individuals and organizations joined together to bring suit against: (1) the Arkansas Board of Education, (2) the director for the Arkansas Department of Education, and (3) the State Textbooks and Instructional Materials Selecting Committee that, collectively, were responsible for translating Act 590 into active educational policy. Among the individuals and organizations that are being represented through the plaintiff side of the case were: The National Association of Biology Teachers, the Arkansas Education Association, the American Jewish Congress, various churches in Arkansas from different denominational backgrounds, as well as a biology teacher from Arkansas and an array of individuals who were parents or friends of students in Arkansas public schools.

The *McLean v. Arkansas Board of Education* trial took place from December 7, 1981 to December 17, 1981. Judge William R. Overton presided over the proceedings and issued his decision on January 5, 1982.

The suit was first filed on May 27, 1981. The complaint maintained that Act 590 was in contravention of the U.S. Constitution because, among other things, that law violated the establishment clause of the First Amendment – which, according to Judge Overton, is made applicable to the states by the way of the 14th Amendment, but, one should point out that the Amendments extend to the people of any given state independently of the 14th Amendment due to the guarantee of a republican form of government in Article IV, Section 4 of the Constitution.

The aforementioned complaint filed by the plaintiffs contained two other charges as well. More specifically, Act 590 denies teachers and students their right to academic freedom by undermining the Free Speech Clause of the 1st Amendment and, in addition, Act 590 is excessively vague and, therefore, violates the Due Process Clause of the 14th Amendment.

In his January 5, 1982 decision, Judge Overton provides a certain amount of legal background to help frame some of the issues in the *McLean v. Arkansas Board of Education dispute*. For instance, he quotes from Justice Black's 1947 decision concerning the *Everson v. Board of Education* case:

"The 'establishment of religion' clause of the First Amendment means at least this: Neither a state nor the Federal Government can set up a church. Neither can pass laws that aid one religion, aid all religions, or prefer one religion over another. Neither can force nor influence a person to go to or to remain away from church against his will or force him to profess a belief or disbelief in any religion ... No tax, large or small, can be levied to support any religious activities or institutions, whatever they might be called, or whatever form they might adapt to teach or practice religion."

The notion of "church" in Justice Black's foregoing statement is used as a representative term that applies to a wide variety of religious institutions that, presumably, is intended to include (despite not being specifically mentioned): Temples, synagogues, mosques, abbeys, cathedrals, meeting halls, houses of worship, spiritual sanctuaries, and the like. The foregoing presumption is strengthened when Justice Black subsequently indicates that the underlying principle extends to: "... religious activities or institutions, whatever they might be called, or whatever form they might adapt to teach or practice religion."

However, although Justice Black seems to assume that everyone will understand what is meant by the idea of a religion or church (including its extended sense noted above), nonetheless, there is considerable vagueness that surrounds and permeates his foregoing statement. As pointed out earlier, the notion of religion might be applicable to almost any conceptual system that involves qualities of: Tying or binding someone to a set of values, teachings, ideas, values, practices, purposes, meanings, methods, understandings, theories, and/or attitudes that are engaged repetitively because they generate a sense of reverence, sacredness, and commitment that orients individuals and/or communities concerning the nature of the truth about an individual's or a community's relation with Being.

Therefore, if a church – irrespective of whatever it might be called or whatever form it might assume – revolves around, in part or in whole, the foregoing set of qualities, properties, and activities, then, Justice Black – possibly without fully understanding the implications of his words -- might be referring to a great deal more than he – or Judge Overton – believes is being claimed in the *Everson v. Board of Education* case. Indeed, any set of practices, ideas, beliefs, values, theories, principles, methods, and so on that one considers to be inviolable, sacrosanct, sacred, and worthy of reverence -- but which cannot necessarily be demonstrated to be true – begins to be indistinguishable from the usual senses associated with terms such as "church" or "religion".

Thomas Jefferson maintained that the "Establishment Clause" of the First Amendment erected a wall of separation between church and State. Yet, depending on what the State holds to be true, one might contend that the policies of the State could give expression to a set of values, ideas, beliefs, principles, methods, and practices that are difficult, if not impossible, to distinguish from religious activities when construed in the broader sense outlined above. If so, then, the so-called wall of separation that, supposedly, was put in place through the "Establishment Clause" of the First Amendment and that was intended to differentiate between church and state tends to dissolve before our eyes.

Judge Overton's decision in *McLean v. Arkansas Board of Education* also cites the words of Justice Felix Frankfurter with respect to the latter's 1948 judgment concerning *McCollum v. Board of Education*. According to Justice Frankfurter:

"Designed to serve as perhaps the most powerful agency for promoting cohesion among a heterogeneous democratic people, the public school must keep scrupulously free from entanglements in the strife of sects. The preservation of the community from divisive conflicts, of Government from irreconcilable pressures by religious groups, of religion from censorship and coercion however subtly exercised, requires strict confinement of the State to instructions other than religious ..."

The idea that public schools should be an agency "for promoting cohesion among heterogeneous democratic people" is put forward as a truism in the foregoing decision. Consequently, Justice Frankfurter does not explore whether, or not, public schools should be an agency "for promoting cohesion", nor does he critically reflect on what might be meant by the notion of cohesion.

Justice Frankfurter wants the instruction that takes place in public schools to be "other than religious," but he doesn't explain precisely what he means by this allusion. Furthermore, although he is clear that public schools should remove themselves "from entanglements in the strife of sects," and although Justice Frankfurter is clear that he is referring to the strife that tends to arise in conjunction with religious sects, he, apparently, fails to consider the possibility that strife also arises in conjunction with all manner of philosophical, scientific, and political sectarian thought and activity, and, as a result, one is thrown deeper into uncertainty concerning the manner of the instruction that is "other than religious" and, therefore, should be adopted by public schools to promote the sort of cohesion he seems to have in mind (at least in a vague sense) for "a heterogeneous democratic people."

During the course of rendering his decision for *McLean v. Arkansas School Board*, Judge Overton makes reference to the opinion of Justice Clark that was issued in conjunction with the 1963 case of *Abbington School District v. Schempp*. In the latter case, Justice Clark maintained that in order to be able to comply with the requirements of the Establishment Clause of the First Amendment, "... there must be a secular legislative purposed and a primary effect that neither advances nor inhibits religion."

The secular constraint upon legislative activity was again affirmed in the 1973 decision concerning *Lemon v. Kurtzman*. In that case, a tripartite set of conditions was established to serve as guidance for trying to parse such matters – namely, (1) the legislation must serve a secular purpose; (2) the primary effect of the legislation must be to neither inhibit nor advance religion, and, finally, (3) such legislation should not encourage or generate excessive government entanglement in religious matters.

Notwithstanding the rather amorphous cloud of meaning in which condition (3) tends to be enveloped as a result of the presence of the term "excessive" (and, therefore, becomes a possible focus for future objections under the Due Process provisions of the 14th Amendment), one might question the requirement that legislation must serve a secular purpose since those purposes not only are fraught with all manner of strife (and, according to Justice Frankfurter, isn't one of the reasons for pursuing secular rather than religious systems of thought is to be able to avoid sectarian strife?) but, perhaps, more importantly, despite the lack of religious vocabulary associated with various notions of secularism, nonetheless, that sort of approach to governance tends to promote views of reality that cannot be proven to be true – anymore than religious models can be proven to be true to everyone's satisfaction – and secular approaches to governance also require citizens to treat legislation as being: Inviolable, sacrosanct, sacred, deserving of reverence, and capable of binding or tying individuals and the community to sectarian theories (of a philosophical kind) concerning the nature of reality?

Is secularism really any less sectarian than overtly religious systems of thought are? Is secularism really any less entangled in issues of strife than are religious sects with respect to disputes about what values, beliefs, ideas, practices, principles, and so on should be treated reverentially and considered to be inviolable, sacrosanct, or sacred and, therefore, worthy of obligating individuals and the community in one way rather than another?

The foregoing considerations are not an attempt to put forth some post-modernist, relativistic deconstruction of the legal system. Rather, an attempt is being made to indicate that there is considerable amorphousness at the heart of the U.S. Constitution as well as in many subsequent judicial decisions concerning the supposed nature of that document.

For instance, if the republican form of government that is guaranteed in Article IV, Section 4 of the U.S. Constitution requires federal government officials – including justices -- to act and make decisions in accordance with republican qualities of: Objectivity, integrity, impartiality, equitability, fairness, independence, disinterestedness, and not being judges in their own affairs, then, why are secular theories of reality being given preference to religious theories of reality? Moreover, displaying a differential preference for secular ideas very likely will not only serve to inhibit the observance, practice, and pursuit of religious values, ideas, practices and so on, but, as well, encourages and promotes secular ideas as if they were religious in nature ... that is, the sort of ultimate views of reality that should be taught in schools and toward which students should develop the requisite reverence and learn how to treat such ideas as being sacred, inviolable, and sacrosanct in nature?

After running through a few relevant aspects of legal history (noted previously in this chapter) in order to provide a context for his decision, Judge Overton's ruling in *McLean v. Arkansas Board of Education* proceeds to offer an extended historical analysis of religious fundamentalism and its decades-long conflict with the theory of evolution. However, Judge Overton does not make any comparable effort to put forth a critical review concerning the theory of evolution and whether, or not, there is a form of fundamentalism to which the theory of evolution might give expression.

Judge Overton does indicate – with a hint of approval -- that the Biological Sciences Curriculum Study (BSCS), which is a non-profit organization that works with scientists and teachers, has developed a series of biology texts that give emphasis to the theory of evolution. He also notes that those texts are being used by 50 percent of the children in American public school systems.

However, Judge Overton, apparently, has nothing to say about whether, or not, requiring school children to use the BSCS books might constitute a contravention of either the Establishment Clause of the First Amendment or the Guarantee Clause of Article IV, Section 4 in the Constitution. After all, the sectarian nature of the theory of evolution and its claim to constitute a scientific portrait concerning the nature of reality has not been proven to be true and, perhaps, can never be shown to be true.

Judge Overton's ruling also makes reference to the history of fundamentalist opposition toward the theory of evolution when he notes that such a history is documented in Justice Fortas' Supreme Court opinion in *Epperson v. Arkansas*. This latter legal decision rescinded the Arkansas legislative Act 1 of 1929 that prohibited the teaching of evolution in public schools.

In each of the foregoing decisions, reasons are given about why fundamentalist views concerning the issue of origins should not be taught in public schools. However, none of those legal decisions explores whether, or not, there might be reasons why the theory of evolution also should not be taught to public school children, and one can't help but wonder whether any of the jurists who were (or are) making decisions concerning the teaching of evolution know much, if anything, about what they are advocating ... or whether their rulings are in compliance with the republican qualities of impartiality, objectivity, integrity, independence, equitability, disinterestedness, and fairness that are guaranteed through Article IV, Section 4 of the Constitution.

After providing an overview of religious fundamentalism and its history of conflict with the theory of evolution, Judge Overton's decision in *McLean v. Arkansas Board of Education* cites some of the evidence that he feels demonstrates the religious intent underlying Act 590 that, supposedly, calls for a balanced treatment of Creation Science and the theory of evolution in the classrooms of public schools. While one is inclined to agree with Judge Overton's assessment of the foregoing evidence, nonetheless, one should keep in mind that there doesn't seem to be any comparable effort on the part of Judge Overton to critically reflect on the possibility that many facets of the theory of evolution also give expression to a religious-like, fundamentalist orientation.

A distinction is made in Judge Overton's decision between, on the one hand, some of the scientific elements that are present in the theory of evolution and, on the other hand, the relative absence of – or the presence of problematic facets of -- scientific rigor in creation science. However, such a distinction tends to obscure the issue that should

| Evolution Unredacted |

have been at the heart of the *McLean v. Arkansas Board of Education* case.

In other words, rather than drawing a distinction between what is science and what is not science, Judge Overton should have better delineated the full nature of the Establishment Clause as well as explored the relevance of Article IV, Section 4 to the matter before his court. As a result, Judge Overton does not appear to issue a ruling that complies with the requirements that are entailed by the guarantee of a republican form of government that is given in the U.S. Constitution.

On the one hand, there is nothing in the Constitution that is functionally dependent on being able to make a distinction between science and non-science. On the other hand, there is a great deal – constitutionally speaking -- that rests on the issue of what constitutes a religion and that rests on the issue of what constitutes establishing a religion.

When the pursuit of scientific methodology leads to the rise of a hermeneutical system like the theory of evolution that has not – and, perhaps, cannot -- be proven to be true (i.e., that the origin of all species is a function of neo-Darwinian dynamics) and that claims that the ultimate nature of reality is both random and material in nature (again, neither of which has been proven to be true, and, perhaps, cannot be proven to be true), then, such a system of hermeneutics becomes indistinguishable from religious systems that seek to impose a sectarian way of thinking on citizens. Consequently, the presence of the foregoing elements in the theory of evolution contravenes both the Establishment Clause of the 1st Amendment, as well as the requirements of Article IV, Section 4 of the Constitution.

According to Judge Overton – and he is basing the following criteria on the testimony of witnesses who participated in the *McLean v. Arkansas Board of Education* trial proceedings – science has five essential properties. (1) Science seeks to discover the nature of the natural laws that govern phenomena; (2) the explanations offered by science are couched in terms of natural laws; (3) the tenets of science can be empirically tested; (4) its conclusions are provisional and, as a result, might change over time; and, (5) the principles of science are capable of being falsified.

Shortly after stating the foregoing characteristics of science, Judge Overton proceeds to point out that Section 4(a) of Act 590 fails to qualify as being scientific because that section depends on the idea that the origin of life arose as a sudden creation "from nothing." Judge Overton claims that such a contention is not scientific because it requires some form of "supernatural intervention that is not guided by natural law", and, consequently, entails an explanation that is not an expression of natural laws, and, in addition, such a thesis is not testable, and cannot be falsified.

In 2012, Lawrence M. Krauss released a book entitled: *A Universe from Nothing*. The author is an atheist, and, therefore, he is not trying to sneak the realm of the supernatural into the discussion by introducing the possibility of something arising from nothing.

The foregoing book is considered to be a book of science. The contents of his book weave together elements from quantum physics, particle physics, astrophysics, thermodynamics, and cosmology to support the idea that the singularity out of which our universe might have arisen could have been an unstable quantum state that spontaneously gave expression to the universe we have inherited and that made life possible.

Of course, whether the foregoing ideas of Lawrence Krauss are correct, or not, is a separate issue. Nonetheless, irrespective of whether his thesis is, or is not, true, the fact that such ideas are considered to be scientific indicates that, contrary to the claim of Judge Overton, the possibility that something might arise out of nothing does not necessarily depend on supernatural intervention.

In any event, insisting on a distinction between natural and supernatural might be something of a snipe hunt. There is nothing that we know of that precludes the possibility that the so-called natural laws of the universe give expression to God's presence in the operations and dynamics that govern that universe, and, as such, God is free to maintain or make exceptions with respect to how those laws unfold in any given case.

If God maintains (or conserves) natural law, this is not supernatural intervention in a natural phenomenon, but, rather, natural law merely becomes a way of marking God's presence in the process of directing physical phenomena. If God makes an exception in the manner in which natural laws are manifested in any given set of circumstances, then, this also would not constitute a supernatural intervention in a natural process but, instead, would merely reflect that God, by virtue of Divine Presence, was modulating the way in which natural law was being manifested in such events.

Judge Overton's perspective concerning the foregoing issues suggests he believes that supernatural events are neither testable nor falsifiable. Notwithstanding the potentially false dichotomy between the natural and the supernatural that is present in Judge Overton's perspective, for thousands of years, mystics from a variety of spiritual traditions have indicated otherwise.

One can elect to dismiss, out of hand, the foregoing claims of the mystics, but doing so seems to exhibit a considerable resonance with the actions of religious clerics who refused to look through Galileo's telescope when given the opportunity to do so. After all, the mystics contend that mysticism is an empirical science in which one is constantly engaged in a process of testing and falsifying various ideas concerning the nature of the mystical path.

One might also point out in passing that, at the present time, the heart of Lawrence Krauss's perspective concerning the possibility of a universe arising from nothing is neither testable nor falsifiable. Yet, he is considered to be a scientist and his ideas are considered to be scientific even as his colleagues understand that the ideas of Lawrence Krauss concerning the possibility of the universe arising from nothing might not be correct.

Also, one might want to keep in mind that like many claims in science, the statements of mystics (as opposed to theologians) also often tend to be tentative in nature. For example, the dissertation that my spiritual guide wrote to satisfy one of the conditions of his doctorate program was considered by A.J. Arberry – an eminent scholar of Islam and the Sufi mystical tradition – to be one of the best treatises on the Sufi path to have been written in the English language.

Early on in his academic career, my spiritual guide would update the foregoing dissertation so that it would better reflect what he experienced and discovered during one, or another, of his 40-day periods of seclusion. However, after a while, he gave up on the idea of modifying the contents of his dissertation because the lived experience generated through his many periods of seclusion were constantly outstripping the written words of his dissertation in too dynamic, rigorous, and ineffable a manner.

The foregoing considerations tend to muddy the waters a little as far as the issue of distinguishing between science and religion is concerned (especially in conjunction with religion's mystical dimension). However, irrespective of whether, or not, one accepts Judge Overton's manner of bringing specific criteria to bear on the problem of distinguishing between science and non-science, none of this is germane to the real issue at the center of *McLean v. Arkansas Board of Education* – namely, whether creation science and the theory of evolution (each in its own way) are, among other things, in contravention of the Establishment Clause of the First Amendment, or the Guarantee Clause of Article IV, Section 4 of the basic Constitution.

Judge Overton provided evidence in his ruling (for example, among, other things, he quoted a statement to this effect from the writing of Duane Gish, a prominent proponent of creation science) that the judge was aware of the claim that the theory of evolution was religious in nature. Yet, he did not seem to pursue this issue and, instead, appeared to accept, at face value, the idea that the theory of evolution was scientific in nature while creation science was not scientific in character.

Conceivably, defense counsel might have done an inadequate job of inducing various witnesses to develop, and elaborate on, the religious-like features that are present in the theory of evolution. Nevertheless, there was enough evidence presented in the *McLean v. Arkansas Board of Education* case to indicate that Judge Overton might not have exercised due diligence with respect to pursuing this facet of the proceedings – especially given that the foregoing issue is far more relevant to the central legal themes of the case (e.g., the Establishment Clause of the First Amendment and Article I, Section 4 of the Constitution) than is the process of trying to differentiate between what is science and what is not science.

Judge Overton was justified in striking down Act 590 of the Arkansas legal code because that piece of legislation clearly violates the prohibitions inherent in the Establishment Clause of the First Amendment, as well as being in contravention of the provisions inherent in Article IV, Section 4 of the Constitution. However, Judge Overton's ruling missed the opportunity to truly deliver a balanced decision (and, therefore, one done in accordance with republican principles) when he failed to overturn the 1968 Supreme Court decision in *Epperson v. Arkansas* that vitiated the Initiated Act of 1929 prohibiting the theory of evolution from being taught in public schools because irrespective of however scientific the theory of evolution might be considered to be, nonetheless, that theory contains an array of elements that render it sectarian in a manner that is indistinguishable from religious theories and, therefore, constitutes a violation of the Establishment Clause of the First Amendment and, in addition, is in contravention of Article IV, Section 4.

Finally, toward the end of his ruling for *McLean v. Arkansas Board of Education*, Judge Overton states:

"Implementation of Act 590 will have serious and untoward consequences for students, particularly those planning to attend college. Evolution is the cornerstone of modern biology ... Any student who is deprived of instruction as to the prevailing scientific thought on these topics will be denied a significant part of science education."

The foregoing warning sounds an awful lot like it is alluding to some sort of a religious-like litmus test for higher education. In other words, Judge Overton's foregoing words seem to be suggesting that unless a person can demonstrate that one is a true believer in the theory of evolution and, as a result, has been thorough indoctrinated into the catechism of evolutionary principles concerning the nature of reality, then that individual risks being thrown into the higher education equivalent of hell or purgatory where such an individual will have to endure boiling in mental anguish for an eternity or, at least, for the duration of one's college career ... and, possibly, longer.

I remember reading Theodosius Dobzhansky's 1973 essay from the *American Biology Teacher* entitled: "Nothing in Biology Makes Sense Except in the Light of Evolution." I thought at the time when I read the foregoing essay that it was an exercise in hyperbole since a great deal of – if not most of – the material in biology makes considerable sense independently of the theory of evolution.

To be sure, the theory of evolution does provide one with a hermeneutical way to tie the phenomena of biology together in a tidy little package that lends more sense to those phenomena than they might have if the theory of evolution is not true. Nevertheless, one can easily jettison the theory of evolution (but not population genetics) and still understand a great deal about the marvelous phenomena to which the study of biology gives expression.

Contrary to what Judge Overton claims in the foregoing quote, evolution is not the cornerstone of biology. The cornerstone of biology is biology.

One doesn't need evolution to understand the principles of photosynthesis, the Krebs cycle, nervous functioning, metabolic pathways, cellular physiology, membrane dynamics, motility, molecular genetics, or a litany of other biological functions and principles. The theory of evolution might tell one – correctly or incorrectly – what purposes and functions are served through various biological processes, but that theory contributes little, or nothing, toward the process of revealing the nuts and bolts of how cells and organisms operate.

At best, the theory of evolution enables biologists to speculate about why cells and organisms might operate in the way they do or why, in certain limited cases, new species might form due to factors such as isolation. But, if someone were to wave a wand that erased the ideas of evolutionary theory from our collective memory banks, human beings would still have discovered a great deal that makes sense with respect to biological processes under a variety of different circumstances.

Nearly a quarter century later, many of the foregoing issues resurfaced again in the 2004-2005 legal proceedings known as *Tammy Kitzmiller, Et Al. v. Dover Area School District Et Al*. The basis for the Pennsylvania case was rooted in an October 18, 2004 memorandum issued by the Dover Area School Board of Directors that announced that students would be required to not only learn about various problems that were entailed by Darwin's theory of evolution, but, as well, students would be required to learn about "other theories of evolution including, but not limited to, intelligent design."

The forgoing resolution was followed a month later by a November 19, 2004 press release from the Dover Area School District stipulating that teachers at Dover High School would be required to read a statement to 9th grade biology students that identified a number of principles. Included in the press release were statements claiming that: There were gaps in the theory of evolution; the theory of evolution was not a fact; the idea of intelligent design provides an account for the origin of life that is different from the theory of evolution, and the book – *Of Pandas and People* – was a resource that students might use in order to learn more about the intelligent design perspective.

A little less than a month later, a suit was filed in U.S. District Court on December 14, 2004. The suit alleged that both the October 18, 2004 resolution of the Dover Area School Board of Directors as well as the November 19, 2004 press release of the Dover Area School District contravened the Establishment Clause of the First Amendment.

The trial began on September 26, 2005. It concluded a little over a month later on November 4, 2005.

The judge presiding over the case was John E. Jones II. He concluded that it was: "…unconstitutional to teach ID [i.e., Intelligent Design] as an alternative to evolution in a public school science classroom."

Like the legal decision in the McLean v. Arkansas Board of Education that was handed down in the 1980s, Judge Jones' judicial decision in the *Kitzmiller, et al v. Dover Area School District et al* case engages in a lengthy discussion that explores a variety of both legal and scientific issues concerning the attempt of Christian fundamentalists to oppose the teaching of the theory of evolution. Such opposition assumed the form of either trying to ban the teaching of the theory of evolution or seeking to have creationist or intelligent design alternatives to the theory of evolution be given equal time in public school classrooms.

During his historical review, Judge Jones II refers to the 1975 Tennessee case of *Daniel v. Waters*. In that dispute, the Sixth Circuit Court of Appeals concluded the legislation at issue gave a "…preferential position for the Biblical version of creation 'over' any account of the development of man based on scientific research and reasoning " and, therefore, was in contravention of the Establishment Clause of the First Amendment.

Although the Sixth Circuit Court of Appeals rightly pointed out that the Tennessee statute that was being explored in the *Daniel v. Waters* case violated the Establishment Clause, the Court failed to indicate that the Tennessee statute also constituted a violation of Article IV, Section 4 of the Constitution because the disputed legislation undermined the principle of republican government that had been guaranteed to each of the states. Extending a preferred position to a Biblical version of creation relative to other non-Biblical accounts concerning the development of human beings that were based on scientific research and reasoning demonstrates that the Tennessee statute was not drawn up in an: Objective, impartial, disinterested, non-partisan, equitable, or fair manner, and, as a result, is inconsistent with the qualities of republicanism.

The Sixth Circuit Court of Appeals does not raise questions in its judicial decision about whether, or not, the theory of evolution should be given a preferred position in public schools. Although the members of the Sixth Circuit Court of Appeals might have felt – if they even considered the matter – that such issues were irrelevant to determining the Constitutional status of the Tennessee statute that was being called into question, the case offered an opportunity for the Court to explore the nature of the Establishment Clause, the Preamble to the Constitution, and Article IV, Section 4 of the Constitution in an equitable, fair, non-partisan, independent, and disinterested fashion, but they failed to do so.

If it is unconstitutional to assign a preferred position to the teaching in public schools of a Biblical account concerning the origins of life or the development of human beings, is it also unconstitutional to assign a preferred position to the teaching of a scientific researched and reasoned theory concerning the evolution of life or the evolution of human beings? Identifying the theory of evolution as being a function of science does not automatically serve to justify why such a theory should be considered to be incumbent on students to learn.

Naturally, those who consider the theory of evolution to be a true account concerning the origins of species believe it is in the best interests of students to be exposed to the research and reasoning that they feel substantiates their evolutionary perspective. However, those who consider the Biblical account concerning the origins of life and the nature of human development also believe the best interests of students are served by exposing students to the research and reasoning that the advocates of creationism feel substantiate their Biblical perspective.

Both the theory of evolution and the creationist approach to origins and human development are sectarian in nature. Why should one suppose that a sectarian position that is claimed to be scientific will be any less likely to violate the Establishment Clause of the First Amendment or to be in contravention of Article IV, Section 4 than is a Biblical approach to those same issues?

By failing to raise the foregoing sort of questions, the Sixth Circuit Court of Appeals is, itself, not only guilty of violating the requirements of Article IV, Section 4 of the Constitution, but, as well, the Court is helping to establish a sectarian framework. As pointed out earlier in this chapter -- and notwithstanding the fact that the theory of evolution does not employ an overtly religious lexicon -- one encounters considerable difficulty avoiding the conclusion that the theory of evolution is, in many ways, virtually indistinguishable from a religious-like framework because the "facts" that it cites are not capable of demonstrating that the theory of evolution is a correct explanation for the origin of all species.

While stating his judicial opinion in the *Kitzmiller et al v. Dover Area School District et al* case, Judge Jones II cites the findings of Judge Overton in *McLean v. Arkansas Board of Education*. More specifically, Judge Jones II summarizes the legal opinion of the earlier case by stating:

"... the United States District Court of Arkansas deemed creation science as merely biblical creationism in a new guise and held that Arkansas's balanced-treatment statute could have no valid secular purpose or effect, served only to advance religion, and violated the First Amendment."

How does one determine what constitutes a "valid secular purpose"? What are the criteria that determine what constitutes a "valid secular purpose"?

More importantly, perhaps, one wonders why secular ideas should be accorded preferential consideration to non-secular ideas in the legal opinion of Judge Jones II. Even if one were to ignore all of the considerations explored earlier in this chapter concerning the religious-like nature of the theory of evolution, as well as ignore the possibility that the theory of evolution might violate the Establishment Cause of the First Amendment when considered from the perspective of a deeper analysis involving a more inclusive notion of religion, nonetheless, the theory of evolution tends to violate the principles inherent in Article IV, Section 4 of the Constitution because that theory cannot necessarily be shown to be true in an objective, impartial, non-partisan, disinterested, equitable, and fair manner by individuals who are not already committed to that theory.

In addition, the District Court of Arkansas seemed to be immune to the irony inherent in their previous quoted words since the theory of evolution serves only to advance the philosophy of evolutionism. This might constitute a secular purpose, but it is not a <u>valid</u> secular purpose because the sectarian nature of the theory of evolution tends to violate the Establishment Clause of the First Amendment as well as contravene the requirements of Article IV, Section 4.

If a person would like to ask whether, or not, the theory of evolution is a scientific theory, then, by all means, ask scientists – and such questions were asked in both *McLean v. Arkansas Board of Education* as well as in *Kitzmiller et al v. Dover School District et al*. However, scientists are not necessarily the people who should be consulted if one is trying to determine the extent to which the theory of evolution constitutes an objective, equitable, fair, independent, impartial, non-partisan, disinterested account of the nature of reality or our relationship to Being and, thereby, is capable of serving a "valid secular purpose" ... that is, one that is capable of satisfying the degrees of freedom and constraints that are set forth in the Constitution (including: The Preamble; the Establishment Clause of the First Amendment; the 9th and 10th Amendment, as well as Article IV, Section 4 of the Constitution).

Judge Jones II commits the same error in his decision concerning *Kitzmiller et al v. Dover Area School District* legal proceedings that Judge Overton committed in the latter's judgment in the *McLean v. Arkansas Board of Education* case. More specifically, each of the foregoing justices spends a great deal of time in their respective decisions making distinctions between science and non-science but spend relatively little time on exploring the nature of the Establishment Clause of the First Amendment, or on analyzing the nature of Article IV, Section 4 of the Constitution, or reflecting on whether, or not -- under the 9th and 10th Amendment -- either secular or non-secular agencies (or neither) should have control of the educational process, or whether, or not, either Federal or State agencies (or neither) should assume control of the educational process.

Both Judge Overton and Judge Jones II make the same point in their respective legal proceedings – namely, that finding fault with the theory of evolution does not necessarily constitute evidence in favor of some edition of creation science or intelligent design. Consequently, each of those judges should have understand that there is a similar logical error present when the two jurists find fault with creationist science or intelligent design and, then proceed to conclude that some form of a secular conceptual system – such as the theory of evolution or science – must, necessarily, constitute the de facto default system that should govern citizens or be taught in public schools.

If Judge Jones II is going to spend an extended period of time pointing out the many problems that permeate the notion of intelligent design and how that notion gives expression to a religious point of view, then, Article IV, Section of the Constitution demands that Judge Jones II also spend an extended period of time exploring the many problems that permeate the theory of evolution and how that theory tends to violate the Establishment Clause of the First Amendment, as well as tends to be in contravention of the 9th and 10th Amendments along with Article IV, Section 4 of the Constitution. By failing to pursue the foregoing sorts of issues in his judicial decision, Judge Jones II was not exhibiting the necessary qualities of: Objectivity, disinterestedness, impartiality, independence, equitability, and fairness that are required by Article IV, Section 4 of the Constitution and that, supposedly, are guaranteed to the people of each of the states.

Judge Jones II describes how five years after the *McLean v. Arkansas Board of Education* decision vacated Act 590 in Arkansas, the Supreme Court of the United States struck down a similar law in Louisiana. The majority opinion in the 1987 decision for *Edwards v.*

Aguillard stipulated that Louisiana's Creationism Act" contravened the Establishment Clause of the First Amendment because the aforementioned Act amounted to "...restructuring the science curriculum to conform with a particular religious viewpoint."

Yet, if one were to retain the logic inherent in the foregoing way of describing the conflict between creationism and evolutionism in *Edwards v. Aguillard*, a person could easily – and justifiably – argue in parallel fashion that the theory of evolution constitutes a restructuring of the science curriculum to conform with a particular sectarian – if not religious-like – viewpoint that seeks to promote an evolutionary philosophy that is dressed up in scientific language. Referring to the theory of evolution as being scientific does not make it any less sectarian, or religious-like in the manner in which it seeks to impose a certain way of thinking on students and, in the process, attempts to induce the latter individuals to consider such a theory to be inviolable, sacrosanct, sacred, and deserving of a reverential-like commitment that should shape a person's understanding and engagement of reality.

Both Judge Overton in *McLean v. Arkansas Board of Education*, as well as Judge Jones II in *Kitzmiller et al v. Dover Area School District et al* seem to be oblivious to the manner in which they each tend to filter the information in their respective cases through the presumptive lenses of science and the theory of evolution rather than filter information through a process of reflecting on that information in a truly objective, impartial, independent, non-partisan, fair, and equitable fashion that tends to lead to the conclusion that, on the one hand, <u>neither</u> creation science or its update counterpart, intelligent design should be taught in public schools, <u>nor</u>, on the other hand, should the theory of evolution be taught in public schools. In fact, the extent to which each of the aforementioned judges seems to be blind to the conceptual dynamic through which their respective cases are being framed and filtered in a manner that give unquestioned priority to science and the theory of evolution indicates just how problematic the issue of establishing a "valid secular purpose" can be if one is going to, simultaneously, try to reconcile such purposes with, say, the requirements of Article IV, Section 4.

Secular purposes are not necessarily the de facto solution for avoiding violations of the Establishment Clause of the First

Amendment or transgressions against the requirements of Article IV, Section 4 of the Constitution. Purposes that are neither secular nor non-secular should be sought ... purposes that require an on-going process of critical reflection intended to ascertain that neither secular nor non-secular perspectives that have sectarian, religious-like features are permitted to be imposed on citizens, and, in addition, to ascertain that the actions and decisions of government officials are in compliance with the requirements of a republican form of government.

During his decision for *Kitzmiller et al v. Dover Area School District et al*, Judge Jones II states:

"We are in agreement with plaintiff's lead expert, Dr. Miller, that from a practical perspective, attributing unsolved problems about nature to causes and forces that lie outside the natural world is a 'science stopper'. As Dr. Miller explained, once you attribute a cause to an untestable supernatural force, a proposition that cannot be disproven, there is no reason to continue seeking natural explanations as we have our answer."

Although the term "natural world" is used in the foregoing excerpt from the legal decision of Judge Jones II, no definition is given for that phrase.

How does one determine what forces and causes lay within, or beyond, the purview of the natural world? How does one prove what forces and causes lay within the boundaries of the natural world?

Just because one has methods at one's disposal that are capable of detecting certain kinds of forces or causal relations in observed phenomena does not mean that other kinds of forces and causes aren't also present that fall beyond the capacity of one's methods for detecting phenomena, forces, and causes. Moreover, forces and causes that cannot be engaged or measured by our current methodology are not necessarily supernatural.

The neutrino is calculated to measure 10^{-24} meters (.000000000000000000000001) or 10 yoctometers. The Planck length is 10^{-35} meters or in the vicinity of .0000000001 yoctometers.

The Planck length tends to mark a boundary for classical ideas concerning the nature of space-time and gravity. Consequently, we have no idea what, if anything, lies on the other side of that boundary marker or how what transpires in that realm of the Universe affects what transpires on the level of the Planck length or larger.

For example, we don't know why constants -- e.g., the mass of an electron that is $9.10938356 \times 10^{-31}$ kilograms -- have the values they do. The Higgs field might have something to do with the mass value of an electron, but if so, at the present time, we do not know what the nature of the dynamics are between the structural properties of the electron and the structural properties of the Higgs field that would result in electrons having such a constant value.

We know that the Higgs field exists because CERN has been able to detect that field through the presence of the Higgs boson. However, we do not know what -- if anything -- makes the Higgs field possible, but irrespective of whatever might make the Higgs field possible and even though we do not, yet, fully understand the properties of that field, we assume that those dynamics are natural in character.

Natural forces and causes are whatever makes observable phenomena possible irrespective of whether, or not, we can detect them, measure them, or understand them. Advances in methodology, measurement, and instrumentation often expand the horizons of the observable and detectible, but, currently, we do not know whether, or not, we will reach a point in the future when we might encounter some sort of inherent limitation to what can be observed or measured through our physical methods and instruments.

If such a limit should be reached, this does not mean that we have exhausted what the natural world has to offer. Instead, what it means is that we will have reached a terminal point for what our methods and instruments can reveal about the character of the natural world.

Conceivably, God operates in the interstitial spaces that cannot be accessed by our methods and instruments. This would not make such dynamics supernatural but, rather, those dynamics would merely give expression to a species of natural phenomena that are beyond our ability to observe, detect, or measure.

Judge Jones II – as well as Dr. Miller, the lead witness for the plaintiff – maintains that: "once you attribute a cause to an untestable supernatural force, a proposition that cannot be disproven, there is no reason to continue seeking natural explanations as we have our answer." Yet, the theory of evolution constantly makes reference to the idea of random, chance events that cannot be proven to be truly – that is, ontologically, rather than just methodologically -- random, chance phenomena, and, as a result, the foregoing perspective has tended to stop scientists from looking for natural explanations that transcend the idea of randomness but still fall within the realm of the natural world even though the properties and characteristics of that natural world might fall beyond the capacity of our present (and, possibly, future) methods, measurements, and instruments to be able to detect.

Neither Judge Jones II nor Dr. Kenneth Miller (the lead witness for the plaintiff) – nor anyone else -- knows how the first protocells came into existence or how the genetic code came into existence. Neither of those individuals knows how consciousness, intelligence, memory, reason, language, or creativity came into being or what made them possible.

They assume that the aforementioned sorts of phenomena are part and parcel of the natural world. Nonetheless, they know almost nothing about the underlying dynamics or causal forces that give expression to those sorts of qualities or properties and, quite possibly, they will never be able to prove or test what, ultimately, is responsible for those phenomena.

In short, neither Judge Jones II nor Dr. Kenneth Miller have defensible grounds for claiming that the natural world is a realm that necessarily excludes the presence of God. Indeed, the nature of God's activity in the natural world might just be among those phenomena that are beyond the capacity of our physical methods and instruments to be able to detect or measure.

When Judge Jones II and Dr. Miller refer to the idea of the supernatural as being a "science stopper", they seem to be blind to the parallel possibility that approaching reality in the way they do could be something of a "soul or spirit stopper". By insisting that: Public schools, their teachers, and their students must adopt a scientific approach to reality that promotes the theory of evolution, they are advocating a policy that, in many respects, cannot be tested or proven to be true, and, therefore, is as much a sectarian system as any religion and, as such, becomes an oppressive force that interferes with the opportunity of individuals to freely seek natural explanations for phenomena – such as life – that fall beyond the limitations of the theory of evolution.

Judge Jones II indicated in his decision that during Dr. Miller's testimony the professor maintained that just because researchers cannot explain all the details of evolutionary theory, this, in an of itself, does not necessarily invalidate the theory of evolution. Perhaps this is true, but, nonetheless, such a claim does tend to lead to the emergence of questions about where and how one should draw the line that enables one to differentiate between problematic speculations and substantiated theories.

The foregoing contention takes place during a section in the judicial decision of Judge Jones II that critically analyzes some of the ideas of Professor Michael Behe concerning the issue of 'irreducible complexity'. Dr. Behe is of the opinion that there are many processes within organisms involving phenomena such as motility, blood clotting, and the immune response that exhibit structural properties of sufficient complexity whose origins, or way of coming together, cannot be explained adequately by the theory of evolution.

Taking issue with the foregoing position of Professor Behe, Judge Jones II cites the testimony of Dr. Miller and Dr. Padian indicating that Dr. Behe's perspective fails to take into consideration well known mechanisms of evolutionary dynamics. For example, Judge Jones II states:

"In fact, the theory of evolution proffers exaptation as a well-recognized, well-documented explanation for how systems with multiple parts could have evolved through natural means."

Exaptation is a process in which biological systems acquire functions that those systems did not originally possess. To illustrate the foregoing issue, Judge Jones II refers to an example provided by Dr. Padian during the latter's testimony indicating that the middle ear bones of mammals arose, over time, from the mammalian jawbone.

Judge Jones II proceeds to claim that the foregoing evidence demonstrates that Professor Behe's notion of 'irreducible complexity' excludes such data from consideration and, therefore, refutes the professor's argument. Yet, Judge Jones II fails to indicate what the set of step-by-step processes was that led the middle ear bones of mammals to arise from and become differentiated from mammalian jawbones.

Consequently, neither Judge Jones II nor Dr. Padian have provided a step-by-step map that plots out how one goes from mammalian jawbones to the emergence of mammalian middle ear bones. Apparently, this is one of the evolutionary details that – according to Judge Jones II and Dr. Kenneth Miller – evolutionary theory is not required to explain but that – quite incredibly -- does not cause the theory of evolution to lose any sense of validity.

Yet, if one were to say that God were responsible for the transition from mammalian jawbones to mammalian middle ear bones, evolutionary scientists would demand that the proponents of that kind of a theory to provide a step-by-step account of how God made such a transition possible. However, if the proponents of that kind of a theory could not provide evidence capable of substantiating their claim, then, evolutionary scientists would very likely argue that the absence of such evidence undermines the validity of a creationist theory of origins.

None of the examples of exaptation that Judge Jones II mentioned in his decision or that Dr. Miller ran through during his testimony provide the step-by-step evidence that is needed to demonstrate that their claims are warranted. They both allude to the possibility of exaptation with respect to the emergence of complex systems of motility, blood clotting, and the immune system, but, apparently, those possibilities are supposed to be accepted without having to present any detailed evidence capable of demonstrating that exaptation correctly (and not just possibly or theoretically) accounts for the emergence of complex systems over time.

Judge Jones writes in his decision that:

"... Dr. Miller presented peer-reviewed studies refuting Professor Behe's claim that the immune system was irreducibly complex. Between 1996 and 2002, various studies confirmed each element of the evolutionary hypothesis explaining the origin of the immune system"

Moreover, on cross-examination Dr. Behe was presented with 58 publications that had been peer-reviewed, along with nine books and a number of chapters from several textbooks on immunology that explored the evolution of the immune system.

To begin with, one might ask if any of the people who were among the peers who reviewed the aforementioned studies on the evolution of the immune system were, or were not, individuals who accepted the theory of evolution. If all of them were proponents of the theory of evolution, then, perhaps, one should not be too surprised that the studies being alluded to might have been acceptable to the peers who reviewed them as long as those studies exhibited the sort of characteristics that would have resonated – to varying degrees -- with the sensibilities of the individuals who were reviewing that material.

Consequently, the foregoing alliance of studies and peers might only indicate that the peers, along with the people who conducted the studies, operated out of a similar world-view. If so, then, the evidence being cited by Judge Jones II or Dr. Miller does not necessarily constitute evidence that the theory of evolution has been shown to be true in some independent fashion.

Secondly, what does it mean to say that a study confirms a given theory? What are the criteria of confirmation? What justifies such criteria?

Since none of the individuals who wrote: Those 58 studies, or nine books, or several textbooks on immunology were present when immune systems began to emerge in various organisms and also were not present when new wrinkles might have been introduced to those systems, I can pretty much guarantee that none of the individuals to whom Judge Jones II or Professor Miller are referring would be able to specify the precise set of steps that led to the appearance of those systems or to their development. Unfortunately, Judge Jones II seems to exhibit little common sense and ask: How do either the authors of those studies and books or the peers who are reviewing that material know that things happened in the way that is being claimed in their studies.

Judge Jones II seems to be treating informed speculation concerning the possible emergence of immune systems as if it were established truth. Furthermore, rather inexplicably, he appears to be claiming that such informed speculation is capable of disproving Dr. Behe's ideas concerning irreducible complexity.

Professor Behe's notion of irreducible complexity might, or might not, be true. However, speculation about what could have happened in the past is not necessarily the same thing as being able to produce step-by-step, verifiable evidence indicating what actually did happen in the past. Therefore, even if all of those 58 studies, 9 books, and assorted chapters that allegedly were considered to confirm the theory of evolution's account concerning the development of immune systems, nevertheless, until one closely and critically examines what is meant by the notion of 'confirmation' and reflects on the criteria that are being used to establish that supposed confirmation (and whether such criteria are justified), one can't really be sure what, if anything, has been demonstrated by the studies and books to which Judge Jones II is alluding.

I'm pretty sure that Judge Jones II did not review the 58 studies, nine books, and chapters in several textbooks of immunology that are being referred to in his legal decision. Instead, he seemed to merely accept, at face value, the testimony of Dr. Miller and several other witnesses for the plaintiff that the foregoing material proved what they claimed it did.

Throughout his decision, Judge Jones II seems to exhibit the same sort of inclination that is being noted above with respect to appearing to be positively deposed toward the idea of the theory of evolution without exhibiting any sort of countering critical reservation concerning that theory. As such, he seems to be in contravention of Article IV, Section 4 of the Constitution because he has failed to act in an: Objective, impartial, non-partisan, independent, equitable, and fair fashion, and, as a result, he is helping to establish the theory of evolution as a sectarian system that is difficult, if not impossible, to differentiate from religious-like systems and, as such, violates the Establishment Clause of the First Amendment.

The way to resolve the issues that arise in *McLean v. Arkansas Board of Education* or in *Kitzmiller et al v. Dover Area School District et al* (or any of the other legal proceedings that have dealt with those issues) is neither to accept the theory of evolution while rejecting some variation on creationist theory, nor should one attempt to resolve the foregoing matters by accepting creation science or intelligent design while rejecting the theory of evolution, nor should one try to resolve those problems by trying to provide a balanced treatment of the two competing visions. Rather, one should proceed with the understanding that creation science, intelligent design, and the theory of evolution all violate the Establishment Clause of the First Amendment, as well as Article IV, Section 4 of the Constitution, and, therefore, should not be permitted to shape educational policy in the public school system.

Bibliography

Books

Halton Arp, *Seeing Red: Redshifts, Cosmology and Academic Science*, Apeiron, 1998.

Peter Atkins, *Four Laws: What Drives the Universe*, Oxford University Press, 2007.

Ian G. Barbour, *Myths, Models and Paradigms: A Comparative Study in Science and Religion*, Harper & Row Publishers, 1974.

John D. Barrow, *New Theories of Everything*, Oxford University Press, 2007.

Alison Bass, *Side Effects: A Prosecutor, A Whistleblower, and a Bestselling Antidepressant on Trial*, Algonquin Books, 2008.

Wayne Becker, Lewis J. Kleinsmith, and Jeff Hardin – with contributions from Gregory Paul Bertoni, *The World of the Cell*, Sixth Edition, Pearson Education, Inc., 2006.

Michael J. Behe, *Darwin's Black Box*, Touchstone, 1996.

Michael J. Benton, *When Life Nearly Died: The Greatest Mass Extinction of All Time*, Thames & Hudson, 2003.

David Bohm, *Wholeness and the Implicate Order*, Ark Paperbacks, 1983.

Peter R. Breggin, *Medication Madness: A Psychiatrist Exposes the Dangers of Mood-Altering Medications*, St. Martin's Press, 2008.

Harold I. Brown, *Perception, Theory and Commitment: The New Philosophy of Science*, The University of Chicago, 1977.

Francis S. Collins, *The Language of God: A Scientist Presents Evidence for Belief*, Free Press, 2006.

Paul Davies, *Cosmic Jackpot: Why our Universe Is Just Right for Life*, Houghton Mifflin, 2007.

Richard Dawkins, *The Selfish Gene*, Paladin, 1978.

Richard Dawkins, *The Blind Watchmaker: Why the Evidence of Evolution Reveals a Universe Without Design*, W.W. Norton & Company, 1987.

Richard Dawkins, *The God Delusion*, Houghton Mifflin, 2006.

Richard Dawkins, *The Greatest Show on Earth: The Evidence for Evolution*, Free Press, 2009.

Daniel C. Dennett, *Darwin's Dangerous Idea: Evolution and the Meaning of Life*, Simon & Schuster, 1995.

Daniel C. Dennett, *Breaking the Spell: Religion as Natural Phenomenon*, Viking Press, 2006.

Tim Friend, *The Third Domain: The Untold Story of Archaea and the Future of Biotechnology*, Joseph Henry Press, 2007.

Douglas Futuyma, *Evolution*, Sinauer Associates Inc, 2005.

Peter Godfrey-Smith, *Theory and Reality: An Introduction to the Philosophy of Science*, University of Chicago Press, 2003.

Robert M. Hazen, *Genesis: The Scientific Quest for Life's Origins*, Joseph Henry Press, 2005.

Nick Herbert, *Quantum Reality: Beyond the New Physics*, Anchor Press/Doubleday, 1985.

John Holland, *Emergence: From Chaos to Order*, Helix Books, 1999.

Robert Kane, *A Contemporary Introduction to Free Will*, Oxford University Press, 2005.

Stuart Kauffman, *Reinventing the Sacred*, Basic Books, 2008.

Manjit Kumar, *Quantum: Einstein, Bohr, and the Great Debate About the Nature of Reality*, W.W. Norton & Company, 2008.

Nick Lane, *Life Ascending*, W.W. Norton & Company, 2009.

David Lindley, *Uncertainty: Einstein, Heisenberg, Bohr and the Struggle for the Soul of Science*, Doubleday, 2007.

Lynn Margulis and Dorion Sagan, *Microcosmos: Four Billion Years of Evolution from Our Microbial Ancestors*, Simon & Schuster, 1986.

Kenneth R. Miller, *Finding Darwin's God: A Scientist's Search for Common Ground Between God and Evolution*, Harper Perennial, 1999.

Kenneth R. Miller, *Only a Theory: Evolution and the Battle for America's Soul*, Viking, 2008.

Melanie Mitchell, *Complexity: A Guided Tour*, Oxford University Press, 2009.

Joanna Moncrieff, *The Myth of the Chemical Cure: A Critique of Psychiatric Drug Treatment – Revised Edition*, Palgrave Macmillan, 2009.

Chris Mooney and Sheril Kirshenbaum, *Unscientific America: How Scientific Illiteracy Threatens Our Future*, Basic Books, 2009.

Debra Niehoff, *The Language of Life: How Cells Communicate in Health and Disease*, Joseph Henry Press, 2005.

Melody Petersen, *Our Daily Meds*, Sarah Crichton Books, 2008.

John W. Santock, *Life-Span Development*, 11th Edition, McGraw-Hill, 2008.

Robert Shapiro, *Origins: A Skeptic's Guide to the Creation of Life on Earth*, Bantam Books, 1986.

Articles

Zvi Bern, Lance J. Dixon and David Kosower, 'Loops, Trees and the Search for New Physics', pp. 34-41, *Scientific American*, May 2012.

Jan Bernauer and Randolf Pohl, 'The Proton Radius Problem', pp. 32-39, *Scientific American*, February 2014.

Kevin L. Campbell and Michael Hofreiter, 'New Life for Ancient DNA', pp. pp. 46-51, *Scientific American*, August 2012.

Theodosius Dobzhansky, 'Nothing in Biology Makes Sense Except in the Light of Evolution', pp. 125-129; *American Biology Teacher*, March 1973.

Alison Gopnik, 'How Babies Think', pp. 76-81, *Scientific American*, July 2010.

Katherine Harmon, 'Shattered Ancestry', pp. 42-49, *Scientific American*, February 2013.

Courtney Humphries, 'Life's Beginnings', pp. 29-33; p. 74, *Harvard Magazine*, September-October 2013.

Ronald Martin and Antonietta Quigg, 'Tiny Plants That Once Ruled the Seas', pp. 40-45, *Scientific American*, June 2013.

Christopher P. McKay and Victor Parro Garcia, 'How to Search for Life On Mars', pp. 44-49, *Scientific American*, June 2014.

Christopher J. Reed, Hunter Lewis, Eric Trejo, Vern Winston and Caryn Evila, 'Protein Adaptations in Archaeal Extremophiles' Hindawai Publishing Corporation, 2013.

Michael Shermer, 'Darwin Misunderstood', page 34, *Scientific American*, February 2009.

Gary Taubes, 'RNA Revolution', pp. 46-52, *Discover*, October 2009.

Bernard Wood, 'Welcome to the Family', pp. 42-47, Scientific American, September 2014.

Kate Wong, 'The Human Saga', pp. 37-39, *Scientific American*, September 2014.

Karen Wright, 'They Came from Outer Space', pp. 46-49, *Discover Magazine – Extreme Universe*, Winter 2010.

Carl Zimmer, 'The Surprising Origins of Life's Complexity', pp. 84-89, *Scientific American*, August 2013.

www.ingramcontent.com/pod-product-compliance
Lightning Source LLC
Chambersburg PA
CBHW070924220526
45471CB00011B/1